Ecotoxicology of Antifouling Biocides

Takaomi Arai • Hiroya Harino
Madoka Ohji • William John Langston
Editors

Ecotoxicology of Antifouling Biocides

Springer

Editors
Takaomi Arai, Ph.D.
Associate Professor
International Coastal Research Center
Ocean Research Institute
The University of Tokyo
2-106-1 Akahama, Otsuchi
Iwate 028-1102, Japan

Madoka Ohji, Ph.D.
Associate Professor
Institute of Symbiotic Science and Technology
Tokyo University of Agriculture and Technology
Fuchu, Tokyo 183-8509, Japan

Hiroya Harino, Ph.D.
Senior Researcher
Osaka City Institute of Public Health and Environmental Sciences
8-34 Tojo-cho, Tennoji
Osaka 543-0026, Japan

William John Langston, Ph.D.
Principal Researcher
Marine Biological Association
Citadel Hill, Plymouth PL1 2PB
UK

ISBN: 978-4-431-85708-2 Springer Tokyo Berlin Heidelberg New York
e-ISBN: 978-4-431-85709-9
DOI: 10.1007/978-4-431-85709-9

Library of Congress Control Number: 2008938544

Printed in Japan

Printed on acid-free paper

Springer is a part of Springer Science+Business Media
springer.com

Foreword

Antifoulants used in paints for marine structures and vessels pose a dilemma for both economic and environmental reasons. Their role is to prevent, or at least reduce, biofouling of submerged structures by colonizing organisms. Considering the financial advantages, savings result from decreased maintenance, usually due to longer lifetimes between servicing and repainting, and from lower fuel consumption by vessels due to less frictional drag on the hull. Antifoulants also generate important environmental benefits. Firstly, better fuel efficiencies produce fewer greenhouse gas emissions. Secondly, the antifoulants reduce the risk of ship hulls acting as a vector for the transport of alien or invasive species. However, antifoulants generally function due to their inherent toxicity. Environmental quality is jeopardized when non-target organisms are adversely affected, with potentially ruinous financial implications if such species are of economic importance, for example, in the case of oyster fisheries. Marine ecosystems can be at risk when ambient concentrations of toxic substances become lethal.

Historically, the economic advantages of antifoulants for maritime transport have driven their usage and technological advances have influenced their development. Records show that the Greeks and Romans used lead sheathing to protect boats. Europeans applied various organic concoctions of grease, pitch, tallow and tar on wooden ships before widely adopting copper sheathing. With the advent of iron ships in the 1800s came the recognition that metallic copper caused corrosion problems, which led directly to the development of marine antifouling paints. Early paints incorporated toxic elements, notably arsenic, copper and mercury. In the mid 1900s, copper-based synthetic paints were predominant until the emergence of tributyltin (TBT) formulations that quickly gained widespread popularity.

TBT has many attributes as an antifoulant. TBT exhibits broad-spectrum biocidal properties and so is effective against a wide range of colonizing organisms. Being colourless, it can be incorporated into coloured paints that find favour in the yachting community. It does not promote galvanic corrosion and so can be used on aluminium. As an organic compound that can be co-polymerised into resin-based paints, TBT was incorporated into self-polishing coatings that remain effective for long periods of time, thereby increasing dry-docking intervals. Consequently, TBT-based paints were applied on vessels of all types and sizes, which served as a very effective means of dispersing the biocide throughout the marine environment, an act not without consequences.

The prevalence of TBT-based paints changed the way antifoulants are viewed and used. Numerous investigations throughout the world confirmed the ubiquitous distribution of TBT in the marine environment (water and sediment) and in organisms at all trophic levels from bacteria and phytoplankton to fish and whales. TBT was quickly recognized as the causative agent for a number of deleterious biological effects, with two impacts of considerable notoriety. Shell thickening of oysters, first recorded in southwest France, threatened a thriving mariculture fishery in Arcachon Bay. Imposex, the masculinisation of female gastropods, led to community level changes in coastal ecosystems; reproductive failure prevented local recruitment and localised extinction of species lacking a motile larval stage. Various ecotoxicological studies demonstrated that, although molluscs were particularly sensitive to TBT, a wide range of organisms exhibited non-lethal effects at concentrations common in the marine environment. Moreover, TBT was shown to cross the placenta in marine mammals and could be measured in the blood of fisher folk. The impact of TBT on marine ecosystems was therefore far greater than had been experienced due to antifoulants previously in service. Thus, environmental and economic consequences, other than the benefits to the shipping industry, on the utilisation of antifoulants had to be considered.

Regulations and laws to restrict the use of TBT as an antifouling agent in marine paints developed piecemeal during the 1980s and 1990s. The first controls were spurred by financial interests and aimed to protect oyster mariculture in France. These regulations banned the application of TBT-based paints on small vessels (i.e. <25 m in length), but were limited regionally. Several countries followed suit with limited controls on the utilisation of TBT. However, as the body of evidence grew regarding the severity of TBT-induced impacts on marine ecosystem health, more comprehensive protection was deemed necessary. Environmental concerns ultimately led to the International Maritime Organization (IMO) adopting an international treaty entitled *The International Convention on the Control of Harmful Anti-Fouling Systems on Ships* in 2001. This Convention will enter into force on September 17, 2008.

Having had the benefits of an antifoulant as effective as TBT for several decades, the various national regulations and the global Convention fostered the search for alternative compounds. Most marine paint formulations have resorted to being copper-based, but incorporating additional toxic substances, generally referred to as biocide boosters, to enhance effectiveness. Several substances have been developed and those most widely marketed are commonly known as Irgarol 1051, diuron, Sea Nine 211, dichlofluanid, chlorothalonil, and zinc pyrithione. The compounds vary in terms of their mode of operation, environmental persistence and toxicological properties, but unlike TBT show no evidence of being endocrine disrupters. Current awareness of their distribution, fate and impact in the marine environment remains limited, prompting many national and regional surveys. Similarly, the full range of ecotoxicological effects is not known, particularly because most studies have been based in Europe and North America.

The importance of this book rests with its timeliness and broad coverage. Firstly, the material presented here provides a comprehensive understanding of TBT in marine ecosystems just as the IMO Convention comes into effect. Noting that TBT is a persistent pollutant, ongoing monitoring is justified with a view to evaluating the efficacy of the global ban on organotin-based paints, and to maintain surveillance of the recovery and eventual remediation of the marine environment. Secondly, this book reviews the current state of knowledge of the biocide booster compounds that are gaining widespread acceptance, a trend that will surely continue as the IMO Convention comes into force. Such biocide boosters, both in current use and under development, need to be fully investigated with regard to their environmental behaviour and ecotoxicological properties. Global studies, especially incorporating tropical regions, must be encouraged. The chief lesson learned from the TBT epoch is that vigilance is essential in order to protect the environment. Future antifouling agents must seek the compromise between the greatest effectiveness for shipping with minimal acceptable danger to the environment.

Stephen J. de Mora
Plymouth Marine Laboratory, UK
August 2008

Preface

Antifouling products play an important role in the shipping industry and are of significant economic importance. It is estimated that, on average, fuel consumption increases 6% for every 100 μm increase in the average hull roughness caused by fouling organisms. For example, at the height of its usage tributyltin (TBT) saved the US Navy an estimated US$150 million annually. Ship fouling is most commonly prevented by the use of antifouling paints. Antifouling paints contain biocides that are released during the lifetime of the coating, creating a concentration of biocide within a surface micro-layer of water adjacent to the paint surface, preventing settlement of juvenile fouling organisms. There are over 4,000 marine fouling species, therefore biocides used in antifouling paints must have a wide spectrum of activity to cover such a diversity of species able to colonise a ships hull.

Due to environmental concerns over the use of TBT as an antifouling biocide and general disquiet over the release of biocides into the environment, new and evolving regulations now demand that antifouling paints must not result in adverse effects on the environment. This has resulted in a significant challenge for the coatings industry, and developments have been underway for many years to emulate the efficiency of antifouling paints based on TBT, while also having significantly reduced environmental impact.

Since the 1970s, widespread use of TBT-based antifouling paints on all vessel types in the commercial and pleasure craft fleets has resulted in elevated ambient concentrations of TBT in water taken from yachting marinas, in busy harbours, and in areas close to ship repair activity. These localised 'hotspots' of TBT, including sediments, were implicated as responsible for damage to cultivated oysters in the vicinity of marinas and to populations of the coastal dogwhelk *Nucella lapillus*, which showed imposex (induction of male characteristics in females) and suppression of breeding activity in extreme cases. In response, the use of TBT on vessels less than 25 m length overall was banned in 1989 throughout the European Union, with similar bans introduced in the USA, Canada, Australia and New Zealand.

In 1990, in response to increasing scientific evidence on the toxicity and occurrence of organotin residues from antifouling paints in the aquatic environment, the IMO-MEPC (International Maritime Organization-Marine Environmental Protection Committee) issued a series of recommendations on the use of TBT antifoulings, including a ban on its use on vessels less than 25 m LOA (length overall).

Japan went further, by firstly restricting applications of TBT usage in Japanese shipyards and subsequently, by banning all applications in 1991. In the mid-1990s, imposex and a reduction in the numbers of the common whelk *Buccinum undatum*, were reported in the North Sea, and TBT implicated as the causative agent. In addition, recent analysis of tissues from sea mammals, fish and some birds have also revealed small but detectable concentrations of TBT. This latest data, although not conclusively demonstrating that TBT is impacting on these species, has caused several governments and environmental organisations to call for a total ban on the use of TBT, throughout the world.

Many governments have cited the precautionary approach, as agreed at the United Nations Rio Convention (and as agreed by IMO-MEPC), as justification to call for a global ban on TBT. This approach permits regulatory action to be taken if there is scientific concern that the use of a material is unsafe to the environment, without the need for scientific proof that actual effects are occurring. In response to calls from several governments to ban the use of TBT as an antifouling biocide throughout the world, the 42nd meeting of the IMO-MEPC (November 1998) unanimously passed a 'draft assembly resolution' calling for a global ban on the application of antifouling products containing organotin compounds which act as biocides, by January 1, 2003, and complete prohibition of their presence on ships' hulls by January 1, 2008.

Antifouling biocides are, by their very nature, toxic to aquatic organisms. An environmental risk assessment is therefore often required before an antifouling paint can be approved. Although some 15 organic chemicals and cuprous oxide have been selected as potential biocides in alternative organotion-free antifouling paints and as antifouling agents for fishing nets, only bis-(1-hydroxy-2(H)-pyridine thionate-O, S) zinc (zinc pyrithion; ZnPT), bis-(1-hydroxy-2(H)-pyridine thionate-O, S) copper (copper pyrithion;CuPT), 4, 5-dichloro-2-n-octy1–3-isothiazolone (Seanine-211), pyridinetripehnylboron (PyB), and 2-methylthio-4-t-butylamino-6-cyclopropylamino-s-triazine (Irgarol 1051) are widely used in the newly developed anti-fouling formulations. These organic chemicals are used with cuprous oxide in order to improve anti-fouling efficiency. However, because the behaviour and toxicity of these compounds have yet to be clarified, compared with organotin compounds, little is known about their effects in the aquatic environment.

The contents of this book reflect recent advances in understanding of antifouling biocides in the environment, including behaviour, toxicity, biological impacts, bioaccumulation and regulation. It is hoped that this book will offer the reader a greater insight into the chemistry and fate of these compounds in aquatic systems.

I would like to thank all our authors for their contributions to this volume and our colleagues who put much effort into reading and commenting on individual chapters.

Takaomi Arai
Iwate, Japan
February 2008

Contents

Section 2 Organotin Compounds as Endocrine Disruptors

Section 3 Toxicity of Organotin Compounds in Aquatic Organisms

Section 4 Genetic and Physiological Impacts of Organotin Compounds

Section 5 Bioaccumulation of Organotin Compounds in Aquatic Organisms

Part III Fundamental Chemistry of Alternative Insecticides

Section 4 Analytical Methods

Contributors

Takaomi Arai
International Coastal Research Center, Ocean Research Institute, The University of Tokyo, 2-106-1 Akahama, Otsuchi, Iwate 028-1102, Japan

Steven J. Brooks
Norwegian Institute of Water Research (NIVA), Gaustadalléen 21, NO-0349 Oslo, Norway

Kazunori Fujii
National Research Institute of Fisheries and Environment of Inland Sea, Fisheries Research Agency, 2-17-5 Maruishi, Hatsukaichi, Hiroshima 739-0452, Japan

Sang Gyoon Kim
Faculty of Agriculture, Graduate School, Kyushu University, 6-10-1 Hakozaki, Higashi-ku, Fukuoka 812-8581, Japan

Takeshi Hano
Faculty of Agriculture, Graduate School, Kyushu University, 6-10-1 Hakozaki, Higashi-ku, Fukuoka 812-8581, Japan

Hiroya Harino
Osaka City Institute of Public Health and Environmental Sciences, 8-34 Tojo-cho, Tennoji, Osaka 543-0026, Japan

Tsuneo Honjo
Faculty of Agriculture, Graduate School, Kyushu University, 6-10-1 Hakozaki, Higashi-ku, Fukuoka 812-8581, Japan

Toshihiro Horiguchi
Research Center for Environmental Risk, National Institute for Environmental Studies, 16-2 Onogawa, Tsukuba, Ibaraki 305-8506, Japan

Ik Joon Kang
Faculty of Agriculture, Graduate School, Kyushu University, 6-10-1 Hakozaki, Higashi-ku, Fukuoka 812-8581, Japan

Kurunthachalam Kannan
Wadsworth Center, New York State Department of Health and Department of Environmental Health Sciences, School of Public Health, State University of New York at Albany, Empire State Plaza, P.O. Box 509, Albany, NY 12201-0509, USA

Shin'ichiro Kawai
Graduate School of Human Environmental Sciences, Kobe College, Hyogo, Japan; Department of Biosphere Sciences, Kobe College, 4-1 Okadayama, Nishinomiya, Hyogo 662-8505, Japan

Katherine H. Langford
Norwegian Institute for Water Research (NIVA), Gaustadalléen 21, NO-0349 Oslo, Norway

William John Langston
Marine Biological Association, Citadel Hill, Plymouth PL1 2PB, UK

Nobuyuki Miyazaki
Center for International Cooperation, Ocean Research Institute, The University of Tokyo, 1-15-1 Minamidai, Nakano-ku, Tokyo 164-8639, Japan

Kazuhiko Mochida
National Research Institute of Fisheries and Environment of Inland Sea, Fisheries Research Agency, 2-17-5 Maruishi, Hatsukaichi, Hiroshima 739-0452, Japan

Tsuyoshi Nakanishi
Laboratory of Hygienics, Gifu Pharmaceutical University, Gifu, Gifu, Japan; Department of Toxicology, Graduate School of Pharmaceutical Sciences, Osaka University, Suita, Osaka, Japan

Ayako Nakayama
Centre for Fish and Wildlife Health, Institute of Animal Pathology, University of Berne, Switzerland;
Graduate School of Human Environmental Sciences, Kobe College, Hyogo, Japan; Department of Biosphere Sciences, Kobe College, 4-1 Okadayama, Nishinomiya, Hyogo 662-8505, Japan

Kei Nakayama
Center for Marine Environmental Studies (CMES), Ehime University, 2-5 Bunkyo-cho, Matsuyama 790-8577, Japan

Jun-ichi Nishikawa
Laboratory of Health Sciences, School of Pharmacy and Pharmaceutical Sciences, Mukogawa Women's University, Nishinomiya, Hyogo, Japan

Madoka Ohji
International Coastal Research Center, Ocean Research Institute, The University of Tokyo, 2-106-1 Akahama, Otsuchi, Iwate 028-1102, Japan

Yuji Oshima
Faculty of Agriculture, Graduate School, Kyushu University, 6-10-1 Hakozaki, Higashi-ku, Fukuoka 812-8581, Japan

Nicholas Dingle Pope
Marine Biological Association, Citadel Hill, Plymouth, UK

Inneke F. M. Rumengan
Faculty of Fisheries and Marine Science, Sam Ratulangi University, Kampus Unsrat Bahu, Manado 95115, Indonesia

Helmut Segner
Centre for Fish and Wildlife Health, Institute of Animal Pathology, University of Berne, Switzerland

Tetsuya Senda
National Maritime Research Institute, 6-38-1 Shinkawa, Mitaka, Tokyo 181-0004, Japan

Yohei Shimasaki
Faculty of Agriculture, Graduate School, Kyushu University, 6-10-1 Hakozaki, Higashi-ku, Fukuoka 812-8581, Japan

Kazunobu Takahashi
Marine Antifouling and Environment Consultant (MAEC): 2-6-13-1008 Shigita-higashi, Jyoto-ku, Osaka 536-0017, Japan;
ISO/TC35/SC9/WG27 Japan Working Group member

Shinsuke Tanabe
Center for Marine Environmental Studies (CMES), Ehime University, 2-5 Bunkyo-cho, Matsuyama 790-8577, Japan

Kevin V. Thomas
Norwegian Institute for Water Research (NIVA), Gaustadalléen 21, NO-0349 Oslo, Norway

Mike Waldock
Cefas Weymouth Laboratory, Fish Diseases Laboratory, The Nothe, Barrack Road, Weymouth, Dorset, DT4 8UB, UK

Part I
Overview of Antifouling Biocides

Chapter 1
Release Rate of Biocides from Antifouling Paints

Kazunobu Takahashi

1.1 Introduction

In October 2001, the International Maritime Organization (IMO) diplomatic conference adopted the draft convention prepared by the Marine Environment Protection Committee (MEPC) of IMO for the "Control of Harmful Anti-Fouling Systems on Ships" (IMO-AFS2001). This international convention banned the application of organotin based antifouling paints by 1 January 2003, with a total ban on the presence of organotin by 1 January 2008. The convention was developed to immediately ban the use of organotin compounds such as tributyltin (TBT) and triphenyltin (TPT) globally in antifouling paints to protect the marine environment. The ban on TBT came about because TBT has extensive detrimental effects on non-target marine organisms. IMO-AFS2001 not only banned organotin, but also encouraged development of the alternative tin-free antifouling systems (i.e. environmentally friendly antifouling systems) (IMO 1999; 2001).

Additionally, the ban to use TBT-antifouling paints has resulted in increased research interest in developing alternative tin-free antifouling paints containing biocides that must be effective to control growth of organisms on submerged ship's hull (Vallee-Rehel et al. 1998; The Japan Shipbuilding Research Association 1993; Omae 2003) The environmental fate and aquatic toxicological profile of these tin-free booster biocides in the marine environment have been studied by many researchers (Okamura et al. 2002; Turley et al. 2000; Callow and Willingham 1996; HSE 2005; Harino 2004; Harino et al. 2005; Konstantinou and Albanis 2004). Here,

K. Takahashi
Marine Antifouling and Environment Consultant (MAEC):
2-6-13-1008 Shigita-higashi, Jyoto-ku, Osaka 536-0017, Japan
ISO/TC35/SC9/WG27 Japan Working Group member

T. Arai et al. (eds.), *Ecotoxicology of Antifouling Biocides*,
DOI: 10.1007/978-4-431-85709-9_1, © Springer 2009

the term 'booster biocides' means a group of compounds normally used in addition to copper compounds such as cuprous oxide (Cu_2O) and cuprous thiocyanate (CuSCN) in antifouling paint formulations. Moreover, the ideal biocides should have the following characteristics (IMO 1999):

1. Broad spectrum activity
2. Low mammalian toxicity
3. Low seawater solubility
4. Low bioaccumulation in the food chain
5. Not persistent in the environment
6. Compatible with paint raw materials
7. Favourable price/performance

As typical candidates of the tin-free booster biocide, Sea-Nine 211 (DCOIT), Irgarol 1051 (CDMTD), Zineb, Ziram (PZ), Preventol A6 (Diuron), Chlorothalonil, Preventol A4-S (Dichlofluanid), Preventol A5-S (Tolylfluanid), Copper Omadine (CuPT), Zinc Omadine (ZnPT) and PK (pyridine-triphenylborane) have been used widely in the commercial TBT-free antifouling paints and copper-free antifouling paints in recent years (Okamura and Mieno 2006).

Therefore, in order to understand both antifouling performance of tin-free antifouling paints and assess the potential risk of copper and booster biocides in the marine environment, it is necessary to determine the release rate or leaching rate of biocides from antifouling paints immersed into seawater (Thouvenin et al. 2002; Samui et al. 1997; European Commission 2002; Takahashi and Ohyagi 1988). This chapter briefly reviews the antifouling paint technology (conventional, ablative, self-polishing copolymer and biocide-free antifouling paints); the methods for determination of release rate biocides from antifouling paints (Ketchum method, ASTM/ISO standard method, US Navy/dome method and CEPE/mass-balance calculation method); the release behavior of biocides from antifouling paints; and the release rate of biocides for conducting the environmental risk assessment.

1.2 Antifouling Paint Technology

In general, effective control of biofouling on ship's hulls has been achieved by means of antifouling paints containing copper compounds and one or more booster biocides. These biocides are present at the paint-water interface and prevent the settling of fouling organisms such as barnacles, mussels, tubeworms and algae (Bowner and Ferrari 1989). Thus, for effective prevention of fouling, a defined and constant concentration of the copper and booster biocide is desirable at the surface of paint, i.e. effective concentration. The release rate of copper and booster biocides from antifouling paints has to remain at this critical value (i.e. Cu: $10\,\mu g/cm^2/day$) for an extended period (Woods Hole Oceanographic Institute 1952). TBT-free antifouling paints are classified into four basic categories by the Japan Paint Manufacturers Association (2005).

1.2.1 Conventional Antifouling Paints

Conventional antifouling paints are composed of a tough insoluble binder such as chlorinated rubber and polyvinyl chloride and rosin within which high concentrations of Cu_2O and booster biocides are physically dispersed in the hard matrix (i.e. free-association antifouling paint). When these paints are newly immersed in seawater, the free associated biocides present near surface of paint are able to diffuse and easily release out from the matrix of the paint film. These paints suffer the inherent defect that the release of biocides decays exponentially with time, resulting in an excessive release of biocides during the early part of the paint's lifetime. Consequently, these paints are effective for only a relatively short lifetime (about 1–1.5 years).

1.2.2 Ablative/Hydration Antifouling Paints

Ablative/hydration antifouling paints differ from conventional hard matrix paints because the paint film matrix is not a hard durable material designed to last long time. In these paints, the film matrix is a mixture of soluble/hydration binder such as vinyl chloride-vinyl acetate copolymer and vinyl chloride-vinyl isobutyl ether copolymer and rosin that are designed to ablate over time, thus allowing the biocides to be released. These paints typically possess a lifetime of 1.5–2 years, but have a major disadvantage of having poor film characteristics, especially after dry out.

1.2.3 Self-polishing Copolymer Antifouling Paints

The release behavior of biocides from self-polishing copolymer (SPC) anti-fouling paints is fundamentally different to that in conventional and ablative antifouling paints. The previous TBT-SPC was a copolymer with tributyltin methacrylate (TBTMA) and methyl methacrylate (MMA) (Atherton et al. 1979). Seawater is able to interact with the hydrophilic copolymer, initiating a hydrolysis reaction that cleaves TBT from the TBT-copolymer (**TBT-acrylate**) and releases it into seawater as shown in Fig. 1.1.

On the other hand, TBT-free SPCs are acrylic copolymers with pendant hydrolysable functions and a variable hydrophobic/hydrophilic balance. Seawater penetration is essential to the biocides release and erosion phenomena. Hydrolysable units lead to a progressive degradation of immersed binder and to initiation and control of erosion rate. The normal lifetime of SPC antifouling paints is typically 3–5 years. The chemical structures of TBT-free SPC such as copper-acrylic copolymer (**Cu-acrylate**), zinc-acrylic copolymer (**Zn-acrylate**)

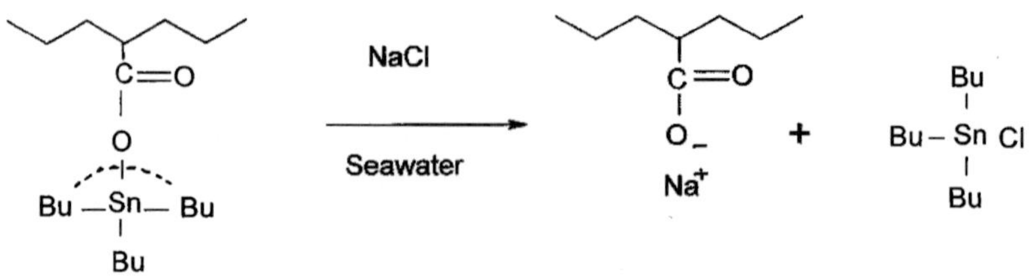

Fig. 1.1 TBT-copolymer (MMA/TBTMA) and hydrolysis reaction of copolymer

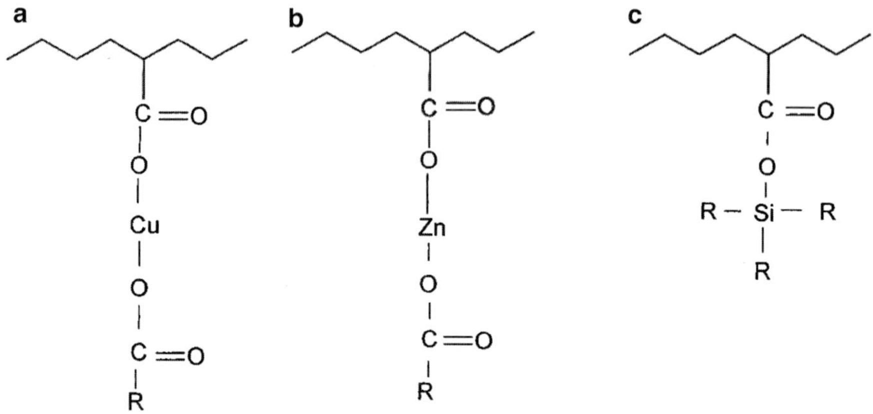

Fig. 1.2 The chemical structures of TBT-free SPC such as Cu-acrylate, Zn-acrylate and Si-acrylate a: Cu-acrylate, b: Zn-acrylate and c: Si-acrylate

and silyl-acrylic copolymer (**Si-acrylate**) are shown in Fig. 1.2. The release behaviour of biocides from conventional, ablative, and SPC antifouling paints are shown diagrammatically in Fig. 1.3. Moreover, Table 1.1 summarizes commercially available TBT-free self-polishing copolymer antifouling paints from around the world.

1.2.4 Biocide-Free Antifouling Paints

Biocide-free antifouling paints (i.e. foul release coatings) are composed of a silicone elastomer and a fluoropolymer that prevent attachment of fouling organisms without the use of biocides. The ultra smooth, low surface energy and slippery surface of these paints allows control of fuel efficiency, speed and maintenance costs since

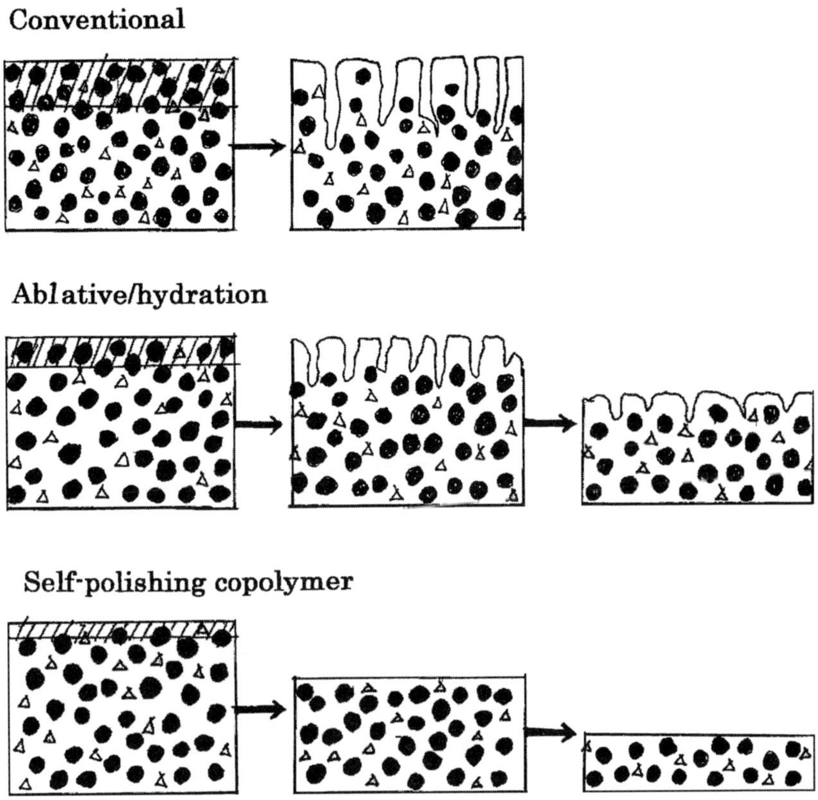

Fig. 1.3 The release behaviors of biocide from conventional, ablative and SPC antifouling paints ●: Cu_2O and Δ: Booster biocide

fouling organism are typically unable to attach, or have difficulty settling on the surface of the ship's hulls (Watermam et al. 2003).

1.3 Methods for the Determination of Biocide Release Rate

1.3.1 Ketchum Method

The Ketchum method was designed at the Woods Hole Oceanographic Institution, USA and measured the release rate of copper from antifouling paints (Ketchum 1952). This method was developed to provide a static laboratory procedure to measure changes in the release rates of biocide that occur during a period of immersion under specified conditions of constant temperature, pH, salinity and

Table 1.1 The commercial products of TBT-free self-polishing copolymer antifouling paint in the world

Company	Product name	Binders	Biocides	Performance	Remark
International Paint	Intersmooth 360/365	Cu-acrylate	Cu_2O/ZnPT	5 Years	Coastal vessel
	460/465				Deep-sea vessel
	Interswift 655	Cu-acrylate/Rosin	Cu_2O/CuPT	3 Years	Coastal vessel
Jotun	Sea Quantum Plus	Silyl-acrylate/Rosin	Cu_2O/CuPT	5 Years	Deep-sea vessel
	Classic				
	Ultra				
Sigma Coatings	Alphagen 50/20/10	Polyvinylpyrrolidone	Cu_2O/Sea-Nine211	5 Years	Deep-sea vessel
	Alpha Trim		Cu_2O/Irgarol1051	3 Years	Coastal vessel
Nippon Paint Marine	Ecoloflex SPC 100/200	Cu-acrylate	Cu_2O/ZnPT	5 Years	Deep-sea vessel
	400/600				Coastal vessel
	Hisol 100/200/300/400	Zn-acrylate	PK/ZnPT	3 Years	Coastal vessel
Hempel	Globic NCT	NAD	Cu_2O/CuPT	5 Years	Deep-sea vessel
	Globic 81900/81950	Fiber Technology	Cu_2O/Sea-Nine211		
Chugoku Marine Paint	Sea Grandprix 1000/	Silyl acrylate/Rosin	Cu_2O/CuPT	5 Years	Deep-sea vessel
	2000				
	Sea Grandprix 500	Zn-acrylate/Rosin	Cu_2O/CuPT	3 Years	Deep-sea vessel/coastal vessel
	Sea CF-10	Zn-acrylate	PK/ZnPT		
	Sea Frontier	Zn-acrylate	PK/ZnPT		Coastal
NKM	Sea Quantum	Silyl-acrylate/Rosin	Cu_2O/CuPT	5 Years	Deep-sea vessel
	Shin LLL			3 Years	Coastal vessel

low biocide concentrations in the surrounding seawater. The outline of modified Ketchum method is as follows:

1.3.1.1 Release Procedure

A flat panel (10 × 15 cm) coated twice with antifouling paint is immersed in a beaker containing 1,500 ml of artificial seawater (pH: 8.0–8.2) placed in a constant temperature room (temperature: 25 ± 2°C). While the contents of the beaker are bubbled by means of an air bubbler (about: 100 ± 10 ml/min), a release test is carried out for 1–2 h. After the release test, the concentration of biocide in artificial seawater is determined by GF-AAS, HPLC and GC. Then, the flat panel used for the release test is immersed again in a holding tank containing artificial seawater. The holding tank for aging of antifouling paint is equipped with units, according to ASTM D5108-90 and ASTM D6442-99. The holding tank is maintained at a temperature between 21°C and 26°C and pH of 8.0–8.2. The flow rate of seawater and the size of the filter are selected to maintain biocide concentration below 100 μg/l.

1.3.1.2 Calculation of the Release Rate of Biocides

The release rate of biocide is calculated from the concentration of the released biocide in the artificial seawater by the following equation:

$$\begin{aligned} R &= (C \times V \times D)/(T \times A) \\ &= (C \times 1.5 \times 24)/(1 \times 200) \\ &= C \times 0.18 \end{aligned}$$

where:

R: the release rate of biocide, in $\mu g/cm^2/day$
C: the concentration of biocide, in μg/l
D: the number of hours per day (= 24)
V: the volume of seawater in the measuring container, in l (=1.5)
T: the time of flat panel immersed in the measuring container (=1 h)
A: the surface area of the paint film (= 200 cm^2)

1.3.2 ASTM/ISO Standard Method

The ASTM/ISO standard method for determination of biocide release rates has been based on the "rotating cylinder method". These methods are the standardized laboratory methods using a rotating cylinder device (diameter: 64 ± 5 mm). Figure 1.4 shows the release rate measuring container and test cylinder device in ISO 15181-1 (ISO 2000) The test cylinders are painted with antifouling paint which, together with an uncoated cylinder are immersed in flowing artificial seawater in a holding tank at (25 ± 1)°C. At specified time intervals, the test cylinders are removed from the holding tank and exposed in measuring containers containing 1,500 ml of artificial

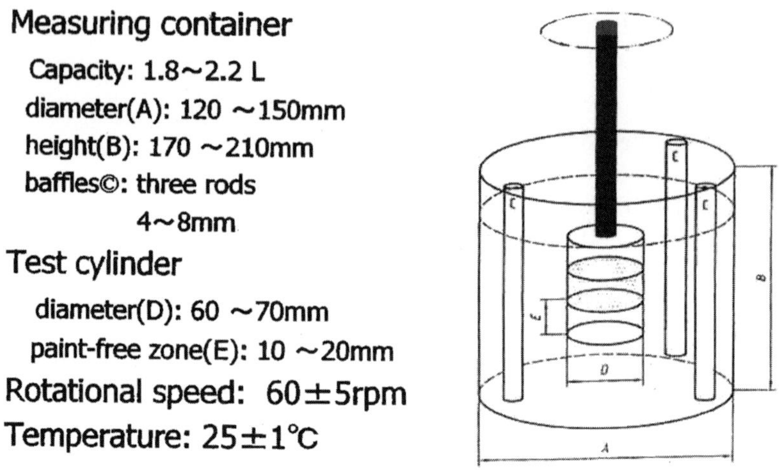

Fig. 1.4 The release rate measuring container and test cylinder in ISO 15181-1 standard method
1. Release rate measuring container
2. Test cylinder
3. Painted zone

seawater before being replaced in the holding tank. The test cylinders in measuring containers are rotated a fixed speed of 60±5 rpm (i.e. about 0.4 knots) for 1 h. The concentration of the biocide released into the artificial seawater of the measuring containers can then be determined. This operation is repeated at defined time intervals (1, 3, 7, 10, 14, 21, 24, 28, 31, 35, 38, 42 and 45 day). Then, the release rate of biocide on each test days, the 14-day cumulative release ($\mu g/cm^2$) and the average release rate ($\mu g/cm^2/day$) from 21 day to the final test day (45 day) are calculated.

ASTM Committee D01 has been establishing the standard test methods as follows:

ASTM D5108-90: Standard Test Method for Organotin Release Rates of Antifouling Coating Systems in Sea Water.

ASTM D6442-99: Standard Test Method for Copper Release Rates of Antifouling Coating Systems in Seawater.

ASTM D6442-06: Standard Test Method for Determination of Copper Release Rate from Antifouling Coatings in Substitute Ocean Water.

ASTM D6903-07: Standard Test Method for Determination of Organic Biocides Release Rates from Antifouling Coatings in Substitute Ocean Water. This method covers the release rate of DCOIT (Sea-Nine211), copper pyrithione (CuPT), zinc pyrithione (ZnPT), and CDMTD (Irgarol 1051).

On the other hand, ISO/TC35/SC9 has been developing the international standard methods for determination of release rate of biocides from antifouling paints and have established as ISO 15181-1, ISO 15181-2, ISO 15181-3, ISO/DIS 15181-4 and ISO/DIS 15181-5 since 1994.

ISO 15181-1: Paints and varnishes – determination of release rate of biocides from antifouling paints – Part 1: General method for extraction of biocides.

ISO 15181-2: Part 2: Determination of copper-ion concentration in extract and calculation of the release rate.

ISO 15181-3: Part 3: Calculation of the zinc ethylene-bis(dithiocarbamate) (Zineb) release rate by determination of the concentration of ethylenethiourea in the extract.

ISO/DIS 15181-4: Part 4: Determination of pyridine-triphenylborane (PTPB) concentration in the extract and calculation of the release rate (in draft).

ISO/DIS 15181-5: Part 5: Calculation of the tolylfluanid and dichlofluanid release rate by determination of the concentration of dimethyltolylsulfamide (DMST) and dimethylphenylsulfamide (DMSA) in extract (in draft).

Table 1.2 shows the operating conditions in ASTM D6442-06 and the revised ISO 15181-1/-2-2007 methods for determination of copper release rate. As shown in Table 1.2, the operating conditions in ASTM D6442-06 and revised ISO 15181-1/-2-2007 are fully harmonized.

Table 1.2 The operating conditions in ASTM D6442-06 and ISO 15181-1/2/2007

	ASTM D6442-06 Copper	ISO15181-1/2/2007 Extraction and copper
Test cylinder:		
Materials	Polycarbonate cylinder (n = 3)	Polycarbonate cylinder (n = 3)
Painted area	200 cm^2	200 cm^2
Diameter	64 mm	59–69 mm
Drying condition	7 ± 1 days/25±2°C	7 ± 1 days/25 ± 2°C
Dry film thickness	Min 100 μm	Min 100–200 μm
Release rate container:		
Materials	Polycarbonate container	Polycarbonate container
Dimension	Diameter: 13.5 cm Height: 19 cm Capacity: 2 l	Diameter: 12–15 cm Height: 17–21 cm Capacity: 1.8–2.2 l
Volume used	1,500 ml	1,500 ml
Temperature	25 ± 1°C	25 ± 1°C
Rotating speed	60 ± 5 rpm/1 h	60 ± 5 rpm/1 h
Test frequency/day	1,3,7,10,14,21,24,28,31,35, 38,42,45	1,3,7,10,14,21,24,28,31,35, 38,42,45
Holding tank:		
Seawater	Substitute ocean water: ASTM D1141	Artificial seawater: ASTM D1141(6)
Temperature	25 ± 1°C	25 ± 1°C
PH	7.9–8.1	7.9–8.1
Salinity	33–34‰	33–34‰
Filter	Carbon filter	Carbon filter + ion exchange filter
Biocide limit:	Cu max 100 ppb	Cu max 100 ppb
Analytical method	GF-AAS	GF-AAS

1.3.3 Others

1.3.3.1 US Navy/Dome Method

The US Navy had developed a field test method for measuring in-situ biocide release rates using a polycarbonate plastic dome (diameter: 30 cm) placed on the painted surface of ship's hull (Seligman and Neumeister 1983). This equipment and technique are similar to those disclosed by Valkirs et al. (2003) (Fig. 1.5). The Space and Naval Surface Warfare Center, San Diego (SSCSD) has recently used the dome method to measure in-situ copper release rate from submerged painted panels and Navy vessel's hull in natural seawater at San Diego Bay, USA (Haslbeck and Ellor 2005).

1.3.3.2 CEPE/Mass-Balance Calculation Method

The mass-balance calculation method for determination of biocide release rate was accepted by regulatory authorities in the European Association of Paint Manufacturers (CEPE) (Hunter 2004). This calculation method is based in a predictive model approach to calculate the average release rate of biocides from antifouling paints (Fig. 1.6). This model was derived from experience of ASTM/ISO method for copper and organotin release rates. The calculation is constructed from the volume of dry paint film applied, its actual loading of biocide and the specified lifetime of the paint. The model assumes that all paint is consumed during the specified lifetime. The high release rate of biocide that typically occurs during the first 14 days of the paint lifetime is also taken into account.

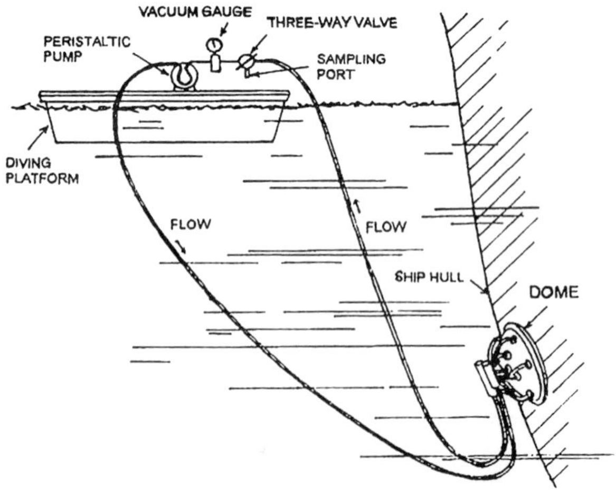

Fig. 1.5 US Navy/dome method system

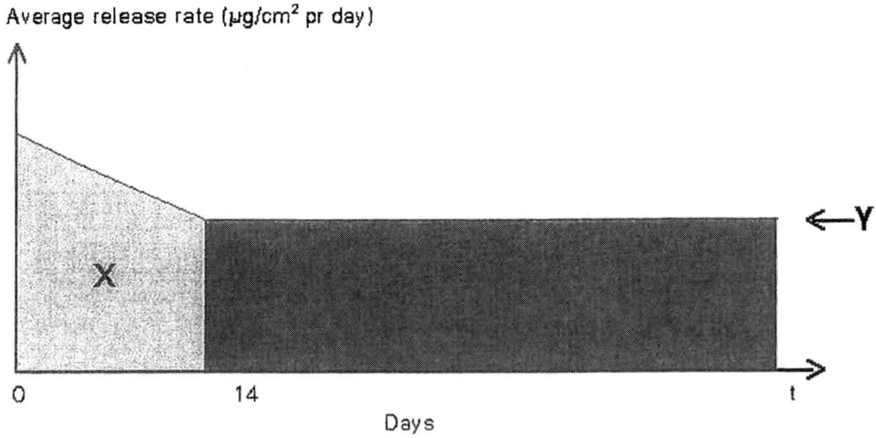

Fig. 1.6 CEPE/mass-balance calculation model X: the amount of biocide released during the first 14 days (µg/cm^2); Y: the average release rate during the rest of the lifetime (µg/cm^2/day)

The calculation of total cumulative amount of biocide released during the lifetime t (µg/cm^2) is made according to the following equations:

1. $X + (t - 1/2) \times 30 \times Y = La \times a \times Wa \times 100/SVR \times SPG \times DFT$
2. $X/Y = 30$

where:

X: the amount of biocide released during the first 14 days (µg/cm^2)
Y: the average release rate during the rest of the lifetime (µg/cm^2/day)
a: the mass fraction of active ingredient in the biocide (organic biocide a = 1; copper in cuprous thiocyanate a = 0.522; copper in cuprous oxide a = 0.86)
Wa: the concentration of biocide in the wet paint in weight %
SPG: the specific gravity of wet paint (g/cm^3)
SVR: the solid volume ratio (volume of dry paint versus volume of wet paint) (in %)
DFT: the dry film thickness specified for the time t (µm)
t: the specified lifetime of the paint (months)
La: the fraction of the active ingredient in the dry film released during the lifetime t
30: 1 month = 30 days
1/2: half a month (14 days)

1.4 Release Behavior of Biocides from Antifouling Paints

1.4.1 Copper Compounds (Cu_2O and CuSCN)

From the results obtained in the round robin test for copper release rate (Cu: µg/cm^2/day) by ISO/WG27 Japan WG (1997), Fig. 1.7 shows the relationship between immersion time and copper release rate from various antifouling paints (US121, RT-1,

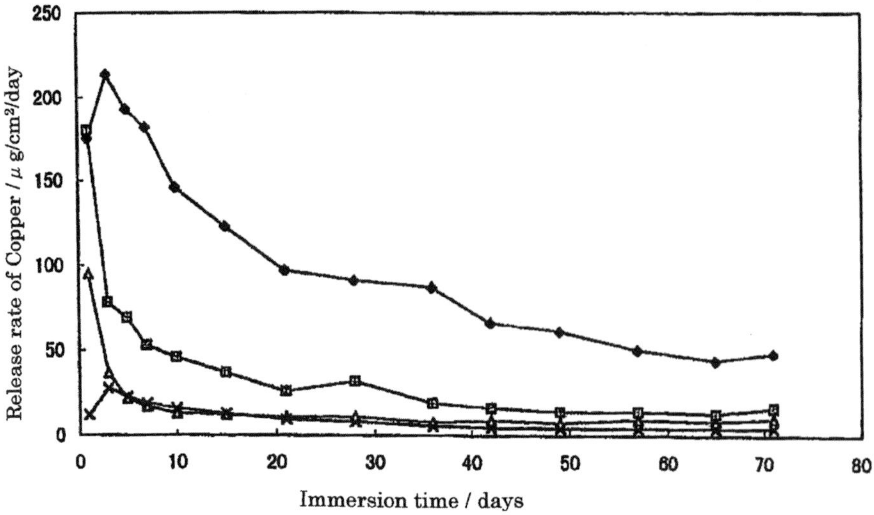

Fig. 1.7 The relationship between immersion time and copper release rate from antifouling paints ♦: US121, ×: RT-1, □: M150 and Δ: F5

M150 and F5). A comparison of copper release rate curves for TBT-SPC antifouling paints (M150 and F5) and ablative antifouling paints (US121 and RT-1) suggests that TBT-SPC paints shows a far more consistent pattern of copper release than ablative antifouling paints. These results show that copper release rates from TBT-SPC antifouling paints reached steady state at about 40 days, while copper release rates from ablative antifouling paints had not reached steady state at about 70 days.

1.4.2 Organotin Compounds (TBT and TPT)

The release rate curves of TBT and TPT from TBT-SPC and TPT-based antifouling paints are shown in Fig. 1.8. As shown in Fig. 1.8, it was found that TPT release rate (TPT^+:μg/cm²/day) from ablative antifouling paint decreased exponentially with time, while TBT release rate (TBT^+:μg/cm²/day) from TBT-SPC antifouling paint reached steady state after about 20 days (Takahashi 1991).

In addition, Fig. 1.9 shows the TBT release rate curves from TBT-SPC antifouling paints using the ASTM/ISO and Ketchum methods (Takahashi and Ikuta 1989). From Fig. 1.9, it was found that TBT initial release rate from TBT-SPC antifouling paint (A) by ASTM/ISO and Ketchum methods were 24.0 and 17.6 μg/cm²/day, respectively and average release rates at steady state were about 1–2 μg/cm²/day. TBT release rates measured by the ASTM/ISO method were 1.3–1.5 times higher than those by Ketchum method. Moreover, these results suggest that TBT release rates from TBT-SPC antifouling paints reached steady state at about 20 days.

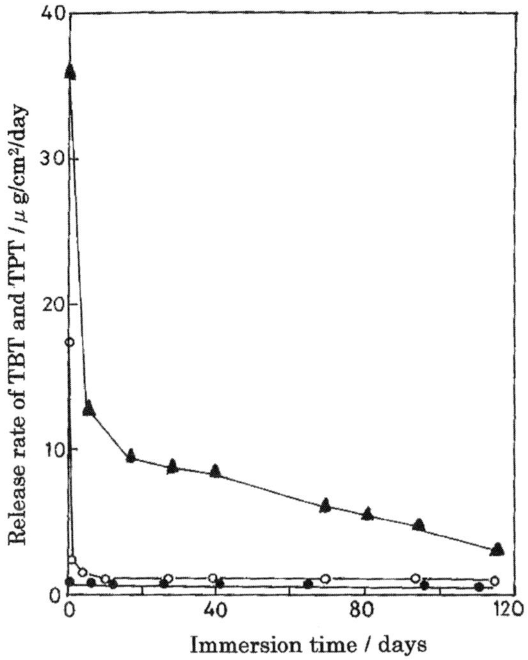

Fig. 1.8 The release rate curves of TBT and TPT from TBT-SPC and TPT-based antifouling paints
○: TBT-SPC AF, ▲: TPTOH-based AF and ●: TPTF-based AF

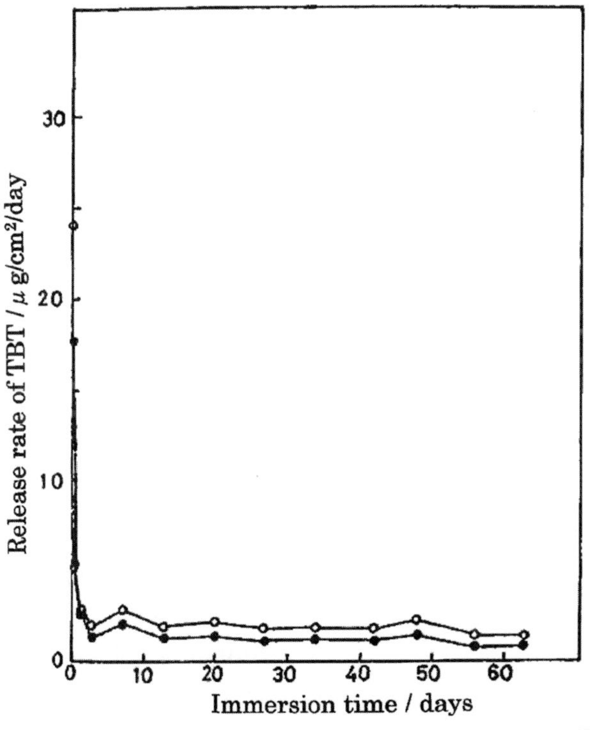

Fig. 1.9 The TBT release rate curves from TBT-SPC antifouling paints
○: ASTM/ISO method and ●: Ketchum method

1.4.3 Sea-Nine211 (DCOIT)

In 1994, the US Environmental Protection Agency (EPA) granted the registration of Sea-Nine211 containing 30% 4,5-dichloro-2-n-octyl-4-isothiazolin-3-one (DCOIT) in xylene as the antifouling biocide. Figure 1.10 shows the relationship between immersion time and release rates of DCOIT from ablative antifouling paints (KR-101 and KR-102) (Takahashi et al. 2002). From Fig. 1.10, the initial release rates of DCOIT from ablative antifouling paints were found to be 38.9–89.6 μg/cm^2/day. After 6 months immersion time, release rates of DCOIT from antifouling paints were about 1.0–2.0 μg/cm^2/day, respectively.

1.4.4 PK (PTPB)

Another tin-free biocide, pyridine-triphenylborane (PTPB) has been widely used in Japan as an alternative, copper-free commercial antifouling paint. Figure 1.11 shows the relationship between immersion time and release rate of PTPB from ablative and SPC copper-free antifouling paints by the ASTM/ISO method (Takahashi et al. 2005).

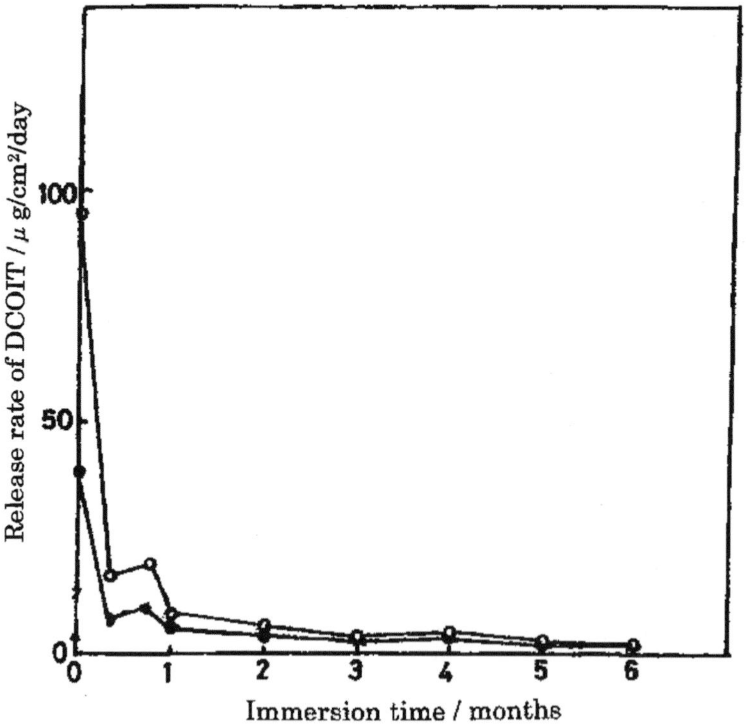

Fig. 1.10 The relationship between immersion time and release rate of DCOIT from ablative antifouling paints
○: KR-101 (AI: 5%) and ●: KR-102 (AI: 3%)

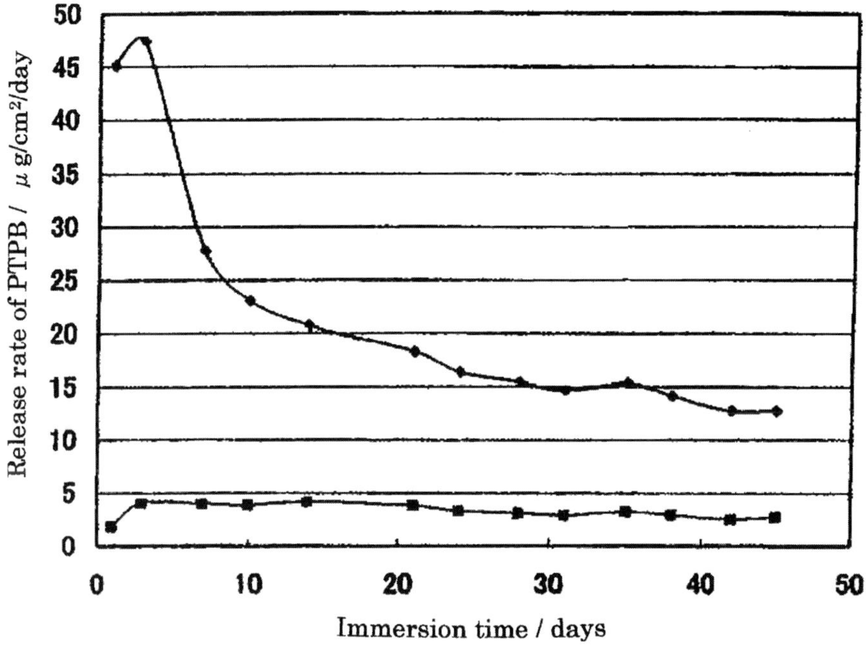

Fig. 1.11 The relationship between immersion time and release rate of PTPB from ablative and SPC copper-free antifouling paints
♦: Ablative copper-free AF and ■: SPC copper-free AF

From Fig. 1.11, the initial release rate of PTPB from ablative and SPC copper-free antifouling paints were found to be 45.1 and 1.8 μg/cm^2/day, respectively.

A comparison of PTPB release rate curves from ablative and SPC copper-free antifouling paints indicates that PTPB release rate from ablative antifouling paint decreased exponentially with time, while PTPB release rate from copper-free SPC antifouling paint reached steady state at about 10 days.

Consequently, it was found that the release rate of biocide from SPC antifouling paints reached steady state after a short time, while the release rate of biocide from free-association (conventional and ablative) antifouling paints decreased exponentially with time and reached steady state at about 4–6 months. In addition, Table 1.3 shows the summary of the release rate of Cu, TBT and booster biocides from antifouling paints.

1.5 Release Rate of Biocides for Environmental Risk Assessment

According to the OECD emission scenario document on antifouling products, the risk assessment for biocides in marine environment can be evaluated by the ratio of the predicted environmental concentration (PEC)/the predicted no-effect concentration

Table 1.3 The summary of release rate of biocides from antifouling paints

Biocides	AF type	Release rate (μg/cm²/day)	Method	Reference
Cu	SPC	15–25	Ketchum	Takahashi (1988)
	SPC	25–40	ASTM/ISO	European Commission (2002)
	Ablative	48–87	ASTM/ISO	Finnie (2006)
	Conventional	131	ASTM/ISO	Finnie (2006)
TBT	SPC	2.0–3.0	ASTM/ISO	Takahashi (1989)
	SPC	1.0–2.0	Ketchum	Takahashi (1989)
	SPC	1.6–5.2	ASTM	US EPA (1987)
TPT	Ablative	7.3	Ketchum	Takahashi (1991)
ZnPT	Ablative	1–3	ASTM/ISO	Turley (2000)
	SPC	3–11	ASTM/ISO	Turley (2000)
Irgarol1051		5.0	ASTM/ISO	van Hattum (2006)
		2.6		van Hattum (2006)
Sea-Nine 211		2.9	ASTM/ISO	van Hattum (2006)
	Ablative	2.0–5.0	Ketchum	Takahashi (2002)
PK	SPC	2.4–3.0	ASTM/ISO	Takahashi (2005)
	Ablative	13.0–15.0	ASTM/ISO	Takahashi (2005)
Diuron		3.3	ISO	van Hattum (2006)

(PNEC) (Fig. 1.12) (OECD 2005). The PEC estimation takes into account emission factors (e.g. release rate, shipping intensities, residence times, ship's hull surface areas), properties and processes of biocide (e.g. Kd, Kow, Koc, volatilization, speciation, hydrolysis, photolysis, biodegradation), and properties and processes related to the specific marine environment (e.g. currents, tides, salinity, pH, temperature, suspended matter load). Therefore, the release rate of biocide is a critical parameter for the environmental risk assessment in calculation of PEC values. MAM-PEC computer model can derive the PEC for biocide in four generic emission scenarios including commercial harbor, marina, shipping lane and open sea (CEPE 1999; Van Hattum et al. 2006).

With regard to the release rate of copper from antifouling paints, Finnie (2006) summarized the release rates of copper from antifouling paints determined by the ASTM/ISO method, CEPE/calculation method and US Navy/dome method as shown in Table 1.4. From Table 1.4, a comparison of the average release rate obtained using the different methods shows that the ASTM/ISO method typically overestimates the environmental release rate of copper (dome method) by factor of 10.4–23.0, while the CEPE/calculation method typically overestimates by factor of 3.9–5.3.

Consequently, the release rates of biocides obtained by the ASTM/ISO method probably do not reflect realistic environmental release rates of biocides for

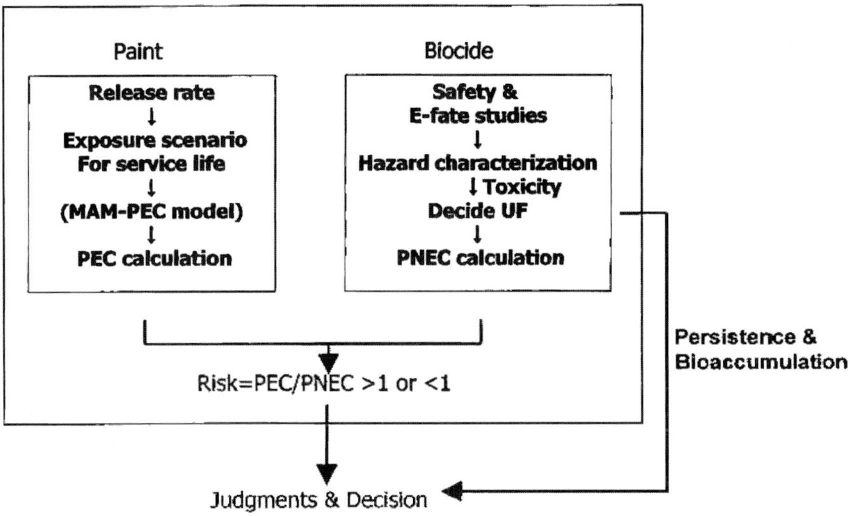

Fig. 1.12 Environmental risk assessment scheme by OECD

Table 1.4 Copper release rates (μg/cm²/day) from each tin-free antifouling paints determined by ASTM/ISO, CEPE/calculation and US Navy/dome methods

Method	BRA640 ablative	BRA540 ablative	Ablative A ablative	Formula 121 conventional	SPC A SPC	SPC B SPC
ASTM/ISO method (21–45 days average)	48.6	58.8	87.0	131.2	100	66.1
CEPE/calculation method	18.4	18.3	15.5	30.4	30.7	30.2
US Navy/dome method (ship's hulls to 758 days)	4.7	4.6	3.9	5.7	–	–
US Navy/dome method (Raft panels to 761 days)	2.2	–	–	–	1.6	1.3
ASTM/ISO: CEPE/calculation ratio	2.6	3.2	5.6	4.3	3.3	3.2
ASTM/ISO: Dome (ship) ratio	10.4	12.8	22.3	23.0	–	–
CEPE/calculation: Dome (ship) ratio	3.9	4.0	4.0	5.3	–	–

antifouling paints. The US Navy/dome method seems to more closely approximate realistic release rates of biocides, but this method is not suitable for widespread use as standard method, since a greater number of people and high costs are required to collect the samples for determination of release rate.

1.6 Conclusions

There are many kinds of methods for the determination of biocide release rates from antifouling paints. The first published ASTM D-5108-90, standard method for organotin release rates was established by ASTM/EPA for regulation of antifouling paints containing organotin compounds. As a result, many governments adopted measures to eliminate the use of antifouling paints with an average TBT release rate of more than 4 $\mu g/cm^2/day$ at steady state (US EPA 1987; WHO 1990). In the case of TBT-SPC antifouling paints, TBT release rates reached steady state in a relatively short time (about 2–3 weeks), so that the average TBT release rate from 21 to 45 days after application are used for regulation in many countries.

However, the release rates of copper from conventional and ablative antifouling paints only reach steady state after a long time (about 4–6 months), such that the average release rate of copper from 21 to 45 days after application, using ASTM/ISO standard methods, do not represent the environmental average release rate of copper over the lifetime of the paints. Especially, it is well known that the release rate of copper from conventional and ablative antifouling paints containing high content of Cu2O decrease exponentially with time. Table 1.5 shows the comparison of parameters in various methods for determination of biocide release rates.

From the discussion document by CEPE AFWG, it is stated in the ASTM/ISO method for determination of copper release rate that 'the test method has not yet been validated to reflect in-situ copper release rates from antifouling paints and therefore should not, at present, be used in the process of generating environmental risk assessment' and 'this test method serves only as a guide for characterization of the early release pattern as well as estimating the steady state release rate of copper from antifouling paints' (OECD 2005).

Hereafter, realistic standard methods for the determination of in-situ release rates of biocides from antifouling paints together with sensitive analytical methods for the monitoring of biocides in coastal seawater samples are necessary in order to

Table 1.5 The parameters in various methods for determination of release rate of biocide

	Laboratory		Field		
Parameters	ASTM/ISO method	Ketchum method	Dome method	Harbor	Sea-lane
pH	7.9–8.1	8.0–8.2	7.6–8.0	7.6–8.0	7.6–8.0
Water movement	Dynamic 0.4 knots	Static Air bubbling	Static	Static	Dynamic 15–20 knots
Seawater	Artificial	Artificial	Natural	Natural	Natural
Temperature (°C)	25 ± 1	25 ± 2	5–28	5–28	5–28
Salinity (‰)	33–34	33–34	–	–	–
Slime layer	No	No	Yes	Yes	Yes
Test period	45 Days	1 Year	1–2 Years	1–2 Years	1–2 Years
Merits/demerits	Standard	Simple	Expensive	–	–

conduct meaningful environmental risk assessments for the regulation of TBT-free antifouling paints.

Acknowledgements The author would like to express sincere thank ISO/TC35/SC9/WG27 Japan WG members for their support and discussions and Dr. Hiroya Harino (Osaka City Institute of Public Health and Environmental Sciences) for his assistance in preparing this manuscript.

References

Atherton D, Verborgt J, Winkeler M A M (1979) New developments in anti-fouling: a review of the present state of the art. J Coat Technol 51: 88–91

Bowner C T, Ferrari G (1989) A new approach to the development and testing of antifouling paints. J Oil Colour Chem Assoc 72: 391–396

Callow M E, Willingham G L (1996) Degradation of antifouling biocides. Biofouling 10: 239–249

CEPE Antifouling WG (1999) Utilization of more environmentally friendly antifouling products-EC project No96/559/3040/DEB/E2, http://www.cepe.org/eas/easpdf/0102B2.pdf

European Commission (2002) Assessment of Antifouling Agents in Coastal Environments (ACE), http://www.pml.ac.uk/ace

Finnie A A (2006) Improved estimates of environmental copper release rates from antifouling products. Biofouling 22: 279–291

HSE (2005) Advisory Committee on Pesticides No223, Evaluation on Copper Pyrithione as a New Active Ingredient in Professional Antifouling Products, p.75

Harino H (2004) Occurrence and degradation of representative TBT free-antifouling biocides in aquatic environment. Coast Mar Sci 29: 28–39

Harino H, Mori Y, Yamaguchi Y et al. (2005) Monitoring of antifouling booster biocides in water and sediment from the port of Osaka, Japan. Arch Environ Contam Toxicol 48: 303–310

Haslbeck E, Ellor J A (2005) Investigating tests for antifoulants: variation between laboratory and in-situ methods for determining copper release rates from navy-approved coatings. J Protective Coating Linings, August, 34–44

Hunter J E (2004) Regulation and registration of anti-fouling coatings in the European Union, in "Proceeding of International Symposium on Antifouling Paint and Marine Environment", p.11, Tokyo

IMO (1999) Focus on IMO. Anti-Fouling Systems, IMO London

IMO (2001) Anti-Fouling Systems, International Convention on the Control of Harmful Anti-Fouling Systems on Ships, IMO London

ISO-15181-1 (2000) Determination of release rate of biocides from antifouling paints – Part 1: general method for extraction of biocides

ISO/TC35/SC9/WG27 Japan WG (1997) ISO ring test for copper biocide release rate determination, p.1–18

Konstantinou I K, Albanis T A (2004) Worldwide occurrence and effects of antifouling paint booster biocides in the aquatic environment: a review. Environ Int 30: 235–248

Ketchum B H (1952) "Marine Fouling and Its Prevention", p.338, US. Naval Institute Press, MD

Omae I (2003) General aspects of tin-free antifouling paints. Chem Rev 103: 3431–3448

Okamura H, Mieno H (2006) Present status of antifouling systems in Japan: tributyltin substitutes in Japan. "Handbook of Environmental Chemistry, Antifouling Paint Biocides." Vol 5, Part O, p.201, Springer

Okamura H, Watanabe T, Aoyama I et al. (2002) Toxicity evaluation of new antifouling compounds using suspension-cultured fish cells. Chemosphere 46: 945–951

OECD (2005) Series on emission scenario documents No.13, emission scenario document on antifouling products, http://www.oecd.org/dataoecd/33/37/34707347.pdf

Samui A B, Hande V R, Deb P C (1997) Synthesis and characterization of copoly(MMA-MA)-Cu complex and study on its leaching behavior. J Coat Technol 69: 67–72

Seligman P A, Neumeister J W (1983) In-Situ. Leach Rate Measuring System. US Patent 4375451

Takahashi K (1991) Measurement of the leaching rates of tributyltin and triphenyltin compounds from antifouling paint by gas chromatography. J Oil Colour Chem Assoc 74: 331–338

Takahashi K, Ikuta H (1989) Comparison of ASTM/EPA and bubbling methods for the determination of tributyltin leaching rate. J Jpn Soc Colour Mater 62: 466–473 (in Japanese)

Takahashi K, Ohyagi Y (1988) The leaching behaviors of copper and tributyltin from self-polishing type antifouling paints. J Jpn Soc Colour Mater 61: 208–214 (in Japanese)

Takahashi K, Ebara M, Mabuchi K et al. (2002) Determination of leaching rate of sea-nine 211 active ingredient, 4,5-dichloro-2-n-octyl-4-isothiazolin-3-one (RH-287) from antifouling paints by gas chromatography. J Jpn Soc Colour Mater 75: 365–370

Takahashi K, Yoshikawa E, Akiyama M et al. (2005) Determination of release rate of pyridine-triphenyborane (PTPB) from copper-free antifouling paints. J Jpn Soc Colour Mater 72: 50–57

The Japan Shipbuilding Research Association. SR-209 meeting (1993) "Technical report of investigation and research for new antifoulants" (in Japanese)

The Japan Paint Manufacturers Association (2005) IMO antifouling system registration list, voluntary control in compliance with IMO AFS Convention, http://www.toryo.or.jp/jp/anzen/imo/index.html

Thouvenin M, Peron J J, Charreteur C et al. (2002) A study of the biocide release from antifouling paints. Prog Org Coat 44: 75–83

Turley P A, Fenn R J, Ritter J C (2000) Pyrithiones as antifoulants, environmental chemistry and preliminary risk assessment. Biofouling 15: 175–182

US EPA (1987) "Tributyltin Technical Support Document Position Document 2/3", Washington, DC

Vallee-Rehel K, Mariette B, Hoarau P A et al. (1998) A new approach in the development and testing of antifouling paints without organotin derivatives. J Coat Technol 70: 55–63

Valkirs A O, Seligman P F, Haslbeck E et al. (2003) Measurement of copper release rates from antifouling paint under laboratory and in situ conditions: implications for loading estimation to marine water bodies. Mar Pollut Bull 46: 763–779

van Hattum B, Baart A, Boon J (2006) Emission estimation and chemical fate modeling of antifoulants. "Handbook of Environmental Chemistry, Antifouling Paint Biocides." Vol 5, part O, p.101, Springer

Watermam B, Daehne B, Michaelis H et al. (2003) "Performance of biocide free antifouling paints – Trials on deep-sea going vessels" Vol 1–3, http://www.limnomar.de/conthtml/eng/mp2_publications_af.htm

WHO (1990) "Environmental Health Criteria 116, Tributyltin compounds", p.40, Geneva

Woods Hole Oceanographic Institute (1952) "Marine Fouling and Its Prevention", p.287, US Naval Institute Press, Annapolis, MD

Chapter 2
International Trends in Regulatory Aspects

Tetsuya Senda

Abbreviations AFS Convention: The International Convention on the Control of Harmful Anti-Fouling Systems on Ships; BPD: Biocidal Products Directive; CG: Correspondence Group; EPA: Environmental Protection Agency; FSUs: Floating storage units; FIFRA: Federal Insecticide, Fungicide, and Rodenticide Act; FPSOs: Floating production storage off-loading units; GESAMP: The Group of Experts on Scientific Aspects of Marine Environmental Protection; GT: Gross Tonnage; IMO: International Maritime Organization; ISO: International Standard Organization; MEPC: Marine Environmental Protection Committee; NOEC: No-Observed-Effect Concentration; OECD: Organization for Economic Co-operation and Development; PEC: Predicted Environmental Concentration; PNEC: Predicted No-Effect Concentration; TBT: Tributyltin; TPT: Triphenyltin; UN: United Nations; WG: Working Group

2.1 Introduction

Fuel consumption of ships is highly dependent on frictional loss occurring between ship's hull and sea water. Fouling of the ship's hull by marine organisms including barnacles, algae and molluscs increases friction, resulting in an increase in fuel consumption and/or a decrease in ship's speed. A number of chemical compounds have been used to prevent those organisms attaching to hulls. A typically used antifouling system involves coating the ship's hull with paint containing substances preventing attachment of organisms. Organotin compounds, tributyltin (TBT) and triphenyltin (TPT) were found to be excellent in efficacy as anti-foulants and also less harmful to paint workers than traditional chemicals including mercury or arsenical compounds. Development of self-polishing organotin co-polymer has produced extremely high performance and long life with an additional effect of keeping the surface smooth by the self-polishing mechanism.

T. Senda
National Maritime Research Institute, 6-38-1 Shinkawa, Mitaka, Tokyo 181-0004, Japan

T. Arai et al. (eds.), *Ecotoxicology of Antifouling Biocides*,
DOI: 10.1007/978-4-431-85709-9_2, © Springer 2009

In the 1970s most of the ships in the world bore organotin based antifouling paints on their hulls. The extent of the use of organotin compounds caused adverse effects on marine organisms particularly on molluscs. Legislative control of antifouling systems was introduced firstly in individual countries and since then there has been a trend towards worldwide regulation. After patient discussion for more than 10 years, an international treaty banning the use of organotin compounds in antifouling systems was adopted at the International Maritime Organization (IMO). In this chapter, the discussion at IMO is firstly reviewed and, subsequently, the treaty controls are described in detail. Finally, regulations for tin-free antifouling systems, currently being implemented in Europe and the United States, are briefly summarized.

2.2 IMO's Effort

In the 1980s, high concentrations of organotin compounds were reported in coastal areas around the world. Because of worrying side-effects of TBT on oysters, France was the first to prohibit the use of organotin-containing paints on ships less than 25 m long, in 1982. TBT-related imposex was then recorded in English coastal waters. Since then, similar legislation was introduced by the United Kingdom, in 1985, and by the United States of America, in 1988. Canada, Switzerland, Austria and Germany subsequently followed this course of action. The Shipbuilders' Association of Japan and the Japanese Shipowners Association collaboratively decided themselves to control the use of TBT based antifouling paints in Japan.

Since shipping is a worldwide activity, controls on antifouling systems by individual states are not effective enough to prevent pollution, even in their own sea area. Therefore, the problem was brought to the International Maritime Organization. IMO is a specialized agency of the United Nations with 167 member states and three associate members, and its purpose is to develop and maintain a comprehensive regulatory framework for shipping including safety and environmental concerns. Environmental issues are normally deliberated by the Marine Environmental Protection Committee (MEPC).

At the 26th session of MEPC held in September 1988, the antifouling paints issue was discussed for first time after the Paris Commission, an organization concerned with prevention of pollution of the North East Atlantic, requested IMO to consider the need for taking measures to restrict the use of TBT compounds on seagoing vessels. Member states were invited to submit information, explaining the ecological effects that TBT compounds might be causing.

At MEPC 29 in March 1990, the antifouling paint issue was included in the agenda for the first time and the United States was appointed the lead country to collect more information. At the following session, MEPC 30, in November 1990, the Resolution MEPC.46(30) entitled "Measures to Control Potential Adverse Impacts Associated with the Use of Tributyl Tin Compounds in Anti-Fouling Paints" was adopted. The resolution included recommendations to eliminate the use of antifouling paints with an average release rate of organotin higher than $4\,\mu g\ cm^{-2}\ day^{-1}$). The resolution also recommended member states of IMO to develop alternative antifouling systems to tin-containing paints and to engage in monitoring to evaluate the effectiveness of control measures that had already been adopted.

At its 38th session in July 1996, MEPC noted monitoring studies undertaken worldwide, the action taken by Japan to ban TBT based antifouling paints, and the progress made in the development of alternative systems. MEPC established a Correspondence Group (CG), consisting of 12 states and four non-governmental organizations (NGO's), coordinated by the Netherlands, to promote further reduction in the use of antifouling paints containing TBT.

The CG submitted the final report to MEPC 41 in March to April 1998 providing conclusions, expressed by the majority of the CG members, that mandatory measures were required to reduce, and eventually eliminate, the use of antifouling systems containing organotin compounds. The CG reports' conclusions are summarized as follows:

(a) Most supported the application of the "precautionary approach" which is based on the idea that, if there is a doubt about harm to the environment from a particular product or action, it should not be done.
(b) Further research on harmful effects of antifouling systems on the marine environment was necessary to focus attention and increase awareness (though many countries saw no need for further evidence of harmful effects of TBT).
(c) Development of alternative, less harmful antifouling systems should be encouraged.
(d) Alternative biocidal systems should be shown to be considerably less harmful to the marine environment than TBT.
(e) Criteria for substances and/or methods used as antifouling systems should be developed.
(f) Interim measures limiting the use of TBT on larger vessels with docking intervals of 2.5 years or less would be needed ahead of a total ban.
(g) As a long-term issue, mandatory measures needed to be developed to bring about a complete ban on TBT, to prevent unfair competition.
(h) An instrument containing mandatory measures for antifouling systems should be developed as a matter of urgency, taking into account enforcement and enforceability of such measures.

After discussion, the MEPC established a Working Group (WG) at the 42nd session and agreed to begin work on drafting regulations to phase out organotin compounds in antifouling systems. Those which opposed or hesitated to establish a treaty for banning TBT raised some viewpoints as follows. Firstly, from a technical viewpoint, it was argue that there were not enough efficient alternative products and therefore uncontrollable hull fouling would lead to potential corrosion and safety hazards. Secondly, from a marine environmental aspect, unknown environmental risks could emerge due to an increased usage of alternative biocides and their metabolites: bioaccumulation potential of organic biocides may become more severe. Thirdly, they pointed out the possibility of some adverse effects on the economy: increased dry-docking for ocean going vessels lowers economic efficiency and, furthermore, unilateral measures could result in losses of business and possible closures of certain shipyards. Finally, concern for global warming was raised, higher heavy oil consumption (resulting from less-efficient antifouling) might contribute to an acceleration of greenhouse gas emission and acid rain effects.

Members states promoting the ban on TBT advocated that there were no essential problems with use of tin-free antifouling systems: alternatives to organotin compounds were already available which could control hull fouling. In fact, the Japanese merchant fleet has coped well with the Japanese ban on TBT (since 1990). Several companies are already selling TBT-free antifouling systems.

Concern over the environmental effects of alternative antifouling products is accepted as a potential problem. The MEPC is looking at how alternative antifouling systems should be assessed, as well as developing regulations to prohibit the use of organotin compounds. Environmental requirements are likely to be included to avoid one harmful substance being replaced with another. Accumulation of alternative antifouling products is unlikely to happen if alternative systems are controlled and the criteria being drawn up for antifouling systems include environmental risks.

Regarding economic disadvantage, some of the new alternative antifouling systems claim to be equally effective as organotin-based systems, allowing dry-docking intervals of up to 5 years. Other systems will require increased frequency of dry-docking, perhaps every 2.5 or 3 years. But for many ships this could fit in with routine surveys or general maintenance. Paint manufacturers are likely to increase research efforts to produce efficient tin-free systems. Figures suggest that increased costs would be small compared to total costs of sea transport.

2.3 AFS Convention

After these discussions, Resolution A.985(21) was adopted at the 21st Assembly (general meeting) of IMO in 1999. It outlined the prospective content of the future convention, subsequently deliberated at the IMO. Following the Resolution, a treaty entitled "The International Convention on the Control of Harmful Anti-fouling Systems on Ships" (AFS Convention) was finally adopted on 5 October 2001 at the Diplomatic Conference of IMO. The AFS Convention, in Article 4, requires the parties (states which accede to the Convention) to prohibit and/or restrict the use of harmful antifouling systems on ships flying their flag, as well as ships not entitled to fly their flag but which operate under their authority and all ships that enter a port, shipyard or offshore terminal of a party. The term "anti-fouling systems" is defined in Article 2(2) as "a coating, paint, surface treatment, surface or device that is used on a ship to control or prevent attachment of unwanted organisms". The specific antifouling systems to be prohibited are given in Annex I to the Convention.

Annex I attached to the Convention presently states that all ships shall not apply or re-apply organotin compounds which act as biocides in antifouling systems by an effective date of 1 January 2003. By 1 January 2008 (effective date), ships either:

(a) Shall not bear such compounds on their hulls or external parts or surfaces or
(b) Shall bear a coating that forms a barrier to such compounds leaching from the underlying non-compliant antifouling systems

The Convention also establishes a mechanism, for future update, to prevent the potential use of other harmful substances in antifouling systems in Article 6. Parties can propose an amendment of Annex I to add any antifouling systems which are believed to be harmful. Receiving such proposals, a "technical group" will be established, according to the Article 7, to include people with relevant expertise. They will review proposals for other substances used in antifouling systems to be prohibited or restricted.

Specific procedures for survey and inspection are given in Annex 4 to the Convention. It requires that ships of 400 GT (gross tonnage) and above, engaged in international voyages (excluding fixed or floating platforms, floating storage units (FSUs) and floating production storage off-loading units (FPSOs)), undergo an initial survey before the ship is put into service or before the International Anti-fouling System Certificate is issued for the first time; and a survey when the antifouling systems are changed or replaced. Ships of 24 m or more in length but less than 400 GT engaged in international voyages (excluding fixed or floating platforms, FSUs and FPSOs) will have to carry a Declaration on Anti-fouling Systems signed by the owner or authorized agent. The Declaration will have to be accompanied by appropriate documentation such as a paint receipt or contractor invoice.

Following Resolution 2, adopted with the Convention, three guidelines have been developed to ensure the implementation of the Convention. These were adopted at the 48th and 49th sessions of MEPC, as follows: Guidelines for survey and certification of anti-fouling systems on ships, Guidelines for brief sampling of anti-fouling systems on ships, and Guidelines for inspection of anti-fouling systems on ships.

Article 18 states that the Convention is to enter into force 12 months after 25 States representing 25% of the world's merchant shipping tonnage have ratified it. Unfortunately, the Convention had not entered into force by the dates described in Annex I, and dates are moved forward to 1 year after the Convention has entered into force. With the accession by Panama on 17 September 2007, the AFS Convention was finally ratified by the required 25 states and a combined 38.1% of world shipping tonnage, and will eventually enter into force on 17 September 2008. The number of states that have subsequently ratified, as of the end of January 2008, now total 28, representing 43.8% of global tonnage. The acceding states are: Antigua and Barbuda, Australia, Bahamas, Bulgaria, Cook Islands, Croatia, Cyprus, Denmark, France, Greece, Iceland, Japan, Kiribati, Latvia, Lithuania, Luxembourg, Mexico, Nigeria, Norway, Panama, Poland, Romania, Saint Kitts and Nevis, Sierra Leone, Slovenia, Spain, Sweden and Tuvalu.

2.4 Antifouling Systems Without Organotin Compounds

2.4.1 Risk Assessment

Since antifouling paints intentionally leach out substances which are active to marine organisms, it is essential to evaluate, prior to use, human health effects (through direct exposure and via food chains) and effects on marine ecosystems, including toxicity to marine organisms. Human and environmental exposure occurs through leaching from coatings on ships' hulls, splashing to air during spraying/painting operations at shipyards, and waste disposal after blasting/scraping of old coatings. Effects on marine ecosystems are a specific issue to antifouling paints, and were a critical factor leading to legislation on organotin compounds. In several countries, assessment of antifouling systems (including both biocides and paint products) requires additional safeguards (above those generally applied to industrial chemicals) to ensure adequate protection for the marine environment. Environmental effects of the test substances are evaluated from properties including persistence in the environment, bioaccumulation and/or bioconcentration, and toxicity to marine organisms.

One method to assess the safety of chemical substances is to give a hazard rating and to apply controls according to this rating. This method is applied to harmful substances intended as cargo, and was proposed by The Group of Experts on Scientific Aspects of Marine Environmental Protection (GESAMP) – an advisory body established in 1969 to advise the United Nations (UN) on scientific aspects of marine environmental protection. In this scheme, the substance is given ratings in terms of various end-points such as bioaccumulation and biodegradation, aquatic toxicity and acute mammalian toxicity. Combination of these ratings determines the hazard profile of the substance which is classified into four categories. Each substance is eventually controlled according to its category, without quantitative considerations.

Another method, that is more frequently used, is the 'environmental risk assessment' which quantitatively evaluates the risk as a product of the hazard and the possibility/probability of exposure. Consideration of the possibility (i.e. concentration) of exposure is essential to this method. Risk assessment of active substances normally consists of the following four processes: hazard identification, dose-response assessment, exposure assessment and risk characterization. In the case of marine ecosystems, for an example, determination of whether or not the substance is likely to be effective to marine organisms is important at the hazard identification stage (to determine the potential hazard of the substance). The dose-response assessment consists of acute and chronic toxicological studies to obtain threshold concentration levels of the substance's activity, such as the no-observed-effect concentration (NOEC). From toxicological data, the predicted no-effect concentration (PNEC) is obtained by applying an appropriate 'safety' factor (determined by considering the extent of uncertainty in the data.

Exposure assessment is carried out to determine the expected concentration of both the active substance and its degradation products (metabolites). Using an appropriate emission scenario, a predicted environmental concentration (PEC) is obtained. Throughout the assessment, a worst-case scenario should be employed – where the lowest toxicity threshold (PNEC) among various tested organisms, and the highest PEC among various emission scenarios, are used. According to the risk characterization process, if the ratio PEC/PNEC is less than unity, the concentration in the environment is likely to be lower than the critical threshold level: risk is then considered as low. If the ratio is more than unity, there may be a risk of deleterious effects and some action to reduce the risk would be recommended.

2.4.2 Regulations in Europe

Registration of antifouling paints for ships has been established in more than ten countries including the United States, Canada, Australia, New Zealand, United Kingdom, Belgium, Ireland, Sweden, Switzerland, Malta and Hong Kong. The European Union has established a unified registration system under the Biocidal Products Directive (BPD), which was adopted in 1998 and entered into force in 2000. Prior to the BPD, many countries in Europe had their own regulations to control the use of antifouling products. In 2005, Belgium, Ireland, Finland, Sweden, The Netherlands, Malta, Denmark and the United Kingdom required registration of antifouling products applied in their shipyards, under their own national laws. These systems were similar, but not identical. All require that both the antifouling products (e.g. paints) and the active ingredients (e.g. biocidal chemicals) should be authorized. All use a risk assessment process to allow use of the product.

The BPD was established as a joint European Council and European Parliament Directive, which introduces an authorization scheme for placing biocidal products on the market, and also for their use. The term "biocidal products", here, is defined as "active substances, and preparations containing one or more active substances, intended to destroy, deter, render harmless, prevent the action of or otherwise exert a controlling effect on any harmful organism by chemical or biological means". The active substance is defined as "a substance or micro-organism, including a virus or a fungus, having a general or specific action on, or against, harmful organisms". This directive applies to those who intend to produce, import and supply the products.

BPD targets include products designed and used for controlling unwanted organisms but excludes plant protection use, medical use, cosmetics and food additives which are regulated by other legislative schemes. The products covered by BPD include 23 types, classified into four groups according to their purpose, as summarized in Table 2.1. Antifouling products are listed as the 21st product in group 4, a group which includes other biocidal products. The BPD requires the producers, importers and suppliers to clarify safety, based on risk assessment methodologies.

In order to have products authorized, applicants should submit a dossier – a set of required data – to one of the member states of European Union (EU). The submitted

Table 2.1 Products covered by BPD

Group	Product type	Products
1	Disinfectants and general biocidal products	
	1	Human hygiene biocidal products
	2	Private and public health area disinfectants and other biocidal products
	3	Veterinary hygiene biocidal products
	4	Food and feed area disinfectants
	5	Drinking water disinfectants
2	Preservatives	
	6	In-can preservatives
	7	Film preservatives
	8	Wood preservatives
	9	Fibre, leather, rubber and polymerised materials preservatives
	10	Masonry preservatives
	11	Preservatives for liquid-cooling and processing systems
	12	Slimicides
	13	Metalworking fluid
3	Pest control	
	14	Rodenticides
	15	Avicides
	16	Molluscicides
	17	Piscicides
	18	Insecticides, acaricides and products to control other arthropods
	19	Repellents and attractants
4	Other biocidal products	
	20	Preservatives for food and feedstocks
	21	Antifouling products
	22	Embalming and taxidermist fluids
	23	Control for other vertebrates

dossier will be evaluated by the state according to the principles described in Annex VI to the Directive. When approved, the products will be included in the Annex I list. Annex IA list the biocides which are evaluated as low risk level. Annex IB is for normally non-biocidal substances which have minor use as biocides (e.g. carbon dioxide and ethanol).

The BPD basically intends to harmonize registration systems for all biocidal products throughout the EU countries and also to ensure a high level of protection for human health and the environment. It should be noted that the BPD introduces a positive list of allowed substances in contrast to the AFS Convention employing a negative list where prohibited substances are listed. If a product is authorized in one member state, mutual recognition should apply throughout the EU countries unless any EU country has special reasons why it should not grant approval. The evaluation by the member state is carried out to determine whether the applied products pose risks for both human health and the environment, or are within allowable thresholds derived from assessment studies. It also requires evidence that the product is effective enough to target organisms at a concentration level which is acceptable, according to the terms of the risk assessment process.

The BPD gives a clear list of all of the data studies required for registration purposes. There are actually two lists for data: one for the active substance and the other for the biocidal product such as antifouling paint for ships, as shown in Fig. 2.1. The data requirement is the same for all active substances and products, and takes no account of the properties or tonnages of the substance on the market.

As for risk assessment of active substances, essential product information is required on the following items: general substance information, effectiveness

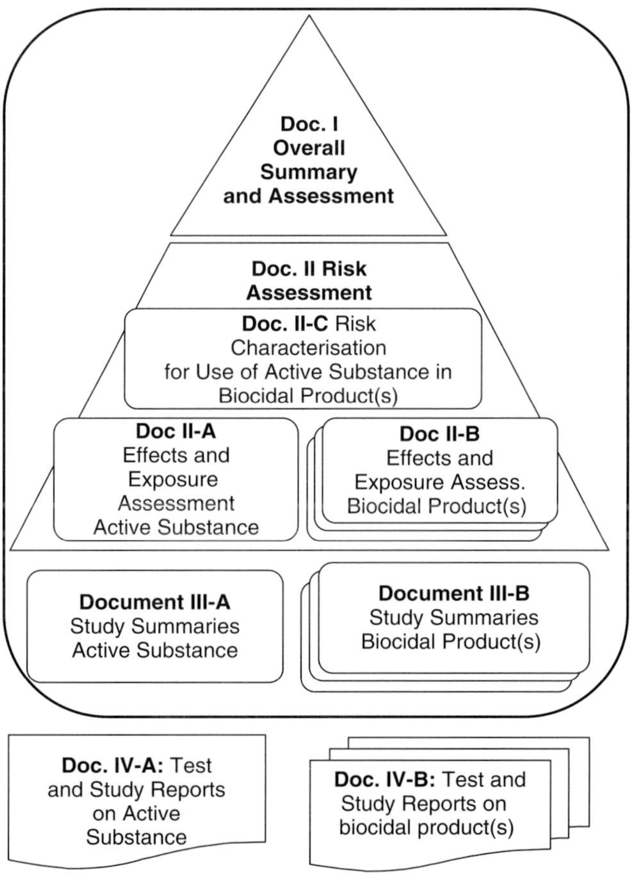

Fig. 2.1 Structure of dossier for BPD application

against target organisms, exposure assessment, human health effects assessment, environmental effects assessment and hazard identification (both physical and chemical). More specifically, data on physical and chemical properties, analytical methods for detection and identification, effectiveness against target organisms, toxicological and metabolic studies, and ecotoxicological profiles (including environmental fate and behavior) are required. For the dossier on active substances, there is a need to demonstrate the use of the active substance in real products, therefore the dossier must contain, within it, a dossier for an antifouling product (e.g. paints).

Risk assessment consists of four stages, as summarized in section 2.4.1. Dose–response analysis should be made for both human health and environmental aspects. Human health effects information requires toxicokinetics and metabolism, acute toxicity, irritation and corrosivity, sensitization, repeated dose toxicity, genotoxicity, carcinogenicity, reproductive toxicity, teratogenicity, fertility, and neurotoxicity.

Data on environmental effects should include information on fate and distribution, degradation, accumulation, and effects on organisms in aquatic, atmospheric and terrestrial fields. Exposure assessment includes intended uses, exposure during manufacture and formulation, and environmental exposure (emission scenarios). To estimate initial input of the substance to the environment, a release rate test is applied to measure the amount leached across a unit surface area per unit time (e.g. $\mu g\ cm^{-2}\ day^{-1}$). In order to estimate how the substance disperses into the environment, an emission scenario has been developed by the Organization for Economic Co-operation and Development (OECD) for estimating the distribution in the environment, under a variety of scenarios, including shipping lanes, harbors, marinas and application and removal of antifouling coatings. By combining this information with data on release rates and degradation properties of the active substance, the predicted environmental concentration (PEC) can be determined.

The risk characterization process considers threats to human health and the environment in order to protect man, animals and their habitat. If the PEC/PNEC ratio is less than unity, the application would be approved and the substance or product will be successfully registered in Annex I. If the ratio is higher than unity, risk/benefit analysis may be made. Even for a product which poses risk, if there are no alternative products, or if significant disadvantage results by eliminating the product from the market, it may still be approved subject to conditions, including a requirement for continuous monitoring of the environment.

2.4.3 Regulations in the United States

In the United States of America, antifouling products are controlled by the Pesticide Registration Program under the Federal Insecticide, Fungicide, and Rodenticide Act (FIFRA) which is governed by the Environmental Protection Agency (EPA). The term pesticide, comparable to biocide in EU's BPD, includes many kinds of ingredients in products such as insect repellants, weed killers, disinfectants, and swimming pool chemicals which are designed to prevent, destroy, repel or reduce pests of any sort.

EPA evaluates and judges pesticides as to whether they meet federal safety standards on both human health and the environmental grounds. Pesticides that meet the requirements are registered to permit their distribution, sale, and use. The examination process investigates the ingredients of the pesticide; the particular site or crop on which it is to be used; the amount, frequency and timing of its use; and storage and disposal practices. From these data, a wide variety of potential human health and environmental effects associated with use of the product are assessed.

The submitted data are evaluated as to whether a pesticide has the potential to cause adverse effects on humans, wildlife, fish, and plants, including endangered species and non-target organisms, as well as possible contamination of surface water or ground water from leaching, runoff, and spray drift. Potential human risks range from short-term toxicity to long-term effects such as cancer and reproductive system disorders. For an application of antifouling paints, the following data are required: data on end use formulation, data on active ingredient(s), product chemistry information, product identity (composition and analysis), physical/chemical properties, acute toxicity data, product performance (efficacy), residue chemistry data, toxicology, terrestrial and aquatic non-target organisms, environmental fate, application exposure monitoring and post-application exposure monitoring.

2.5 Future Prospect

The AFS Convention that controls the use of antifouling paints will enter into force in September 2008. It will be an effective tool to phase out the use of organotin compounds such as TBT and TPT all over the world. The next stage for marine environmental protection would be the control of other antifouling biocides which may be harmful to marine ecosystem. Resolutions adopted by IMO along with the AFS Convention, at the Diplomatic Conference in 2001, emphasize the importance of establishing an international framework for further assessment. Resolution 3 invites member states of IMO to approve, register or license antifouling systems applied in their territories. It also urges to continue the work, in appropriate international fora, for the harmonization of test methods and performance standards for antifouling systems containing biocides. Resolution 4 requests member states to promote and provide directly, or through IMO, information and support, in particular to developing states that request technical assistance for:

(a) The assessment of the implications of ratifying, accepting, approving, or acceding to, and complying with, the Convention
(b) The development of national legislation to give effect to the Convention and
(c) The introduction of other measures, including the training of personnel, for the effective implementation and enforcement of the Convention

It also requests all relevant states and organizations to promote co-operation for scientific and technical research on the effects of antifouling systems as well as monitoring these effects.

Many countries have their own law to protect human health from chemical products, however, a limited number of countries have established regulation focusing on marine environmental protection. A proposal has been brought to the International Standard Organization (ISO), at its Technical Committee 8 (TC8), that provides a draft risk assessment protocol for antifouling systems. This prospective protocol intends to provide a standard method to evaluate the safety of antifouling chemicals and products and also to promote self-management by industries – to show that their products are safe and of minimal risk to the marine environment. The proposed methodology employs an environmental risk assessment scheme similar to the one used in BPD, to harmonize with preceding evaluation systems.

While antifouling systems often release substances which may exhibit adverse effects on human health and/or marine ecosystems, they also represent a substantial economical benefit. Reducing fuel consumption provides not only savings in transportation costs but also substantial environmental advantages, such as suppressing carbon dioxide emission and preventing unwanted translocation of ‘alien species’ on ships’ hulls. To help overcome this dilemma, appropriate regulations, supported by strong scientific evidence, should be established as soon as possible.

Part II
Behavior of Organotin Compounds and Their Effects on Aquatic Organisms

Section 1
Distribution of Organotin Compounds in Aquatic Environments

Chapter 3
Global Contamination by Organotin Compounds

Kurunthachalam Kannan and Shinsuke Tanabe

3.1 Introduction

Organotin compounds were first developed as moth-proofing agents in the 1920s, and were only later used more widely as bactericides and fungicides (WHO 1980). Organotin compounds produced for commercial applications include methyltins, butyltins, phenyltins, octyltins, and cyclohexyltins. The major uses of organotins are as polyvinyl chloride (PVC) heat stabilizers, catalysts (for silicone and polyurethane production), biocides, agrochemicals, and glass coatings. The use of tributyltin (TBT) in marine antifouling paints dates from the 1960s, initially as a booster biocide in copper-based formulations. As a result of TBT's efficacy over copper, the use of TBT-based paints accelerated greatly in the 1970s. Annual production of organotin compounds increased from <5,000 t in 1955 to >50,000 t in 1995 (Fent 1996; OECD 2001), with 15–20% of the production accounted for by triorganotins (Bennett 1996). The global annual production of TBT alone was estimated to be 4,000 t in the late 1990s (OECD 2001). In addition to antifouling uses, TBT was used in wood and material preservatives, and slimicides. The use of TBT in antifouling paints applied to hulls of ships and boats, fish-nets, crab pots, docks, and water cooling towers contributed to the direct release of organotins into the aquatic environment. These antifouling usages have caused the greatest environmental concern, because of TBT's high aquatic toxicity. Since the widespread use of TBT-based paints began in the early 1970s, several researchers have reported the harmful effects of TBT on economically important marine food species such as

K. Kannan
Wadsworth Center, New York State Department of Health and Department of Environmental Health Sciences, School of Public Health, State University of New York at Albany, Empire State Plaza, P.O. Box 509, Albany, NY 12201-0509, USA

S. Tanabe
Center for Marine Environmental Studies (CMES), Ehime University, 2-5 Bunkyo-cho, Matsuyama 790-8577, Japan

T. Arai et al. (eds.), *Ecotoxicology of Antifouling Biocides*,
DOI: 10.1007/978-4-431-85709-9_3, © Springer 2009

oysters and mussels (for reviews, see Fent 1996; Champ and Seligman 1996 and other chapters in this volume).

One of the first documented instances of TBT toxicity was in Pacific oysters, *Crassostrea gigas,* in France's Arcachon Bay. Abnormal spatfall, decrease in larval survival rates, and shell malformations were observed as early as 1974. By the early 1980s, effects on oysters in Arcachon Bay had been linked to TBT, and in 1982 France banned the use of TBT-containing antifouling paints on vessels less than 25 m in length (Alzieu 1991). Many other countries adopted similar regulations from the late 1980s, e.g., the UK, the USA, Australia, Canada, The Netherlands, Switzerland, Japan, Denmark, and Hong Kong. The toxic effect of butyltins was also recognized in other bivalves, especially mussels, and in gastropods. Several studies in the 1980s established a link between TBT exposure and 'imposex' (the imposition of male sexual characteristics on females) in certain neogastropods, and the decline of populations in the waters off southwestern England (Smith 1981; Bryan et al. 1986). Imposex can be initiated in mollusks at water TBT concentrations in the low nanogram per liter range (i.e., <10 ng/l) (Bryan et al. 1986), also the concentration range at which shell deformities and larval mortalities occur (Alzieu 1991). In the mid 1980s, bioaccumulation of TBT in farmed salmon held in net-pens that had been treated with TBT-based antifouling paints was reported (Short and Thrower 1986; Davis and McKie 1987).

Butyltin pollution is not limited to the coastal marine environment; it also extends to the freshwater environment. Austria and Switzerland banned the use of TBT, even though these two countries are land-locked. Nevertheless, as stated earlier, the regulations on TBT-based antifouling paints are only partial in most of the countries that limit usage applying to recreational boats and vessels <25 m in length. The International Maritime Organization (IMO) adopted a global treaty to ban the application of TBT-based paints from January 2003, and total prohibition by January 2008 (IMO 2001). Details of the worldwide regulatory strategies for organotin compounds have been reviewed by Champ (2000).

In general, the regulations have resulted in reduced TBT contamination in water and some organisms, and recovery of mollusc populations, particularly those close to marinas. However, researchers have noted that there has been little or no reduction in TBT concentrations in sediments, even several years after the regulations were enacted (e.g. Quevauviller et al. 1994; Fent and Hunn 1995; Chau et al. 1997). Persistence of TBT in sediments is indicated by a half-life of between 2 and 30 years in temperate regions (Dowson et al. 1996; Maguire 2000; WHO 2006), whilst continued occurrence of imposex at sites near shipping activities (e.g. Fent 2004; Santos et al. 2004), indicates that the legacy of TBT is likely to be long-lasting: appropriate management of potential impacts is an important consideration. In addition, unregulated use of TBT on vessels >25 m in length and on aluminum-hulled boats will continue to be sources of release in large harbors, anchorages, and shipping lanes. Although the IMO's global treaty on the use of TBT is applicable to member countries of the organization, TBT will continue to be produced and used as biocides, especially in developing countries and those countries that do not join the organization. In addition, TBT continues to be used in other applications such as material and wood preservatives and in slimicides.

While studies in the 1970s focused primarily on the biological effects of TBT, only in the 1980s were analytical methods developed, enabling compound-specific analysis of butyltins in marine and fresh waters, in sediments, and in biological samples (Meinema et al. 1978; Hodge et al. 1979; Maguire 1982; Matthias et al. 1986; Sullivan et al. 1988). However, more sensitive analytical methods continued to evolve through the 1990s enabling detection at the sub-parts-per-trillion level and compound-specific determination of organotins in environmental and biological matrices. The U.S. EPA's threshold criterion for protection of saltwater aquatic life from chronic toxic effects of TBT is 7.4 ng/l (USEPA 2003) and some states in the USA have even lower threshold values. Furthermore, analytical methods for lipid-rich biological matrices were not available until the early 1990s (Harino and Fukushima 1992; Iwata et al. 1994; Abalos et al. 1997). Several earlier publications have reviewed these analytical methods, as well as the environmental distribution, fates, and ecotoxicological effects of organotin compounds (e.g. Champ and Seligman 1996; Fent 1996; Champ 2000; Maguire 2000; Antizar-Ladislao 2007).

In this chapter, we describe the global distribution of organotin compounds, with an emphasis on butyltin compounds. The distributions of other organotin compounds such as phenyltin, methyltin, and octyltin compounds have not yet been studied in detail. Mono- and dialkyltin compounds, especially methyl- and octyl-tin compounds, are used as additives/stabilizers in PVC. These compounds are chemically bound to the plastic matrix and, consequently, migration or leaching from rigid PVC is thought to be insignificant. A few studies have reported the occurrence of phenyltin compounds arising from antifouling usage of triphenyltin (TPT) and its usage as a fungicide in agriculture (Kannan et al. 1995b; Kannan and Lee 1996). Although phenyl-, octyl-, and methyl-tin compounds have been reported to occur in environmental and biological matrices from sites worldwide, the magnitude and frequency of distribution of these compounds are low compared to the situation for butyltins (Champ and Seligman 1996; Attar 1996; Ebdon et al. 1998; Champ 2000). Contamination of the waters of the open ocean is an indication of the global distribution of butyltins. Thus, we have here compiled data from monitoring studies that reported concentrations of butyltin compounds in mussels and oysters, pelagic squid, pelagic and coastal fish, oceanic birds, and marine mammals, in various locations, to illustrate the distribution of butyltin compounds on a global scale. TBT degrades by a stepwise dealkylation pathway to dibutyltin (DBT), monobutyltin (MBT), and finally inorganic tin, in sediments and biota. In this review, ΣBT refers to the sum of MBT, DBT, and TBT concentrations. Butyltin concentrations are presented as cations throughout the chapter.

Unlike persistent organic pollutants (POPs) such as polychlorinated biphenyls (PCBs), which are subjected to long-range atmospheric transport, butyltin compounds have other mechanisms of dispersion in the environment. Given the strong affinity of TBT for suspended particles and sediments, benthic sediments are regarded as the major sink for TBT in the environment (Clark et al. 1988; Hoch 2001). Butyltin compounds have low vapor pressure (e.g. the vapor pressure of *bis*-tributyltin oxides ranges from $6.4\ 10^{-7}$ to $1.2\ 10^{-4}$ mmHg at room temperature) suggesting that these compounds are less volatile than several POPs. The aqueous solubilities of butyltin compounds are greater than that for POPs and decrease from mono- to trialkyl-tin species depending on the anion, temperature, and salinity of the water. For example, published values for TBT solubility range from 17 to 256 mg l^{-1} in distilled water, and from 1 to 50 mg l^{-1} in seawater (WHO 1980). Because of the low vapor pressure, and high particle affinity, atmospheric transport of butyltins to remote locations is expected to be negligible. Furthermore, organotin compounds can be subjected to photodegradation in the atmosphere (Kannan and Lee 1996). Transport by water currents and international ship traffic, including commercial shipping activities, are considered to be the major pathways of global dispersion of butyltin compounds.

3.2 Organotins in Polar Regions

The occurrence of POPs in environmental matrices and biota in polar regions has been known for over 4 decades. Nevertheless, few studies have reported on the occurrence of butyltins in subpolar regions (AMAP 2004). TBT and its degradation products were measured in blue mussels collected in the mid to late 1990s from Greenland, the Faroe Islands, northern Norway, and Iceland. TBT concentrations in mussels from the Faroe Islands ranged from 24 to 186 ng/g, wet wt (Følsvik et al. 1998; values have been converted from dry weight to wet weight, based on a moisture content of 80%). In northern Norway, concentrations of TBT in mussels collected in 1993 and 1994 were generally high (Berge et al. 1997), in particular near harbors, where concentrations as high as 880 ng/g, wet wt (dry weight values have been converted to wet weight basis) were reported. In Iceland, the mean concentration of TBT in blue mussels collected in the early 1990s was 123 ng/g, wet wt, near large harbors (Skarphédinsdóttir et al. 1996).

Livers of minke whales (*Balaenoptera acutorostrata*) collected from the Antarctic Ocean in 1985 did not contain measurable concentrations of butyltins (Tanabe 1999). However, more recently, TBT was found in Antarctic marine sediments (Negri et al. 2004), possibly originating from antifouling paints applied to the research vessels that operate in polar latitudes. Within the Arctic circle, ΣBT was found in the livers of harbor porpoises (*Phocoena phocoena*) collected in 1999 from the west coast of Greenland at concentrations ranging from 2 to 18 ng/g, wet wt (Strand et al. 2005). Butyltins were found in livers of male Dall's porpoises sampled in 1992 from the Aleutian Islands (Alaska), the Bering Sea, and the northwestern North Pacific Ocean (ΣBTs = 41–180 ng/g, wet wt) (Tanabe et al. 1998). Livers of pilot whales (*Globicephala melas*) collected from the Faroe Islands in the North Atlantic contained low, but detectable, concentrations of ΣBTs (<0.3–10 ng/g, wet wt) (AMAP 2004).

There is growing concern regarding the toxic effects of TBT in sub-Arctic and Arctic waters (Svavarsson 2000; Svavarsson et al. 2001). In 1995, occurrence of imposex in dogwhelk (*Nucella lapillus*) in Icelandic waters indicated that even sub-Arctic organisms are affected by TBT exposures (Skarphédinsdóttir et al. 1996). Imposex has been documented in mud snails in harbors of northern Norway, Svalbard, Iceland, and Alaska. In Greenland, high TBT concentrations in mussels were found near the capital Nuuk, where most of the ship traffic occurs (Jacobsen and Asmund 2000). TBT concentrations in mussels from Greenland were lower than in mussels from Iceland and the Faroe Islands (Jacobsen and Asmund 2000).

3.3 Mussels as Indicators of Coastal Pollution

Bivalves, especially mussels, have been widely used as sentinel organisms for marine and estuarine pollution, due to their propensity to bioaccumulate and bioconcentrate organic and metallic pollutants (including butyltins); thus, these organisms provide an indication of temporal and spatial distribution of contamination in surrounding waters (see also Chapter 18). Bioconcentration factors of TBT in mussels and oysters ranged from 2,300 to 11,400 (WHO 1980; OECD 2001). The analysis of bivalves collected in the Mussel Watch Program of the U.S. National Oceanic and Atmospheric Administration (NOAA) from the West, East, and Gulf coasts during 1987–1990 showed a ubiquitous distribution for butyltins (Wade et al. 1988; Uhler et al. 1993). Although the bivalve samples were collected from locations not influenced by local sources of pollution (e.g., marinas and harbors were avoided), occurrence of butyltins in all of the bivalves indicated the widespread nature of contamination. Nevertheless, temporal trends of butyltins in bivalve mollusks collected along the US coasts from 1987 to 1993 suggested a general decrease in concentrations in most of the sites (O'Connor 1996). There are no available recent reports of trends in the concentrations of butyltins in the US coastal waters.

A comprehensive and systematic survey of butyltin contamination along the coasts of several Asia-Pacific countries was conducted by Tanabe and co-workers at Ehime University, Japan (Fig. 3.1). Green mussels (*Perna viridis*) from Cambodia, Hong Kong, India, Indonesia, Korea, Malaysia, Philippines, South Korea, South China, and Vietnam, collected during 1997–1999 as a part of the Asia-Pacific Mussel Watch Program, were analyzed for butyltin concentrations and results showed widespread contamination along these coasts (Kan-atireklap et al. 1997; Sudaryanto et al. 2000; 2002, 2004, 2005; Hong et al. 2002; Shim et al. 2005a, b). Some of the highest concentrations of butyltins were found in mussels collected from Japan and Korea (Shim et al. 2005a, b), reflective of the high levels of ship building and repair activities, and shipping traffic, in these two countries. ΣBT concentrations as high as 3,000 ng/g wet wt, were found in mussels from Korea (Shim et al. 2005a, b; Hong et al. 2002). Marinas, harbors, and mariculture activities were shown to contribute to butyltin contamination in coastal areas. The concentration of butyltins in mussels, expressed as a national average, was significantly correlated with the per capita gross national product (GNP) of the country, in these surveys of coastal waters. Although the concentrations of butyltins in mussels from developing countries in Asia were lower than the concentrations reported for developed nations such as the USA and Japan, the lack of regulation of the production and use of butyltins in most of the Asian developing countries is a cause for concern. Butyltin concentrations did

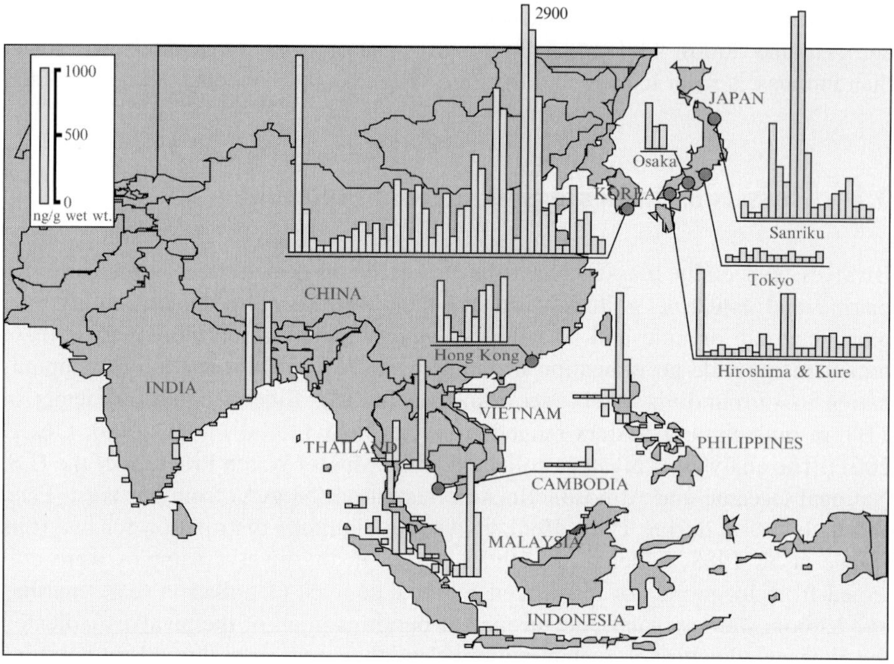

Fig. 3.1 Distribution of ΣBT (MBT + DBT + TBT) concentrations in mussels collected from coastal waters of Asian countries (Data from Sudaryanto et al. 2002; Hong et al. 2002)

not decrease in mussels collected from 1989 to 1999 in Hong Kong (Sudaryanto et al. 2002). Similarly, concentrations of butyltins in mussels collected from India during 1994 and 1999 did not show an evidence for declining concentrations (Sudaryanto et al. 2002). Furthermore, butyltin concentrations in mussels from several harbors did not decline after the partial ban on the application of TBT to the hulls of small boats and to fish-nets in Korea (Hong et al. 2002; Shim et al. 2005b). Considering the high rates of economic growth, the continuing lack of regulation of TBT, and the ever-increasing demand for marine-coating paints in Asian countries, butyltin contamination is expected to continue, and probably to increase, in the short term.

3.4 Squid as Indicators of Butyltin Pollution in the Oceans

Pelagic squid have been used as bioindicators, to trace organotin pollution in open ocean waters (Yamada et al. 1997). Livers of 13 species of squid collected in 1989–1993 from 77 locations across various oceans were analyzed for butyltin and phenyltin concentrations (Fig. 3.2). Butyltins were found in

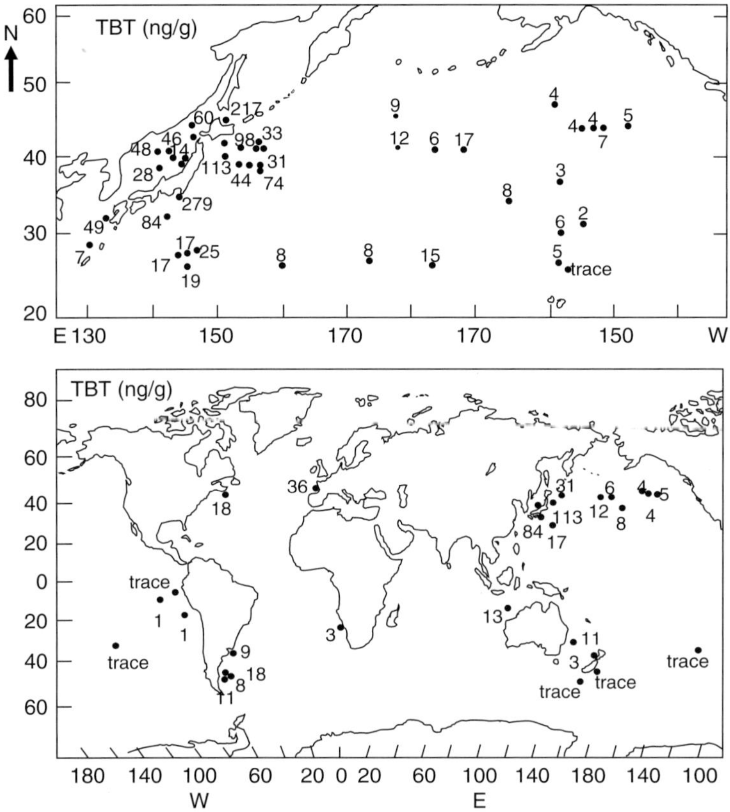

Fig. 3.2 Concentrations of tributyltin (TBT) in livers of squid from the North Pacific Ocean (*top*), and from the oceans worldwide (*bottom*). Numbers indicate the concentrations (ng/g, wet wt) (Adapted from Yamada et al. 1997 with some modifications)

squid collected from all of the locations sampled; the concentrations of both butyltins and phenyltins were greater in individuals from coastal waters than in individuals from offshore waters. Concentrations of ΣBT in the livers of squid from Japanese coastal waters were 17–279 ng/g, wet wt. Concentrations of ΣBT in livers of squid from the East China Sea (7–49 ng/g), western North Pacific Ocean (17–25 ng/g), central North Pacific Ocean (6–17 ng/g), and northeastern North Pacific Ocean (<8 ng/g) were lower than the concentrations found in individuals from coastal locations. The highest concentrations of ΣBT were found in individuals from coastal waters off Japan, France, and Canada. The concentrations of butyltins and phenyltins were higher in squid collected from the Northern Hemisphere than in those from the Southern Hemisphere (see also Chapter 18).

3.5 Coastal and Pelagic Fish Species as Indicators of Butyltin Pollution

TBT accumulates in fish with bioconcentration factors of between 100 and 10,000 (OECD 2001). The first report of the occurrence of butyltins in fish species was by Short and Thrower (1986), who described the accumulation of TBT in Chinook salmon raised in marine net-pens in Alaska, USA. The growth of fouling organisms on netting cages had been a widespread problem for marine salmon farming units, so that TBT was applied as an antifoulant to the aquaculture nets. Muscle tissues of Chinook salmon reared in nets accumulated TBT at concentrations as high as 900 ng/g, wet wt. Widespread occurrence of butyltins in fish was later shown by surveys conducted in several countries in Asia and Oceania (Batley et al. 1992; Kannan et al. 1995a; Sudaryanto et al. 2004). Butyltin compounds were also shown to occur in fish collected from offshore and open oceans. De Brito et al. (2002) reported the occurrence of TBT in muscle of walleye pollock (*Theragra chalcogramma*) from the Bering Sea, Gulf of Alaska, and Japan Sea, at ΣBT concentrations ranging from 1.1 to 5.5 ng/g, wet wt. Butyltin concentrations in fish from offshore locations were one to two orders of magnitude lower than the concentrations found in fish from coastal locations such as Otsuchi Bay (Harino et al. 1998; Takahashi et al. 1999), Tokyo Bay (Takayama et al. 1995), the Baltic Sea off Poland (Kannan and Falandysz 1997; Senthilkumar et al. 1999a), the Mediterranean Sea off Italy (Kannan et al. 1996; Morcillo et al. 1997), the North Sea off Germany (Shawky and Emons 1998), the Mediterranean Sea off Greece (Tselentis et al. 1999), the Pacific and Gulf coastal waters off the USA (Krone et al. 1996; Kannan et al. 1997a), coastal waters off countries of the Middle East (de Mora et al. 2003), and coastal waters of Taiwan (Hung et al. 1998). Consumed fish and shellfish are the major sources of human exposure to butyltins (Belfroid et al. 2000). A compilation of concentrations of butyltins reported for fish samples from a number of countries is illustrated in Fig. 3.3. There was a general pattern of higher concentration of butyltins in fish from developed countries in Europe and North America than in developing countries in Asia.

The distribution of butyltins in offshore and open-ocean waters was investigated using skipjack tuna (*Katsuwonus pelanus*) as a bioindicator of contamination (Ueno et al. 2004). Skipjack tuna are principally distributed in offshore to open-ocean waters in tropical and temperate regions around the world, such as the Pacific, Atlantic, and Indian Oceans. Ueno et al. (2004) collected skipjack tuna from offshore waters of various Asian countries (Japan, Taiwan, the East China Sea, the Philippines, Indonesia, the Bay of Bengal), the Seychelles; Brazil and the North Pacific Ocean during 1996–2001. Butyltins were found in all liver samples analyzed, at concentrations ranging from 3.8 to 400 ng/g, wet wt (Ueno et al. 2004). The distribution of ΣBT concentrations in livers of skipjack tuna from several oceans is shown in Fig. 3.4. In skipjack tuna, as for squid (see above), some of the highest ΣBT concentrations were in specimens collected from offshore waters around Japan (the Japan Sea and the East China Sea). The results of this skipjack

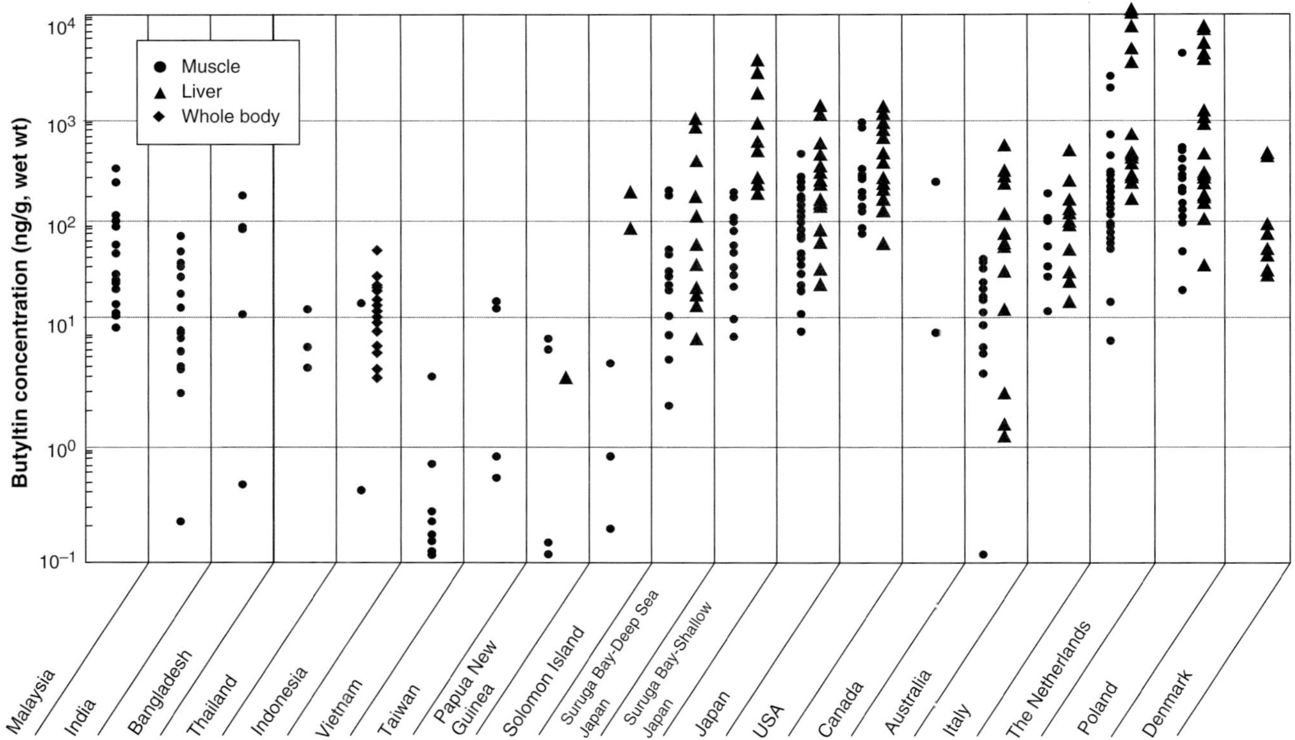

Fig. 3.3 Concentrations of total butyltins (ΣBT) in fish from several countries (Data from Kannan et al. 1995a; Stab et al. 1996; Takahashi et al. 1999; Senthilkumar et al. 1999a; Strand and Jacobsen 2005)

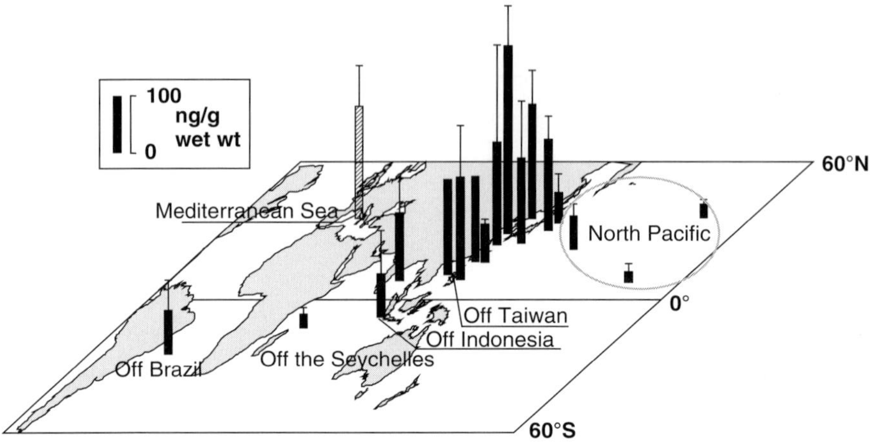

Fig. 3.4 Distribution of total butyltins (ΣBT) in livers of skipjack tuna collected from coastal locations and open seas (From Ueno et al. 2004). Patterned bar indicates data for tuna from Kannan et al. (1996)

tuna survey suggested both a worldwide distribution of butyltins and continuing inputs of butyltins to offshore waters. TBT-based antifouling paints used on large (unregulated) vessels, and on vessels from countries that have not established any regulation of TBT are thought to be the sources of butyltins in open oceans. Skipjack tuna collected from offshore waters around Asian developing countries, such as the South China Sea, the Philippines, Indonesia and the Bay of Bengal contained ΣBT concentrations comparable to the concentrations found for tuna from the waters around Japan. The concentrations of ΣBT in tuna collected in waters near developing countries have been increasing, owing to rising demand for antifouling paints in Asia and to the lack of regulations on the use. In contrast, concentrations of ΣBT in skipjack tuna collected from open-ocean locations such as in the North Pacific Ocean were low. Again, similar to the pattern found for squid, skipjack tuna from the Southern Hemisphere showed lower ΣBT concentrations than did tuna from the Northern Hemisphere.

While the sources of butyltins in coastal areas relate to commercial shipping activities, the significance of such sources to the contamination of remote offshore areas remains a subject of debate (Ten Hallers-Tjabbes et al. 1994). There is an increasing body of evidence to support a link between the density of ship traffic and occurrence of biological effects (predominantly imposex) in organisms inhabiting offshore waters. Contamination of offshore waters and pelagic organisms by butyltin compounds has been reported by several investigators (Ten Hallers-Tjabbes et al. 1994; Hashimoto et al. 1998; Gómez-Ariza et al. 2006; Viglino et al. 2006). Water, sediment, and gastropods collected in the Atlantic Ocean, off the Iberian Peninsula, contained measurable concentrations of butyltins (Gómez-Ariza et al. 2006). The occurrence of butyltins in offshore locations has been related to the intensity of ship traffic.

3.6 Coastal and Oceanic Birds as Indicators of Butyltin Contamination

The first report of the occurrence of butyltins in waterbirds was by Osborn and Leach (1987), who reported butyltins in livers of oystercatchers (*Haematopus ostralegus*) from the Exe Estuary in the UK. Later studies revealed the occurrence of butytlins in aquatic birds from various inland and coastal locations. Studies from Lake Biwa, Japan (Guruge et al. 1996, 1997), Lake Westeinder, the Netherlands (Stab et al. 1996), Gdansk Bay, Poland (Kannan and Falandysz 1997), the Great Lakes, USA (Kannan et al. 1998b) and Vancouver Harbor, Canada (Kannan et al. 1998b; Elliott et al. 2007) confirmed the presence of ΣBT at a few hundreds of nanograms per gram concentrations in various species of waterbirds. Butyltins were also reported to occur in several species of coastal birds from Danish coastal waters (Strand and Jacobsen 2005). The highest concentration of ΣBT reported thus far for a waterbird species, 4,600 ng/g, wet wt, was measured in the liver of a long-tailed duck (*Clangula hyemalis*) collected from Gdansk Bay, Poland (Kannan and Falandysz 1997). Recently, ΣBT concentrations on the order of a few thousand nanograms per gram, on a wet weight basis, were found in livers of surf scoter (*Melanitta perspicillata*) collected from Vancouver Harbor, on the west coast of Canada (Elliott et al. 2007).

Studies have also reported the occurrence of butyltins in pelagic birds from remote oceanic locations, again underscoring the worldwide distribution of these contaminants (Table 3.1). Oceanic birds, including several species of albatross from the Indian Ocean, and albatross, fulmar, shearwater, and puffin from the North Pacific Ocean, collected in the early to mid 1990s, contained measurable levels of butyltins (Guruge et al. 1997). Concentrations of ΣBT in open-ocean birds from the North Pacific Ocean were an order of magnitude lower than the concentrations found in inland and coastal birds (Guruge et al. 1997). Among the oceanic birds sampled, livers of Laysan albatross contained the highest concentration of ΣBT ranging from 7 to 110 ng/g, wet wt. Concentrations of ΣBT in birds from the Indian Ocean were lower than the concentrations found in birds from the North Pacific Ocean (Guruge et al. 1997). Glaucous gulls (*Larus hyperboreus*) collected from Bear Island in the Barents Sea (northern Norway) in 1998 contained several tens of nanograms per gram concentrations of ΣBT (Berge et al. 2004). ΣBT concentrations in the range of several tens to a few hundreds of nanograms per gram were reported in livers of eider duck, black-backed gull, and cormorant from Danish coastal waters (Strand and Jacobsen 2005).

ΣBT concentrations in feathers of migrant birds from South India were higher than the concentrations in the birds' soft tissues (Senthilkumar et al. 1999b). Concentrations of butyltins in birds collected from inland waterbodies in South India and the Philippines were lower than concentrations found in waterbirds from coastal regions of Japan (Senthilkumar et al. 1998). Feathers accounted for 20–30% of the ΣBT total body burden in cormorants (Guruge et al. 1996). Thus, molting of feathers is likely to be an important route of excretion of butyltins in birds (Senthilkumar et al. 1998; Guruge et al. 1996).

Table 3.1 Concentrations (ng/g, wet wt) of butyltins in kidney and liver of adult open ocean birds (From Guruge et al. 1997)

Species	Location	Organ	No of sample	MBT	DBT	TBT	Total BTs
Laysan albatross (1985)	North Pacific Ocean	Kidney	4 (2♂, 2♀)	6.1	2.0	1.0	8.5
				(<5–10)	(<3–3)	(<1–2)	(3–13)
Laysan albatross (1986)	North Pacific Ocean	Liver	5 (♂)	28	15	0.4	43
				(<5–71)	(7–31)	(<1–4)	(7–110)
Black-footed albatross	North Pacific Ocean	Kidney	3 (♂)	5.4	3.2	<1	8.8
				(<5–11)	(2–5)		(2–16)
Sooty shearwater	North Pacific Ocean	Kidney	3 (♂)	<5	7.3	<1	7.3
					(5–9)		(5–9)
Northern fulmar	North Pacific Ocean	Kidney	4 (♂)	4.2	3.0	0.3	7.5
				(<5–9)	(<3–5)	(<1–2)	(4–12)
Tuffed puffin	North Pacific Ocean	Kidney	4 (♂)	<5	1.8	<1	1.8
					(<3–7)		(ND–7)
Light-mantled sooty albatross	South Indian Ocean	Kidney	1 (♂)	2.7	4.3	3.9	11
		Liver	1 (♂)	4	1.7	4.1	9.8
Royal albatross	South Indian Ocean	Kidney	1 (♂)	<5	4.5	1.7	6.2
		Liver	1 (♂)	<5	4.2	1.6	5.8
White-capped albatross	South Indian Ocean	Kidney	2 (♂)	4.3	11	1.3	17
				(4–5)	(10–12)	(1–2)	17
Black-browed albatross	South Indian Ocean	Kidney	6 (4♂, 2♀)	<5	12	3	15
					(<3–29)	(<1–5)	(ND–29)
Gray-headed albatross	South Indian Ocean	Kidney	3 (♂)	<5	3	0.7	3.7
					(1–4)	(<1–2)	(1–5)
Yellow-nosed albatross	South Indian Ocean	Kidney	3 (♂)	3.4	4.1	2.0	9.3
				(1–5)	(3–6)	(<1–6)	(7–13)
Northern giant petrel	South Indian Ocean	Kidney	3 (♂)	4.5	7.2	1.5	13
				(2–8)	(3–12)	(1–2)	(6–22)

ND: Not detected

3.7 Marine Mammals as Indicators of Global Pollution

Marine mammals, because of their long life spans and migratory behaviors, embody an integration of the temporal and spatial aspects of contamination by bioaccumulative organic compounds. For instance, marine mammals have been used as indicators of coastal and oceanic pollution by POPs (Tanabe et al. 1994). The first report of the occurrence of butyltins in the blubber of marine mammals was by Iwata and co-workers in 1994 (Iwata et al. 1994). Prior to that study, no method for the determination of butyltins in lipid-rich biological matrices was available. Furthermore, butyltins were thought not to bioaccumulate in organisms at higher trophic levels. Following the finding of butyltins in marine mammal tissues in 1994, several studies examined aspects of the accumulation of butyltins in dolphins, whales, porpoises, seals, sea lions, and sea otters from oceans and seas including the Pacific coast of Japan (Iwata et al. 1995), Asian coastal waters including the East China Sea, the Sulu Sea, and the Bay of Bengal (Tanabe et al. 1998; Le et al. 1999; Liu et al. 2003; Harino et al. 2007), the Bering Sea off the Aleutian Islands (Alaska), and the North Pacific Ocean (Tanabe 1999), the Mediterranean Sea off Italy and Greece (Kannan et al. 1996; Focardi et al. 2000), the Atlantic and the Pacific coasts of the USA (Kannan et al. 1997a, 1998a; Kajiwara et al. 2001), the Norwegian Arctic (Berge et al. 2004), the Black Sea (Madhusree et al. 1997), the Caspian Sea (Kajiwara et al. 2002), the Baltic Sea (Kannan and Falandysz 1997; Ciesielski et al. 2004), coastal waters of the UK (Law et al. 1998, 1999), and coastal waters of Denmark and Greenland (Strand and Jacobsen 2005; Strand et al. 2005). In addition, butyltin compounds have been reported to occur in freshwater aquatic mammals such as the Ganges River dolphin from India (Kannan et al. 1997b), river otters from the states of Washington and Oregon in the USA (Kannan et al. 1999), and beluga whales from the St. Lawrence River estuary in Canada (Yang et al. 1998).

Butyltins have been found in marine mammals collected in the late 1980s and the 1990s from almost all of the global oceans and seas surveyed, including remote marine locations (Fig. 3.5). Concentrations of ΣBT in the livers of marine mammals were in the range of several hundreds of nanograms per gram to a few thousand nanograms per gram, on a wet weight basis (Fig. 3.5). Some of the highest concentrations, exceeding 10,000 ng/g, wet wt, for ΣBT were reported in livers of bottlenose dolphins from the Atlantic coast of the USA (Kannan et al. 1997a), livers of finless porpoise from Seto Inland Sea, Japan (Iwata et al. 1995), and livers of Indo-Pacific humpback dolphins from the Hong Kong coast (Takahashi et al. 2000). Similarly, livers of sea otters collected along the coastal waters of California, USA, contained ΣBT concentrations as high as 9,200 ng/g, wet wt (Kannan et al. 1998a). Livers of striped dolphins collected from the Baltic Sea off Poland contained ΣBT concentrations as high as 7,700 ng/g, wet wt (Ciesielski et al. 2004). Although the concentrations in liver are higher than the concentrations found in other body tissues of marine mammals, a few studies have reported higher concentrations of ΣBT in kidneys than in other tissues. The concentration of ΣBT in the kidney of a striped dolphin from the Mediterranean Sea was as high

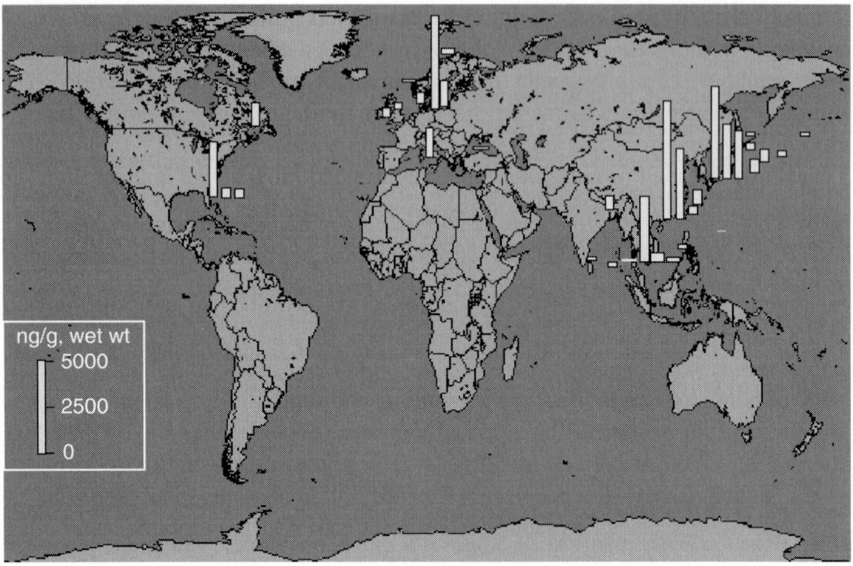

Fig. 3.5 Mean concentrations of total butyltins (ΣBT = MBT + DBT + TBT) in livers of cetaceans from coastal and offshore locations. More than one bar per location indicates analysis of several species for that location (Adapted from Tanabe 1999 with additional data points)

as 8,000 ng/g, wet wt (Focardi et al. 2000). Overall, concentrations of butyltins in marine mammals that inhabit coastal waters are higher than the concentrations found in pelagic species. For example, the highest concentrations of ΣBT were found in four coastal species of dolphins, finless porpoise, bottlenose dolphin, harbor porpoise, and Indo-Pacific humpback dolphin. ΣBT concentrations were in the range of a few tens to a few hundreds of nanograms per gram in livers of harbor porpoises from the Barents Sea, northern Norway (Berge et al. 2004). Harbor porpoises collected from more industrialized coastal areas, such as the Baltic Sea off Poland contained ΣBT concentrations on the order of several hundreds to thousands of nanograms per gram, on a wet weight basis (Ciesielski et al. 2004). Similarly, harbor porpoises from coastal waters of the UK and Denmark contained ΣBT concentrations in the range of several hundreds of nanograms per gram to a few thousands of nanograms per gram in livers (Strand et al. 2005; Law et al. 1998). Dall's porpoises from the Bering Sea (Tanabe et al. 1998) and beluga whales from the St. Lawrence River estuary in Canada (Yang et al. 1998), both of which are waterbodies away from local sources of contamination, contained ΣBT concentrations on the order of a few tens to a few hundreds of nanograms per gram, on a wet weight basis. Butyltins have been found in livers of sea otters from Kamchatka, Russia (18–79 ng/g, wet wt) and the Aleutian Islands, Alaska (6.5–270 ng/g, wet wt) (Murata et al. 2008).

Marine mammals collected from coastal waters of developed countries contained higher concentrations of butyltins than did mammals from waters near less developed countries (Tanabe 1999; Takahashi et al. 2000). For example, concentrations of ΣBT in livers of marine mammals collected from coastal waters of India, the Philippines, and Taiwan were on the order of a few tens to a few hundreds of nanograms per gram, on a wet weight basis (Tanabe 1999; Liu et al. 2003). However, studies have reported the occurrence of high concentrations of ΣBT, on the order of a few hundred to a few thousand nanograms per gram (58–4,860 ng/g, wet wt), in livers of several species of whales collected from coastal waters of Thailand (Harino et al. 2007) and in livers of dolphins collected from coastal waters of Hong Kong (Takahashi et al. 2000). These results suggest that the contamination by butyltins persists in coastal areas of developing countries where no regulations on the use of TBT exist.

Several studies have also reported the occurrence of butyltins in pinnipeds such as seals and sea lions from various oceans and seas (Fig. 3.6). Concentrations of ΣBT in livers of seals were lower than the concentrations reported for livers of cetaceans. The highest concentration of ΣBT reported for a pinniped species was 300 ng/g, wet wt, in the liver of a Steller sea lion (*Eumetopias jubatus*) collected from Hokkaido, Japan (Kim et al. 1996a). Concentrations of ΣBT in livers of Caspian seals from the Caspian Sea were between 0.49 and 17 ng/g (Kajiwara et al. 2002),

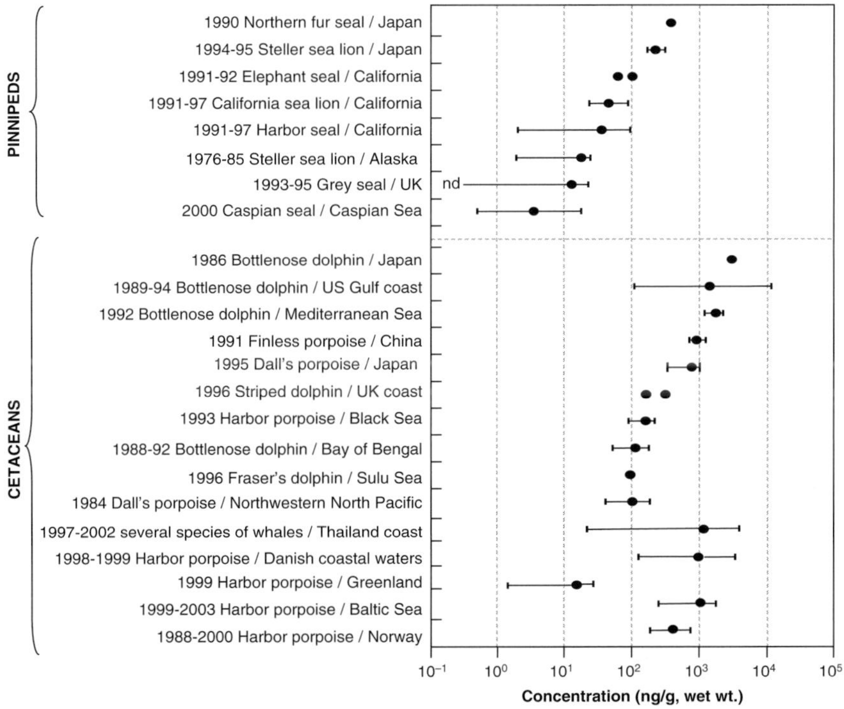

Fig. 3.6 Total butyltin (ΣBT) concentrations (mean and range) in livers of pinnipeds and cetaceans from various seas and oceans (Adapted from Kajiwara et al. 2002 with additional data points)

values comparable to the concentrations reported for Steller sea lions from Alaska (1.9–24 ng/g; Kim et al. 1996a) and grey seals from the UK (1–22 ng/g; Law et al. 1998). Lower concentrations of ΣBT in seals than in dolphins were considered to be due to the greater capacity of seals to metabolize and excrete butyltin compounds (Kim et al. 1996a, b; Takahashi et al. 2000; Kajiwara et al. 2002). Butyltins are bound to keratinaceous proteins found in the hair/fur of pinnipeds. For example, concentrations of ΣBTs in the hair of Steller's sea lions from Hokkaido, Japan, were in the range of 490–2,700 ng/g, wet wt (Kim et al. 1996a, b), values higher than the concentrations found in the animals' livers. The annual moult – a route of elimination of butyltins in pinnipeds – does not occur in cetaceans (which lack hair).

Butyltin compounds are potent immunotoxic compounds (Whalen et al. 1999). Mitogen-induced proliferation of lymphocytes was significantly suppressed in the blood of several species of marine mammals, at a concentration of 77 ng/ml for DBT and at a concentration of 89 ng/ml for TBT (Nakata et al. 2002). Other studies have also shown that DBT is a more potent immunotoxicant than TBT (Snoeij et al. 1987). The liver-concentration to blood-concentration ratio of DBT in marine mammals was three (Iwata et al. 1995). Based on this ratio, the threshold concentration of DBT in livers of marine mammals was calculated to be 231 ng/g, wet wt. Although this threshold value does not incorporate safety or uncertainty factors, marine mammals from several locations contain DBT concentrations greater than 231 ng/g, and may be at risk from exposures. Therefore, further studies are needed to monitor the concentrations of butyltins in marine mammals; only then can we evaluate the risks, and ascertain the efficacy of regulations on the use of TBT.

3.8 Conclusions

In the late 1980s, TBT was viewed as a transient environmental problem, since the assumption was that a partial ban would, over time, reduce butyltin levels in the environment. The regulations imposed in the 1980s were based on a rationale that, since shipping and boating activities contribute to the contamination in marinas and harbors, the ban on use of TBT-based paints for recreational boats and vessels less than 25 m in length would effectively reduce environmental levels. When the regulations were imposed in the late 1980s, the persistence of these compounds, which contributes to their bioaccumulation and wider dissemination in the environment, was underestimated. Nevertheless, studies collectively documenting the worldwide distribution of butyltins and bioaccumulation in organisms at higher tropic levels appeared after the partial regulations had been imposed in several countries. TBT is now known to meet the criteria of persistence, bioaccumulation, and toxicity under Canada's Toxic Substances Management Policy, which mandates measures for TBT's virtual elimination (Maguire 2000; Loganathan et al. 2001). As was pointed out by Goldberg (1986), TBT is perhaps the most toxic substance to have been deliberately introduced into the marine environment.

The results of butyltin-contamination surveys of the oceans clearly suggest that butyltin contamination is a persistent and pervasive problem. Concentrations of butyltins in marine organisms, including dolphins and oceanic birds from open ocean waters, are at or above the threshold for adverse effects. The use of TBT-based antifouling paints has resulted, and continues to result in a global environmental distribution. Concentrations of butyltins in the marine environment around developing countries, for which no regulations exist, are expected to increase yet further in the near future. The continued, unregulated use of butyltins as additives in a wide range of consumer products will ensure additional contamination sources in the environment. Ongoing monitoring of butyltin compounds in the environment is needed, with an emphasis on elucidation of temporal trends, especially in the oceans and in coastal waters off developing countries. Whether efforts to address the emerging challenges will draw on the lessons of the past remains to be seen.

References

Abalos M, Bayona J-M, Compañó R et al. (1997) Granados M, Leal C, Prat M-D (1997) Analytical procedures for the determination of organotin compounds in sediment and biota: a critical review. J Chrom A 788: 1–49

Alzieu C (1991) Environmental problems caused by TBT in France: assessment, regulations, prospects. Mar Environ Res 32: 7–17

AMAP (2004) AMAP Assessment 2002: Persistent Organic Pollutants (POPs) in the Arctic. Arctic Monitoring and Assessment Programme (AMAP), Oslo, Norway. xvi+310 pp.

Antizar-Ladislao B (2007) Environmental levels, toxicity and human exposure to tributyltin (TBT)-contaminated marine environment: a review. Environ Int 34: 292–308

Attar KM (1996) Analytical methods for speciation of organotins in the environment. Appl Organomet Chem 10: 317–337

Batley GE, Scammell MS, Brockbank CI (1992) The impact of the banning of tributyltin-based antifouling paints on the Sydney rock oyster, *Saccostrea commercialis*. Sci Total Environ 122: 301–314

Belfroid AC, Purperhart M, Ariese F (2000) Organotin levels in seafood. Mar Pollut Bull 40: 226–232

Bennett RF (1996) Industrial manufacture and applications of tributyltin compounds. In: de Mora SJ (ed) Tributyltin: case study of an environmental contaminant. Cambridge University Press, Cambridge, pp 21–61

Berge JA, Berglind L, Brevik EM et al. (1997) Levels and environmental effects of TBT in marine organisms and sediments from the Norwegian coast. Norwegian State Pollution Monitoring Programme, Report 6393/97.

Berge JA, Brevik EM, Bjørge A et al. (2004) Organotins in marine mammals and seabirds from Norwegian territory. J Environ Monit 6: 108–112

Bryan GW, Gibbs PE, Hummerstone LG et al. (1986) The decline of the gastropod *Nucella lapillus* around southwest England: evidence for the effect of tributyltin from antifouling paints. J Mar Biol Assoc UK 66: 611–640

Champ MA (2000) A review of organotin regulatory strategies, pending actions, related costs and benefits. Sci Total Environ 258: 21–71

Champ MA, Seligman PF (1996) An introduction to organotin compounds and their use in antifouling coatings. In: Champ MA, Seligman PF (eds) Chapter 1. Organotin: environmental fate and effects. Chapman & Hall, London, pp 601–614

Chau YK, Maguire RJ, Brown M et al. (1997) Occurrence of organotin compounds in the Canadian aquatic environment five years after the regulation of antifouling uses of tributyltin. Water Qual Res J Canada 32: 453–521

Ciesielski T, Wasik A, Kuklik I et al. (2004) Organotin compounds in the liver tissue of marine mammals from the Polish coast of the Baltic Sea. Environ Sci Technol 38: 1415–1420

Clark EA, Sterritt RM, Lester JN (1988) The fate of tributyltin in the aquatic environment. Environ Sci Technol 22: 600–604

Davis IM, McKie JC (1987) Accumulation of total tin and tributyltin in muscle tissue of farmed Atlantic salmon. Mar Pollut Bull 18: 405–407

De Brito APX, Ueno D, Taakahashi S et al. (2002) Organochlorine and butyltin residues in walleye Pollock (*Theragra chalcogramma*) from Bering Sea, Gulf of Alaska and Japan Sea. Chemosphere 46: 401–411

De Mora SJ, Fowler SW, Cassi R et al. (2003) Assessment of organotin contamination in marine sediments and biota from the Gulf and adjacent region. Mar Pollut Bull 46: 401–409

Dowson PH, Bubb JM, Lester JN (1996) Persistence and degradation of pathways of tributyltin in freshwater and estuarine sediments. Estuar Coast Shelf Sci 42: 551–562

Ebdon L, Hill SJ, Rivas C (1998) Organotin compounds in solid waste: a review of their properties and determination using high-performance liquid chromatography. TRAC-Trends Anal Chem 17: 277–288

Elliott JE, Harris ML, Wilson LK et al. (2007) Butyltins, trace metals and morphological variables in surf scoter (*Melanitta perspicillata*) wintering on the south coast of British Columbia, Canada. Environ Pollut 149: 114–124

Fent K (1996) Ecotoxicology of organotin compounds. Crit Rev Toxicol 26: 1–117

Fent K (2004) Ecotoxicological effects at contaminated sites. Toxicol 205: 223–240

Fent K, Hunn J (1995) Organotins in freshwater harbours and rivers: temporal distribution, annual trends and fate. Environ Toxicol Chem 14: 1123–1132

Focardi S, Corsolini S, Aurigi S et al. (2000) Accumulation of butyltin compounds in dolphins stranded along the Mediterranean coasts. Appl Organomet Chem 14: 48–56

Følsvik N, Brevik EM, Berge JA et al. (1998) Organotin and imposex in the littoral zone in the Faroe Islands. Fróðskaparrit 46: 67–80

Goldberg ED (1986) TBT: an environmental dilemma. Environment 22: 42–44

Gómez-Ariza JL, Santos MM, Morales E et al. (2006) Organotin contamination in the Atlantic Ocean off the Iberian Peninsula in relation to shipping. Chemosphere 64: 1100–1108

Guruge KS, Tanabe S, Iwata H et al. (1996) Distribution, biomagnification, and elimination of butyltin compound residues in common cormorants (*Phalacrocorax carbo*) from Lake Biwa, Japan. Arch Environ Contam Toxicol 31: 210–217

Guruge KS, Iwata H, Tanaka H et al. (1997) Butyltin accumulation in the liver and kidney of seabirds. Mar Environ Res 44: 191–199

Harino H, Fukushima M (1992) Simultaneous determination of butyltin and phenyltin compounds in the aquatic environment by gas chromatography. Anal Chim Acta 246: 91–96

Harino H, Fukushima M, Yamamoto Y et al. (1998) Contamination of butyltin and phenyltin compounds in the marine environment of Otsuchi Bay, Japan. Environ Pollut 101: 209–214

Harino H, Ohji M, Wattayakorn G et al. (2007) Accumulation of organotin compounds in tissues and organs of stranded whales along the coasts of Thailand. Arch Environ Contam Toxicol 53: 119–125

Hashimoto S, Watanabe M, Noda Y et al. (1998) Concentration and distribution of butyltin compounds in a heavy tanker route in the Strait of Malacca and in Tokyo Bay. Mar Environ Res 45: 169–177

Hoch M (2001) Organotin compounds in the environment: an overview. Appl Geochem 16: 719–743

Hodge VF, Seidel SL, Goldberg ED (1979) Determination of tin(IV) and organotin compounds in natural waters, coastal sediments, and macro algae by atomic absorption spectrometry. Anal Chem 51: 1256–1259

Hong H-K, Takahashi S, Min B-Y et al. (2002) Butyltin residues in blue mussels (*Mytilus edulis*) and arkshells (*Scapharca broughtonii*) collected from Korean coastal waters. Environ Pollut 117: 475–486

Hung T-C, Lee T-Y, Liao T-F (1998) Determination of butyltins and phenyltins in oysters and fishes from Taiwan coastal waters. Environ Pollut 102: 197–203

IMO (2001) International Maritime Organization. International convention on the control of harmful antifouling systems on ships, http://www.imo.org/Conventions/mainframe.asp?topic_id=529

Iwata H, Tanabe S, Miyazaki N et al. (1994) Detection of butyltin compound residues in the blubber of marine mammals. Mar Pollut Bull 28: 607–612

Iwata H, Tanabe S, Mizuno T et al. (1995) High accumulation of toxic butyltins in marine mammals from Japanese coastal waters. Environ Sci Technol 29: 2959–2962

Jacobsen JA, Asmund G (2000) TBT in marine sediments and blue mussels (*Mytilus edilus*) from central-west Greenland. Sci Total Environ 245: 131–136

Kajiwara N, Kannan K, Muraoka M et al. (2001) Organochlorine pesticides, polychlorinated biphenyls, and butyltin compounds in blubber and livers of stranded California sea lions, elephant seals and harbour seals from coastal California, USA. Arch Environ Contam Toxicol 41: 90–99

Kajiwara N, Niimi S, Watanabe M et al. (2002) Organochlorine and organotin compounds in Caspian seals (*Phoca caspica*) collected during an unusual mortality event in the Caspian Sea in 2000. Environ Pollut 117: 391–402

Kan-atireklap S, Tanabe S, Sanguansin J et al. (1997) Contamination by butyltin compounds and organochlorine residues in green mussel (*Perna viridis* L.) from Thailand coastal waters. Environ Pollut 97: 79–89

Kannan K, Falandysz J (1997) Butyltin residues in sediment, fish, fish-eating birds, harbour porpoise and human tissues collected from the Polish coast of the Baltic Sea. Mar Pollut Bull 34: 203–207

Kannan K, Lee RF (1996) Triphenyltin and its degradation products in foliage and soils from sprayed pecan orchards and in fish from adjacent ponds. Environ Toxicol Chem 15: 1492–1499

Kannan K, Tanabe S, Iwata H et al. (1995a) Butyltins in muscle and liver of fish collected from certain Asian and Oceanian countries. Environ Pollut 90: 279–290

Kannan K, Tanabe S, Tatsukawa R (1995b) Phenyltin residues in horseshoe crabs, *Tachypleus tridentatus* from Japanese coastal waters. Chemosphere 30: 925–932

Kannan K, Corsolini S, Focardi S et al. (1996) Accumulation pattern of butyltin compounds in dolphin, tuna and shark collected from Italian coastal waters. Arch Environ Contam Toxicol 31: 19–23

Kannan K, Senthilkumar K, Loganathan BG et al. (1997a) Elevated levels of tributyltin and its breakdown products in bottlenose dolphins (*Tursiops truncatus*) found stranded along the US Atlantic and Gulf coasts. Environ Sci Technol 31: 296–301

Kannan K, Senthilkumar K, Sinha RK (1997b) Sources and accumulation of butyltin compounds in Ganges river dolphin, *Platanista gangetica*. Appl Organomet Chem 11: 223–230

Kannan K, Guruge KS, Thomas NJ et al. (1998a) Butyltin residues in sea otters (*Enhydra lutris nereis*) found dead along California coastal waters. Environ Sci Technol 32: 1169–1175

Kannan K, Senthilkumar K, Elliott JE et al. (1998b) Occurrence of butyltin compounds in tissues of waterbirds and seaducks from the United States and Canada. Arch Environ Contam Toxicol 35: 64–69

Kannan K, Grove RA, Senthilkumar S et al. (1999) Butyltin compounds in river otters (*Lutra canadensis*) from the Northwestern United States. Arch Environ Contam Toxicol 36: 462–468

Kim GB, Lee JS, Tanabe S et al. (1996a) Characteristics of butyltin and organochlorine compounds accumulation in various organs and tissues including hair in Steller sea lion (*Eumetopias jubatus*). Mar Pollut Bull 32: 558–563

Kim GB, Tanabe S, Tatsukawa R et al. (1996b) Characteristics of butyltin accumulation and its biomagnifications in Steller sea lion (*Eumetopias jubatus*). Environ Toxicol Chem 15: 2043–2048

Krone CA, Stein JE, Varanasi U (1996) Butyltin contamination of sediments and benthic fish from the East, Gulf and Pacific coasts of the United States. Environ Monit Assess 40: 75–89

Law RJ, Blake SJ, Jones BR et al. (1998) Organotin compounds in liver tissue of harbour porpoises (*Phocoena phocoena*) and grey seals (*Halichoerus grypus*) from the coastal waters of England and Wales. Mar Pollut Bull 36: 241–247

Law RJ, Blake SJ, Spurrier CJH (1999) Butyltin compounds in liver tissues of pelagic cetaceans stranded on the coasts of England and Wales. Mar Pollut Bull 38: 1258–1261

Le LTH, Takahashi S, Saeki K et al. (1999) High percentage of butyltin residues in total tin in the livers of cetaceans from Japanese coastal waters. Environ Sci Technol 33: 1781–1786

Liu LL, Chen SJ, Chou LS et al. (2003) Accumulation of butyltin compounds in the pantropical spotted dolphin, *Stenella attenuata*, from the coasts of Taiwan. J Food Drug Anal 11: 320–323

Loganathan BG, Kannan K, Owen DA et al. (2001) Butyltin compounds in freshwater ecosystem. In: Lipnick RL, Hermens JLM, Jones KC, Muir DCG (eds) Persistent, bioaccumulative and toxic chemicals I. American Chemical Society, Washington, DC, pp 134–149

Madhusree B, Tanabe S, Öztürk AA et al. (1997) Contamination by butyltin compounds in harbour porpoise (*Phocoena phocoena*) from the Black Sea. Fresenius J Anal Chem 359: 244–248

Maguire RJ (1982) Occurrence of organotin compounds in lakes and rivers. Environ Sci Technol 16: 698–702

Maguire RJ (2000) Review of the persistence, bioaccumulation and toxicity of tributyltin in aquatic environments in relation to Canada's toxic substances management policy. Water Qual Res J Canada 35: 633–679

Matthias CL, Olson GJ, Brinckman FE et al. (1986) Comprehensive method for the determination of aquatic butyltin and butylmethyltin species at ultratrace levels using simultaneous hybridization/extraction with gas chromatography-flame photometric detection. Environ Sci Technol 20: 609–615

Meinema HA, Burger-Wiersma T, Versluis-deHan G et al. (1978) Determination of trace amounts of butyltin compounds in aquatic ecosystems by gas chromatography/mass spectrometry. Environ Sci Technol 12: 288–293

Morcillo Y, Borghi V, Porte C (1997) Survey of organotin compounds in the western Mediterranean using mollusks and fish as sentinel organisms. Arch Environ Contam Toxicol 32: 198–203

Murata S, Takahashi S, Agusa T et al. (2008) Contamination status and accumulation profiles of organotins in sea otters (*Enhydra lutris*) found dead along the coasts of California, Washington, Alaska (USA), and Kamchatka (Russia). Mar Pollut Bull 56: 641–649

Nakata H, Sakakibara A, Kanoh M et al. (2002) Evaluation of mitogen-induced responses in marine mammal and human lymphocytes by in-vitro exposure of butyltins and non-ortho coplanar PCBs. Environ Pollut 120: 245–253

Negri AP, Hales LT, Battershill C et al. (2004) TBT contamination identified in Antarctic marine sediments. Mar Pollut Bull 48: 1142–1144

O'Connor TP (1996) Trends in chemical concentrations in mussels and oysters collected along the US coast from 1986 to 1993. Mar Environ Res 41: 183–200

OECD (2001) Report on the test results of endocrine disrupting effects of tributyltin (TBT) on fish. http://www.oecd.org/dataoecd/7/1/2461193.pdf

Osborn D, Leach DV (1987) Organotin in birds: pilot study. Final report to the UK department of the environment. Institute of Terrestrial Ecology, Huntingdon, UK, 15 pp

Quevauviller P, Donard OFX, Etcheber H (1994) Butyltin distribution in a sediment core from Arcachon Harbour (France). Environ Pollut 84: 89–92

Santos MM, Vieira N, Reis-Henriques MA et al. (2004) Imposex and butyltin contamination off the Oporto Coast (NW Portugal): a possible effect of the discharge of dredged material. Environ Int 30: 793–798

Senthilkumar K, Kannan K, Tanabe S et al. (1998) Butyltin compounds in resident and migrant birds collected from Philippines. Fresenius Environ Bull 7: 561–571

Senthilkumar K, Duda CA, Villeneuve DL et al. (1999a) Butyltin compounds in sediment and fish from the Polish coast of the Baltic Sea. Environ Sci Pollut Res 6: 200–206

Senthilkumar K, Tanabe S, Kannan K et al. (1999b) Butyltin residues in migratory and resident birds collected from South India. Toxicol Environ Chem 68: 91–104

Shawky S, Emons H (1998) Distribution pattern of organotin compounds at different trophic levels of aquatic ecosystems. Chemosphere 36: 523–535

Shim WJ, Hong SH, Kim NS et al. (2005a) Assessment of butyl- and phenyltin pollution in the coastal environment of Korea using mussels and oysters. Mar Pollut Bull 51: 922–931

Shim WJ, Yim UH, Kim NS et al. (2005b) Accumulation of butyl- and phenyltin compounds in starfish and bivalves from the coastal environment of Korea. Environ Pollut 133: 489–499

Short JW, Thrower FP et al. (1986) Accumulation of butyltins in muscle tissue of Chinook salmon reared in sea pens treated with tri-n-butyltin. Mar Pollut Bull 17: 542–545

Skarphédinsdóttir H, Ólafsdóttir K, Svavarsson J et al. (1996) Seasonal fluctuations of tributyltin (TBT) and dibutyltin (DBT) in the dogwhelk, *Nucella lapillus* (L.) and the blue mussel, *Mytilus edulis* L., in Icelandic waters. Mar Pollut Bull 32: 358–361

Smith BS (1981) Male characteristics on female mud snails caused by antifouling paints. J Appl Toxicol 1: 22–25

Snoeij NJ, Penninks AH, Seinen W et al. (1987) Biological activity of organotin compounds – an overview. Environ Res 44: 335–353

Stab JA, Traas TP, Stroomberg G et al. (1996) Determination of organotin compounds in the foodweb of a shallow freshwater lake in the Netherlands. Arch Environ Contam Toxicol 31: 319–328

Strand J, Jacobsen JA (2005) Accumulation and trophic transfer of organotins in a marine foodweb from the Danish coastal waters. Sci Total Environ 350: 72–85

Strand S, Larsen MM, Lockyer C (2005) Accumulation of organotin compounds and mercury in harbour porpoises (*Phocoena phocoena*) from the Danish waters and West Greenland. Sci Total Environ 350: 59–71

Sudaryanto A, Takahashi S, Tanabe S et al. (2000) Occurrence of butyltin compounds in mussels from Indonesian coastal waters and some Asian countries. Water Sci Technol 42: 71–78

Sudaryanto A, Takahashi S, Monirith I et al. (2002) Asia-Pacific mussel watch: monitoring of butyltin contamination in coastal waters of Asian developing countries. Environ Toxicol Chem 21: 2119–2130

Sudaryanto A, Takahashi S, Iwata H et al. (2004) Contamination of butyltin compounds in Malaysian marine environments. Environ Pollut 130: 347–358

Sudaryanto A, Takahashi S, Iwata H et al. (2005) Organotin residues and the role of anthropogenic tin sources in the costal marine environment of Indonesia. Mar Pollut Bull 50: 227–236

Sullivan JJ, Torkelson JD, Wekell MW et al. (1988) Determination of tri-n-butyltin and di-n-butyltin in fish as hydride derivatives by reaction gas chromatography. Anal Chem 60: 626–630

Svavarsson J (2000) Imposex in dogwhelk (*Nucella lapillus*) due to TBT contamination: improvement at high latitudes. Mar Pollut Bull 40: 893–897

Svavarsson J, Gramno Å, Ekelund R et al. (2001) Occurrence and effects of organotins on adult common whelk (*Buccinum undatum*; Mollusca, Gastropoda) in harbours and in a simulated dredging situation. Mar Pollut Bull 42: 370–376

Takahashi S, Tanabe S, Takeuchi I et al. (1999) Distribution and specific bioaccumulation of butyltin compounds in a marine ecosystem. Arch Environ Contam Toxicol 37: 50–61

Takahashi S, Le LTH, Saeki H et al. (2000) Accumulation of butyltin compounds and total tin in marine mammals. Water Sci Technol 42: 97–108

Takayama T, Hashimoto S, Tokai T et al. (1995) Measurements of organotin compounds in fishes and crustaceans of Tokyo Bay. Environ Sci 8: 1–9 (in Japanese)

Tanabe S (1999) Butyltin contamination in marine mammals – a review. Mar Pollut Bull 39: 62–72

Tanabe S, Iwata H, Tatsukawa R (1994) Global contamination by persistent organochlorines and their ecotoxicological impact on marine mammals. Sci Total Environ 154: 163–177

Tanabe S, Prudente M, Mizuno T et al. (1998) Butyltin contamination in marine mammals from North Pacific and Asian coastal waters. Environ Sci Technol 32: 193–198

Ten Hallers-Tjabbes CC, Kemp JF, Boon JP (1994) Imposex in whelks (*Buccinum undatum*) from the Open North Sea: Relation to shipping traffic intensities. Mar Pollut Bull 28: 311–313

Tselentis BS, Maroulakou M, Lascourreges JF et al. (1999) Organotins in sediments and biological tissues from Greek coastal areas: preliminary results. Mar Pollut Bull 38: 146–153

Ueno D, Inoue S, Takahashi S et al. (2004) Global pollution monitoring of butyltin compounds using skipjack tuna as a bioindicator. Environ Pollut 127: 1–12

Uhler AD, Durell GS, Steinhauer WG et al. (1993) Tributyltin levels in bivalve mollusks from the east and west coasts of the United States: results from the 1988–1990 national status and trends mussel watch project. Environ Toxicol Chem 12: 139–153

USEPA (2003) Ambient Aquatic Life Water Quality Criteria for Tributyltin (TBT) – Final. Office of Water, EPA 822-R-03-031, United States Environmental Protection Agency, Washington, DC http://www.epa.gov/waterscience/criteria/tributyltin/tbt-final.pdf

Viglino L, Pelletier E, Lee LEJ (2006) Butyltin species in benthic and pelagic organisms of the Saguenay Fjord (Canada) and imposex occurrence in common whelk (*Buccinum undatum*). Arch Environ Contam Toxicol 50: 45–59

Wade TL, Garcia-Romero B, Brooks JM (1988) Tributyltin contamination in bivalves from United States coastal estuaries. Environ Sci Technol 22: 1488–1493

Whalen MM, Loganathan BG, Kannan K (1999) Immunotoxicity of environmentally relevant concentrations of butyltins on human natural killer (NK) cells *in vitro*. Environ Res 81: 108–116

WHO (1980) Tin and organotin compounds. International Programme on Chemical Safety. Environmental Health Criteria 15, World Health Organization, Geneva

WHO (2006) Mono- and disubstituted methyltin, butyltin, and octyltin compounds. Concise international chemical assessment document 73, http://www.who.int/ipcs/publications/cicad/cicad73.pdf

Yamada H, Takayanagi K, Tateishi M et al. (1997) Organotin compounds and polychlorinated biphenyls of livers in squid collected from coastal waters and open ocean. Environ Pollut 96: 217–226

Yang F, Chau YK, Maguire RJ (1998) Occurrence of butyltin compounds in beluga whales (*Delphinapterus leucas*). Appl Organomet Chem 12: 651–656

Chapter 4
Contamination by Organotin Compounds in Asia

Takaomi Arai and Hiroya Harino

4.1 Introduction

Tributyltin (TBT) compounds have been used extensively as a biocide in marine antifouling paints. These compounds are persistent in the marine environment, especially in sediments, due to slow degradation rates and consistent flux (Stewart and de Mora 1990; Michel and Averty 1999). Further, they can accumulate in a variety of marine organisms, from plankton and fish, to various marine mammals (Harino et al. 1999, 2003, 2007a, b, c). Numerous deleterious biological effects of TBT on non-target organisms have been observed (Fent 1996). The most obvious manifestations of TBT contamination have been shell deformation in Pacific oysters (Alzieu 1996) and the development of imposex/intersex in gastropods (Gibbs and Bryan 1996). BT compounds may potentially affect human health through consumption of contaminated seafood.

Owing to the widespread deleterious effects on non-target organisms, the use of TBT as an antifouling agent has been regulated in developed countries for over 20 years (Bosselmann 1996). France was the first country to implement a ban on the use of TBT antifouling paints on ships of less than 25 m at the beginning of 1982. Most European countries, the USA, Canada, Australia, and New Zealand implemented similar limited legislation. However, only a few countries or regions in Asia have such regulations, although Japan banned the use of TBT on all vessels in 1991. Considering these facts, it is necessary to understand the present status of BT contamination in Asian coastal waters before implementation of the global ban to evaluate the effectiveness in the future.

T. Arai
International Coastal Research Center, Ocean Research Institute, The University of Tokyo, 2-106-1 Akahama, Otsuchi, Iwate 028-1102, Japan

H. Harino
Osaka City Institute of Public Health and Environmental Sciences, 8-34 Tojo-cho, Tennoji, Osaka 543-0026, Japan

T. Arai et al. (eds.), *Ecotoxicology of Antifouling Biocides*,
DOI: 10.1007/978-4-431-85709-9_4, © Springer 2009

In this chapter, the levels of contamination of TBT compounds, and their distribution in coastal waters of Asian countries are overviewed.

4.2 Status of Butyltin Contamination

4.2.1 China

In China, industrial and commercial activities are concentrated on the east coast, particularly the southeastern coast. The coastal environment of China may experience adverse effects from intense industrial activities. Until recently, environmental protection of near shore water was not a high priority in China (Yuan et al. 2001). Concentrations of monobutyltin (MBT), dibutyltin (DBT), and TBT in sediments ranged from ND to 12.62, from ND to 6.99, and from ND to 24.03 ng g^{-1} dry wt, respectively (Yuan et al. 2001) (Table 4.1). Although TBT, DBT and MBT are found in varying amounts, the concentrations of BT in China are within the range of levels found throughout the world. Sediment samples from Victoria Harbor, Hong Kong have relatively high concentration of BTs, particularly DBT and MBT. Victoria Harbor is one of the largest seaports in the world in terms of tonnage of shipping handled. It was previously reported that sediment from Victoria Harbor had very high concentrations of aliphatic hydrocarbons and PAHs, attributable to heavy shipping activity (Connell et al. 1998). Shipping activity is probably also the cause of the BT contamination in Victoria Harbor as well. The relatively low TBT to MBT ratio in the sediment from Victoria Harbor indicates that the BTs have undergone significant extent of degradation.

Table 4.1 BT levels in coastal sediment in Asian countries reported after 2000

		Concentration (ng g^{-1} dry wt)			
Country	Survey period	MBT	DBT	TBT	Reference
China	1996, 1999	ND–12.62	ND–6.99	ND–24.03	Yuan et al. (2001)
India	2002–2003	–	ND–469	5–2,384	Bhosle et al. (2004)
	2000–2002	–	ND–469.9	4.5–16,816	Bhosle et al. (2006)
Indonesia	1998	<2.5–37	<2.7–87	0.51–190	Sudaryanto et al. (2005)
Japan	2003	4.3–22	2.3–23	1.2–19	Ohji et al. (2007)
	2005	<1–3,300	2–3,400	<1–14,000	Harino et al. (2007a)
Korea	1996	<2–1,390	<1–5,090	7–13,300	Shim et al. (2002)
Malaysia	1997–1998	5.0–360	3.8–310	2.8–1,100	Sudaryanto et al. (2004)
	2006	4–542	1–232	0.7–492	Harino et al. (2008b)
Thailand	2004	1–293	1–368	2–1,246	Harino et al. (2006)
Vietnam	2002	<0.04–11	0.64–3.2	0.89–34	Midorikawa et al. (2004)
	2003	3.9–29.7	8.1–42.7	8.3–50.5	Nhan et al. (2005)

In bivalve samples from Chinese Bohai coastal waters, however, the detection rate was about 90%: 101 out of 113 of the samples were found to contain detectable BTs, which indicated a wide incidence of BTs pollution along this coastline (Yang et al. 2006) (Table 4.2). Concentrations of BTs ranged from the detection limit (<2.5 ng g^{-1} dry wt) to 397.6 ng g^{-1} dry wt (mean 63 ng g^{-1} dry wt). The concentrations of TBT, DBT and MBT in individual bivalve samples ranged from <2.8 to 383.9, <3.2–158.1 and <2.5–52.2 ng g^{-1} dry wt, respectively (Yang et al. 2006). Among BTs, TBT was detected at relatively high concentrations in most of samples, whereas the concentrations of DBT and MBT were low (Yang et al. 2006). The high level of BT compounds and high percentage of TBT implied fresh input of TBT in Chinese coastal sites, which is probably attributable to the continuous usage of TBT-based antifouling paints in China. BT residue levels in the bivalve samples varied significantly between different sampling sites. The highest mean BTs level (108 ng g^{-1}) was found in the samples collected from Dalian, which is one of the important seaports of China and a world-famous tourist destination with consequently shipping density is high here (Yang et al. 2006). TBT released from antifouling paints applied these vessels probably accounts for the high BT contents in Dalian bivalve samples. The lowest residue level of BTs (mean ΣBTs 17.2 ng g^{-1} wet wt) was detected in Shouguang samples, which were collected at

Table 4.2 BT levels in bivalves in Asian countries reported after 2000

Country	Survey period	Concentration (ng g^{-1} wet wt) MBT	DBT	TBT	Reference
Cambodia	1998	<2.0–25	<0.98–37	2.4–88	Sudaryanto et al. (2002)
China	2003	ND–52.2	ND–158.1	ND-383.9	Yang et al. (2006)
India	1998	<2.2–66	<0.86–150	0.83–570	Sudaryanto et al. (2002)
	2001	64–732	ND–358	64–732	Bhosle et al. (2004)[a]
Indonesia	1997–1998	<1.5–13	<0.58–14	2.2–38	Sudaryanto et al. (2000)
Japan	2003	0.83–2.9	0.83–3.1	0.77–11	Ohji et al. (2007)
	2005	4–32	3–92	3–287	Harino et al. (2007a)
Korea	1995–1998, 2001	<5–1,270	27–1,950	48–2,800	Shim et al. (2005)[a]
Malaysia	1997–1998	<2.6–74	<1.0–160	3.5–730	Sudaryanto et al. (2002)
	2006	41–102	3–5	8–32	Harino et al. (2008b)
Philippines	1994, 1997	<3–51	<1–100	<1–640	Tanabe et al. (2000)
	1998	<2.0–15	<1.3–19	0.8–47	Sudaryanto et al. (2002)
Thailand	1994–1995	<3–45	1–80	3–680	Tanabe et al. (2000)
	2004	8–20	4–9	4–45	Harino et al. (2006)
Vietnam	2002	0.1–44	0.5–10	1.4–56	Midorikawa et al. (2004)
	2003	2.8–18.4	4.4–26.6	3.8–15.1	Nhan et al. (2005)

[a]Data are shown as ng g^{-1} dry wt.

relatively clean sites. Noticeably, most of the samples from Yingkou, a marine aquaculture area away from the seaport, were also characterised by high contents of BT compounds (mean ΣBTs 37.8 ng g^{-1} wet wt). It was found that TBT-based coatings on the marine aquaculture facilities, such as fishnets and sea pens, could be one of the sources for BTs in this area. High levels of BTs were also found in bivalve near other port cities possessing a high density of ships coated with TBT-based antifouling paints.

4.2.2 India

India is one of the most rapidly developing countries in South Asia. TBT compounds have been used as antifouling agents in marine paints. There is as yet no legislation regulating the usage of organotins.

The concentration of BTs in coastal waters varied spatially (Bhosle et al. 2004). The stations located in the close vicinity of shipyard and dry dock areas had relatively higher concentrations of both DBT and TBT. The concentration of BTs in sediment samples also varied spatially (Bhosle et al. 2004). In Marmugao harbour, one of the major ports on the west coast of India (Bhosle et al. 2004), the concentrations of DBT and TBT ranged from undetected to 469 ng g^{-1} dry wt and 5–2,384 ng g^{-1} dry wt of sediment, respectively (Table 4.1). In addition to being a busy commercial port, Marmugao is home to a large fleet of fishing boats, a ship building unit and a dry dock facility involved in the construction, repair and painting of various commercial and naval vessels. Relatively high concentrations of BTs in the sediment collected from the harbour probably reflects the combination of these sources and particularly the effect of dry dock activities. Biological samples such as mussel, clam and fish *Mugil* sp. also contained high concentrations of BTs with DBT being particularly abundant in these animals. The relatively higher abundance of DBT indicates that abiotic and/or biotic degradation processes of TBT were occurring in these animals.

A wide range of concentrations of DBT and TBT were observed in the samples collected from various locations along the coast of India (Sudaryanto et al. 2002; Bhosle et al. 2004). The concentration levels of BTs found in sediment and marine organisms are comparable with those found in different parts of the world, whereas the detected water concentrations are in the range of values recorded during the pre-ban period (Hoch 2001). The BT levels in the water samples of the west coast of India were relatively higher than those reported for the east coast of India (Bhosle et al. 2004; Rajendran et al. 2001). The distribution pattern indicates localized areas of contamination. The mussels from some rural areas found TBT at lower concentrations and proportions, indicating that TBT usage as antifouling agents in the rural areas seems to be minimal. For the sediment and water samples, TBT was generally the most abundant constituent of BTs (Bhosle et al. 2004). Relatively higher abundance of TBT suggests fresh inputs and/or less degradation of TBT.

4.2.3 *Indonesia*

Indonesia is the largest archipelago in the world, situated at the confluence of the Pacific and the Indian Oceans, and bridging two continents, Asia and Australia. Indonesia shares borders with Singapore, Malaysia, and the Philippines to the north and Australia to the south across narrow straits of water. The country relies on shipping as an important mode of transportation of natural resources, goods and people, with about 538 ports of which 131 are open for international trade (Rumengan et al. 2008). Therefore, there is great concern over the potential risk from organotin-based paints derived from ships. Large quantities of TBT and other organotins still enter the sea directly from the anti-fouling paints on ships' hulls, especially where there are no restrictions.

In sediments from the Indonesian Archipelago, concentrations of total BT (ΣBTs = TBT + DBT + MBT) ranged from 0.51 to 320 ng g^{-1} dry wt. Among the location sampled, elevated concentrations of ΣBTs (23–320 ng g^{-1} dry wt) were found in sediments collected from areas of higher maritime activities, such as Belawan (Medan City, North Sumatra Province), Jakarta Bay (Jakarta Province) and Surabaya (East Java Province). Jakarta Bay and Surabaya, where the biggest ports are located, can be designated as hot-spots of BT contamination in Indonesia. Higher concentrations of BTs found in sites of intensive maritime activities suggest that TBT-based antifouling paints are used on ship hulls and confirmed the fact that port activities are the main sources of BTs in Indonesia. According to Sudaryanto et al. (2005), however, at half of the sites surveyed, BT concentrations were in trace amounts (0.51–3.2 ng g^{-1} dry wt), including at intensive marine aquaculture areas. Concentration of BTs found in sediments and mussels (Sudaryanto et al. 2000, 2002) in such areas were in fact much lower than those at other marine culture sites of Southeast Asian countries (Sudaryanto et al. 2004). This result supports to the previous suggestion of minimal or no usage of BTs on fish culture nets in Indonesia (Sudaryanto et al. 2000, 2002). The residue levels of BT found were lower than those reported in many coastal and harbour areas of the world (Tables 4.1 and 4.2) including several other developing Asian countries, as well as urbanized industrial nations such as the USA, Europe, Australia and Japan (Table 4.3). These results clearly indicate low BT contamination in the marine environment of Indonesia.

4.2.4 *Japan*

In Japan, bis(tributyltin)oxide (TBTO) was designated a Class 1 Specified Chemical Substance in 1989, and TBT species excluding TBTO (13 chemicals) were designated as Class 2 Specified Chemical Substances in 1990. In spite of these regulations, it is reported that the elevated concentrations of OTs still exist in the coastal waters of Japan (Harino et al. 2004, 2007a, 2008a), though concentrations of TBT in water and sediment have generally decreased between 1990 and 2002. The Environmental Agency has also surveyed TBT in mussels, fish and birds between

Table 4.3 BT levels in coastal sediment in the world except for Asian countries reported after 2000

Country	Survey period	Concentration (ng g^{-1} dry wt)			Reference
		MBT	DBT	TBT	
Australia	1999	<1–161	<1–71	4.2–1,275	Haynes and Loong (2002)
Canada	1998	4–989	45–997	97–888	Regoli et al. (2001)
Germany	1997–1998	10–1,300	30–14,000	80–17,000	Biselli et al. (2000)
Italy	1999–2000	<1	<1	3–27	Chiavarini et al. (2003)
Spain	1993–1994	<0.8–140	<0.7–100	<0.6–140	Gomez-Ariza et al. (2000)
Sweden	2000	6.5–44.4	16.7–98.0	16.5–366.0	Brack (2002)
UK	1998	–	<0.001–790	<2–5,800	Thomas et al. (2000)
USA	1991–1997	–	–	<2.9–68,613	Elgethun et al. (2000)

1985 and 2002 and although the concentrations in these animals decreased at the start of the survey, concentrations of these chemical compounds have remained constant in recent years.

According to evidence from several sources, the mean concentrations of TBT at seven sites of Japanese coastal waters appear to be constant in recent years at 0.011 mg l^{-1}, in spite of the differences in the utilization of these coastal areas (Inoue et al. 2002; Harino et al. 2003; Ohji et al. 2007). TBT concentrations were similar in the littoral region around Japan. The proportion of BT present as TBT in these coastal waters is <50%, implying that the degradation rate of TBT may be sufficient to overcome fresh input of TBT to water. Spatial variation has been observed, however, by Takeuchi et al. (2004) who compared the BT concentrations in water samples among 18 areas along the Japanese coast, and indicated that TBT concentrations in the western Japan were higher than those from the Pacific coast of northern Japan, the coast along the Sea of Japan and Tokyo Bay and the adjacent area.

Average TBT concentrations in sediments from Ofunato Bay, the Port of Osaka, Tanabe Bay and Hakata Bay, where there are small or large ports, shipyards, docks and fishing nets, were over 0.1 mg kg^{-1} dry (Harino et al. 2008a). Furthermore, extremely high TBT concentration (14,000 ng g^{-1} dry) can still be found close to shipyards (Harino et al. 2007a) (Table 4.1). These results suggest increased use of this compound recently, because sediments tend to integrate TBT loadings over the longer term. High concentrations of TBT have been observed in mussel samples from Yamada and Otsuchi Bays in the Sanriku coastal area, suggesting that the loading of TBT in these bays has also continued OTs contaminations in mussels tend to integrate and reflect those in water during the previous 2 or 3 months exposure (Short and Sharp 1989).

Generally, TBT concentrations in water and biological samples have declined in comparison to those in 1990 when OT regulation was introduced in Japan. However significant changes of TBT have not been observed in sediments. There are still a number of TBT hot-spots in the country.

4.2.5 *Korea*

In Korea, the application of trialkyltin-based antifouling has been banned since 2000 except on oceangoing vessels, nets, and immersed structures. However, there are no regulations for releasing wastewater containing TBT from shipyards in the country. BTs in surface sediments from a shipyard area were detected at high levels (Shim et al. 2002). TBT, DBT, and MBT concentrations in the surface sediments ranged between 7–13,300, <1–5,090, and <2–1,390 ng g^{-1} dry wt, respectively (Table 4.1). The highest BTs concentrations were found in the front of shipyard. TBT accounted for the largest portion (69%) of total BTs, followed by DBT (20%) and MBT (11%) (Shim et al. 2002). These results suggest that shipyards are one of the major sources of TBT in the coastal environmental of Korea as in other coastal areas.

Mussel and oyster samples analyzed from the coast of Korea contained detectable amounts of BTs (Shim et al. 2005). TBT, DBT, and MBT concentrations were ranged from 48 to 2,800, from <3 to 1,950, and from <4 to 1,270 ng g^{-1} dry wt, respectively (Table 4.2). Organotin residue levels in the coastal environment of Korea are therefore comparable to those in developed countries (Shim et al. 2005). However, TBT concentrations in Korean mussels and oysters were much higher than those in other Asian countries (Shim et al. 2005) (Table 4.2). The highest TBT (2,800 ng g^{-1} dry wt), DBT (1,950 ng g^{-1} dry wt), and MBT (1,270 ng g^{-1} dry wt) concentrations were observed in front of a large maintenance shipyard (Shim et al. 2005). Relatively high TBT concentrations were also recorded in the innermost sites of the enclosed bays where harbors or shipyards were located, consistent with notion of antifouling from ship's hulls being primary source of TBT, as is the case in other coastal areas.

4.2.6 *Malaysia*

In Malaysia, as in other Asian countries except Japan and Hong Kong, there is no specific legislation controlling the usage of TBT (Champ 2000). Malaysia is surrounded by the busy waterways such as the Malacca Strait and South China Sea, which are recognized as major shipping routes in Southeast Asia. This country has better economic status than other developing Asian nations (Sudaryanto et al. 2002). Therefore, this situation may lead to heavy contamination by organotins in the environment from a variety of sources. Hashimoto et al. (1998) has reported high levels of BT compounds in seawater from a busy tanker route in the Strait of Malacca. Recent measurements reveal concentrations of MBT, DBT and TBT in sediment from the coastal water of Peninsular Malaysia along the Strait of Malacca were in the range of 4–242, 1–186 and 0.7–228 ng g^{-1} dry wt, respectively (Harino et al. 2008b) (Table 4.1). The concentrations of MBT, DBT and TBT in sediment from the coastal water of Johor Strait were in the range of 83–542, 30–232 and 41–492 ng g^{-1} dry wt, respectively (Table 4.1) and were higher than those in Peninsular Malaysia. MBT, DBT and TBT concentrations in mussels sampled along the Strait of Malacca were detected in the range of 41–102, 3–5 and 8–32 ng g^{-1} dry wt, respectively (Table 4.2). Combining these results with values reported previously for sea water, sediments, mussels and cockles (Tong et al. 1996; Hashimoto et al. 1998; Sudaryanto et al. 2002, 2004), it is apparent that BT contamination is widely distributed along coastal waters of Malaysia and has spread to a wide range of environmental media and biota. Sudaryanto et al. (2004) reported that the concentration of BTs in sediments were in the range of 2.8–1,100 ng g^{-1} dry wt for TBT, 3.8–310 ng g^{-1} dry wt for DBT and 5.0–360 ng g^{-1} dry wt for MBT (Table 4.1). The high proportion of TBT found in these samples suggests fresh inputs of TBT and the presence of recent sources along coastal waters of Malaysia. Lack of any TBT legislation has clearly resulted in significant contamination of BTs in Malaysia.

Sites with consistently elevated concentrations of BTs in fish, sediment and mussels were located along the coast of the Malacca Strait, particularly in locations with intensive maritime activities such as Penang, Johore and Johor Bahru) (Sudaryanto et al. 2004; Harino et al. 2008b). Notably, highest concentration of BTs were found in sediments (1,400 ng g^{-1} dry wt) and fish (210 ng g^{-1} wet wt) in samples collected in the narrowest area of the Johor Strait and Malacca Strait where large harbors and major shipping routes are located. Hashimoto et al. (1998) also observed higher concentrations of TBT in seawater near to major shipping lanes in the Malacca Strait. Lowese levels of BTs were recorded in mussels from rural sites in Sabah at Kalimantan Island (Sudaryanto et al. 2004).

Potential sources of BTs contamination in Malaysian coastal waters are thus similar to those in developed countries in which significant TBT contamination was associated with harbors, marinas, shipyards and high boating activities (Fent 1996). Usage of TBT as biocide in antifouling paints on ship hulls and/or other marine structures in harbors clearly contributes to heavy contamination of BTs in Malaysia. However, intensive aquaculture activities may also contribute to BTs contamination sources in this country: relatively high concentrations of BTs have been recorded in mussels collected from an aquaculture site in Langkawi (140 ng/g wet wt) (Sudaryanto et al. 2002).

4.2.7 Philippines

At most sampling sites, TBT in molluscan bioindicators was detected at relatively high concentrations, compared with DBT and MBT (Tanabe et al. 2000) (Table 4.2). Concentrations of MBT, DBT, and TBT in green mussels were in the range of <3–55, <1–100, and <1–640 ng g^{-1} wet wt, respectively (Tanabe et al. 2000). Relatively high TBT concentrations were observed in samples collected from areas with high boating activities suggest that the source of TBT was antifouling paints. However, BT levels were found to be low in green mussels collected from aquaculture areas, implying minimal usage of BTs for aquaculture activities in the Philippines. Continuing inputs of TBT into the coastal waters of the Philippines may provide a plausible explanation for the higher ratio of TBT found in mussel samples analysed. It is interesting to note however, that mussels from rural areas contained TBT in lower concentrations and proportions, indicating that TBT usage as antifouling agents in these rural areas seems to be minimal. A possible pollution source of BTs in these locations may be the usage of MBT and DBT in plastic products as stabilizers and catalysts. Plastic litter has often been found in abundance along the coasts of Asian developing countries, even at rural sites (Sudaryanto et al. 2002).

Relatively low concentrations of BTs were found in the liver of cetaceans from the Sulu Sea, which ranged at 42–98 ng g^{-1}, wet wt (Prudente 2008). These findings were similar to the low BT residues (ranging <1–30 ng g^{-1} wet wt) found in green mussels from the waters in the Visayas region, which is close to the site where these marine mammals were collected. Cetaceans inhabiting waters adjacent to developing countries in the tropics and subtropics (including the Philippines) generally contain significantly lower hepatic BT concentrations compared with those inhabiting temperate waters near developed nations such as Japan (Tanabe et al. 1998). This could be indicative of significant and continuing inputs of BTs in the coastal waters of these developed countries: comparatively lower levels of BT usage in the Philippines, is implied at present.

4.2.8 Thailand

Thailand is one of the fastest growing developing countries in Southeast Asia. Human and industrial activities in Thailand have recently increased, and trading is flourishing because of the increase of economic activity. The sailing of foreign flag vessels is, therefore, increasing in the Gulf of Thailand. This imposes a potential risk from antifouling biocides. Kan-atireklap et al. (1997a, b) monitored the contamination by BT compounds in sediment and green mussels (*Perna viridis*, L.) from coastal areas of Thailand in 1994 and 1995, and found these to be heavily contaminated. More recently (Harino et al. 2006), the concentrations of MBT, DBT, and TBT in sediment from Thailand reported to be in the range of 1–293, 1–368, and 2–1,246 ng g^{-1} dry wt, respectively (Table 4.1). Because the use of OTs is not regulated in this country, it can be assumed that the country is still subjected to BT pollution. The levels of OTs in Thailand are, in fact, similar to those in Malaysia.

Highest concentrations of TBT in sediment were observed at industrial areas with shipyards, suggesting the entrainment of paint chips containing TBT in sediment. The proportion of TBT relative to other BTs was over 60% (Harino et al. 2006). The higher concentration and ratio of TBTs in sediment indicate continuing inputs from shipping. BT compounds have also been detected in green mussels from Thailand with a mean TBT concentration of 13.6 ng g^{-1} wet wt (Harino et al. 2006). Sudaryanto et al. (2002) surveyed the BT concentrations in green mussels from various developing Asian countries between 1997 and 1999. TBT was detected at mean wet weight concentrations of 2.38, 26.1, 11.9, 2.14, 18.3, 1.95, and 2.61 ng g^{-1} in samples from Cambodia, China, India, Indonesia, Malaysia, the

Philippines, and Vietnam, respectively. Although the level of TBT in green mussels from Thailand was in the range of detected values from other countries, it was at the higher end of this range. The highest concentration was found in an aquaculture area subjected to heavy cargo shipping. It can be predicted that the TBT concentration in water in such an area is also higher in comparison with that of the other sites, because the concentrations of BTs in mussels probably reflect the BT concentrations in water (Harino et al. 1999). Higher concentrations of TBT in water suggest that TBT has either been used for fishing equipment, or is derived from the large commercial vessels docking in the offshore area. Harino et al. (2007b) reported that concentrations of BTs in organs and tissues of various species of stranded whales from Thailand were at >10 times greater than those from green mussels. The BT concentrations in the whales stranded in Thailand were higher compared with other reported values for cetaceans. In general, the concentrations of BTs tend to decrease toward the open sea compared with the coastal area because of the lower density of ships and better exchange of water. Nevertheless, BT concentrations in whales in open sea were higher than those in mussels distributed in the coastal area. The phenomena suggests extensive spread of BTs pollution from Thailand's coastal towards the open ocean.

4.2.9 Vietnam

Vietnam's ongoing economic liberalization has seen the rapid development of industry, and vigorous international trade, imposing a potential risk for BT contamination, which in turn could exert stress on the aquatic environment (Kannan and Falandysz 1997). Concentrations of MBT, DBT, and TBT in sediments ranged from <0.04 to 29.7, from 0.64 to 42.7, and from 0.89 to 50.5 ng g^{-1} dry wt, respectively (Midorikawa et al. 2004; Nhan et al. 2005) (Table 4.1). Though the contamination status of BTs in sediment varied depending on countries, the concentration of BT in Vietnam is within the range of levels found throughout the world. Further, it appears that the level of BTs in sediment from Vietnam is lower in comparison with those in non-regulated countries.

The concentrations of BT in clams from Vietnam were within the range of those reported from other countries. Concentrations of MBT, DBT, and TBT in clam *Meretrix* spp. were in the range of 0.1–44, 0.5–26.6, and 1.4–56 ng g^{-1} wet wt, respectively (Midorikawa et al. 2004; Nhan et al. 2005) (Table 4.2). The highest levels of TBT in sediments from Vietnamese coastal waters occurred in trading ports, in which many small vessels moor, and from areas where shipyards or vessel repair facilities are located. The higher proportions of TBT in sediment from fishing ports and trading ports indicate a continuous input of BTs in these areas. In contrast, when comparing the mean relative composition of the BT compounds, clams contain less TBT, but more DBT and MBT than the sediments from the same locations. This presumably reflects the ability of clams to debutylate TBT and DBT.

4.3 International Comparison

Bivalves such as mussels have been used as bioindicators for monitoring BTs contamination in coastal waters, for example in numerous 'mussel watch' programmes. Sudaryanto et al. (2002) surveyed the concentrations of BTs in green mussel (*Perna viridis*) from Asian countries. Median concentrations of TBT were 0.41, 25.4, 0.29, 0.64, 7.78, 0.62 and 1.33 ng g^{-1} wet wt in Cambodia, China, India, Indonesia, Malaysia, Philippines and Vietnam, respectively. Harino et al. (2007a) reported that concentrations of TBT ranged from 3 to 287 ng g^{-1} wet wt (median 30 ng g^{-1} wet wt) in Japan, whilst TBT was detected over the range 0.95–2.7 ng g^{-1} wet wt in common mussels (*Mytilus edulis*) from North and Baltic seas. (Rüdel et al. 2003). The concentrations of TBT in mussels from Japan were generally much higher than the other countries. Mussels sampled in China, Korea, Malaysia, and Thailand also contained relatively high levels of BTs compared with those from other Asian countries. Sudaryanto et al. (2002) suggested that BT contamination was likely to be correlated with the industrial and human activities shown as per capita GNP. The contamination increases in accordance with high economic growth rate in those countries because the usage of TBT for aquaculture and shipping activities tends to increase depending on economic status. This suggests that coastal pollution by BTs in Asian countries would become more serious in the future without a global ban on the use of these chemicals. Among the Asian countries, only Japan (since 1990) has regulated the usage of TBT. In Korea, partial restrictions on application on small boats and fish nets have been implemented since 2000. Although reductions in TBT concentrations have been recorded in many developed countries following regulation (Champ 2000), there are still reports od significant contamination (Table 4.3). This indicates continuing pollution by BTs derived from heavy ship traffic, which may originate in countries having no regulations on TBT-based antifouling paints. Thus, continuing monitoring and investigations on BT contamination are required in worldwide.

4.4 Conclusions

The present status of BT residues in Asian coastal waters reveals widespread contamination. Highest concentrations of BTs were found at locations with intensive maritime activities, implying the usage of TBT as a biocide in antifouling paints and in marine aquaculture. During the International Maritime Organization Assembly conducted in November 1999, the Environmental Protection Committee proposed a global prohibition on the application of organotin compounds as biocides in antifouling systems on ships by January 1, 2003, and a complete prohibition on the presence of organotin compounds on ships' hulls by January 1, 2008 (Champ 2000). In Japan a ban on the use of TBT antifouling paints has been in place for 20 years that. In spite of the regulation, however elevated concentrations of BTs have been detected in coastal waters, with extremely high levels reported as hot-spots (Harino et al. 2004, 2007a). This implies that BT contamination in the aquatic environment of developing Asian countries may become a serious issues in future due to the unregulated usage of organotins and the increasing demand for antifouling paints in these regions. Thus, it is important to continue monitoring for BT contamination and to conduct ecotoxicological risk assessments to further our understanding of the fate of contamination and the impacts on organisms in Asian countries.

References

Alzieu C (1996) Biological effects of tributyltin on marine organisms. In: de Mora SJ (ed) Tributyltin: Case Study of an Environmental Contaminant, Cambridge University Press, Cambridge

Bhosle NB, Garg A, Jadhav S et al. (2004) Butyltins in water, biofilm, animals and sediments of the west coast of India. Chemosphere 57:897–907

Bhosle NB, Garg A, Harji R et al. (2006) Butyltins in the sediments of Kochi and Mumbai harbours, west coast of India. Environ Int 32:252–258

Biselli S, Bester K, Huhnerfuss H et al. (2000) Concentrations of the antifouling compound Irgarol 1051 and of organotins in water and sediments of German North and Baltic Sea marinas. Mar Pollut Bull 40:233–243

Bosselmann K (1996) Environmental law and tributyltin in the environment. In: de Mora SJ (ed) Tributyltin: Case Study of an Environmental Contaminant, Cambridge University Press, Cambridge

Brack K (2002) Organotin compounds in sediments from the Gota alv estuary. Water Air Soil Pollut 135:131–140

Champ MA (2000) A review of organotin regulatory strategies, pending actions, related costs and benefits. Sci Total Environ 258:21–71

Chiavarini S, Massanisso P, Nicolai P et al. (2003) Butyltins concentration levels and imposex occurrence in snails from the Sicilian coasts (Italy). Chemosphere 50:311–319

Connell DW, Wu RSS, Richardson BJ et al. (1998) Fate and risk evaluation of persistent organic contaminants and related compounds in Victoria Harbour, Hong Kong. Chemosphere 36: 2019–2030

Elgethun K, Neumann C, Blake P (2000) Butyltins in shellfish, finfish, water and sediment from the Coos Bay estuary (Oregon, USA). Chemosphere 41:953–964

Fent K (1996) Ecotoxicology of organotins compounds. Crit Rev Toxicol 26:1–117

Gibbs PE, Bryan GW (1996) TBT-induced imposex in neogastropod snails: masculinization to mass extinction. In: de Mora SJ (ed) Tributyltin: Case Study of an Environmental Contaminant, Cambridge University Press, Cambridge

Gomez-Ariza JL, Giraldez I, Morales E (2000) Temporal fluctuations of tributyltin in the bivalve Venerupis decussata at five stations in southwest Spain. Environ Pollut 108:279–290

Harino H, Fukushima M, Kawai S (1999) Temporal trends in organotin compounds in the aquatic environment of the Port of Osaka, Japan. Environ Pollut 105:1–7

Harino H, Yamamoto Y, Kawai S et al. (2003) Butyltin and phenyltin residues in water, sediment and biological samples collected from Otsuchi Bay, Japan. Otsuchi Mar Sci 28:84–90

Harino H, Yamamoto Y, Kawai S et al. (2004) Distribution of organotin compounds in aquatic environment – The Port of Osaka, Japan. In: Shibata K, Senda T (eds) Proceedings of International Symposium on Antifouling Paint and Marine Environment. pp. 61–69. Tokyo.

Harino H, Ohji M, Wattayakorn G et al. (2006) Occurrence of antifouling biocides in sediment and green mussels from Thailand. Arch Environ Contam Toxicol 51:400–4007

Harino H, Yamamoto Y, Eguchi S et al. (2007a) Concentrations of antifouling biocides in sediment and mussel samples collected from Otsuchi Bay, Japan Arch Environ Contam Toxicol 52:179–188

Harino H, Ohji M, Wattayakorn G et al. (2007b) Accumulation of organotin compounds in tissues and organs of stranded whales along the coasts of Thailand. Arch Environ Contam Toxicol 53:119–125

Harino H, Ohji M, Wattayakorn G et al. (2007c) Concentrations of organotin compounds in tissues and organs of dugongs from Thai coastal waters. Arch Environ Contam Toxicol 53:495–502

Harino H, Eguchi S, Yamamoto Y et al. (2008a) Distribution of Organotin Compounds in Representative Coastal Areas from Japan. Coast Mar Sci 32:88–95

Harino H, Arai T, Ohji M et al. (2008b) Occurrence of Antifouling biocides in Southeast Asia- Malaysia, Thailand and Vietnam, In Miyazaki M, Wattayakorn G (ed): The ASEAN International Conference: Conservation on the Coastal Environment, Shinjusha, Tokyo

Hashimoto S, Watanabe M, Noda Y et al. (1998) Concentration and distribution of butyltin compounds in a heavy tanker route in the Strait of Malacca and in Tokyo Bay. Mar Environ Res 45:169–177

Haynes D, Loong D (2002) Antifoulant (butyltin and copper) concentrations in sediments from the Great Barrier Reef World Heritage Area, Australia. Environ Pollut 120:391–396

Hoch M (2001) Organotin compounds in the environment-an overview. Appl Geochem 16:719–743

Inoue S, Oshima U, Imada N et al. (2002) Tributyltin pollution in coastal areas around northern Kyushu. Jpn J Environ Toxicol 5:43–50

Kannan K, Falandysz J (1997) Butyltin residues in sediment, fish, fish eating birds, harbor porpoise and human tissues from the Polish coast of the Baltic Sea. Mar Pollut Bull 34:203–207

Kan-atireklap S, Tanabe S, Sanguansin J (1997a) Contamination by butyltin compounds in sediments from Thailand. Mar Pollut Bull 34:894–899

Kan-atireklap S, Tanabe S, Sanguanisin J et al. (1997b) Contamination by butyltin compounds and organochlorine residues in green mussel (*Perna viridis*, L.) from Thailand coastal waters. Environ Pollut 97:79–89

Michel P, Averty B (1999) Distribution and fate of tributyltin in surface and deep waters of the northwestern Mediterranean. Environ Sci Technol 33:2524–2528

Midorikawa S, Arai T, Harino H et al. (2004) Concentrations of organotin compounds in sediment and clams collected from coastal areas in Vietnam. Environ Pollut 131:401–408

Nhan DD, Loan DT, Tolosa I et al. (2005) Occurrence of butyltin compounds in marine sediments and bivalves from three harbour areas (Saigon, Da Nang and Hai Phong) in Vietnam. Appl Organomet Chem 19:811–818

Ohji M, Arai T, Midorikawa S et al. (2007) Distribution and fate of organotin compounds in Japanese coastal waters. Water Air Soil Pollut 178:255–265

Prudente (2008) Monitoring of butyltin compounds in the aquatic environments of the Philippines. Coast Mar Sci 32:108–115

Rajendran RB, Tao H, Miyazaki A et al. (2001) Determination of butyl-, phenyl-, octyl- and tributylmonomethyltin compounds in a marine environment (Bay of Bengal, India) using gas chromatography-inductively coupled plasma mass spectrometry. J Environ 3:627–634

Regoli L, Chan HM, de Lafontaine Y et al. (2001) Organotins in zebra mussels (Dreissena polymorpha) and sediments of the Quebec City Harbour area of the St. Lawrence River. Aquat Toxicol 53:115–126

Rüdel H, Lepper P, Steinhanses J (2003) Retrospective monitoring of organotin compounds in marine biota from 1985 to 1999: results from the German Environmental Specimen Bank. Environ Sci Technol 37:1731–1738

Rumengan IFM, Ohji M, Arai T et al. (2008) Contamination status of butyltin compounds in Indonesian coastal waters. Coast Mar Sci 32:116–126

Shim WJ, Hong SH, Yim UH et al. (2002) Horizontal and vertical distribution of butyltin compounds in sediments from shipyards in Korea. Arch Environ Contam Toxicol 43:277–283

Shim WJ, Hong SH, Kim NS et al. (2005) Assessment of butyl- and phenyltin pollution in the coastal environment of Korea using mussels and oysters. Mar Pollut Bull 51:922–931

Short JW, Sharp JL (1989) Tributyltin in bay mussel (*Mytilus edulis*) of the Pacific Coast of the United States. Environ Sci Technol 23:740–743

Sudaryanto A, Takahashi S, Tanabe S et al. (2000) Occurrence of butyltin compounds in mussels from Indonesian coastal waters and some Asian countries, Water Sci Technol 42:71–79

Sudaryanto A, Takahashi S, Monirith I et al. (2002) Asia-Pacific mussel watch:monitoring of butyltin contamination in coastal waters of Asian developing countries. Envrion Toxicol Chem 21:2119–2130

Sudaryanto A, Takahashi S, Iwata H et al. (2004) Contamination of butyltin compounds in Malaysian marine environment. Environ Pollut 130:347–358

Sudaryanto A, Takahashi S, Iwata H et al. (2005) Organotin residues and the role of anthropogenic tin sources in the coastal marine environment of Indonesia. Mar Pollut Bull 50:227–236

Stewart C, de Mora SJ (1990) A review of the degradation of. tri(n-butyl)tin in the marine environment. Environ Technol 11:565–570

Takeuchi I, Takahashi S, Tanabe S et al. (2004) Butyltin concentrations along the Japanese coast from 1997 to 1999 monitored by *Caprella* spp. (Crustacea: Amphipoda). Mar Environ Res 57:397–414

Tanabe S, Prudente M, Mizuno T et al. (1998) Butyltin contamination in marine mammals from North Pacific and Asian coastal waters. Environ Sci Technol 32:193–198

Tanabe S, Prudente M, Kan-atireklap S et al. (2000) A Mussel watch: marine pollution monitoring of butyltins and organochlorines in coastal waters of Thailand, Philippines and India. Ocean Coast 43:819–839

Thomas KV, Blake SJ, Waldock MJ (2000) Antifouling paint booster biocide contamination in UK marine sediments. Mar Pollut Bull 40:739–745

Tong SL, Pang FY, Phang SM et al. (1996) Tributyltin distribution in the coastal environment of Peninsular Malaysia. Environ Pollut 91:209–216

Yang RQ, Zhou QF, Liu JY et al. (2006) Butyltins compounds in molluscs from Chinese Bohai coastal waters. Food Chem 97:637–643

Yuan DX, Yang DN, Wade TL et al. (2001) Status of persistent organic pollutants in the sediment from several estuaries in China. Environ Pollut 114:101–111

Chapter 5
Behaviour of Organotins in the Coastal Environment

William John Langston, Hiroya Harino, and Nicholas Dingle Pope

5.1 Introduction

The use of TBT and TPT as constituents of antifouling compounds, over more than three decades, has left many coastal environments with a longstanding legacy of contamination. Inputs to water should no longer be an acute threat. However, despite recent actions by IMO to ban the use of these compounds, recovery will not be instantaneous and contamination could even increase at some locations where the legislation is ineffective, or where coatings are replaced, or sediments re-mobilised. Partitioning to solids is reversible and hence sediments may act as a persistent sink and secondary source of adsorbed organotins, as well as those residues entrained as paint flakes from boatyards and docks. Estimates of TBT half-times in sediments range from a few months to decades (in anoxic sediments), indicating that this 'reservoir' of organotins is likely to remain biologically relevant, and will require management, for a considerable period. For some ports and harbours, appropriate dredging and disposal of TBT-enriched sediments represents an extremely costly option to maintain viability, and it will be important to ensure that further harm to the environment does not ensue from remobilization of these residues. Based on example in the UK, we review here some of the factors which influence long-term partitioning behaviour and persistence of organotions (predominantly TBT) – and the likely timescales for recovery.

W.J. Langston and N.D. Pope
Marine Biological Association, Citadel Hill, Plymouth PL1 2PB, UK

H. Harino
Osaka City Institute of Public Health and Environmental Sciences,
8-34 Tojo-cho, Tennoji, Osaka 543-0026, Japan

T. Arai et al. (eds.), *Ecotoxicology of Antifouling Biocides*,
DOI: 10.1007/978-4-431-85709-9_5, © Springer 2009

5.2 Persistence of Organotins in Water and Sediments

Persistence of any chemical in the environment is a function of biological, chemical and physical degradation, together with physical removal mechanisms. Photolysis and chemical degradation of tri-substituted organotins (mainly TBT) has been demonstrated in the laboratory though, in nature, particularly in turbid estuaries where attenuation of UV light occurs within the upper few centimeters, microbially-mediated debutylation to DBT and, to a lesser extent MBT is the predominant mechanism by which these compounds are broken down (Seligman et al. 1996a). Only in the photic zone of clearer freshwater systems is abiotic photolysis likely to be of significance (half time ~3 months).

In seawater and estuarine water, half-times for TBT, derived from short-term laboratory experiments, are rather variable, but generally observe zero-order kinetics, with values ranging from 4 to 19 days (reviewed by Seligman et al. 1996a). However, longer term mesocosm studies suggest rather longer degradation (months), together with first-order kinetics and rate constants in the range 0.1–0.2 day^{-1}. There are some indications that the rate of degradation, expressed as the proportion of metabolites in the sample, is a function of residence time of the water body, and is also a function of microbial populations in the water column and temperature (low in winter). Relationships have been found between degradation and the presence of photosynthetic organisms such as diatoms and dinoflagellates. It is also possible that degradation rates may be enhanced by prior exposure of the microbial population to TBT or even other xenobiotics, though adaptation to TBT has not been widely demonstrated.

The fate of TBT is more complex once incorporated into sediment. Seligman and co-authors (1996a) have described rather rapid, sediment-enhanced degradation of spiked TBT in aerated conditions (half times 2–15 days); however TBT bound tightly to sediment or paint particles appears to be considerably more persistent. It has been suggested that TBT in paint particles may be considerably less bioavailable than that adsorbed to sediment particles and if so, may represent a much more persistent legacy. The proportion of paint particles in sediment could conceivably explain some of the variation in estimates of half lives seen in nature (see section 5.4), though as in water, there are numerous characteristics which influence behaviour.

Measured half times for TBT in experiments with unspiked sediments are invariably longer than in water and typically are of the order of several months (sometimes longer), with both bacteria and fungi contributing significantly to measured degradation rates through dealkylation to both DBT and, particularly, hydrophilic MBT (Seligman et al. 1996a). Several authors have indicated that sediment anoxia is a principal factor determining half times; in extreme cases estimates of half lives in benthic muds are in excess of 10 years, as suggested by Maguire (2000). Sediments where biological degradation was suppressed by KCN showed no degradation after ~1 year (Maguire and Tkacz 1985).

Our own laboratory experiments conducted in flow-through aquaria (complete water exchange daily), containing undisturbed sediments labeled with ^{14}C-TBT (0.3 mg kg^{-1}, as Sn), indicate very low rates of degradation. Half-times increased with depth in the sediment and ranged from 2.6 years at the surface (0–5 cm) to

3.5 years (at 5–10 cm), and to 8.5 years at the more anoxic bottom of the cores (10–15 cm). Degradation decreased with depth and the proportion remaining as TBT, after 4 years was 72%, 85% and 89% in upper middle and lower parts of the core, respectively. Loss rates via this mechanism were estimated to vary from 0.013% to 0.028% day^{-1}. Diffusive losses were also relatively small in these undisturbed cores (0.0015% day^{-1}), representing a leach rate of 1.37 $\mu g\ m^{-2}\ day^{-1}$. Clearly, in undisturbed sites in estuaries marinas and docks, sediments could retain loadings for perhaps a decade or more (see section 5.4). Not surprisingly, however, release rates increased as a function of flow rate over these cores.

Interestingly, the estimated TBT release rates from sediment cores are comparable to *in situ* measurements made at the sediment-water interface (Stang and Seligman 1987), and are of the same order as many estimates of leach-rates from anti-fouling paints. Seligman et al. (1996b) have shown the latter may vary considerably between coatings (0.1–2.8 $\mu g\ cm^{-1}\ day^{-1}$) but it would appear that sediments operate as a slow-release mechanism for TBT which is analogous to many paint matrices.

Experiments with sediments in mesocosms have confirmed that once incorporated into sediment, degradation is dramatically slowed to a rate of $\sim 1 \times 10^{-4}\ day^{-1}$, most notably during a declining temperature regime (to 5°C) designed to simulate winter conditions (Seligman et al. 1996a). These authors have suggested a short-lived chemical (possibly catalytic) component to degradation of OTs at the sediment-water interface (with MBT as a predominant product) followed by a much longer period of microbial degradation lasting several months or more. Sediments thus act as a store for any TBT that is not degraded and may act as a slow release mechanism, depending on conditions.

Potentially, the biological activity of infaunal macro-organisms (bioturbation) could enhance the process of mobilization of organotins in sediments. In laboratory leaching studies with ^{14}C-TBT labeled sediment (0.3 $mg\ kg^{-1}$, as Sn), in flow through aquaria (complete water exchange daily), the presence of both worms *Nereis diversicolor* (at a density of 550 m^{-2}) and burrowing clams *Scrobicularia plana* (at a density of 165 m^{-2}) were shown to enhance release of TBT to overlying water compared to sediments devoid of macrofauna. This was particularly evident in the initial few days after fauna were added, as they established burrow systems. Average daily net losses due to bioturbation (over 69 and 128 day periods) were 0.0012% and 0.0008% day^{-1} for worms and clams, respectively. This is roughly equivalent in magnitude to losses by diffusion/desorption and represents a relatively small fraction of the sediment budget (<5%). Comparable behaviour has also been observed following colonisation of contaminated sediment by the mud-snail *Hinia reticulata* (Pope 1998), though losses were higher (>0.1% day^{-1}) due to the larger number of animals (300 m^{-2}) and their greater mobility. Figure 5.1 shows the influence of addition and removal of these snails on the release of TBT from sediment (10.2 $\mu g\ g^{-1}$) to overlying water. It is possible that major bioturbator species, such as burrowing shrimps belonging to the family Callianassidae, could have even larger effects on release and redistribution of TBT from, and within, sediment (through burial and re-working). The scale of these biological processes in different types of bioturbating organism, and their influence on the fate of organotins, are interesting topics for further study.

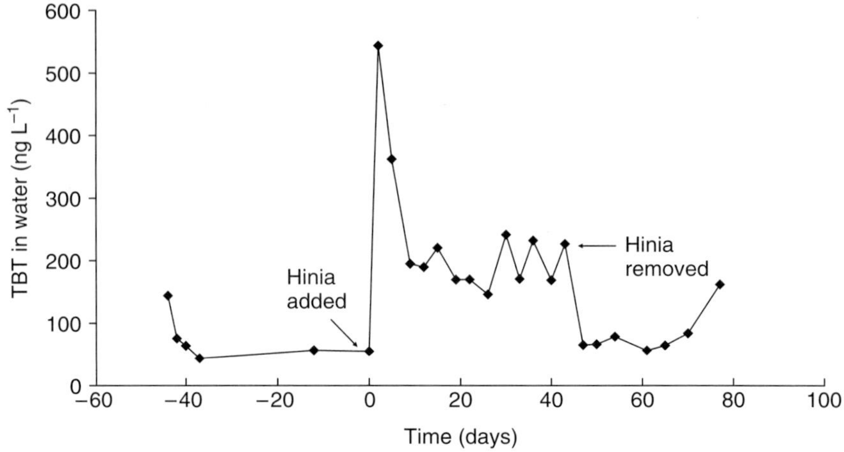

Fig. 5.1 Influence of bioturbation on the release of TBT to overlying water following the introduction (day 0) and removal (day 43) of the mud-snail *Hinia reticulata* to a flow-through sediment-water system equilibrated with ^{14}C TBT

Thus, whilst the rapid degradation of TBT observed in incubated water samples has led some authors to label TBT as relatively non-persistent, the fate in coastal ecosystems can be modified by a wide range of specific properties. The most meaningful evaluations of behaviour are therefore based on site-specific field observations (see section 5.4). Flushing rates, microbial populations and temperature are important variables. For example, the degradation of TBT is minimal below ~5°C and will consequently be very slow in cold climates and vulnerable polar areas. However, the influence of particulates, coupled with hydro- and sediment dynamics, is probably the most critical factor affecting the fate of TBT residues in coastal habitats. Many sediments will represent a substantial legacy of TBT use and may provide continuing inputs to the water column. Crucially, such sediments may also represent disturbed biological habitats and furthermore, delay the TBT degradation and loss processes significantly.

5.3 Determinants of Sediment-Water Partitioning

Given that the behaviour of organotins differs significantly in sediment and water, and that degradation rates are at least an order of magnitude longer in sediments, the long-term fate of these compounds will depend heavily on their relative partitioning between aqueous and particulate phases. Factors that promote adsorption of organotins to particulates will tend to increase their long-term retention, and in sheltered coastal and estuarine habitats, sediments will act as a reservoir and secondary source of contamination long after new inputs have ceased as a result of paint bans. Several models have been developed to describe the adsorption process, however, at low concentrations of solute, all sorption isotherm models predict linear sorption, leading to widespread acceptance of the linear (constant partitioning) model (Eq. 5.1) under these conditions:

$$q = K_p C \tag{5.1}$$

Where, at equilibrium, q is the mass of solute sorbed per unit mass of the solid phase, C is the concentration in solution and K_p is a constant termed the *partition coefficient*.

Most sorption isotherm models assume a condition of dynamic equilibrium and reversibility, suggesting that sorbed compounds will readily become desorbed when concentrations in the surrounding water drop below equilibrium levels. Experimental studies indicate this is the case for TBT in freshly spiked systems, at low concentrations (Langston and Pope 1995), though in organic rich benthic muds it is possible that a proportion of the compound may be bound in an irreversible state which in some cases may also be related to the "age" of the solute/sorbent combination. At higher concentrations, TBT sorption exhibits a Freundlich-type (non-linear) isotherm whereby the value of the partition coefficient varies with TBT concentration (K_p decreases as TBT increases). This is a feature of ionisable organic solutes and reflects decreasing solute-sediment affinities as more adsorption sites become filled.

A number of characteristics (of solute and sorbent) influence the sorption process. Firstly there is the hydrophilic-hydrophobic balance of the solute, usually measured as the degree to which the solute partitions between n-octanol and water – the octanol-water partition coefficient (K_{ow}). TBT is sequestered readily by suspended particulates (usually equilibrium occurs within a few hours), due to its low solubility and its hydrophobicity (log K_{ow} ~3.8). Therefore it has a relatively strong tendency to become incorporated into estuarine sediments where it may remain long after new inputs have ceased.

Secondly, because of the affinity of TBT for hydrophobic organic phases, the fractional organic carbon content (f_{oc}) of the solid phase can be important in determining concentration, such that an organic carbon normalised partition coefficient (K_{oc}) may be a useful way of comparing sediment contamination (Eq. 5.2), explaining some but not all of the variation.

$$K_{oc} = \frac{K_p}{f_{oc}} \tag{5.2}$$

However, organic content alone seldom explains partitioning behaviour totally. Several studies have observed inverse relationships between K_p and both particle concentration, and particle size; the latter can be partly explained by the total surface area available for sorption, although often this may be attributable to higher carbon content of smaller particles. It is also important to note that partitioning data derived from experiments can be influenced artificially by drying, oxidation or freezing of the sediment, unrealistic concentrations, inadequate phase separation and inadequate recovery. Together these factors explain some of the variability for reported K_p values for TBT (Table 5.1). Using radiolabelled ^{14}C-TBT, we have attempted to avoid these pitfalls and have determined equilibrium K_p values of 248–24,6771 kg^{-1} in batch experiments with 16 natural sediments from the south coast of England (Langston and Pope 1995; Pope 1998). Highest K_p values were associated with organic-rich silty sediments. Published *apparent* sorption coefficients, based on *in situ* field measurements of sediment and overlying water, exhibit a wide variety values (over three orders of magnitude), though often in the range 10^3–10^4 l kg^{-1}, at the high end of the scale found in laboratory-based equilibrium experiments. Harris et al. (1996) and others have suggested that, in the field, some of the higher values may signify the presence of antifoulant paint chippings, near boat maintenance facilities, though this has yet to be verified. It may be that higher K_p values occur *in situ* because the system is not at equilibrium: long-term desorption experiments support the hypothesis that sub-equilibrium levels of TBT are desorbed from benthic sediments which are not well-mixed (Pope 1998).

The nature of the aqueous phase may influence sorption in several ways. pH is one of the master variables for surface coordination reactions, reflecting solution hydrolysis, protonation of adsorbing ions and surface charge properties of the adsorbent. The effect of pH on TBT partitioning revealed highest values for K_p (~107,5891 kg^{-1}) at neutral pH (~7) (Fig. 5.2). The K_p decreased sharply with both increasing or decreasing pH (~19,243 and 17,6671 kg^{-1} at pH 4.65 and 9.19 respectively), reflecting increased solubility of TBT under these conditions. In estuarine environments, pH is usually buffered via the seawater carbonate system to near-neutral conditions (pH ~8) but it is feasible that under low salinity conditions in the upper reaches of estuaries, or under high freshwater flow, pH could become a relevant factor in TBT partitioning (Langston and Pope 1995).

Next to pH, competitive adsorption is probably the second most important way in which dissolved solutes affect adsorption of cations – notably bulk inorganic constituents (e.g. Ca, Mg) that mutually adsorb onto the same sites on the solid phase. The extent of competition depends upon relative concentrations of these solutes and affinities for the same sites. Ionic strength (or salinity) therefore influences sorption, affecting activities in solution and surface charge on particles. Increasing salinity also increases the aggregation of suspended particulates, which may in turn influence the kinetics of sorption – by necessitating solute diffusion into aggregate pore waters. Concomitantly the total surface area, and therefore the rapidly reversible component of sorption, decrease. Given this complexity it is not surprising that studies on the effects of salinity on TBT partitioning often showed conflicting results (Randall and Weber 1986; Unger et al. 1987, 1988; Harris et al. 1996); interpretation is confounded in some of these studies by use of unrealistic TBT levels, artificial substrates, processed sediments or non-natural waters. In attempting to avoid some of these inconsistencies, investigations with two natural sediments and estuarine water (Langston and Pope 1995; Pope 1998) showed highest K_p values in fresh water (60,613 and 30,3331 kg^{-1}) reducing to 15,000 and 17,2001 kg^{-1}, respectively, at salinities between 25% and 100% seawater. Concentrations in water were at a maximum between 25% and 75% seawater, perhaps reflecting the effect

Table 5.1 Reported sediment-water partition coefficients for TBT

Partition coefficient ($l\ kg^{-1}$)	Suspended sediment load ($mg\ l^{-1}$)	Salinity ($g\ l^{-1}$)	pH	Notes	Reference
3,278–3,918	5.8–6.7	Sea water			Valkirs et al. (1986)
929	14	Sea water			
340	50	Sea water			
3,000	10	Sea water			M & T Chemicals (unpublished). Referred to in Valkirs et al. (1986)
1,500–1,900,000	10–1,000	Artificial sea water	6.2–8.2	Fulvic acid-coated hydrous iron oxide used as the solid phase	Randall and Weber (1986)
2,180	5,000	Sea water		Natural harbour sediment, unfiltered water	Maguire and Tkacz (1985)
111–8,200	No data	0–35 (Artificial seawater)		Only aqueous phase analysed	Unger et al. (1988)
6,250–55,439	No data	Sea water		*In situ* measurements in Pearl Harbor mesocosms	Stang and Seligman (1987)
4,608–39,352	1.6–8.6	Sea water			Valkirs et al. (1987)
200–1,400	60–100	0–32 (filtered seawater)		Sediments were freeze dried, ground, and sieved (100 mesh)	Harris and Cleary (1987)
17–4,500	1,000	Seawater		Experimentally determined values	Kram et al. (1989)
32–292,000				Calculated from field analyses of sediments and overlying water	
248–24,677	<1,300	Seawater		Extensive studies on TBT partitioning using radiolabelled ^{14}C-TBT	Langston and Pope (1995); Pope (1998)
30,333–60,613		River water		Natural sediments and water, effects of pH, salinity, sediment load and type	
19,243		Seawater	4.65		
107,589		Seawater	7.0		
17,667		Seawater	9.19		

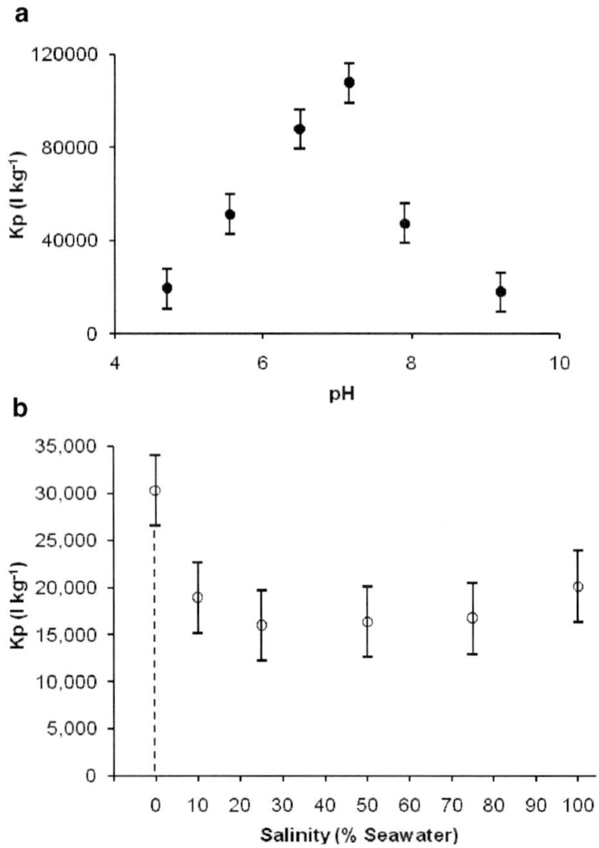

Fig. 5.2 Effect of pH (a) and salinity (b) on TBT partitioning (Adapted from Langston and Pope 1995)

of chloride ion concentration on TBT speciation in solution: above 75% seawater there was a 'salting-out' effect due to lower solubility, consistent with most organic (hydrophobic) contaminants. The increasing removal to particulates at salinities below 25% seawater may reflect the growing influence of the ionic components of the TBT molecule and shifts in speciation, coupled with modifications to charges on the sediment surface, in water of low ionic strength.

5.4 Natural Resuspension, Dredging and Disposal of Contaminated Spoils – The Continuing Legacy of TBT

The partition coefficient does not appear to change markedly (<2-fold) over a wide range (30-fold) of suspended solids concentrations (Langston and Pope 1995), though obviously, as TBT partitions in favour of sediment, the amount of sediment in suspension will influence the transport budget of TBT, particularly in turbid estuaries and coastal systems. The fate and persistence of TBT in the water column and sediments is therefore dependent on a combination of hydrodynamic and biogeochemical factors (Ruiz et al. 1996). Localised circulation patterns and tidal flows usually determine the residence time for TBT and hence the ultimate loadings. Where water movements are strong, distributions of TBT may be fairly homogeneous, however in less dynamic environments, heterogenous distributions in sediments are more likely. Sediment sources will be of more significance where contamination levels, and K_p, are high. Seasonal trends can also markedly influence residence times for TBT, and hence the influence of sediment release on concentrations in water. A discussion of the dynamics of release and how sediment loadings might be simulated under estuarine conditions is provided by Harris et al. (1996). Although reasonable estimates of partition coefficients and degradation rates are now available to incorporate into such simulations, estimates of the total mass of exchangeable sediment and its exchange rate with the water column are often more difficult to parameterise. If mixing times are very rapid, relative to equilibration times for TBT partitioning, laboratory- derived K_p values may be less representative of natural partitioning.

Behaviour of TBT and the role of sediments as a secondary source will clearly be site specific. As TBT inputs to water decline there will be a tendency, over time, for K_p values to increase according to a logarithmic relationship favouring retention in sediments (Langston and Pope 1995). Thus, as sources of 'new' TBT are eliminated indications are that releases from relatively undisturbed sediments at most locations are likely to be slower, in relation to the original inputs. Nevertheless, any process that results in increased sediment/water mixing (tidal resuspension, dredging, bioturbation) would invariably be expected to remobilise and release some of the sediment-bound TBT. The potential for desorption of TBT, as sediments age, does not appear to vary greatly over relatively long periods (Pope 1998).

An example of natural tidal resuspension affecting transport and distribution of TBT is shown in the axial profile (summer) for the macro-tidal Tamar Estuary, UK, in Fig. 5.3, which encompasses sites from freshwater to the sea. Prior to TBT legislation TBT was dominated by inputs from the dockyards and marinas towards the mouth of the Estuary (concentrations were sometimes in excess of 100 ng l^{-1} here). These inputs have now all but ceased as a result of control measures and TBT concentrations are much lower whilst the estuarine profile is governed by natural physical and chemical processes. A dominant aspect is the sharp change occurring upstream (0–10 km from the tidal weir) near the freshwater-seawater interface. This region coincides with the region of maximum turbidity (note high suspended solids loads) caused by tidal-resuspension of particles at the salt wedge. Since there is no evident freshwater input for TBT, origins are presumed to be the legacy of sources near the mouth, described above. The type of profile exhibited in the Tamar is generated internally by interactions within the estuary, with the TBT maximum at the head resulting from upstream transport of tidally re-suspendable sediment, coupled with remobilization and desorption at low salinities (consistent with partitioning results, described above). This occurs primarily in the calmer low-flow summer period: in periods of high-flow (usually winter) distributions are more homogeneous as material

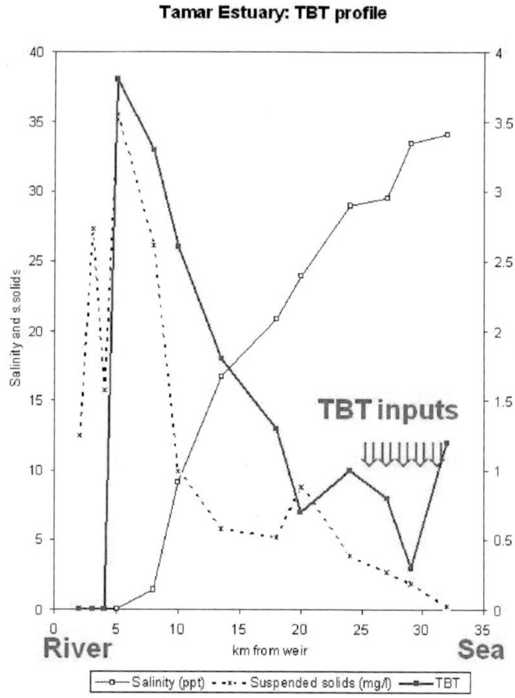

Fig. 5.3 Profiles showing TBT concentrations, salinity and suspended solids in the water column of the Tamar Estuary, UK

is transported back down estuary, completing the process of internal cycling. In the absence of significant net export from the system this pattern of cycling could mean that estuaries retain TBT burdens for a number of years.

Maintenance of the seasonal TBT 'inputs' at the turbidity maximum, typified in Fig. 5.3, can thus be envisaged as a continuous balance between recruitment from and deposition in estuarine sediment. This advective process is a more effective means of transferring TBT from sediment to the water column than diffusion alone and will be subjected not only to climatic variation but also to oscillations in spring/neap tidal energy inputs. Such pronounced reactivity may not be evident in less dynamic systems. Nevertheless, this example highlights the fact that estuarine particulates can become the dominating force determining distributions in the water column, following the removal of primary sources.

For port authorities, regulators and conservation agencies, concern arises where the re-mobilization of TBT from sediments is artificially enhanced following dredging or general disturbance (Brack 2002) and difficulties in managing this problem have frequently been highlighted (e.g. Svavarsson et al. 2001; Santos et al. 2004). For example, dredging of parts of Southampton Water during the last decade, to improve access to ports and marinas, has coincided with a halt in the decline of TBT levels in water at some sites. The time course of TBT reduction in waters can be substantially altered by prolonged activity, as indicated for the Hamble, Southampton, in Fig. 5.4; continuing use of TBT on shipping using the commercial

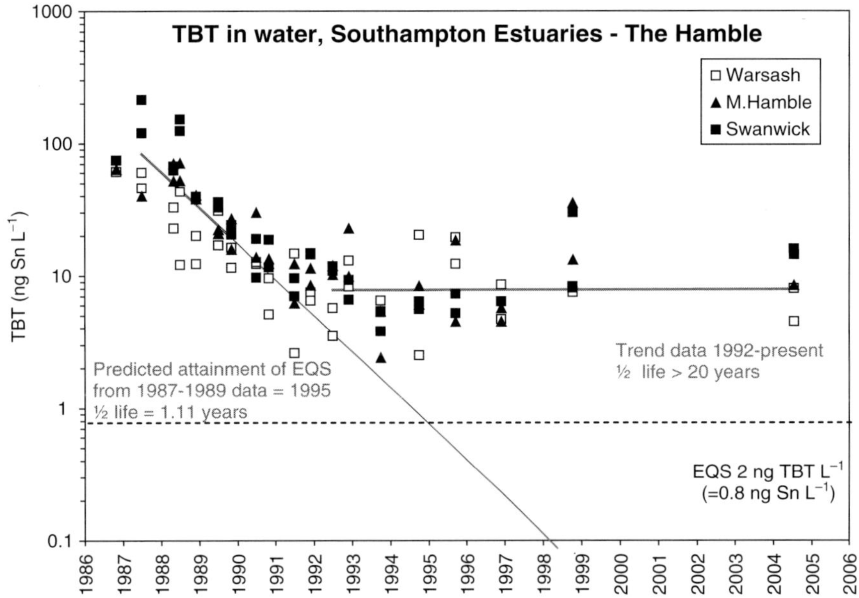

Fig. 5.4 TBT in water (ng l^{-1} as Sn), Hamble Estuary. Initial recovery, following legislation in 1987 has been halted, probably due to a combination of sediment re-release (including dredging) and some continuing use of TBT coatings

port, nearby, may have also contributed to the inputs from sediment disturbance. As a result, the system has been at steady-state in recent years – at concentrations substantially above the Environmental Quality Standard (EQS) of 0.002 μg l^{-1}.

Maritime ports are constantly under pressure to intensify the volume of containerised transport vessels to meet competition and, in Europe, there are plans to meet this demand by developing some ports as regional hubs (Marcadon 1999). Expansion could lead to enhanced remobilization of sediment-bound TBT and possibly even further recharge of sediments in some locations, depending on conditions. In addition, as channels may have to be extended and maintained to greater depths, to allow access to larger vessels (Side and Jowitt 2002), the impact of resuspension events and dredge spoil disposal may increase at sites which are currently unaffected by TBT (Svavarsson et al. 2001). This is particularly contentious where there is potential for resuspended material containing TBT to be transported extensively into adjacent statutory conservation areas. Recently for example, expansion of the Port of Southampton, UK, has been refused because of such concerns over designated sites. Given such important economic and conservation issues, there are strong arguments to establish better cause-effect relationships between dredged spoil disposal activities and the impacts of TBT on protected species and habitats. Without such evidence there is a risk of poor decisions being made: at one extreme this could result in failure to protect vulnerable species, at the other, over-precautionary actions could lead to unnecessary increases in disposal costs and possibly even port closure.

Clearly however, given the reversible nature of TBT adsorption in sediment, it is important to ensure that levels do not exceed thresholds for the most sensitive species as a result of remobilization events. Even where re-release is negligible, there may still be a significant threat from particulates enriched with TBT, since for infaunal species such as the clam *Scrobicularia plana* and mud snails *Hinia reticulata*, sediments are an important vector for bioaccumulation (Langston and Burt 1991; Pope 1998; *see* Chapter 16). Bioavailability and toxicity of sediment-bound TBT, like partitioning, will vary according to sediment characteristics of which organic carbon content is a primary determinant. Organic carbon appears particularly important in modifying desorption and bioavailability in species such as *Hinia* which derive body burdens through contact with sediment rather than ingestion, perhaps through its regulation of pore water concentrations. Thus, desorption tends to be highest where sediments are low in organics (low K_p and K_{oc}), and, correspondingly, aqueous routes of uptake – and toxicity – become more prominent (Meador et al. 1997; Pope 1998). There are undoubtedly other, more subtle, sediment influences, such as particle size, humic and colloidal content, which have yet to be explained.

5.5 Monitoring Temporal Trends in Coastal Systems

After initial legislation prohibiting the use of organotin antifouling on most leisure craft (1987 in the UK), TBT and TPT continued to be used on larger vessels, on the assumption that levels in offshore seas would be diluted sufficiently to prevent deleterious effects. Around the UK, on open coastlines, reduced contamination has been demonstrated by declining imposex severity and the recovery of dogwhelk *Nucella lapillus* populations (Fig. 5.5; Hawkins et al. 2002). Even so, this recovery process has been relatively slow (10+ years) and there are still populations of neogastropods which are heavily impacted close to ports, both in the UK and elsewhere (Galante-Oliveira et al. 2006; Gibbs 2009). Similarly, in Arcachon Bay, France, despite rapid reductions in seawater concentrations of between five- and ten-fold in the 3 years after legislation (1982), and accompanying improvements in *Crassosstrea gigas,* sediment contamination was still considered responsible for effects in oysters some 20 years after TBT legislation (Alzieu 2000). Furthermore, the assumption that removing TBT from the leisure market would be sufficient protection for the marine environment was not upheld by the evidence from a number of post-ban studies, notably those which demonstrated the continuing presence of imposex in snails close to offshore shipping routes and anchorages (Hallers-Tjabbes et al. 2003; Rato et al. 2006).

In a number of inshore and estuarine areas where small boats are predominant, such as the Crouch Estuary (Waldock et al. 1999) and Poole Harbour, UK (Langston et al. 1987, 1994) sediment-dwelling macrofauna have responded successfully to legislation and the earlier decline of populations of many species has been halted, as levels of bioavailable TBT are reduced. In other areas, however, reductions of TBT levels, and recovery of biota has been slower, perhaps because of the combination

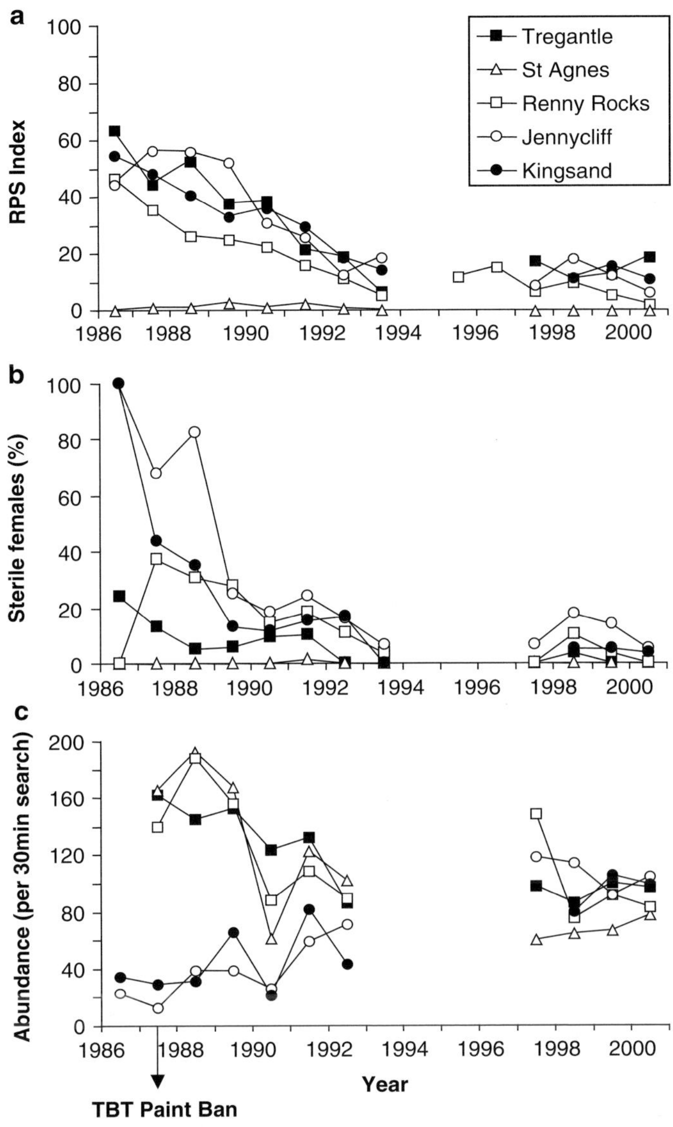

Fig. 5.5 Recovery of dogwhelks (*Nucella lapillus*) from TBT pollution following UK ban in 1987 on boats <25 m (Reproduced from Hawkins et al. 2002, with acknowledgements to Elsevier)

of protracted release of organotin bound up in sediments, and continuing inputs from the commercial fleet (Langston et al. 1990, 1994; Langston and Burt 1991; St.-Jean et al. 1999). This is evident for example, in parts of Southampton Water and its estuaries – a major port and sailing centre in the UK.

Table 5.2 compares half times for TBT in different environmental compartments at different sites in Poole Harbour and Southampton (for data up to 2004).

Table 5.2 Poole Harbour and Southampton Estuaries: Summary of TBT environmental half-times (months) in water, sediment, clams *Scrobicularia plana* (post-1987)

POOLE HARBOUR

Site	Harbour area	Boat traffic (mainly leisure)	Water	Half-times sediment	(months) *Scrobicularia*
Jerry's Point	South	Low	29	122	–
Brands Bay	South	Negligible	52	117	55
Wytch Farm	South	Negligible	39	85	57
Parkstone	North	Moderate	45	62	47
Power Station	North	Moderate	36	101	–
Sterte (Holes Bay)	North	Moderate-high	41	95	47
Lytchett Bay	North	Low-moderate	39	(NS)	68
Marina	North	High	18	–	–

SOUTHAMPTON ESTUARIES

Site	Estuary	Boat traffic	Water	Half-times sediment	(months) *Scrobicularia*
Totton	Test	Upstream of commercial port	NS	NS	NS
Cracknore	Test	Commercial and Naval	NS	NS	NS
Swanwick	Hamble	Leisure-high	23	55	44
Mid Hamble	Hamble	Leisure-high	31	55	44
Warsash	Hamble	Leisure-high	49	51	55
Upper Itchen	Itchen	Leisure low	30	89	91
St.Denys	Itchen	Leisure – moderate	36	NS	106
Northam	Itchen	Leisure – high	51	111	207
Woolston	Itchen	Leisure and commercial	96	NS	204

NS – Slope of regression not significantly different from zero.

Half-times are predictably influenced by the density and type of shipping/boating activity and also by nature of the site/sample type.

Poole Harbour, with its narrow entrance and limited tidal range and currents, was originally selected for study as a possible worst-case situation with regard to contamination from tin-based anti-fouling paints. This large natural lagoon is utilised by several thousand leisure vessels (and a relatively small number of commercial ships) and was therefore considered an ideal location to examine contamination in relation to boating patterns. Because of the predominance of small-vessels (prohibited from using TBT paints in 1987), it was also considered ideal for judging the effectiveness of legislation on reducing pollution; hence monitoring of water, sediment, and a variety of bioindicator species (clams *Scrobicularia plana*, *Mya arenaria*; worms, *Nereis diversicolor*; seaweed *Fucus vesiculosus* and snails *Littorina littorea*) has taken place at intervals over the last two decades (Langston et al. 1987, 1994). Elevated TBT residues were encountered in various parts of the Harbour during the 1980s, often in excess of 100 ng l^{-1} and sometimes in excess of 1 µg l^{-1}, notably near marinas and high-density moorings situated along the northern shoreline. In contrast, the southern shoreline is relatively undeveloped and contamination levels were correspondingly lower (several nanograms per liter) though still often exceeding the EQS (Langston et al. 1987). The success of TBT regulations, in reducing inputs to the water column, can be assessed from the summary of environmental half-lives in Table 5.2 and was most effective at marina sites. Following TBT legislation in 1987, environmental 'half lives' for TBT in Poole waters ranged between 18 and 52 months – presumably extended in comparison with laboratory estimates by continuing residual inputs from boats and sediment desorption (Table 5.2). The rate of decline in TBT concentrations in water may be slowing down at some sites and, occasionally, reversed. It is possible that small amounts of 'new' TBT are still being introduced illegally into the system, or that maintenance dredging is causing transient pulses of TBT in water.

The relatively high affinity of TBT for the particulate phase partly explains why high concentrations tend to be retained close to the major TBT sources (marinas) in the sheltered environment of Poole Harbour. Some entrainment of paint particles is also likely. The distribution of TBT in sediments closely resembles that described for water, with severest contamination being restricted to areas associated with high boating activity. Mean concentrations of TBT in surface sediments in the years immediately following the ban ranged from 0.014 μg g^{-1}(as Sn) at southern sites, to 0.52 μg g^{-1} at Parkstone in the north. Sediments in boatyards and marinas can be more than an order of magnitude higher. The environmental half-lives for TBT concentrations in sediment are longer than the equivalent values for TBT in overlying water, mostly ranging from 5 to 10 years, though at one site, Lytchett Bay, no significant change could be detected (Table 5.2). Despite reductions, TBT contamination throughout the Harbour has remained closely related to the densities of moorings and routes of major boat traffic.

TBT levels in bioindicators such as clams *S. plana* display comparable spatial and temporal trends to those described above. As with sediments, concentrations of TBT in *S. plana* varied by more than an order of magnitude between "clean" sites in the south (0.08 μg g^{-1}) and the more contaminated sites in the north (1.35 μg g^{-1}). Half-times for TBT in *S. plana*, post 1987, were of the order of 5–6 years.

Populations of *S. plana,* were in decline at heavily contaminated sites in Poole during the 1980s and exhibited an inverse correlation between TBT levels and the occurrence and abundance of adult clams (Langston et al. 1990). Concentrations of sediment TBT in the range 0.1–0.3 μg g^{-1} (equivalent to levels in the more contaminated Poole sediments) would almost certainly result in poor survival of juveniles and contribute to a decline in clams numbers (Langston and Burt 1991): after more than a decade this trend has now been reversed as TBT levels fall. More rapid recovery has been observed in epifaunal pacific oysters *Crassostrea gigas* in the outer harbour, where previous attempts at culturing proved abortive, as TBT-induced shell-thickening resulted in abnormal and unmarketable oysters (Dyrynda 1992). In 1989, 2 years after the TBT ban, there was evidence of reversion to more normal growth both in the remnants of this stock, and in freshly laid oysters. However, animals laid in Holes Bay (an area of higher boat traffic, poor flushing and high sediment TBT) continued to show abnormal shell-thickening after legislation, consistent with delayed return of TBT concentrations to no-effects levels. The classic TBT indicator *Nucella lapillus* is not a native of Poole Harbour (unsuitable substrates and other physical and chemical constraints). Close to the entrance to Poole Harbour, in Poole Bay, the species appears to have been eliminated by TBT in the late 1980s.

Comparable monitoring of water, surface sediment and benthic biota has been undertaken over more than two decades in the three major estuaries which feed into Southampton Water, namely the Test (commercial and naval shipping), Itchen (mainly small boats, with the influence of commercial shipping at the mouth) and Hamble (large concentration of yachts) (Langston et al. 1994). In the Test Estuary, there were large differences in TBT concentrations in water depending on proximity to shipping (~1 ng/l upstream to ~100 ng/l, as Sn, near docks and boatyards). Despite spatial variation in concentration, temporal trends for TBT levels at individual sites consistently indicate that the 1987 legislation (banning TBT use on small vessels) has had little impact: at hotspots, concentrations may still exceed EQS by two orders of magnitude. The dominant factor here is the presence of commercial and naval shipping, and boatyards. Trends in sediment and biota from the Test are similar to that in water (half-lives cannot be determined, as regression slopes are not significantly different from zero, Table 5.2). In contrast, legislation was effective initially, in reducing TBT levels in water, sediment and biota in the Hamble Estuary, presumably because small boats predominate at this site. However, initial projections that the EQS would be attained by 1995 were not fulfilled due to a slow down in the reduction in levels (see Fig. 5.4). Delays to further improvements in water quality could be related to sedimentary sinks of the compound, to dredging, and to the presence, nearby (Test/Southampton Water), of commercial vessels (discussed above). For similar reasons this pattern is also observed for the Lower Itchen Estuary (Woolston), where attainment of EQS was initially predicted for 1998. Concentrations here also appear to be close to steady state and still exceed the EQS. In contrast, there have been rather more consistent improvements in water quality at sites in the upper Itchen Estuary (remote from TBT sources and considerably influenced by uncontaminated freshwater).

The summary of half lives for TBT in sediments and clams at each of the Southampton sites (Table 5.2) shows a similar spatial pattern to that in water, though often demonstrating greater persistence. Environmental half-times vary from 4 years upwards. Shortest times were in the small boat dominated Hamble, but in parts of the Itchen and particularly the Test, losses can scarcely be detected; half lives under such conditions are likely to be of the order of decades. TBT concentrations in sediments close to dockyards, marinas, and hull cleaning facilities often lie in the range 0.1–1 μg TBT g^{-1}, whilst chronic contamination of a few nanograms TBT per gram may be detected in deposits at considerable distance upstream from TBT sources (e.g. Upper Itchen and Totton in the Upper Test). As in Poole Harbour, populations of deposit-feeding clams *Scrobicularia plana,* have been in decline at several of the more TBT-contaminated sites around Southampton Water.

Thus, long-term trends in sediment TBT retention are reflected in slow recovery of benthic clams at some sites, notably at Cracknore in the Test Estuary, which is close to a ship repair facility and opposite the Port of Southampton. Here clams were eliminated for much of the 1980s (they were relatively abundant in 1978 prior to popular usage of TBT). Despite a successful settlement of larvae in the 1990s a sustainable population has not developed due to the persistence of high levels in sediment and overlying waters (Fig. 5.6). This contrasts with the site at Totton, upstream of sources, where the clam has remained viable, apparently because of low TBT levels (Fig. 5.6).

Progress towards recovery at sites in Poole Harbour and Southampton Water, as in other estuarine and coastal areas, is clearly variable and dependent on a range of site specific factors. The long lived nature of sediment-bound TBT and the nature and severity of (historical) inputs are amongst the most important of these. The presence of paint particles, hydrology and geochemistry of sediments, salinity and

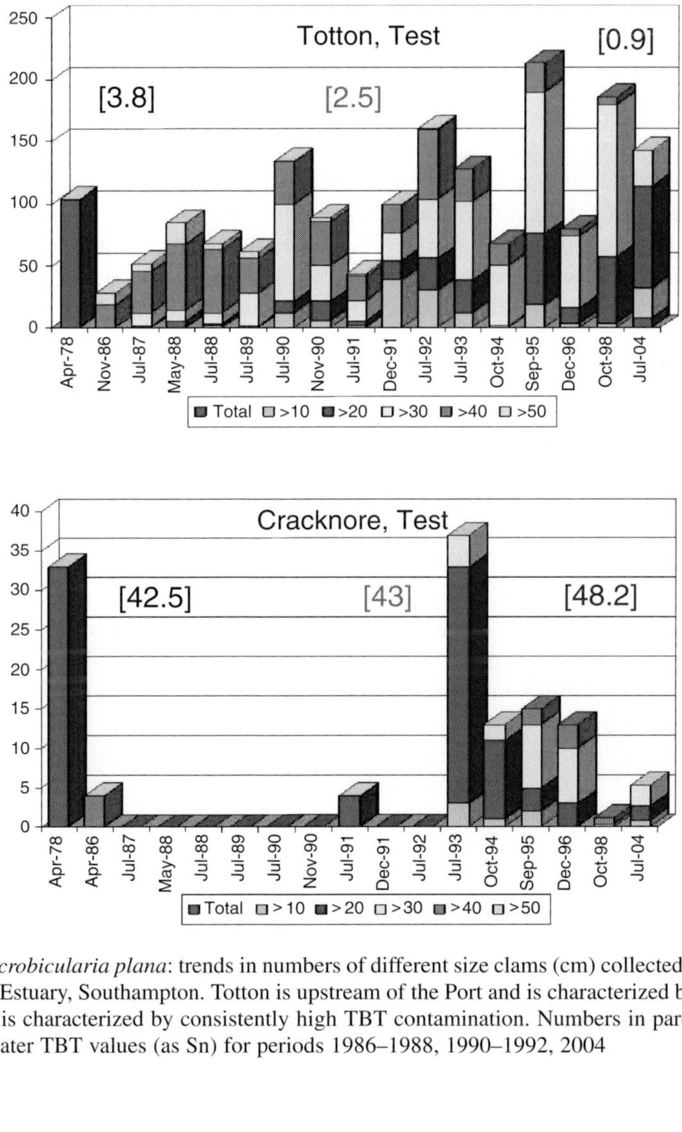

Fig. 5.6 *Scrobicularia plana*: trends in numbers of different size clams (cm) collected at two sites in the Test Estuary, Southampton. Totton is upstream of the Port and is characterized by low TBT. Cracknore is characterized by consistently high TBT contamination. Numbers in parenthesis are mean seawater TBT values (as Sn) for periods 1986–1988, 1990–1992, 2004

temperature regime, and the indigenous micro- and macro-biotic communities will contribute towards site-specific variations in the pattern of recovery. The possibility of illegal usage might also be a confounding feature at some sites in the immediate future, before the IMO recommendations begin to take hold.

5.6 Conclusions

Understanding the partitioning and persistence of organotins in sediments and water is paramount in terms of predicting fate, effects and recovery in coastal and estuarine systems. In the UK there have been widespread improvements in open coastal areas, following the introduction of legislation on small vessels in 1987, however response in estuarine locations in the UK such as Poole and Southampton has been variable, depending on conditions of the site and the relative influence of large ships. Uncontained discharges from boatyards have, until recently, posed a substantial threat to the local aquatic environment. Consequently, TBT levels remain biologically significant in some areas as a result of usage on commercial vessels and sedimentary sinks/sources. Even in some small-boat dominated areas in the UK, persistence in sediments is evident more than a decade after legislation and this secondary source may delay recovery for the foreseeable future. The longevity of effects will be affected by a combination of physicochemical parameters which determine partitioning behaviour, the hydrodynamic energy of the environment and bioturbator activity. Furthermore, longevity of impact will vary between taxonomic groups, with benthic molluscs amongst those slowest to return to normal. In light of the anticipated removal of TBT inputs from much of the global fleet in 2008, surveys of organotin concentrations and accompanying biological records should continue for benchmarking purposes and to broaden our understanding of long-term environmental trends and to assess the rates of recovery. There are still many lessons to be learned, particularly with regard to timescales, susceptibility, reversibility of damage, and longevity in sediments. The TBT ‘model’ represents a unique opportunity to chart the entire lifespan of a major pollutant, from its first introduction in the environment to its (anticipated) harmless conclusion. This may be invaluable for predicting and managing emerging chemical threats associated with alternative biocides.

References

Alzieu C (2000). Impact of tributyltin on marine invertebrates. Ecotoxicology 9:71–76

Brack K (2002). Organotin compounds in sediments from the Gota Alv estuary. Water Air Soil Pollut 135:131–140

Dyrynda EA (1992). Incidence of abnormal shell thickening in the Pacific Oyster Crassostrea gigas in Poole Harbour (UK), subsequent to the 1987 TBT restrictions. Mar Pollut Bull 24(3): 156–163

Galante-Oliveira S, Langston WJ, Burt GR et al. (2006). Imposex and organotin body burden in the dog-whelk (Nucella lapillus L.) along the Portuguese coast. Appl Organomet Chem 20:1–4

Gibbs PE (2009). Long-term tributyltin (TBT)-induced sterilization of neogastropods: persistence of effects in Ocenebra erinacea over 20 years in the vicinity of Falmouth (Cornwall, UK). J Mar Biol Assoc UK, in press

Hallers-Tjabbes CC Ten, Wegener JW, Van Hattum B et al. (2003). Imposex and organotin concentrations in *Buccinum undatum* and *Neptunea antiqua* from the North Sea: Relationship to shipping density and hydrographical conditions, Mar Environ. Res. 55(3)203–233

Harris JRW, Cleary JJ (1987). Particle-water partitioning and organotin dispersal in an estuary. Oceans '87, Institute of Electrical and Electronics Engineers, New York.

Harris JRW, Cleary JJ, Valkirs AO (1996). Particle-water partitioning and the role of sediments as a sink and secondary source of TBT. In: Champ MA, Seligman PF (eds) Organotin. Chapman & Hall, London: 459–473

Hawkins SJ, Gibbs PE, Pope ND et al. (2002). Recovery of polluted ecosystems: the case for long-term studies. Mar Environ Res 54(3–5):215–222

Kram ML, Stang PM, Seligman PF (1989). Fate and distribution of organotin in sediments of four U.S. harbors. Tech. Rep. U.S. Nav. Ocean Syst. Cent. 88

Langston WJ, Burt GR (1991). Bioavailability and the effects of sediment bound TBT in deposit feeding clams Scrobicularia plana. Mar Environ Res 32:61–77

Langston WJ, Pope ND (1995). Determinants of TBT adsorption and desorption in estuarine sediments. Mar Pollut Bull 31(1–3):32–43

Langston WJ, Burt GR, Zhou M (1987). Tin and organotin in water, sediments, and benthic organisms of Poole Harbour. Mar Pollut Bull 18:634–639

Langston WJ, Bryan GW, Burt GR et al. (1990). Assessing the impact of tin and TBT in estuaries and coastal regions. Funct Ecol 4:433–443

Langston WJ, Bryan GW, Burt GR et al. (1994). Effects of sediment metals on estuarine benthic organisms, R&D Note 203, National Rivers Authority

Maguire RJ (2000). Review of the persistence, bioaccumulation and toxicity of tributyltin in aquatic environments in relation to Canada's toxic substances management policy. Water Qual Res J Can 35(4):633–679

Maguire RJ, Tkacz RJ (1985). Degradation of the tri-normal-butyltin species in water and sediment from Toronto Harbor. J Agric Food Chem 33(5):947–953

Marcadon J (1999). Containerisation in the ports of Northern and Western Europe. GeoJournal 48(1):15–20

Meador JP, Krone CA, Dyer DW et al. (1997). Toxicity of sediment-associated tributyltin to infaunal invertebrates: species comparison and the role of organic carbon. Mar Environ Res 43:219–241

Pope ND (1998). The bioavailability of sediment-bound tributyltin (TBT), Ph.D. thesis, University of Plymouth, Plymouth

Randall L, Weber JH (1986). Adsorptive behavior of butyltin compounds under simulated estuarine conditions. Sci Total Environ 57:191–203

Rato M, Sousa M, Quinta R et al. (2006). Assessment of inshore/offshore tributyltin pollution gradients in the northwest Portugal continental shelf using Nassarius reticulatus as a bioindicator. Environ Toxicol Chem 25(12):3213–3220

Ruiz JM, Bachelet G, Caumette P et al. (1996). Three decades of tributyltin in the coastal environment with emphasis on Arcachon Bay France. Environ Pollut 93(2):195–203

Santos MM, Viera N, Reis-Henriques MA et al. (2004). Imposex and butyltin contaminants off the Oporto Coast (NW Portugal): a possible effect of the discharge of dredged material. Environ Int 30:793–798

Seligman PF, Maguire RJ, Lee RF et al. (1996a). Persistence and fate of organotins in aquatic ecosystems. In: Champ MA, Seligman PF (eds) Organotin. Chapman & Hall, London: 429–457

Seligman PF, Adema J, Grouvhoug J et al. (1996b). Environmental loading of tributyltin from drydocks and ship hulls. In: Champ MA, Seligman PF (eds) Organotin. Chapman & Hall, London: 405–427

Side J, Jowitt P (2002). Technologies and their influence on future UK marine resource development and management. Mar Policy 26:231–241

St.-Jean SD, Courtenay SC, Pelletier E et al. (1999). Butyltin concentrations in sediments and blue mussels (*Mytclus edules*) of the Southern Gulf of St Lawrence, Canada. Environ. Technol 20:181–189

Stang PM, Seligman PF (1987). In situ adsorption and desorption of butyltin compounds from Pearl Harbor, Hawaii sediment. Oceans '87, Institute of Electrical and Electronics Engineers, New York

Svavarsson J, Granmo A, Ekelund R et al. (2001). Occurrence and effects of organotins on adult common whelk (Buccinum undatum) (Mollusca, Gastropoda) in harbours and in a simulated dredging situation. Mar Pollut Bull 42(5):370–376

Unger MA, Maclntyre WG, Huggett RJ (1987). Equilibrium Sorption of Tributyltin Chloride by Chesapeake Bay Sediments. Oceans '87 International Organotin Symposium, Halifax, Nova Scotia, Institute of Electrical and Electronics Engineers, New York

Unger MA, Maclntyre WG, Huggett RJ (1988). Sorption behavior of tributyltin on estuarine and fresh-water sediments. Environ Toxicol Chem 7(11):907–915

Valkirs AO, Seligman PF et al. (1986). Butyltin partitioning in marine waters and sediments. Oceans '86 Organotin Symposium, Washington DC, Institute of Electrical and Electronics Engineers

Valkirs AO, Stallard MO, Lee RF (1987). Butyltin Partitioning in Marine Waters. Oceans '87 International Organotin Symposium, Halifax, Nova Scotia, Institute of Electrical and Electronics Engineers, New York

Waldock R, Rees HL, Matthiessen P et al. (1999). Surveys of the benthic infauna of the Crouch Estuary (UK) in relation to TBT contamination. J Mar Biol Assoc UK 79(2):225–232

Chapter 6
Organotin Contamination in Deep Sea Environments

Hiroya Harino, Takaomi Arai, Madoka Ohji, and Nobuyuki Miyazaki

Abbreviations BMF: Bio magnification factor; CHL: Chlordane compounds; DBT: Dibutyltin; DDTs: DDT and its metabolites; DPT: Diphenyltin; HCH: Hexachlorobenzene; MBT: Monobutyltin; MPT: Monophenyltin; OT: Organotin; PAH: Polycyclic aromatic hydrocarbon; PCB: Polychlorinated biphenyl; POPs: Persistent organic pollutants; ROVs: Remotely operated vehicles; SPM: Suspended particulate material; TBT: Tributyltin; TPT: Triphenyltin

6.1 Introduction

Deep sea environments are divided into the bathyal zone (200–2,000 m), the abyssal (2,000–6,000 m) and the hadal zone (over 6,000 m). These zones cover the largest part of the ocean biome (more than 80%). Until now, it was considered that these zones were deserts because sunlight could not reach to such depths and pressures were too high for biota. Advances in deep sea submersibles and image capturing technologies are now increasing the opportunities for marine biologists to observe

H. Harino
Osaka City Institute of Public Health and Environmental Sciences,
8-34 Tojo-cho, Tennoji, Osaka 543-0026, Japan

T. Arai and M. Ohji
International Coastal Research Center, Ocean Research Institute, The University of Tokyo,
2-106-1 Akahama, Otsuchi, Iwate 028-1102, Japan

N. Miyazaki
Center for International Cooperation, Ocean Research Institute, The University of Tokyo,
1-15-1 Minamidai, Nakano, Tokyo 164-8639, Japan

T. Arai et al. (eds.), *Ecotoxicology of Antifouling Biocides*,
DOI: 10.1007/978-4-431-85709-9_6, © Springer 2009

and uncover the mysteries of the deep ocean realm. Remotely operated vehicles (ROVs) have been used underwater since the 1950s. ROVs are basically unmanned submarine robots with umbilical cables used to transmit data between the vehicle and researcher for remote operation in areas where diving is constrained by physical hazards. ROVs are often fitted with video, cameras, mechanical tools for specimen retrieval and measurements. Subsequently, manned deep sea submersibles have been developed and research has progressed. Although the deep sea is in total darkness, is extremely cold, and subjection to great pressure, the marked development of bathyscaphes has revealed the presence of many deep-sea organisms such as bivalves and gastropods in water depths of 3,000 m and more, and has permitted the collection of sediment and marine organisms (e.g. Endo et al. 1999; Okutani et al. 2002; Okutani and Iwasaki 2003).

The contamination of deep-sea ecosystems by man-made chemicals has also been clarified by progress in diving technology. Organochlorine insecticide residues were measured in the livers of *Antimore rostrata*, a deep sea fish collected from 2,500 m in 1972, 1973, and 1974 off the east coast of the United States (Berber and Warlen 1979). Subsequently, metals, polycyclic aromatic hydrocarbons (PAHs), polychlorinated biphenyls (PCBs) and chlorinated pesticides in were determined in tilefish (*Lopholatilus chamaeleonticeps*) collected from Lydonia Canyon (on the Georges Bank) in 1981–1982 (Steimle et al. 1990). Persistent organochlorines such as PCBs, DDT and its metabolites (DDTs), chlordane compounds (CHLs), and hexachlorobenzene (HCH) were detected in deep-sea organisms from a water depth of 180–980 m in Suruga Bay (Lee et al. 1997), whilst PCBs have also been detected (22 mg kg^{-1}) along with DDTs (13 mg kg^{-1}) in amphipods collected from a water depth of 2,075 m in the Arctic Ocean (Hargrave et al. 1992). More recently, detection of persistent organic pollutants has been reported in many kind of samples from various water depths (e.g., Takahashi et al. 1997a). It has thus been concluded that persistent organic pollutants (POPs) and various other contaminants can be transferred to deep-sea areas where they may be accumulated by deep-sea organisms.

It is well known that organotins (OTs) cause advertise effects in marine organisms (Gibbs and Bryan 1986; Gibbs et al. 1991; Ohji et al. 2002) and may persisted in sediments for a long time (Dowson et al. 1993; Harino et al. 1999; see also Chapter 6, this volume). Therefore the current status of OTs and their effects in deep-sea organisms are of much concern. In this chapter, recent findings on the contamination of deep-sea environments by OTs are summarized. Throughout, concentrations of OTs are expressed as Sn^{4+}.

6.2 Details of Deep-Sea Areas

There are few papers on the contamination of OTs in deep sea environments, because of the difficulty and costs in taking samples in the deep sea To our knowledge, only six such surveys (of OTs) have been performed in the Mediterranean Sea

(Michel and Averty 1999; Borghi and Porte 2002), and at five sites around Japan – Yamato Bank (Ikeda et al. 2002), Tosa Bay (Takahashi et al. 2001), Suruga Bay (Takahashi et al. 1997b), off-Tohoku (De Brito et al. 2002) and the Nankai Trough (Harino et al. 2005). The Mediterranean Sea bordered many countries each housing considerable numbers of ports, harbours and marinas. The latter have increased substantially in recent years, especially, in the northwestern Mediterranean Sea. There is also considerable commercial and naval shipping traffic.

The Yamato Bank is located in the central part of the Sea of Japan between China and Japan, and has become an eminent fishing ground. The central part of the Yamato Bank is divided by a ravine of 2,000 m. Suruga Bay is located near the central part of the Pacific coast of Honshu and has a deep and narrow trough along its long axis. The bathyal zone (200–1,000 m) of the eastern continental slope of the bay has a steep but simple physiography. There are many open ocean fisheries bases and some paper mill industries in the inner part of the bay. Tosa Bay is located along the Pacific coast of Kochi Prefecture and has an entrance which is deep and opens into the Pacific Ocean. The central part of the continental slope of the bay is at a depth of 25–800 m. The hydrological circulation within the bay is strongly influenced by the inflow of the oceanic water, the Kuroshio current, resulting in effective water exchange in and out of the bay. Human influences are not intensive along the coast of Tosa Bay, except for the harbor of Kochi Port which houses a significant number of maritime and industrial activities. Off-Tohoku is located in the northwestern Pacific and human activities are relatively low along the adjacent coastal area of northern Japan: here there are major ocean currents, Oyashio (northward cold water) and Kuroshio (southward warm water); which together contribute to, Kuroshio – Oyashio interfrontal zone along the Pacific coast of northern Japan. The continental slope in this region is at a depth of 150–1,300 m. The Nankai Trough, referring to the western edge of the smaller Philippine Sea Plate which is in contact with Japan, is located 100 km off the southern coast of Japan and has a water depth of 3,000–4,000 m. Fishing is the main industry of Cape of Muroto, the closest point to Nankai Trough, and whilst many fishing boats from other countries also frequent these waters.

6.3 The Concentrations of Organotins in Deep Sea Environments

The concentrations of OTs at these deep sea sites are summarized in Table 6.1. Three vertical profiles between 25 and 2,500 m in water columns from the northwestern Mediterranean Sea were surveyed by Michel and Averty (1999). The vertical profile of salinity and tributyltin (TBT) concentrations at the three stations are shown in Fig. 6.1. Highest levels occurred in high salinity surface water. The inflow of sea water at a depth of around 500 m is confirmed by the vertical profile of salinity and the lowest TBT concentrations occurred at this level. Below 500 m, a secondary peak in TBT concentrations was observed at 1,200 m, and thereafter,

Table 6.1 The concentrations of OTs in deep sea environment

Location	Water depth	Sample	MBT	DBT	TBT	MPT	DPT	TPT
Northwestern Mediterranean[a]	500–2,500	Water	–	–	<06–16	–	–	–
Northwestern Mediterranean[b]	1,008–1,816	Fish (liver)	<1.0–72	6.1–82	2–64	<1–190	<1–110	<1.4–240
		Fish (muscle)	<1.0	<1–9.2	<1	<1	<1	<1–4.3
		Fish (gill)	<1.0–1.9	<1–6.8	<1–18	<1	<1	<1.4–71
		Fish (disgestive grand)	<1.0–2.2	<1–3.5	<1–20	<1	<1–18	<1.4–49
Yamato Bank, Japan[c]	350–400	Water	–	–	246–328	–	–	<0.3
	350–401	Sediment	–	–	2.3–6.6	–	–	1.3–2.3
	200–740	Fish	–	–	1.5–34	–	–	14–75
	350–980	Crustaceans	–	–	0.74–32	–	–	44–156
	200–540	Cephalopods	–	–	28–43	–	–	27–48
Suruga Bay, Japan[d]	200–740	Fish	<10–160	<2.0–350	<0.8–280	–	–	–
	350–980	Crustaceans	<10–17	<2.0–47	3.2–150	–	–	–
	200–540	Cephalopods	<10–61	<2.0–82	1.8–98	–	–	–
	135–350	Echinoderms	<2.7–14	2.8–5.6	2.3–53	–	–	–
	135–350	Gastropods	<10	4.2	5.3	–	–	–
North Pacific, off-Tohoku, Japan[e]	250–1,000	Fish	<1.2–28	<0.2–27	0.1–180	–	–	–
	150–1,000	Cephalopods	1.0–7.5	0.5–4.4	0.9–13	–	–	–
	250–2,000	Crustaceans	<1.2–18	<0.2–1.4	<0.1–5.3	–	–	–
	250–400	Gastropods	<1.2–5.4	<0.2–4.6	0.6–1.9	–	–	–
	2,000	Sea star	<1.2	<0.2	0.8	–	–	–

Water; ng l^{-1}, sediment; μg kg^{-1} dry, biological sample; μg kg^{-1} wet
[a]Michel and Averty (1999)
[b]Borghi and Porte (2002)
[c]Ikeda et al. (2002)
[d]Takahashi et al. (1997b)
[e]De Brito et al. (2002)

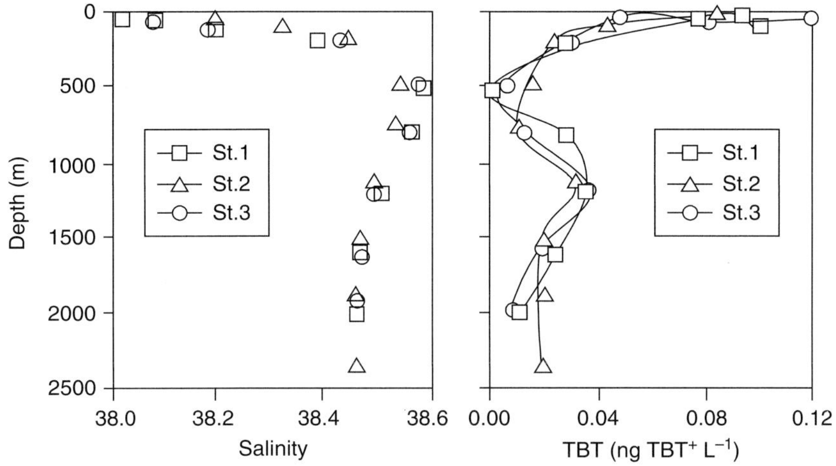

Fig. 6.1 The vertical profile of TBT and salinity in Mediterranean Sea (Borghi and Porte 2002)

contamination declined progressively down to 2,500 m. Concentrations of TBT in deep sea fish tissues (liver, muscle, gill and digestive grand) from the northwestern Mediterranean Sea were in the range <1–64 μg kg^{-1} (Borghi and Porte 2002). Despite the lower concentrations of TBT in these deep sea fish, in comparison with shallow water fish collected in harbors and coastal areas, these results highlight the transport of BT contamination to deep-sea organisms. Furthermore, the TBT residues recorded in deep-sea organisms were within the estimated effective concentrations (8–40 μg kg^{-1}) based on chronic toxicity endpoints in molluscs and were close to concentrations which exert cytotoxic effects in fish hepatoma cells. Interestingly, triphenyltin (TPT) concentrations in *A. rostratus* and *M. moro* from deep waters were much higher than those reported in coastal fish from the area.

The locations of the five areas near Japan in which surveys of OT contamination in the deep sea were conducted are shown in Fig. 6.2. Kono et al. (2004) measured TBT and TPT in water from the Yamato Bank at a depth of 350–400 m. The concentrations of TBT were in the range 0.2–0.3 ng l^{-1}, which were lower than in coastal waters; TPT was not detected in this area (<0.3 ng l^{-1}). TBT in sediment was detected in the range 2.3–6.6 μg kg^{-1} dry weight, again lower than those in coastal area however, TPT concentrations in deep sea sediments ranged between 1.3 and 2.3 μg kg^{-1} dry weight, comparable to values in coastal areas. TBT and TPT concentrations in sediment were 10^{5} times higher than those in seawater.

OT concentrations have been measured in marine organisms from Suruga Bay, off-Tohoku and Tosa Bay. In the bathyal region of Suruga Bay, maximum TBT concentrations in the tissues of deep sea fish, crustaceans, cephalopods, echinoderms, and gastropods were 280, 150, 98, 53 and 5.3 μg kg^{-1}, respectively. Off Tohoku, the maximum concentrations of TBT were 180, 13, 5.3, 1.9 and 0.8 μg kg^{-1} in fish, cephalopods, crustaceans, gastropods and sea star, respectively. Maximum concentrations of TBT in fish, crustaceans, cephalopods and echinoderms from Tosa Bay

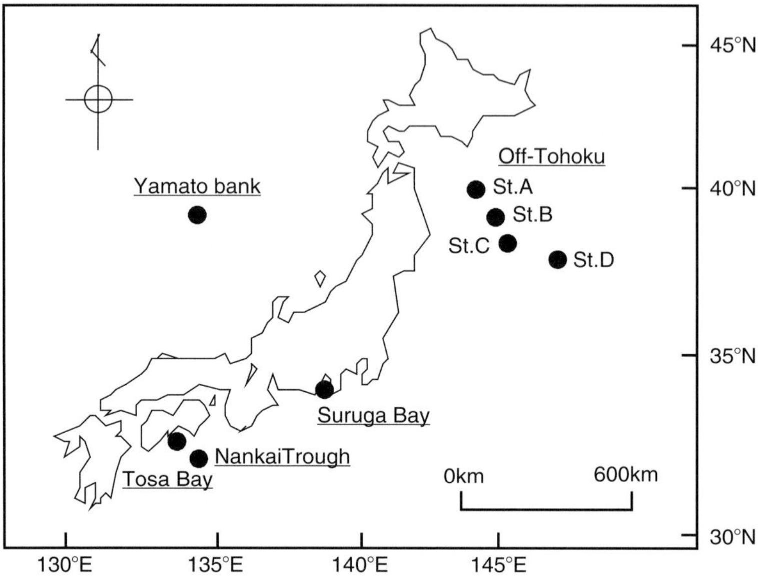

Fig. 6.2 Deep sea sampling sites for organotins, Japan

were 16, 4.5, 2.9 and <0.8 μg kg^{-1}, respectively. In these deeper bays, TBT concentrations in aquatic organisms were apparently lower than those in various coastal waters. However, their presence in tissues indicates the influence of industrial and human activities and that OT contamination is exported to the deep sea environment. Differences in BTs concentrations in myctophid fish, which undergo species-specific diel vertical migration between mesopelagic (200–1,000 m) and epipelagic (<100 m) zones, have been described by Takahashi et al. (2000). Myctophid fishes can be grouped into vertical migratory species, semi - migratory species, and non-migratory species. BTs concentrations in migratory and semi-migratory species were higher than those in non-migratory species, which may be due to higher exposure of BTs in surface water.

In the Naikai Trough, the deepest region sampled, gastropods (*Colliloconcha nankaiensis*), sea cucumbers (*Psychropotes verrucosa*), galatheid crabs (*Munidopsis albatrossae* and *Munidopsis subsquamosa*), bivalves (*Clyptogena tsubasa* and *Clyptogena nautilei*) were collected, along with sediments (from a depth of up to 1 cm and a depth of up to 15 cm below the sediment surface). BT concentrations in the 0–1 cm and 0–15 cm sediment cores were 41 and 21 μg kg^{-1} dry, respectively. The sedimentation rate near the sampling point is 1.09 mm year^{-1} (Iwai et al. 2004). This result therefore indicates a recent increase in BT input. The compositions of BTs in the 0–1 cm surface samples were similar to those in sub-surface samples from a depth of up to 15 cm in the sediment core (Fig. 6.3). The PT concentration in the 0–1 cm surface sediment (0.028 mg kg^{-1} dry) was lower than at 0–15 cm depths (0.052 mg kg^{-1} dry). Moreover, the proportion of DPT and MPT, which are degradation products of TPT, was higher in surface layers (Fig. 6.3). This profile

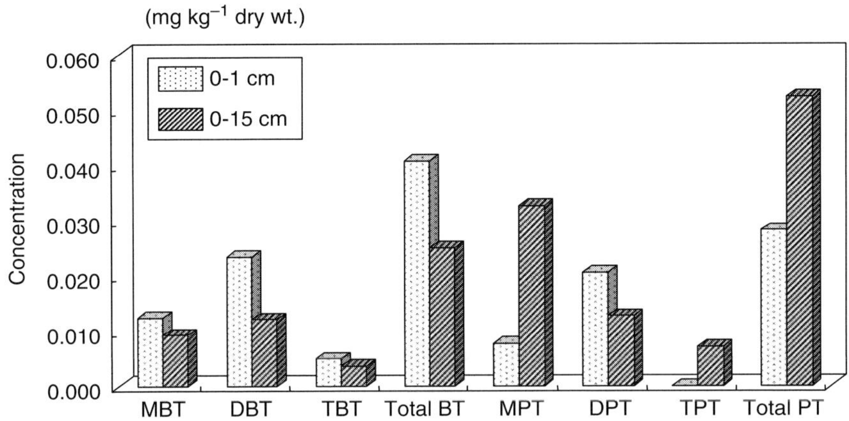

Fig. 6.3 The composition of BTs and PTs in sediment from Nankai Trough

implies a recent reduction of PT concentrations in deep-sea areas, consistent with observations in coastal areas. TBT and TPT in sediment from Nankai Trough were detected in the range 4–5 μg kg^{-1} dry weight and <1–7 μg kg^{-1} dry weight, respectively. It was reported by Ministry of the Environment, Japan (2003) that the geometric means of TBT and TPT concentrations in coastal areas of Japan were 9.4 and 1.2 μg kg^{-1} dry, respectively. The levels of TBT and TPT in sediment from deep-sea areas were thus similar to those in coastal areas, confirming the spread of OT contamination from coastal areas.

OT concentrations measured in the six species of deep-sea organism from Nankai Trough are shown in Table 6.2. Means of BT concentrations in *C. nankaiensis*, *P. verrucosa*, *M. albatrossae*, *M. subsquamosa*, *Cl. tsubasa*, and *Cl. nautilei* were, respectively, 0.089, 0.057, 0.018, 0.016, 0.019 and 0.026 mg kg^{-1} wet weight. The corresponding PT concentrations in those animals were 0.212, 0.363, 0.166, 0.186, 0.030 and 0.025 mg kg^{-1} wet weight, respectively. High concentrations were thus observed in gastropods and sea cucumbers.

6.4 Bio-Concentrations of Organotin Compounds in Deep Sea Organisms

Borghi and Porte (2002) estimated the food chain magnification of OTs in *M. moro* and *L. lepidon*. The biomagnification factor (BCF) is obtained by dividing the concentration in the whole body by the concentration in contents of their digestive tube. BCF for TBT ranged from 0.02 in *M. moro* to 0.04 in *L. lepidion*, whereas TPT ranged from 0.27 in *M. moro* to 0.93 *L. lepidion*. They clearly indicate the biodegradable nature of TBT in comparison with TPT, though on this basis neither appear to be biomagnified in the food chain.

Table 6.2 The concentrations of OTs in deep sea environment

Location	Water depth	Sample	MBT	DBT	TBT	MPT	DPT	TPT
Tosa Bay, Japan[a]	150–400	Fish	<1.4–8.2	<0.8–4.2	0.9–16	–	–	–
	150–400	Crustaceans	<1.4–2.4	<0.8–1.0	<0.8–4.5	–	–	–
	100–400	Cephalopods	<1.4–7.5	<0.8–1.8	2.0–2.9	–	–	–
	100–800	Echinoderms	<1.4–2.2	<0.8–0.8	<0.8	–	–	–
Nankai Trough, Japan[b]	3,553–3,571	Sediment	9–12	12–23	4.0–5.0	8–33	13–21	<1.0–7.0
	3,536–3,587	Bivalves	1–52	2–35	7.0–89	13–48	<1.0	<1.0
	3,540–3,575	Gastropods	10–110	5–56	2.0–34	73–330	<1.0–2.0	<1.0–13
	3,571–3,574	Galatheid	10	6	3.0	140	<1.0	26
	3,571–3,575	Crab	10	6	1.0–3.0	140–180	<1–1	6.0–26
	3,655	Sea cucumber	44	9	3.0	360	5.0	3.0

Water; ng l^{-1}, sediment; μg kg^{-1} dry, biological sample; μg kg^{-1} wet
[a]Takahashi et al. (2001)
[b]Harino et al. (2005)

Kono et al. (2004) reported the concentration of OTs in relation to trophic level in deep sea organisms from Yamato Bank (Fig. 6.4). Trophic level was classed by stomach contents of the predators: Trophic level 1 was composed of organisms ingesting mainly detritus organisms; Trophic level 2 was composed of organisms ingesting polychaeta,

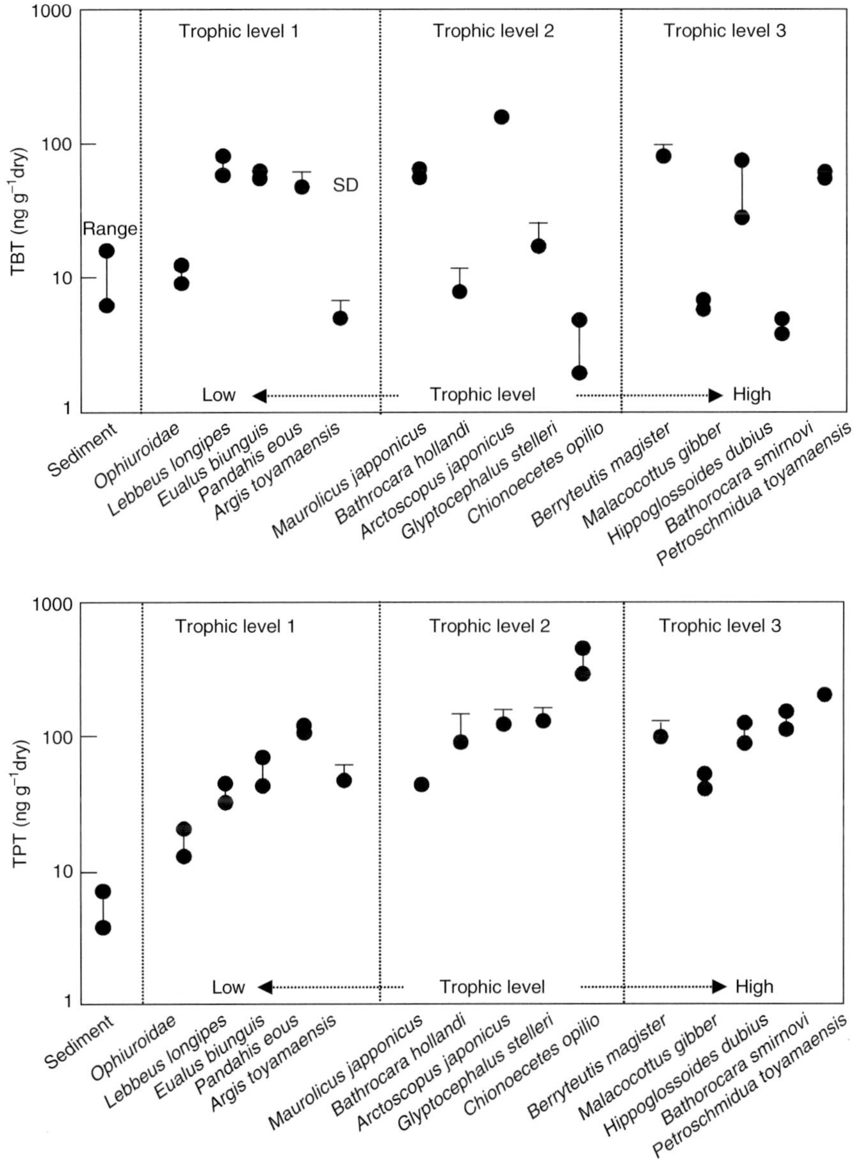

Fig. 6.4 The relationship between TBT or TPT concentrations and trophic level in Yamato Bank (Ikeda et al. 2002)

amphipoda, euphausia, and ophiuroidae; whilst Trophic level 3 was composed of organisms ingesting fish, decapods and squid. According to this classification, TBT concentration was apparently unaffected by trophic level. In contrast, TPT concentration was higher in organisms belonging to higher trophic levels in the deep water, suggesting the bioconcentration of TPT through the food web.

Takahashi et al. (1997b) have previously proposed the concentration of BTs along food chains. Bigger carnivorous fish and scavengers that are at the higher trophic levels of the food web contained relatively elevated levels of BT residues in their tissues. The bio magnification factor (BMF tissues relative to their stomach contents) was estimated roughly for BTs. These values ranged from 0.8 (in dories) to 1.8 (in dogfish sharks). Although the increased exposure by feeding on contaminated diet would be expected, generally, to result in accumulation in higher trophic species, BMFs for BTs were estimated to be less than 10 in most species including deep-sea fish. This may be due to the biodegradable nature of TBT in higher organisms, and/or lower assimilative potency as compared to other persistent contaminants such as PCBs and methylmercury.

TBT was the predominant OTs compounds in almost all deep sea biota (Takahashi et al. 2001). This suggests a continuing input of TBT into the deep-sea environment and a relatively low capacity to degrade TBT in these organisms.

Carbon and nitrogen isotopes are useful to characterize the trophic level of an organism (Broman et al. 1992). Iwasaki et al. 2005 (personal communication) has classified various species of deep-sea organisms by $\delta^{13}C$ value into two groups. Group A includes *C. nankaienssi*, *Cl. tsubasa*, and *Cl. nautilei* and group B *P. verrucosa*, *M. albatrossae*, and *M. subsquamosa*. The organisms in group A use the organic matter chemosynthesized by symbiotic bacteria as nutrient source, while those in group B depend on photosynthesis carried out near the surface by phytoplankton. The relationship between $\delta^{13}C$ and OT concentration is shown in Fig. 6.5. No difference in BT or TBT concentration was observed between the two groups, but PT and TPT concentrations were significantly higher in group B than in group A ($p < 0.001$). Trophic levels in the food chain are estimated using $\delta^{15}N$ values (Broman et al. 1992). The relationship between $\delta^{15}N$ and OT concentration is shown in Fig. 6.6. No change in BT or TBT concentration was observed to accompany the increase of $\delta^{15}N$ values, across trophic levels, but PT and TPT concentration generally increased with increasing of $\delta^{15}N$ values. These findings suggested that TPT concentration rises as trophic level increases, but no such trend is observed for TBT.

The partition coefficient of OT compounds between deep-sea organisms and sediment, were calculated by dividing the TBT concentration in each deep-sea organism by the concentration in sediment. The average partition coefficients of TBT for *C. nankaiensis*, *P. verrucosa*, *M. albatrossae*, *M. subsquamosa*, *Cl. tsubasa* and *Cl. nautilei* were 2.6, 0.72, 0.63, 0.19, 0.43, 0.30 and 0.46, respectively. *C. nankaiensis* is a carrion feeder as well as a deposit feeder. The high partition coefficient for *C. nankaiensis* may therefore be due to bioavailability if TBT in its prey.

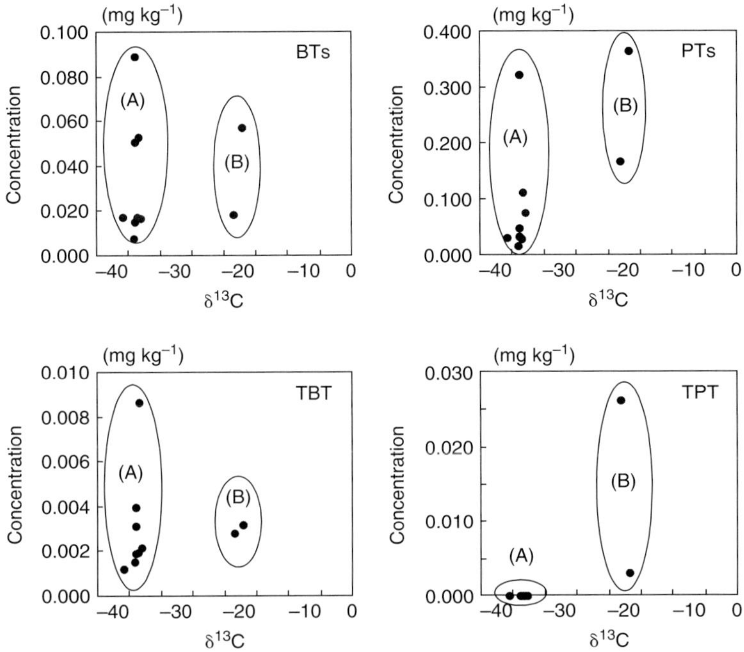

Fig. 6.5 The relationship between BTs or PTs and $\delta^{13}C$

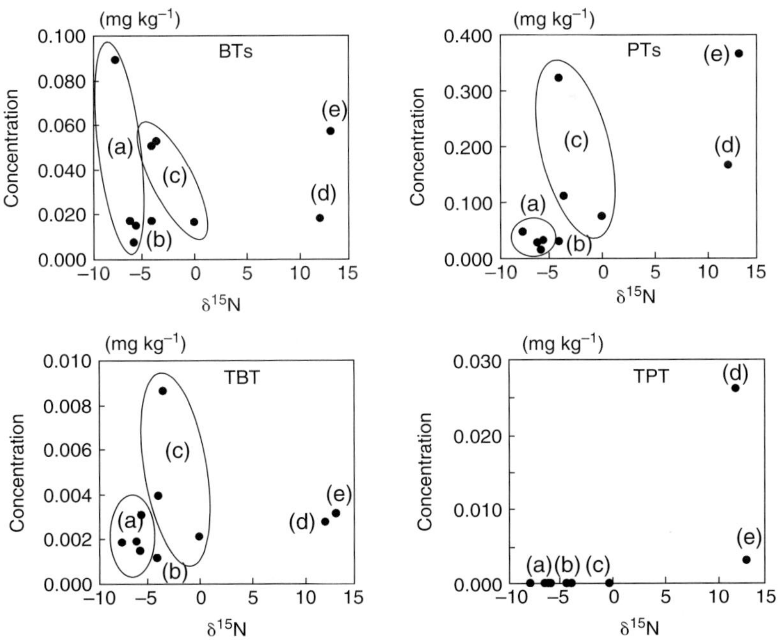

Fig. 6.6 The relationship between BTs or PTs and $\delta^{13}N$

6.5 Transportation of Organotin Compounds to Deep Sea Area

The transportation process whereby harmful chemical compounds reach deep sea environments is of great concern. Many artificial chemical compounds such as organochlorine insecticide residues have been detected in the deep sea environment. However, about the process behind transport of these chemicals are uncertain. In the case of TBT, the degradation rate following release into water from vessel hulls is relatively fast (half-lives; 7–19 days). TBT is thus less persistent in the aquatic environment than many organochlorines. Nevertheless, BT and PT residues have been detected in deep-sea organisms and sediments. Various hypotheses have been proposed concerning the process of OT transportation to bathyal regions. Michel and Averty (1999) estimated the vertical transport of OTs in Mediterranean Sea. The fraction of TBT absorbed by suspended particulate material (SPM) is less than 10^{-3}, because Mediterranean Sea waters are not very turbid: the mean SPM is of the order of 0.28 mg l^{-1} and particulate carbon content is between 0.024 and 0.036 mg l^{-1} for waters below 200 m. Furthermore, the sedimentation rate on the highest abyssal plains in 0.01 cm $year^{-1}$. Considering the scientific data, it is hard to imagine that TBT which adheres to SS will deposited to deep sea environment in significant quantities. Dissolved TBT may, however, have been transported vertically down to the 1,200 m level. The presence of TBT in deep waters in this region is therefore more likely to be due to vertical movements of water masses of surface origin.

Harino et al. (1998) reported that OTs are concentrated or absorbed readily on suspended substances such as plankton and OTs which are absorbed to sediment and suspended substance, may persist for a considerable period (Dowson et al. 1993). Where present in significant densities, such suspended material may be responsible for transport and deposition of absorbed OTs in the bathyal region. OTs also accumulate in biota such as marine mammals and fish and the bioconcentration factor for these biota are reported to be very high (Le et al. 1999; Yamada and Takayanagi 1992; Harino et al. 1998, 2000): The carcasses of these organisms are also deposited in the bathyal region and represent a further transport route.

6.6 Conclusions

As the deep sea is remote from human activity, it has been assumed that contamination in the bathyal region will not have an influence on our lives, even if polluted. Therefore, in the past, parts of the deep sea have been used for disposal of waste. However, this is a false assumption. The deep sea has a close relationship with human life. Recently, deep sea water has been sold as for consumption and deep sea fish are being taken, increasingly, as food. Furthermore, the deep sea may have an influence on the coastal environment by determining ocean currents and transportation of aquatic organisms. With the development of the submersible research vessels, the pollution situation of the deep sea is beginning to be unravelled. It is regrettable that the deep sea environment has been found to be polluted by OT compounds, especially, TPT which was banned in Japan at an early stage. In order to help preserve the deep sea environment, the top research priority is to elucidate the transport process of OTs. The transportation process of OTs into deep sea environment is largely unknown at present, although some hypotheses have been proposed. Furthermore, it is also important to survey the persistence of these compounds in the bathyal region. Clearly there are a number of unresolved issues surrounding, the fate of OT contamination in deep sea environment.

References

Berber R T, Warlen S M (1979) Organochlorine insecticide residues in deep sea fish from 2500 m in the Atlantic Ocean. Environ Sci Technol 13:1146–1148

Borghi V, Porte C (2002) Organotin pollution in deep-Sea fish from the Northwestern Mediterranean. Environ Sci Technol 36: 4224–4228

Broman D, Naf C, Rolff C et al. (1992) Using ratios of stable nitrogen isotopes to estimated bioaccumulation and flux of polychlorinated dibenzo-p-dioxins (PCDDs) and dibenzofurans (PCDFs) in two food chains from the northern Baltic. Environ Toxicol Chem 11:331–34

De Brito A P X, Takahashi S, Ueno D et al. (2002) Organochlorine and butyltin residues in deep-sea organisms collected from the western North Pacific, off Tohoku, Japan. Mar Pollut Bull 45:348–361

Dowson P H, Bubb J M, Williams T P et al. (1993) Degradation of tributyltin in freshwater and estuarine marina sediments. Water Sci Technol 28:133–137

Endo H, Iwasaki N, Machida Y et al. (1999) Preliminary research on the deep-sea demersal fishes and crustaceans off Cape Muroto using deep-tow camera. JAMSTEC J Deep Sea Res 14:411–420

Gibbs P E, Bryan G W (1986) Reproductive failure in populations of the dog-whelk, *Nucella lapillus* caused by imposex induced by tributyltin from antifouling paints. J Mar Biol Assoc UK 66:767–777

Gibbs P E, Spencer B E, Pascoe P L (1991) The American oyster drill, *Urosalpinx cunerea* (gastropoda): evidence of decline in an imposex – affected population (R. Blackwater, Essex). J Mar Biol Assoc UK 71:827–838

Hargrave B T, Harding G C, Vass W P et al. (1992) Organochlorine pesticides and polychlorinated biphenyls in the Arctic Ocean food web. Arch Environ Contam Toxicol 22:41–54

Harino H, Fukushima M, Yamamoto Y et al.(1998) Organotin compounds in water, sediment and biological samples from the Port of Osaka, Japan. Arch Environ Contam Toxicol 35:558–564

Harino H, Fukushima M, Kawai S (1999) Temporal trends of organotin compounds in the aquatic environment of the Port of Osaka, Japan. Environ Pollut 105:1–7

Harino H, Fukushima M, Kawai S (2000) Accumulation of butyltin and phenyltin compounds in various fish species. Arch Environ Contam Toxicol 39:13–19

Harino H, Iwasaki N, Arai T et al. (2005) Accumulation of organotin compounds in the deep – sea environment of Nankai Trough, Japan. Arch Environ Contam Technol 49:1–8.

Ikeda K, Minami T, Yamada H et al. (2002) Bioaccumulation of organotin compounds through the food web developed in the deep water of the Japan Sea. J Environ Chem 12:105–114 (in Japanese)

Iwai M, Fujiwara O, Momma H et al. (2004) Holocene seismoturbidites from the Tosa bay Trough a landward slope basin of Nankai Trough off Muroto: Core KR9750P1. Mem Geol Soc Japan 58:137–152 (in Japanese)

Kono K, Minami T, Yamada H et al. (2004) Bioaccumulation of organotin compounds through the food web in deep water of the Japan Sea. In: Shibata K and Senda T (eds) Proceedings of International Symposium on antifouling paint and marine environment. Tokyo, pp. 45–51

Le L T H, Takahashi S, Saeki K et al. (1999) High percentage of butyltin residues in total tin in the livers of cetaceans from Japanese coastal waters. Environ Sci Technol 33:1781–1786

Lee J S, Tanabe S, Takemoto N et al. (1997) Organochlorine residues in deep-sea organisms from Suruga Bay, Japan. Mar Pollut Bull 34:250–258

Michel P, Averty B (1999) Distribution and fate of tributyltin in surface and deep waters of the northwestern Mediterranean. Environ Sci Technol 33:2524–2528

Ministry of the Environment, Japan (2003) Chemicals in the Environment, 247–265 (in Japanese)

Ohji M, Arai T, Miyazaki N (2002) Effects of tributyltin exposure in the embryonic stage on sex ratio and survival rate in the caprellid amphipod *Caprella danilevskii*. Mar Ecol Prog Ser 235:171–176

Okutani T, Iwasaki N (2003) Noteworthy abyssal molluscs (excluding vesicomyid bivalves) collected from the Nankai Trough off Shikoku by the ROV Kaiko of the Japan Marine Science & Technology Center. VENUS 62:1–10

Okutani T, Kojima S, Iwasaki N (2002) New and known vesicomyid bivalves recently collected from the western and central Nankai Trough off Shikoku and Honshu, by deep sea research systems of Japan Marine Science and Technology Center. VENUS 61:129–140

Steimle F W, Zdanowicz V, Gadbois D (1990) Metals and organic contaminants in Northwest Atlantic deep-sea tilefish tissues. Mar Pollut Bull 21:530–535

Takahashi S, Tanabe S, Takemoto N et al. (1997a) Organochlorine residues in deep-sea organisms from Suruga Bay, Japan. Mar Pollut Bull 34:250–258

Takahashi S, Tanabe S, Kubodera T (1997b) Butyltin residues in deep-sea organisms collected from Suruga Bay, Japan. Environ Sci Technol 31:3103–3109

Takahashi S, Tanabe S, Kawaguchi K (2000) Organochlorine and butyltin residues in mesopelagic Myctophid fishes form the Western North Pacific. Environ Sci Technol 34:5129–136

Takahashi S, Hayashi S, Kasai R et al. (2001) Contamination of deep-sea organisms from Tosa Bay, Japan by organochlorine and butyltin compounds. Nat Sci Mus Monogr 20:363–380

Yamada H, Takayanagi K (1992) Bioconcentration and elimination of bis(tributyltin)oxide (TBTO) and triphenyltin chloride (TPTC) in several marine fish species. Water Res 26:1589–1595

Section 2
Organotin Compounds as Endocrine Disruptors

Chapter 7
Mechanism of Imposex Induced by Organotins in Gastropods

Toshihiro Horiguchi

Abbreviations APGWamide: Alanine-proline-glycine-tryptophan amide; AR: Androgen receptor; ER: Estrogen receptor; GC-MS: Gas chromatography with mass spectrometry; hRXRs: Human retinoid X receptors; ILO: International Labour Organization; IPCS: International Programme on Chemical Safety; 9CRA: 9-*cis* retinoic acid; TBT: Tributyltin; TPT: Triphenyltin; UNEP: United Nations Environment Programme; VDS: Vas deferens sequence; WHO: World Health Organization

7.1 Introduction

Certain environmental chemicals cause feminization of males and/or masculinization of females, and such phenomena are generally called endocrine disruption (Colborn et al. 1996). The current status of studies of endocrine disruption both in wildlife and humans is reviewed by the International Programme on Chemical Safety (IPCS) under the joint work of the United Nations Environment Programme (UNEP), the International Labour Organization (ILO) and the World Health Organization (WHO) (International Programme on Chemical Safety 2002). Here, the author will review the masculinization of female gastropod mollusks, called imposex, in terms of the basic biology and induction mechanism of imposex.

The first report of masculinized female gastropods was made by Blaber (1970), describing a penis-like outgrowth behind the right tentacle in spent females of the dog-whelk, *Nucella lapillus* around Plymouth, UK. The term *imposex*, however, was coined by Smith (1971) to describe the syndrome of a superimposition of male type genital organs, such as the penis and vas deferens, on female gastropods. Imposex is thought to be irreversible (Bryan et al. 1986). Reproductive failure may occur in females with severe imposex, resulting in population decline or even

T. Horiguchi
Research Center for Environmental Risk, National Institute for Environmental Studies, 16-2 Onogawa, Tsukuba, Ibaraki 305-8506, Japan

T. Arai et al. (eds.), *Ecotoxicology of Antifouling Biocides*,
DOI: 10.1007/978-4-431-85709-9_7,

mass extinction (Gibbs and Bryan 1986, 1996). In some species, imposex is typically induced by tributyltin (TBT) and triphenyltin (TPT), chemicals released from antifouling paints used on ships and fishing nets (Bryan et al. 1987, 1988; Gibbs et al. 1987; Horiguchi et al. 1995, 1997a).

As of 2004, approximately 150 gastropod species have been reported to be affected by imposex worldwide (Bech 2002a, b; Fioroni et al. 1991; Horiguchi et al. 1997b; Marshall and Rajkumar 2003; Sole et al. 1998; ten Hallers-Tjabbes et al. 2003; Terlizzi et al. 2004); many of these gastropod species belong to the families Muricidae (e.g., *N. lapillus*, *Ocenebra erinacea*, *Thais clavigera,* and *Urosalpinx cinerea*), Buccinidae (e.g., *Babylonia japonica*, *Buccinum undatum,* and *Neptunea arthritica arthritica*), Conidae (e.g., *Conus marmoreus bandanus* and *Virroconus ebraeus*), and Nassariidae (e.g., *Ilyanassa obsoleta* and *Nassarius reticulatus*) of the Neogastropoda (Fioroni et al. 1991; Horiguchi et al. 1997b).

Regarding Japanese gastropods, at least 39 species (seven mesogastropods and 32 neogastropods) have been found to be affected by imposex among 69 species examined (Horiguchi et al. 1997b; Horiguchi 2000). Although imposex has been observed mostly in shallow-water species in previous surveys, detailed studies of species living at depths of 200 m or more should also be considered, because of the discovery of imposex in Alabaster False Tun (*Galeoocorys leucodoma*) trawled from the depths of 200–250 m off the Atsumi Peninsula, Japan in 1999 (Horiguchi 2000).

The endocrine-disrupting effect of organotins on aquatic organisms is described in Chapter 8, in terms of anatomical, histopathological and ecological aspects (individual- and population-level effects).

7.2 Mode of Action of Organotin Compounds on Development of Imposex in Gastropods

7.2.1 Endocrinology of Gastropod Mollusks

Because of the lack of information on the basic biology of mollusks, understanding of reproductive physiology and/or endocrinology of gastropods is very limited. Knowledge has been mainly obtained from research on Opisthobranchia (e.g., *Aplysia californica*) and Pulmonata (e.g., *Lymnaea stagnalis*); in these gastropods several neuropeptides released from the visceral ganglia, cerebral ganglia, or the prostate gland act as hormones to promote ovulation, egg-laying, or egg-release (Chiu et al. 1979; Ebberink et al. 1985; Joosse and Geraerts 1983). There is very little understanding of the reproductive physiology and/or endocrinology of Prosobranchia (including Archaeo-, Meso- and Neogastropoda). A review by LeBlanc et al. (1999) has suggested that gastropods have both peptide and steroid hormones, however it remains unclear exactly what type of sex hormone gastropods have (see below).

Because sex steroid hormones, such as testosterone and 17β-estradiol, play physiologically important roles in the development of sex organs and the maturation of gonads (i.e., oogenesis and spermatogenesis) in vertebrates, similar sex steroid hormones might also regulate the reproduction of invertebrates, such as gastropods (LeBlanc et al. 1999). After the removal of the hermaphroditic organ, oogenesis and spermatogenesis were observed respectively in the gonads of 17β-estradiol-treated females and testosterone-treated males of the slug *Limax marginatus*; egg-laying was also induced by 17β-estradiol in female slugs, implying the existence of vertebrate-type sex steroid hormones in this species (Takeda 1979, 1983). The *in vitro* metabolism of androstenedione and the identification of endogenous steroids (androsterone, dehydroepiandrosterone, androstenedione, 3α-androstanediol, estrone, 17β-estradiol and estriol) by gas chromatography with mass spectrometry (GC-MS) were reported for *Helix aspersa* (Le Guellec et al. 1987). Several vertebrate-type sex steroids (androsterone, estrone, 17β-estradiol and testosterone) and the synthetic estrogen (ethynylestradiol) were also identified by high resolution GC-MS in the gonads of *T. clavigera* and *B. japonica*. The detection of the synthetic estrogen, ethynylestradiol, in the gonads, presumably represents environmental, rather than endogenous origins – indicating that contamination of the habitat of *B. japonica* had occurred (Lu et al. 2001). It is therefore likely that the presence of other vertebrate-type sex steroids in *T. clavigera* and *B. japonica* may have been due to environmental exposure as opposed to synthesis *in vivo*.

In contrast, the biotransformation of testosterone has been characterized in the mud snail (*I. obsoleta*) (Gooding and LeBlanc 2001). However, as there has been no scientific verification on the presence of AR in gastropods (see below), we should perhaps interpret the biological significance of the transformation of testosterone in the *I. obsoleta* exposed at a relatively high dose (1.0 μM (150,000 DPM) [^{14}C] testosterone), with caution (Gooding and LeBlanc 2001).

Further evidence of steroid-producing cells and synthetic/metabolic enzymes for steroid biosynthesis needs to be obtained to clarify the existence of vertebrate-type sex steroid hormones in gastropods. Aromatase-like activity has been measured and reported in several gastropod species (Morcillo and Porte 1999; Santos et al. 2002), however, the measured aromatase-like activity does not necessarily confirm the existence of vertebrate-type aromatase in gastropods. To the best of our knowledge, there has been no scientific report that has elucidated the successful isolation of aromatase protein from invertebrates.

Although an estrogen receptor (ER)-like cDNA has been isolated from *A. californica* (Gastropoda: Opisthobranchia), and the protein it encodes functions as a constitutively activated transcription factor, estrogen cannot bind this protein (Thornton et al. 2003). Similarly, an ER-like protein has also been isolated from *T. clavigera*; though this too is not bound by estrogen (Kajiwara et al. 2006; Iguchi et al. 2007). This *T. clavigera* protein is also a constitutively activated transcription factor (Iguchi et al. 2007). To the best of our knowledge, no scientific report has described the successful cloning of an androgen receptor (AR) from the tissues of invertebrates, including gastropods. In the absence of direct evidence for ER and AR, their physiological role in mollusks remains in doubt – even if estrogens and androgens are detected in tissues. Based on a study of fully sequenced invertebrate genomes, homologues of ER and AR have yet to be found in invertebrates (Escriva et al. 1997). Thus, it remains unclear whether gastropods have AR and ER. Further studies are necessary to identify steroid receptors and clarify their functions in gastropods.

7.2.2 Mode of Action of Organotin Compounds on the Development of Imposex

Regarding the induction mechanism of imposex, several hypotheses have been proposed and they can be summarized as follows: (1) increased androgen levels, such as testosterone, due to aromatase inhibition by TBT (Bettin et al. 1996; Santos et al. 2005; Spooner et al. 1991); (2) inhibition of the excretion of sulfate conjugates of androgens by TBT (Ronis and Mason 1996); (3) disturbance of the release of penis morphogenetic/retrogressive factor from pedal/cerebropleural ganglia by TBT (Féral and Le Gall 1983); and (4) increase in a neuropeptide, alanine-proline-glycine-tryptophan amide (APGWamide) level caused by TBT (Oberdörster and McClellan-Green 2000, 2002).

Experimental evidence, however, is weak for these four hypotheses. There is a lack of correlation between the time course of the increase in testosterone titres and penis growth in females in the aromatase inhibition hypothesis (Bettin et al. 1996; Spooner et al. 1991). Regarding the hypotheses (1) and (2), Spooner et al. (1991) reported that testosterone levels were significantly elevated in TBT-exposed dog-whelks (*Nucella lapillus*) on days 28 and 42 when compared to the control, although the penis length of female *Nucella lapillus* started to increase on day 14. In another study, a combination of the aromatase inhibitor fadrozole (5 μg/g wet wt) and testosterone (0.1 μg/g wet wt) had little effect on the induction and/or promotion of imposex in *T. clavigera*, as indicated by the incidence of imposex and penis growth (Iguchi et al. 2007). Consequently, there seems uncertain about the mechanism by which organotins induce imposex in gastropods, assuming that vertebrate-type steroid hormones are involved.

It is unknown whether aromatase-like activity is actually inhibited by TBT concentrations in tissues of gastropods collected at natural sites slightly contaminated by TBT. There is also contradictory evidence of the relationship between reduced aromatase-like activity and advance imposex symptoms in the gastropod, *Bolinus brandaris* (Morcillo and Porte 1999). Santos et al. (2005) suggested the involvement of AR, besides aromatase inhibition, in the development of imposex in *N. lapillus*, although gastropods may not inherently have AR (Escriva et al. 1997). If gastropods also have AR similar to vertebrates, it may be profitable to consider the possible activation of androgen receptor-mediated responses caused by TBT or TPT in gastropods, as the enhancements of androgen-dependent transcription and cell proliferation by TBT and TPT have been reported in human prostate cancer cells (Yamabe et al. 2000).

There is a possibility that the results given in support of the 'inhibition of testosterone excretion' hypothesis (Ronis and Mason 1996) may reflect a phenomenon that is at least partly short-term and/or associated with acutely toxic TBT concentrations (Matthiessen and Gibbs 1998).

Several neuropeptides released from the visceral ganglia, cerebral ganglia, or the prostate gland of gastropods (e.g., *A. californica* and *L. stagnalis*) act as ovulation, egg-laying, or egg-releasing hormones (Chiu et al. 1979; Ebberink et al. 1985). Féral and Le Gall (1983) suggested that TBT-induced imposex in *O. erinacea* might be related to the release of neural morphogenetic controlling factors. Their study used *in vitro* tissue cultures derived from a presumed penis-forming area of the immature slipper limpet, *Crepidula fornicata*, and the isolated nervous systems of male or female *O. erinacea* in the presence/absence of TBT (0.2 µg/l) (Féral and Le Gall 1983). The accumulation of TBT or TPT in the central nervous systems of *H. gigantea* (Horiguchi et al. 2002), *N. lapillus* (Bryan et al. 1993), and *T. clavigera* (Horiguchi et al. 2003) indicates the potential for the toxic effects of TBT and TPT on neuroendocrine systems. Oberdörster and McClellan-Green (2000) reported that APGWamide, a neuropeptide released from the cerebral ganglia of gastropods such as *L. stagnalis,* markedly induced the development of imposex in female *I. obsoleta*. The effect of APGWamide in the induction and/or promotion of the development of imposex, however, appears weak based on the experimental results of the incidences of imposex and penis growth (Oberdörster and McClellan-Green 2000, 2002), because the incidences of imposex and penis growth were higher and much longer in gastropods exposed to TBT and/or TPT in the laboratory, respectively (Horiguchi 2006).

Thus, at present, four hypotheses regarding the induction mechanism of imposex in gastropods cannot be fully supported.

There are several characteristics in the development of imposex induced by organotin compounds, such as TBT and TPT in gastropods. At the initial stage of imposex development, the differentiation and growth of male type genital organs (i.e., penis and vas deferens) occur and lead to ovarian spermatogenesis at the severely affected stage, involving oviduct blockage due to the proliferation of epidermal tissues surrounding the vas deferens (Gibbs and Bryan 1986; Gibbs et al. 1988, 1990, 1991; Horiguchi 2000; Horiguchi and Shimizu 1992; Horiguchi et al. 1994, 2000, 2002, 2005, 2006; Oehlmann et al. 1996; Schulte-Oehlmann et al. 1997). Therefore, the author considers that the true mechanism of action of TBT or TPT in the development of imposex in gastropods must encompass an explanation of each of the characteristics mentioned above (Horiguchi 2000).

Nishikawa et al. (2004) proposed a unique mechanism of action of TBT or TPT on the development of imposex in gastropods, which was completely different from other hypotheses already proposed as the imposex induction mechanism. Nishikawa et al. (2004) showed that organotins (both TBT and TPT) bound to the human retinoid X receptors (hRXRs) with high affinity and the injection of 9-*cis* retinoic acid (9CRA), the natural ligand of hRXRs, into female rock shells (*T. clavigera*) induced the development of imposex (Figs. 7.1 and 7.2). The cloning of an RXR homologue from *T. clavigera* revealed that the ligand-binding domain of the rock shell RXR was very similar to that of the vertebrate RXR and bound to both 9CRA and organotins (Nishikawa et al. 2004). Horiguchi et al. (2007b) treated female rock shells (*Thais clavigera*) with three different concentrations (0.1, 1, or 5 µg/g wet wt) of 9CRA or with a single concentration (1 µg/g wet wt) of TBT, TPT (as positive controls), or fetal bovine serum (as a negative control) to confirm the effectiveness of 9CRA in inducing the development of imposex in *T. clavigera*.

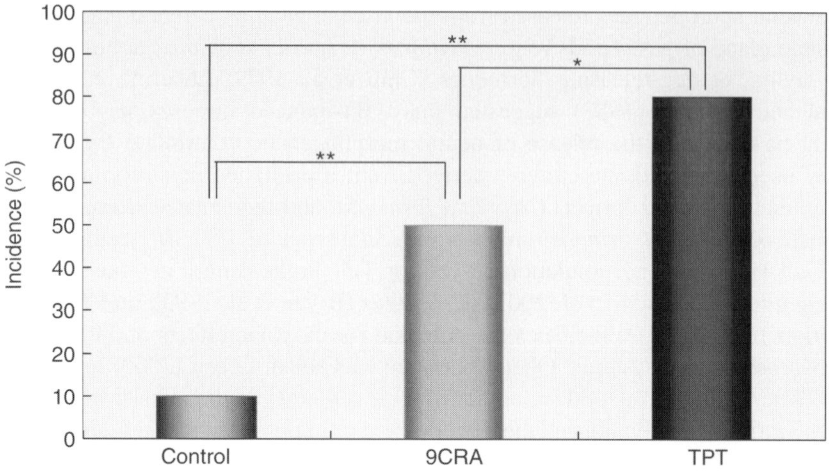

Fig. 7.1 Incidence of imposex in female rock shells (*Thais clavigera*) 1 month after treatment with fetal bovine serum (control), 1 μg/g (wet wt) of 9-*cis*-retinoc acid (9CRA), or 1 μg/g (wet wt) of triphenyltin chloride (TPT)
$^{*}P < 0.05$; $^{**}P < 0.01$

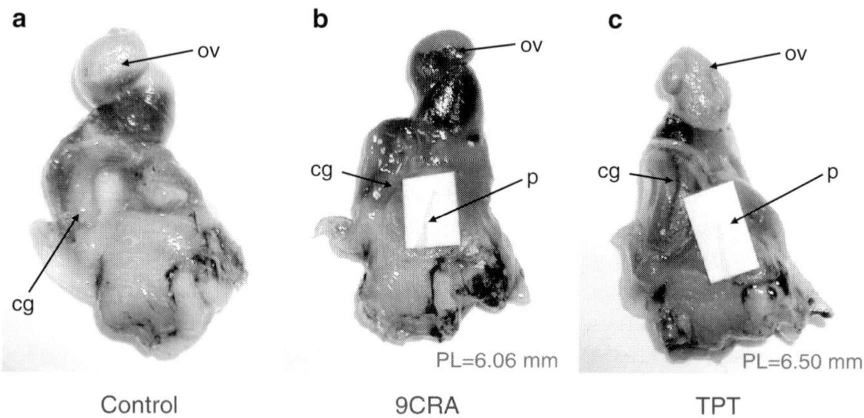

Fig. 7.2 Substantial penis growth observed in the female rock shells after a month of 9CRA injections. cg: capsule gland, ov: ovary, p: penis. (a) Neither penis nor vas deferens observed in the control female (after shell removal). (b) Substantial penis growth as well as vas deferens development observed in the female which received 1 μg/g (wet wt) of 9CRA injection (after shell removal; penis length: 6.06 mm). (c) Substantial penis growth as well as vas deferens development observed in the positive control female which received 1 μg/g (wet wt) of TPT injection (after shell removal; penis length: 6.50 mm). Imposex symptoms based on penis length and vas deferens sequence (VDS) index of the females which received 9CRA injections were clearly promoted, similar to those of females receiving TPT injections

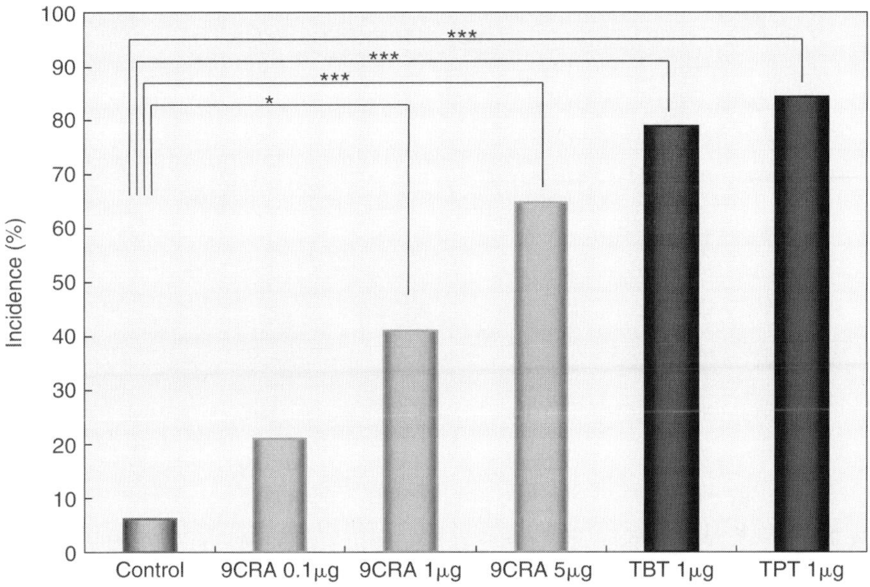

Fig. 7.3 Incidence of imposex in female rock shells (*T. clavigera*) 1 month after treatment with fetal bovine serum (control), three different concentrations of 9-*cis*-retinoc acid (9CRA), tributyltin chloride (TBT), or triphenyltin chloride (TPT)
$^{*}P < 0.05$; $^{***}P < 0.001$

9CRA induced imposex in a dose-dependent manner (Fig. 7.3); imposex incidence was significantly higher in the rock shells that received 1 μg ($P < 0.05$) or 5 μg ($P < 0.001$) 9CRA than in the controls. After 1 month, the rock shells treated with 5-μg 9CRA exhibited substantial growth of the penis-like structure. The length of the structure differed between the 0.1 and 5 μg 9CRA treatment groups ($P < 0.05$) but not between the 1 μg and 5 μg 9CRA treatment groups ($P > 0.05$). Compared with the control, the vas deferens sequence (VDS) index increased significantly in the 1 μg ($P < 0.05$) and 5 μg ($P < 0.001$) 9CRA groups. A light microscopic histological observation revealed that the penis-like structures behind the right tentacle in female rock shells treated with 5 μg 9CRA were essentially the same as the penises and vasa deferentia of normal males and of TBT-treated or TPT-treated imposexed females (Fig. 7.4; Horiguchi et al. 2007b).

Horiguchi et al. (2007a) investigated RXR gene expression and measured the RXR protein content in various tissues of wild male and female rock shells (*T. clavigera*) to further elucidate the role of RXR in the development of organotin-induced imposex in gastropod mollusks. By using the methods of quantitative real-time polymerase chain reaction, Western blotting, and immunohistochemistry with a commercial antibody against human RXR alpha (α), they revealed that RXR gene expression was significantly higher in the penises of males ($P < 0.01$) and in imposexed females ($P < 0.05$) than in the penis-forming areas of normal females (Fig. 7.5). Western blotting demonstrated that the antibody could detect rock shell

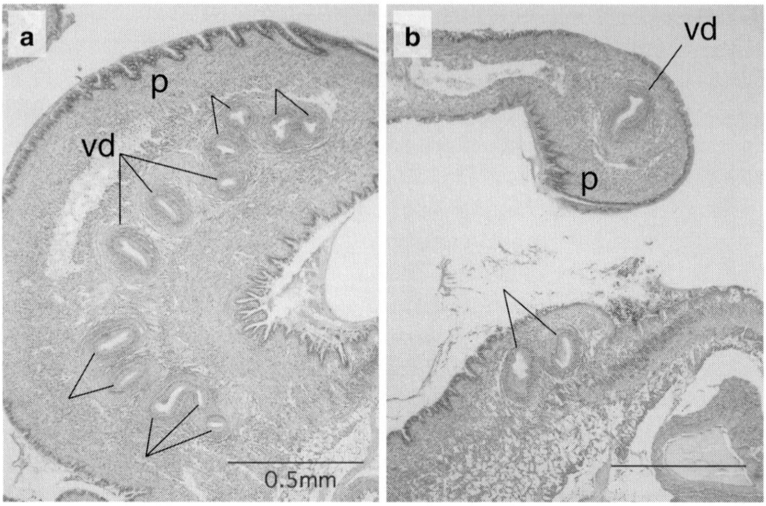

Fig. 7.4 Histology of the penis-like structure (7.00 mm in length) that developed behind the right tentacle of a female rock shell (*T. clavigera*) 1 month after treatment with 5 µg/g (wet wt) of 9CRA. The sections in a and b were stained with hematoxylin and eosin. The scale bars represent 0.5 mm. p, penis; vd, vas deferens

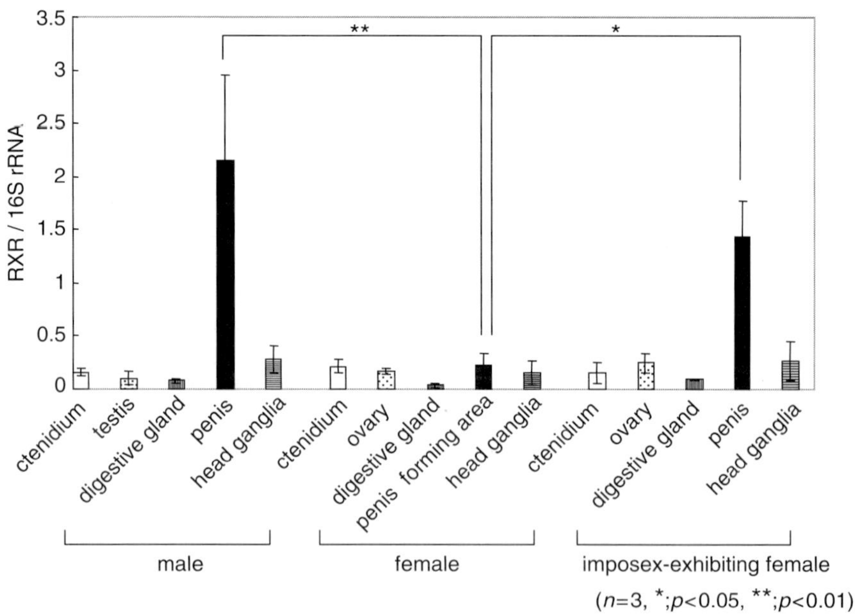

Fig. 7.5 RXR gene expression in various tissues of male, normal female, and imposex-exhibiting female rock shells (*T. clavigera*)

RXR and showed that the male penis had the highest RXR protein content among the analyzed tissues of males and morphologically normal females. Moreover, immunohistochemical staining revealed nuclear localization of RXR protein in the epithelial and smooth muscle cells of the vas deferens and in the interstitial or connective tissues and epidermis of the penis in males and in imposexed females (Fig. 7.6). Based on the results of this study as well as their previous studies, Horiguchi et al. (2007a) suggested that RXR could be involved in organotin-mediated induction of male-type genitalia (penis and vas deferens) in female rock shells.

Castro et al. (2007) also reported that imposex in the dog-whelk (*N. lapillus*) could be mediated by RXR, although they observed the highest expression level of

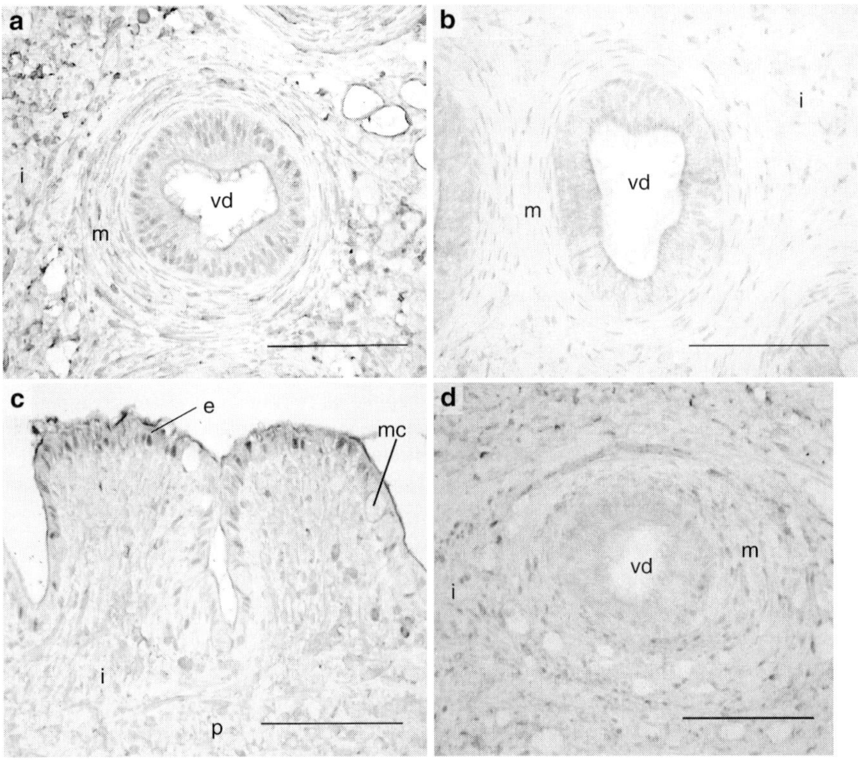

Fig. 7.6 Immunohistochemical expression of RXR in male and imposex-exhibiting female rock shells (*T. clavigera*). Males and severely imposexed females collected at Jogashima (a contaminated site) in December 2003 (a–e). Normal females collected at Hiraiso (a reference site) in December 2003 (h). Slightly affected imposex-exhibiting females collected at Hiraiso in December 2004 (f and g). (a) Penis of a male, showing positive nuclear staining in the epithelial cells of the vas deferens and the surrounding smooth muscle. (b) Penis of a male stained with antibody neutralized with blocking peptide, showing no staining. Non-specific staining was visible in the superficial region of the epithelium of the vas deferens. (c) Epidermal region of penis in a male showing positive staining of epithelial cells. Staining of the superficial layer of the epidermis was non-specific. (d) Penis of a severely affected imposex-exhibiting female, showing similar RXR expression to that in the male (a).

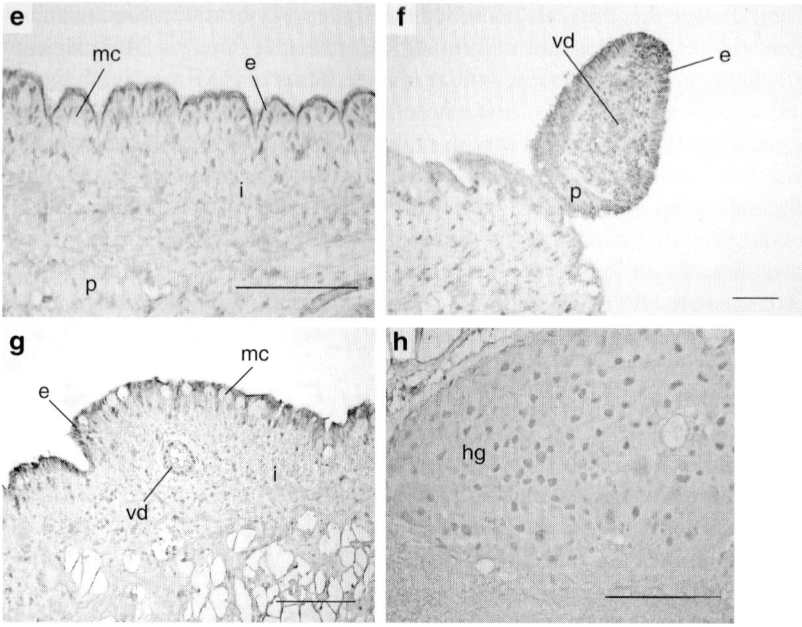

Fig. 7.6 (continued) (e) Epidermal region of penis in a severely imposexed female, showing positive staining in the epidermal cells. (f) Tiny penis of a slightly affected imposex-exhibiting female, showing positive staining in the epithelial cells of the vas deferens, and in the epidermal and interstitial cells. (g) Epidermal region behind the right tentacle of a slightly affected imposex-exhibiting female, showing positive staining in the epithelial cells of the developing vas deferens, and in the epidermal and interstitial cells. (h) Head ganglia of a normal female, showing positive staining in the nerve cells. Sections were counterstained with Hematoxylin. Scale bars indicate 100 µm. e, epithelium (or epithelial cell); hg, head ganglia; i, interstitial tissue (or interstitial cell); m, muscle layer (or smooth muscle cell); mc, mucous cell; p, penis; vd, vas deferens

RXR gene in gonads, which was different from the rock shell (*T. clavigera*). They discussed the induction mechanism of imposex caused by organotins in gastropods, on the basis of a complicated scenario that integrates the interaction between three cascades (retinoic, neuroendocrine and steroid).

Overall, however, these findings suggest that RXR plays an important role in inducing the development of imposex, namely the differentiation and growth of male type genital organs in female gastropods.

Preliminary experimental results on RXR gene expression and induction of imposex after 3-month flow-through exposure to TPT with the rock shell (*T. clavigera*) further support the hypothesis that RXR plays an important role in inducing the development of imposex caused by organotins in female gastropods (Horiguchi et al., manuscript in preparation).

Further studies involving histological, immunohistochemical, biochemical and molecular biological techniques are needed to elucidate the complete mechanism of action of TBT or TPT on the development of imposex in gastropods. This may involve the clarification of a natural ligand and target gene(s) of the rock shell RXR, and when and how the differentiation and proliferation of the stem cells of the penis and vas deferens in a female rock shell are initiated and promoted, which could lead to the epidermal differentiation and proliferation of the penis and vas deferens formation. Morphogenetic factors could be involved in the formation of the curved penis and vas deferens. It is also possible that other factors, such as certain neuropeptides induced in the head ganglia by exposure to organotins, might be associated with the RXR gene-mediated development of imposex – if these factors are induced downstream of the RXR cascade (Morishita et al. 2006).

7.3 Conclusions

It appears that the physiological regulatory system of reproduction may be different in gastropods, compared to that of vertebrates. The retinoid X receptor (RXR) has an important role in the TBT and/or TPT-induced development of imposex, and the subsequent differentiation and growth of male type genital organs, in female gastropods.

References

Bech M (2002a) A survey of imposex in muricids from 1996 to 2000 and identification of optimal indicators of tributyltin contamination along the east coast of Phuket Island, Thailand. Mar Pollut Bull 44:887–896.

Bech M (2002b) Imposex and tributyltin contamination as consequence of the establishment of a marina and increasing yachting activities at Phuket Island, Thailand. Environ Pollut 117:421–429.

Bettin C, Oehlmann J, Stroben E (1996) TBT-induced imposex in marine neogastropods is mediated by an increasing androgen level. Helgol Meeresunters 50:299–317.

Blaber SJM (1970) The occurrence of a penis-like outgrowth behind the right tentacle in spent females of *Nucella lapillus* (L.). Proc Malacol Soc Lond 39:231–233.

Bryan GW, Gibbs PE, Hummerstone LG et al. (1986) The decline of the gastropod *Nucella lapillus* around south-west England: evidence for the effect of tributyltin from antifouling paints. J Mar Biol Assoc UK 66:611–640.

Bryan GW, Gibbs PE, Burt GR et al. (1987) The effects of tributyltin (TBT) accumulation on adult dog-whelks, *Nucella lapillus*: long-term field and laboratory experiments. J Mar Biol Assoc UK 67:525–544.

Bryan GW, Gibbs PE, Burt GR (1988) A comparison of the effectiveness of tri-*n*-butyltin chloride and five other organotin compounds in promoting the development of imposex in the dog-whelk, *Nucella lapillus*. J Mar Biol Assoc UK 68:733–744.

Bryan GW, Bright DA, Hummerstone LG et al. (1993) Uptake, tissue distribution and metabolism of ^{14}C-labelled tributyltin (TBT) in the dog-whelk, *Nucella lapillus*. J Mar Biol Assoc UK 73:889–912.

Castro LFC, Lima D, Machado A et al. (2007) Imposex induction is mediated through the Retinoid X Receptor signalling pathway in the neogastropod *Nucella lapillus*. Aquat Toxicol 85:57–66.

Chiu AY, Hunkapiller MW, Heller E et al. (1979) Purification and primary structure of neuropeptide egg-laying hormone of *Aplysia californica*. Proc Natl Acad Sci USA 76:6656–6660.

Colborn T, Dumanoski D, Myers JP (1996) Our Stolen Future. Dutton, New York: 306p.

Ebberink RHM, Loenhout H, van Geraerts WPM et al. (1985) Purification and amino acid sequence of the ovulation neurohormone of *Lymnaea stagnalis*. Proc Natl Acad Sci USA 82:7767–7771.

Escriva H, Safi R, Hänni C et al. (1997) Ligand binding was acquired during evolution of nuclear receptors. Proc Natl Acad Sci USA 94:6803–6808.

Féral C, Le Gall S (1983) The influence of a pollutant factor (TBT) on the neurosecretory mechanism responsible for the occurrence of a penis in the females of *Ocenebra erinacea*. In: Lever J, Boer HH (eds) Molluscan Neuro-Endocrinology. North-Holland, Amsterdam: pp.173–175.

Fioroni P, Oehlmann J, Stroben E (1991) The pseudohermaphroditism of prosobranchs; morphological aspects. Zool Anz 226:1–26.

Gibbs PE, Bryan GW (1986) Reproductive failure in populations of the dog-whelk, *Nucella lapillus*, caused by imposex induced by tributyltin from antifouling paints. J Mar Biol Assoc UK 66:767–777.

Gibbs PE, Bryan GW (1996) TBT-induced imposex in neogastropod snails: masculinization to mass extinction. In: de Mora SJ (ed) Tributyltin: Case Study of an Environmental Contaminant. Cambridge University Press, Cambridge: pp.212–236.

Gibbs PE, Bryan GW, Pascoe PL et al. (1987) The use of the dog-whelk, *Nucella lapillus*, as an indicator of tributyltin (TBT) contamination. J Mar Biol Assoc UK 67:507–523.

Gibbs PE, Pascoe PL, Burt GR (1988) Sex change in the female dog-whelk, *Nucella lapillus*, induced by tributyltin from antifouling paints. J Mar Biol Assoc UK 68:715–731.

Gibbs PE, Bryan GW, Pascoe PL et al. (1990) Reproductive abnormalities in female *Ocenebra erinacea* (Gastropoda) resulting from tributyltin-induced imposex. J Mar Biol Assoc UK 70:639–656.

Gibbs PE, Spencer BE, Pascoe PL (1991) The American oyster drill, *Urosalpinx cinerea* (Gastropoda): evidence of decline in an imposex affected population (R. Blackwater, Essex). J Mar Biol Assoc UK 71:827–838.

Gooding MP, LeBlanc GA (2001) Biotransformation and disposition of testosterone in the eastern mud snail *Ilyanassa obsoleta*. Gen Comp Endocrinol 122:172–180.

Horiguchi T (2000) Molluscs. In: Kawai S, Koyama J (eds) Problems of Endocrine Disruptors in Fisheries Environment. [in Japanese] Koseisha-Koseikaku, Tokyo: pp.54–72.

Horiguchi T (2006) Masculinization of female gastropod molluscs induced by organotin compounds, focusing on mechanism of actions of tributyltin and triphenyltin for development of imposex. Environ Sci 13:77–87.

Horiguchi T, Shimizu M (1992) Effects on aquatic organisms, mainly on molluscs. In: Satomi Y, Shimizu M (eds) Organotin Pollution and Its Effects on Aquatic Organisms. [in Japanese] Koseisha-Koseikaku, Tokyo: pp.99–135.

Horiguchi T, Shiraishi H, Shimizu M et al. (1994) Imposex and organotin compounds in *Thais clavigera* and *T. bonni* in Japan. J Mar Biol Assoc UK 74:651–669.

Horiguchi T, Shiraishi H, Shimizu M et al. (1995) Imposex in Japanese gastropods (Neogastropoda and Mesogastropoda): effects of tributyltin and triphenyltin from antifouling paints. Mar Pollut Bull 31:402–405.

Horiguchi T, Shiraishi H, Shimizu M et al. (1997a) Effects of triphenyltin chloride and five other organotin compounds on the development of imposex in the rock shell, *Thais clavigera*. Environ Pollut 95:85–91.

Horiguchi T, Shiraishi H, Shimizu M et al. (1997b) Imposex in sea snails, caused by organotin (tributyltin and triphenyltin) pollution in Japan: a survey. Appl Organomet Chem 11:451–455.

Horiguchi T, Takiguchi N, Cho HS et al. (2000) Ovo-testis and disturbed reproductive cycle in the giant abalone, *Haliotis madaka*: possible linkage with organotin contamination in a site of population decline. Mar Environ Res 50:223–229.

Horiguchi T, Kojima M, Kaya M et al. (2002) Tributyltin and triphenyltin induce spermatogenesis in ovary of female abalone, *Haliotis gigantea*. Mar Environ Res 54:679–684.

Horiguchi T, Shiraishi H, Morita M (2003) Specific tissue distributions of organotin compounds in prosobranch gastropods. [Abstract] SETAC 24th Annual Meeting Abstract Book p.290.
Horiguchi T, Kojima M, Takiguchi N et al. (2005) Continuing observation of disturbed reproductive cycle and ovarian spermatogenesis in the giant abalone, *Haliotis madaka,* from an organotin-contaminated site of Japan. Mar Pollut Bull 51:817–822.
Horiguchi T, Kojima M, Hamada F et al. (2006) Impact of tributyltin and triphenyltin on ivory shell (*Babylonia japonica*) populations. Environ Health Persp 114:Supplement 1:13–19.
Horiguchi T, Nishikawa T, Ohta Y et al. (2007a) Retinoid X receptor gene expression and protein content in tissues of the rock shell *Thais clavigera*. Aquat Toxicol 84:379–388.
Horiguchi T, Ohta Y, Nishikawa T et al. (2007b) Exposure to 9-*cis* retinoic acid induces penis and vas deferens development in the female rock shell, *Thais clavigera*. Cell Biol Toxicol doi:10.1007/s10565-007-9051-9.
Iguchi T, Katsu Y, Horiguchi T et al. (2007) Endocrine disrupting organotin compounds are potent inducers of imposex in gastropods and adipogenesis in vertebrates. Mol Cell Toxicol 3:1–10.
International Programme on Chemical Safety (2002) Global Assessment of the State-of-the-Science of Endocrine Disruptors (Damstra T, Barlow S, Bergman A, Kavlock R, Kraak GVD (eds)). WHO/PCS/EDC/02.2, 180 p.
Joosse J, Geraerts WPM (1983) Endocrinology. In: Saleuddin ASM, Wilbur KM (eds) The Mollusca Vol. 4 Physiology Part 1. Academic, New York: pp.317–406.
Kajiwara M, Kuraku S, Kurokawa T et al. (2006) Tissue preferential expression of estrogen receptor gene in the marine snail, *Thais clavigera*. Gen Comp Endocrinol 148:315–26. Epub 2006 Jun 19.
LeBlanc GA, Campbell PM, den Besten P et al. (1999) The endocrinology of invertebrates. In: deFur PL, Crane M, Ingersoll C, Tattersfield L (eds) Endocrine Disruption in Invertebrates: Endocrinology, Testing, and Assessment. SETAC Press, Pensacola, FL: pp.23–106.
Le Guellec D, Thiard MC, Remy-Martin JP et al. (1987) *In vitro* metabolism of androstenedione and identification of endogenous steroids in *Helix aspersa*. Gen Comp Endocrinol 66:425–433.
Lu M, Horiguchi T, Shiraishi H et al. (2001) Identification and quantitation of steroid hormones in marine gastropods by GC/MS. [in Japanese] Bunseki Kagaku 50:247–255.
Marshall DJ, Rajkumar A (2003) Imposex in the indigenous *Nassarius kraussianus* (Mollusca: Neogastropoda) from South African harbours. Mar Pollut Bull 46:1150–1155.
Matthiessen P, Gibbs PE (1998) Critical appraisal of the evidence for tributyltin-mediated endocrine disruption in mollusks. Environ Toxicol Chem 17:37–43.
Morcillo Y, Porte C (1999) Evidence of endocrine disruption in the imposex-affected gastropod *Bolinus brandaris*. Environ Res A 81:349–354.
Morishita F, Minakata H, Takeshige K et al. (2006) Novel excitatory neuropeptides isolated from a prosobranch gastropod, *Thais clavigera*: the molluscan counterpart of the annelidan GGNG peptides. Peptides 27:483–492.
Nishikawa J, Mamiya S, Kanayama T et al. (2004) Involvement of the retinoid X receptor in the development of imposex caused by organotins in gastropods. Environ Sci Technol 38:6271–6276.
Oberdörster E, McClellan-Green P (2000) The neuropeptide APGWamide induces imposex in the mud snail *Ilyanassa obsoleta*. Peptides 21:1323–1330.
Oberdörster E, McClellan-Green P (2002) Mechanism of imposex induction in mud snail, *Ilyanassa obsoleta*: TBT as a neurotoxin and aromatase inhibitor. Mar Environ Res 54:715–718.
Oehlmann J, Fioroni P, Stroben E et al. (1996) Tributyltin (TBT) effects on *Ocinebrina aciculata* (Gastropoda: Muricidae): imposex development, sterilization, sex change and population decline. Sci Total Environ 188:205–223.
Ronis MJJ, Mason AZ (1996) The metabolism of testosterone by the periwinkle (*Littorina littorea*) *in vitro* and *in vivo*: effects of tributyltin. Mar Environ Res 42:161–166.
Santos MM, ten Hallers-Tjabbes CC, Vieira N et al. (2002) Cytochrome P450 differences in normal and imposex-affected female whelk *Buccinum undatum* from the open North Sea. Mar Environ Res 54:661–665.

Santos MM, Castro LF, Vieira MN et al. (2005) New insights into the mechanism of imposex induction in the dogwhelk *Nucella lapillus*. Comp Biochem Physiol C Toxicol Pharmacol 141:101–109.

Schulte-Oehlmann U, Oehlmann J, Fioroni P et al. (1997) Imposex and reproductive failure in *Hydrobia ulvae* (Gastropoda: Prosobranchia). Mar Biol 128:257–266.

Smith BS (1971) Sexuality in the American mud snail, *Nassarius obsoletus* Say. Proc Malacol Soc Lond 39:377–378.

Sole M, Morcillo Y, Porte C (1998) Imposex in the commercial snail *Bolinus brandaris* in the northwestern Mediterranean. Environ Pollut 99:241–246.

Spooner N, Gibbs PE, Bryan GW et al. (1991) The effect of tributyltin upon steroid titres in the female dogwhelk, *Nucella lapillus*, and the development of imposex. Mar Environ Res 32:37–49.

Takeda N (1979) Induction of egg-laying by steroid hormones in slugs. Comp Biochem Physiol 62A:273–278.

Takeda N (1983) Endocrine regulation of reproduction in the snail, *Euhadra peliomphala*. In: Lever J, Boer HH (eds) Molluscan Neuro-endocrinology. North-Holland, Amsterdam: pp.106–111.

ten Hallers-Tjabbes CC, Wegener JW, Van Hattum BA et al. (2003) Imposex and organotin concentrations in *Buccinum undatum* and *Neptunea antiqua* from the North Sea: relationship to shipping density and hydrographical conditions. Mar Environ Res 55:203–233.

Terlizzi A, Delos AL, Garaventa F et al. (2004) Limited effectiveness of marine protected areas: imposex in *Hexaplex trunculus* (Gastropoda: Muricidae) populations from Italian marine reserves. Mar Pollut Bull 48:188–192.

Thornton JW, Need E, Crews D (2003) Resurrecting the ancestral steroid receptor: ancient origin of estrogen signaling. Science 301:1714–1717.

Yamabe Y, Hoshino A, Imura N et al. (2000) Enhancement of androgen-dependent transcription and cell proliferation by tributyltin and triphenyltin in human prostate cancer cells. Toxicol Appl Pharmacol 169:177–184.

Chapter 8
The Endocrine-Disrupting Effect of Organotin Compounds for Aquatic Organisms

Toshihiro Horiguchi

Abbreviations CRM: Certified Reference Material; DBT: Dibutyltin; DPT: Diphenyltin; GC-FPD: Gas chromatography with flame photometric detection; AFS Convention: International Convention on the Control of Harmful Anti-fouling System on Ships; IMO: International Maritime Organization; MBT: Monobutyltin; MPT: Monophenyltin; RPL: Relative penis length; ΣBTs: Sum of butyltins; ΣPTs: Sum of phenyltins; TBT: Tributyltin; TPT: Triphenyltin; VDS: Vas deferens sequence; wet wt: Wet weight

8.1 Introduction

The first report of masculinized female gastropod mollusks was made by Blaber (1970), describing a penis-like outgrowth behind the right tentacle in spent females of the dog-whelk, *Nucella lapillus* around Plymouth, UK. The term, "imposex", however, was defined by Smith (1971), meaning imposed sexual organs, to describe the syndrome of a superimposition of male genital tracts, such as penis and vas deferens, on female gastropods. Imposex is thought to be an irreversible syndrome (Bryan et al. 1986). Reproductive failure may be brought about in severely affected stages of imposex, resulting in population decline and/or mass extinction (Gibbs and Bryan 1986, 1996). Imposex is known to be induced in many species by tributyltin (TBT), and also by triphenyltin (TPT) released from antifouling paints on ships and fishing nets (Bryan et al. 1987, 1988; Gibbs et al. 1987; Horiguchi et al. 1995, 1997a).

Up until July 2004, approximately 150 gastropod species worldwide have been reported to be affected by imposex (Bech 2002a, b; Fioroni et al. 1991; Horiguchi et al. 1997b; Marshall and Rajkumar 2003; Sole et al. 1998; ten Hallers-Tjabbes et al. 2003;

T. Horiguchi
Research Center for Environmental Risk, National Institute for Environmental Studies, 16-2 Onogawa, Tsukuba, Ibaraki 305-8506, Japan

T. Arai et al. (eds.), *Ecotoxicology of Antifouling Biocides*,
DOI: 10.1007/978-4-431-85709-9_8, © Springer 2009

Terlizzi et al. 2004). Regarding Japanese gastropods, 39 species, including seven mesogastropods and 32 neogastropods, have been found to be affected by imposex (Horiguchi 2000). Several of these belong to the genus Muricidae (e.g., *Lepsiella scobina, Nucella lapillus, N. lima, Ocenebra erinacea, Ocinebrina aciculata, Rapana venosa venosa, Thais clavigera, Urosalpinx cinerea* etc.); Buccinidae (e.g., *Babylonia japonica, Buccinum undatum, Neptunea arthritica arthritica* etc.); Conidae (e.g., *Conus marmoreus bandanus, Virroconus ebraeus* etc.) and Nassariidae (e.g., *Ilyanassa obsoleta, Hinia reticulata* etc.). Numerous studies have examined the incidence or severity of imposex; investigated the use of certain gastropod species as biological indicators of TBT contamination, and surveyed TBT contamination also using gastropods. Only a few reports, however, have presented evidence for population-level effects of reproductive failure due to imposex, based on either morphological or histological methods (Bryan et al. 1986; Gibbs and Bryan 1986; Gibbs et al. 1988, 1990, 1991; Horiguchi 2000; Horiguchi et al. 1994, 2000, 2006; Oehlmann et al. 1996; Schulte-Oehlmann et al. 1997).

Here, the author will summarize current status of imposex and contamination by organotins in the rock shell *Thais clavigera* in Japan, and then focus on two case studies of imposex in the ivory shell *Babylonia japonica* and masculinization of female abalone *Haliotis madaka* and *H. gigantea*, which is similar to imposex in other gastropods (Meso- and Neogastropoda), with special reference to possible linkages with declining populations.

8.2 Current Status of Imposex and Contamination by Organotins in the Rock Shell *Thais clavigera* from Japan

Among rock shell (*T. clavigera*) samples collected between January 1999 and November 2001 from 174 locations along the Japan coast, imposex was observed at 166 locations, whereas no, or rare, cases were found at the remaining eight locations. The percentage occurrence of imposex was as high as or close to 100% in approximately half of the affected locations surveyed. It is expected that spawning obstruction occurs in more than half the population of females when relative penis length (RPL) index exceeds 40, on the basis of the relationship between RPL index, vas deferens sequence (VDS) index and the percentage occurrence of oviduct (vulva) blockage in females. Among the 174 locations, RPL index values exceeding 40 were found in 41 locations. High values of RPL and VDS indices were generally observed in the western part of Japan. Compared with the results of a previous survey (conducted between 1996 and 1999), the indices seemed to have decreased, but remained almost unchanged in some locations (Horiguchi 2004).

TPT concentrations in tissues of the rock shell showed a decrease over time but varied distinctly between locations; relatively high pollution levels in a few locations were detected. Decreases in TBT concentrations were also distinct in general but the degree of decrease was lower than those in TPT concentrations.

Changes in concentrations over time were not observed in several locations. An increase in the concentrations of TBT was observed in two locations near fishing ports (Horiguchi 2004).

8.3 Collapse of Commercial Fisheries for the Ivory Shell *Babylonia japonica* in Japan: Did Reproductive Failure Caused by Imposex Bring About Drastic Population Decline?

The ivory shell *Babylonia japonica* (Neogastropoda: Buccinidae), which inhabits sandy or muddy sediments in shallow water (approximately 10–20 m in depth) from the south of Hokkaido to Kyushu, Japan, is a scavenger in inshore ecosystem, and traditionally a target species of commercial fisheries in Japan. Imposex seems to have been observed in the ivory shell since the 1970s (Kajikawa and Hamada, personal communication, Tottori, Japan, July 1991), and the total catch drastically decreased all over Japan in the late 1970s or early 1980s (Horiguchi and Shimizu 1992).

Much effort has been made to enhance the ivory shell stocks: Seed production using adult ivory shells reared in hatcheries, with subsequent release of seeds/juveniles into the sea. Most of seeds/juveniles of the ivory shells released into the sea (approximately 90% of total production in Japan) have been produced at a hatchery in Tomari, Tottori Prefecture, located in the western part of Japan (Horiguchi et al. 2006). In Tottori Prefecture, however, not only the total catch but also the number of egg capsules spawned by adult shells at the hatchery and seeds/juveniles artificially produced/released into the sea has decreased since the mid-1980s (Horiguchi et al. 2006; Fig. 8.1). The total catch has drastically decreased since 1984, 2 years after the first observation of imposex-affected female ivory shells from Tottori Prefecture, involving the increase of both percentage occurrence of imposex individuals and mean penis length in females (Hamada et al. 1988, 1989; Kajikawa 1984; Kajikawa et al. 1983; Fig. 8.1). The number of egg capsules spawned by adult ivory shells at the hatchery, as well as the number of seeds/juveniles artificially released into the sea, has also decreased since the mid-1980s (Fig. 8.1). Introduction of adult ivory shells from another prefecture (Niigata Prefecture, Japan) to compensate for insufficient numbers of the normal brood stock also resulted in failure of the release of seeds/juveniles into the sea due to their high mortality at the hatchery prior to release (Fig. 8.1). Recovery of total catch of the ivory shell has not been observed in spite of such efforts to enhance the ivory shell stocks (Fig. 8.1). Finally, operation of the ivory shell hatchery for stock enhancement in Tottori had to be stopped, and the hatchery was closed in 1996 (Fig. 8.1). Therefore, possible reproductive failure caused by imposex in the ivory shell was suspected.

Horiguchi et al. (2006) examined the incidence of reproductive failure accompanied by imposex in the ivory shell, based on the histopathological observation of gonads, and investigated the relationship between organotin compounds and imposex in the ivory shell, based on chemical analysis of organotin concentrations

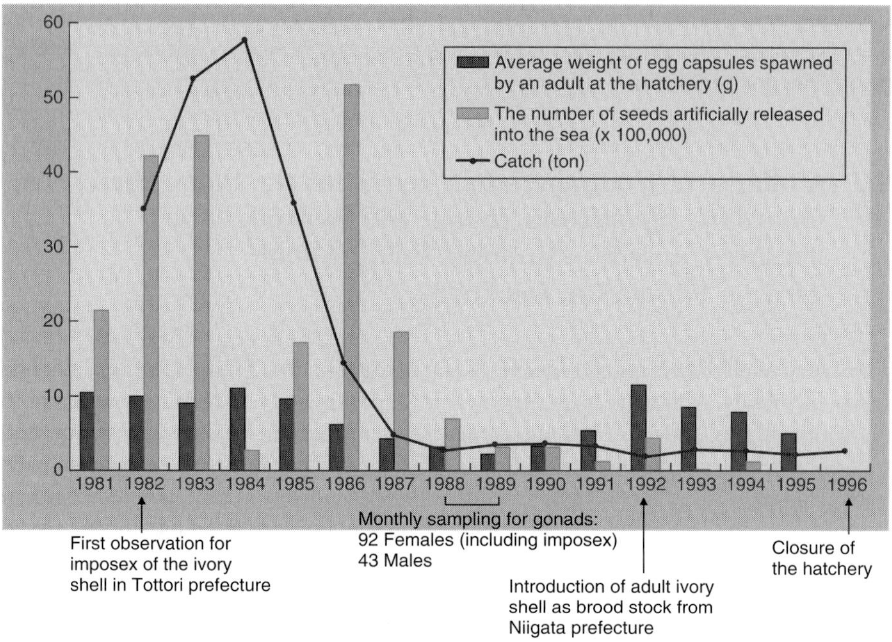

Fig. 8.1 Temporal trends for average weight of egg capsules spawned by adults at the hatchery, the number of seeds/juveniles released into the sea and the total catch of the ivory shell, *Babylonia japonica* in Tottori Prefecture, Japan (Horiguchi et al. 2006)

in tissues of the ivory shell. Horiguchi et al. (2006) also discussed the possibility that the marked decline in the ivory shell (*Babylonia japonica*) populations from Japan could have been brought about mainly by reproductive failure accompanied by imposex, induced by TBT and TPT from antifouling paints.

Horiguchi et al. (2006) performed histopathological examination of gonads in the ivory shell: Adult *B. japonica* reared in the hatchery of the Tottori Prefectural Sea Farming Association were sampled monthly from December 1988 to November 1989, when the number of egg capsules spawned by adult *B. japonica* at the hatchery had reached a minimum. Gonad samples of 10–15 *B. japonica* specimens were fixed in Bouin's fluid, embedded in paraffin, and stained with hematoxylin-eosin for histopathological examination under a light microscope. In total, 135 *B. japonica* specimens were examined (43 males and 92 females, consisting of 16 normal females and 76 imposex-exhibiting individuals). To quantitatively evaluate the gonadal maturation of *B. japonica,* female and male reproductive cells were scored based on developmental stages, similar to those described in Takamaru and Fuji (1981) and Horiguchi et al. (2000). The individual reproductive developmental score was the mean value of these scores for the reproductive cells of each *B. japonica*. The population reproductive developmental score was the monthly mean value of the individual reproductive developmental scores (Horiguchi et al. 2000).

Horiguchi et al. (2006) also carried out chemical analysis of organotin compounds in tissues of the ivory shell: Adult *B. japonica* specimens collected at Yodoe, Tottori

Prefecture, in June 1991 were used for chemical analysis of organotin (butyltin and phenyltin) compounds. The specimens were dissected for original sex and imposex determination. A total of 52 *B. japonica* specimens were used for chemical analyses (25 males and 27 females, consisting of three normal females and 24 imposex-exhibiting individuals). Chemical analyses of butyltin and phenyltin compounds in tissues (muscle [foot], head with tentacle, radula with sac, oesophagus with crop, stomach, digestive gland, kidney, rectum, ovary or testis, oviduct, siphon, ctenidium, heart, osphradium, and mantle) of *B. japonica* specimens were conducted with composite samples and quantified by gas chromatography with flame photometric detection (GC-FPD), using the method described in Horiguchi et al. (1994).

The percentages of occurrence of imposex were 82.6% and 88.9% in *B. japonica* specimens collected from December 1988 to November 1989 and in June 1991, respectively. Both penis and vas deferens were found to be well developed in imposex-exhibiting females (Horiguchi et al. 2006). No oviduct blockage (i.e., occlusion of the vulva) by vas deferens formation, however, was observed in imposex-exhibiting female *B. japonica* (Horiguchi et al. 2006), a finding that differs from the imposex symptoms observed in *N. lapillus*, *Ocinebrina aciculata*, and *T. clavigera* (Gibbs and Bryan 1986; Gibbs et al. 1987; Horiguchi et al. 1994; Oehlmann et al. 1996).

Temporal variations in the reproductive developmental score of the *B. japonica* population differed between females (including imposex-exhibiting females) and males (Horiguchi et al. 2006; Fig. 8.2). Although the spawning season for *B. japonica*

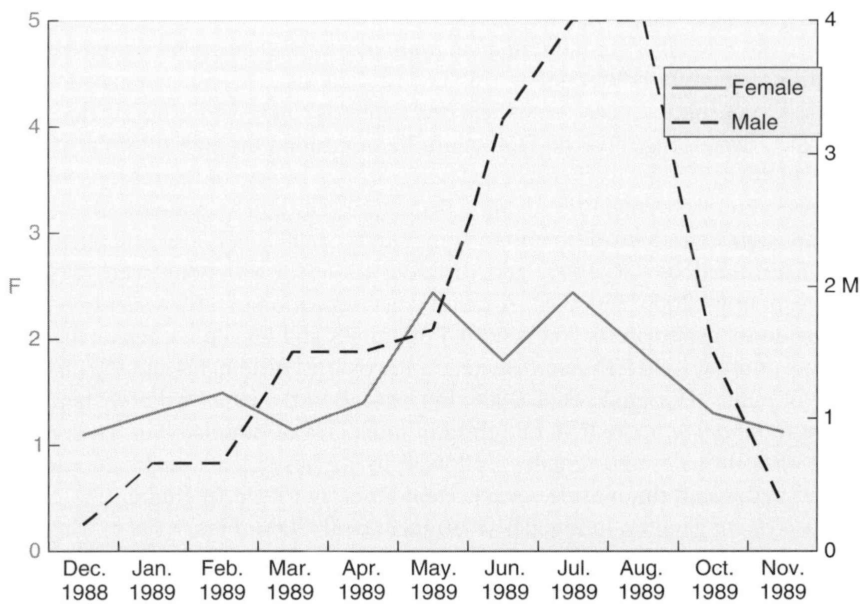

Fig. 8.2 Reproductive cycle of the ivory shell (*Babylonia japonica*) represented by the population reproductive developmental scores. Female reproductive cells were scored based on five categories, and those of males were based on four categories (Horiguchi et al. 2006). The female curve includes imposex-exhibiting females

is late June to early August (Kajikawa et al. 1983), ovarian maturation seemed to be suppressed in females, compared to testicular maturation in males (Horiguchi et al. 2006; Fig. 8.2), which is probably due to the presence of immature females throughout the spawning season. During the spawning season, clearer ovarian maturation and spawning of many more egg capsules were observed in *B. japonica* females in a population from Teradomari, Niigata Prefecture, Japan, compared to those from Tottori (Hamada and Inoue 1993, 1994, 1995). Testicular maturation in males from Tottori was clear in July and August, the spawning season for *B. japonica* (Horiguchi et al. 2006; Fig. 8.2). Thus, the reproductive cycle was unclear in females but it was clearly observed in males (Horiguchi et al. 2006; Fig. 8.2). This suppressed ovarian maturation during the spawning season could be the direct reason for the decreased number of egg capsules spawned by adult *B. japonica* at the hatchery and might accompany imposex in *B. japonica* (Gibbs et al. 1988).

Ovarian spermatogenesis (i.e., an ovo-testis) was observed in 6 (one normal female and five imposex individuals) of 92 female or imposex *B. japonica* specimens examined, a frequency of about 6.5% (Horiguchi et al. 2006; Fig. 8.3). It is well known that most prosobranchs (including *B. japonica*) are dioecious although there are relatively few hermaphroditic prosobranchs in which the gonad produces eggs and sperm simultaneously (Fretter 1984; Uki 1989). Ovarian spermatogenesis has been observed in neogastropods (e.g., *N. lapillus*, *O. aciculata,* and *T. clavigera*) and archaeogastropods (e.g., *Haliotis madaka* and *H. gigantea*) exposed to TBT or TPT, although no penis formation is involved in spermatogenesis in ovaries of female abalone (see below) (Gibbs et al. 1988; Horiguchi and Shimizu 1992; Horiguchi et al. 2000, 2002, 2005; Oehlmann et al. 1996). Ovarian spermatogenesis was even observed in a normal female *B. japonica* without any penis or vas deferens formation, although the frequency was low (one of six, 16.7%). The development of male-type genital organs (penis and vas deferens) and ovarian spermatogenesis in females exposed to TBT or TPT might be controlled through different physiological pathways. This ovarian spermatogenesis may be one of the reasons why the spawning ability of female *B. japonica* decreased (Horiguchi et al. 2006).

Tissue concentrations of organotin compounds, such as butyltins and phenyltins, were determined by GC-FPD, and different tissue distributions were observed (Horiguchi et al. 2006; Fig. 8.4). A marked accumulation of TBT was observed in the ctenidium, osphradium, and heart in both males and females, whereas the highest concentrations of TPT were detected in the ovaries of females and the digestive glands of males (Horiguchi et al. 2006; Fig. 8.4). Based on the total body burden of TBT in *B. japonica*, more than one-third of total TBT accumulated in the digestive glands of both males and females, followed by the testis, ctenidium, muscle, and heart in males and the muscle, ovary, ctenidium, and head (including the central nervous system ganglia) in females (Horiguchi et al. 2006; Figs. 8.5a, c). Based on the total body burden of TPT, approximately three-quarters and more than one-half of total TPT accumulated in the digestive glands of males and females, respectively. The second highest tissue burden of TPT was observed in the gonads of both males and females, followed by the muscle, ctenidium, and heart in males and the muscle, oviduct, and head in females (Horiguchi et al. 2006; Figs. 8.5b, d).

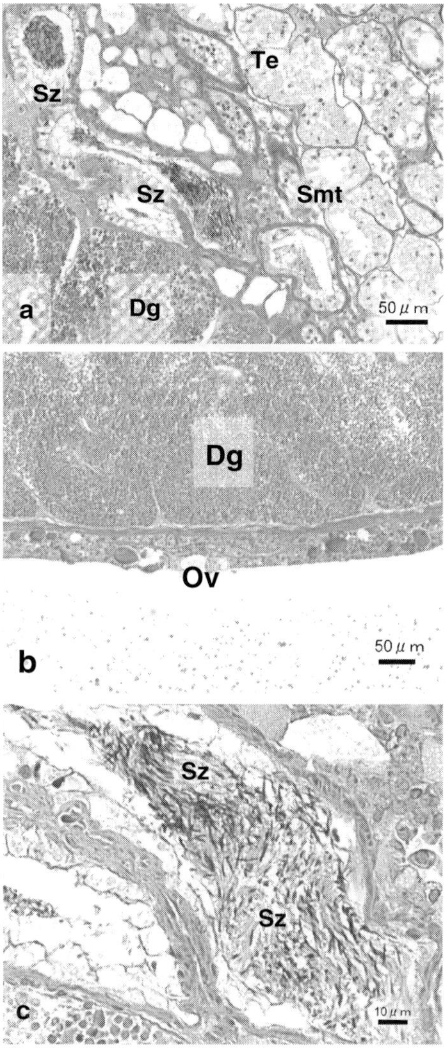

Fig. 8.3 Spermatogenesis in the ovary of a normal female *Babylonia japonica* (i.e., without penis and vas deferens). Testicular (a) and ovarian (b) tissues (i.e., ovo-testis) were observed in the gonad of a female *B. japonica*, which was classified originally as a female because of the presence of female accessory sex organs (e.g., a capsule gland) with neither penis nor vas deferens. Spermatogenesis was also observed in seminiferous tubules of the ovo-testis (c). Dg, digestive gland; Ov, ovary; Smt, seminiferous tubule; Sz, spermatozoon; Te, testis (Horiguchi et al. 2006)

A similar accumulation pattern was also observed in *T. clavigera* (Horiguchi et al. 2003), whereas a slightly different pattern was found in *O. erinacea*, in which approximately half of the total body TBT burden accumulated in the capsule gland (Gibbs et al. 1990), possibly suggesting a difference in organotin accumulation patterns among species. Although concentrations of TBT and TPT in ganglia were quite high in *T. clavigera*, the total tissue burden of those organotins was not high because of the relatively small ganglia tissue in that species (Horiguchi et al. 2003); this may also be the case with *B. japonica* in this study. Similar concentrations of TBT and TPT were also detected in ganglia of *B. undatum* (Mensink et al. 1997).

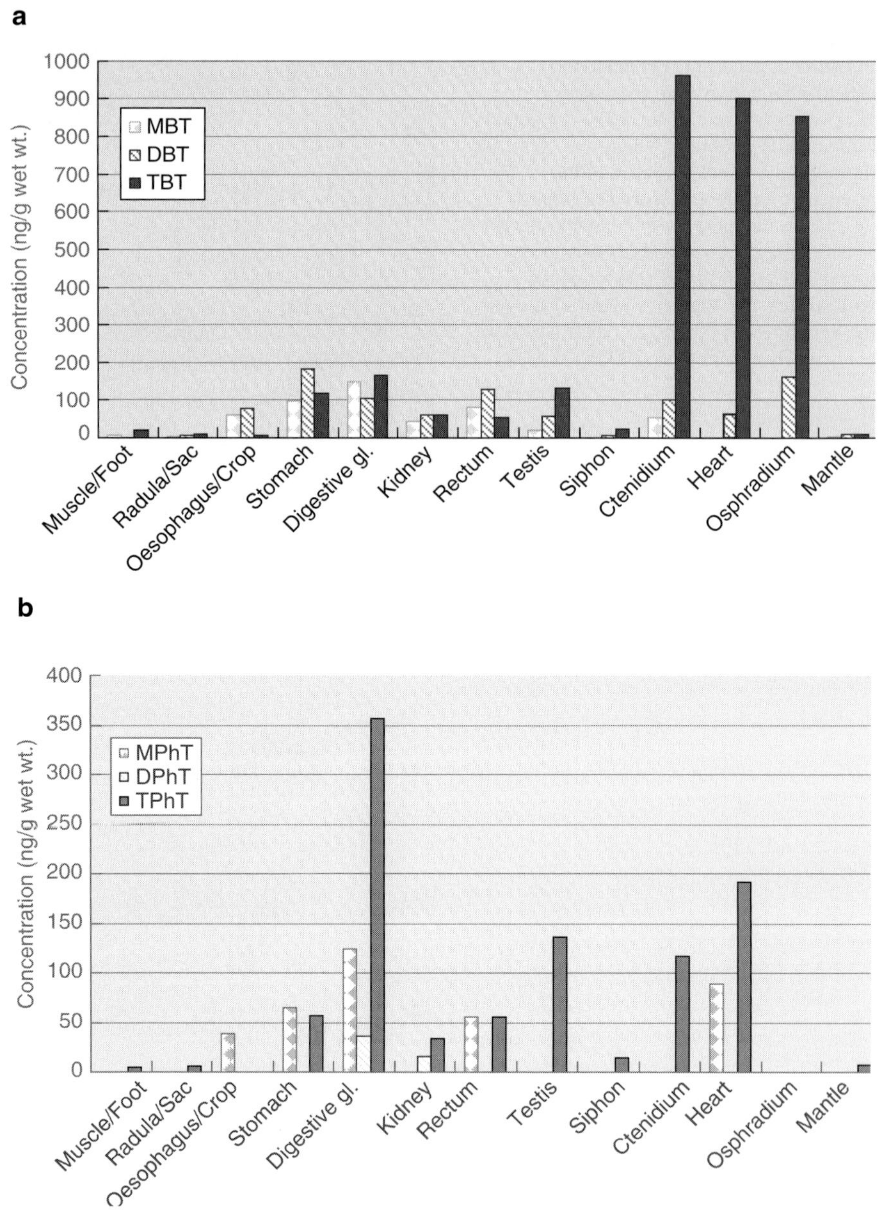

Fig. 8.4 Tissue distribution of organotin compounds in the ivory shell (*Babylonia japonica*) from Yodoe, Tottori, Japan (June 1991): (**a**) butyltins in males; (**b**) phenyltins in males;

Mortality of larvae and seeds or juveniles might also be due to the accumulation of TPT and TBT in ovaries as well as contamination of seawater with TPT or TBT (Coelho et al. 2001; Inoue et al. 2004; Lapota et al. 1993; Li et al. 1997; Nakayama et al. 2005; Ruiz et al. 1995; Treuner et al., unpublished manuscript). Based on a

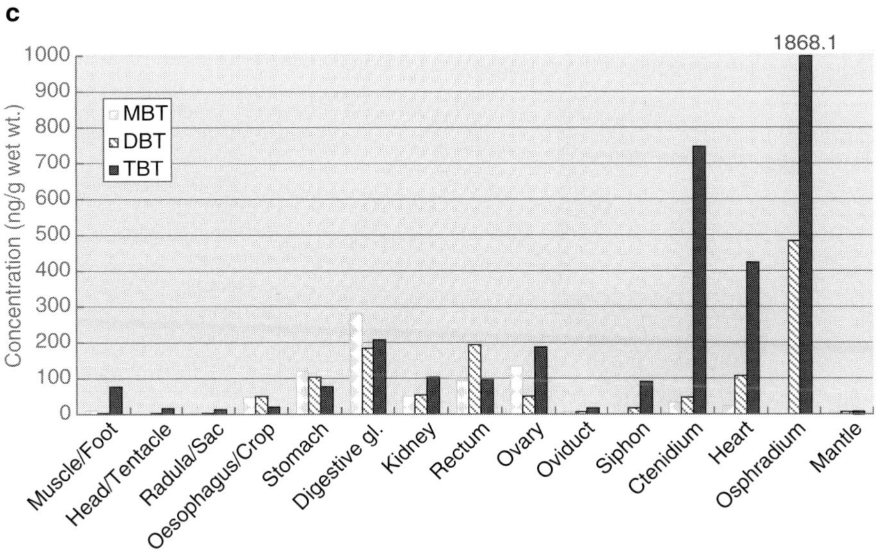

Fig. 8.4 (continued) (**c**) butyltins in females (including imposex individuals);

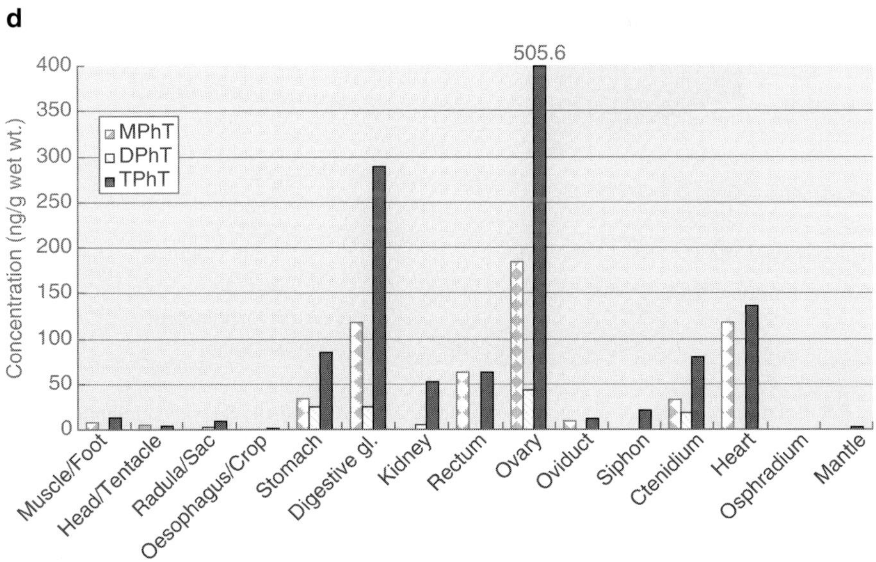

Fig. 8.4 (continued) (**d**) phenyltins in females (including imposex individuals) (Horiguchi et al. 2006)

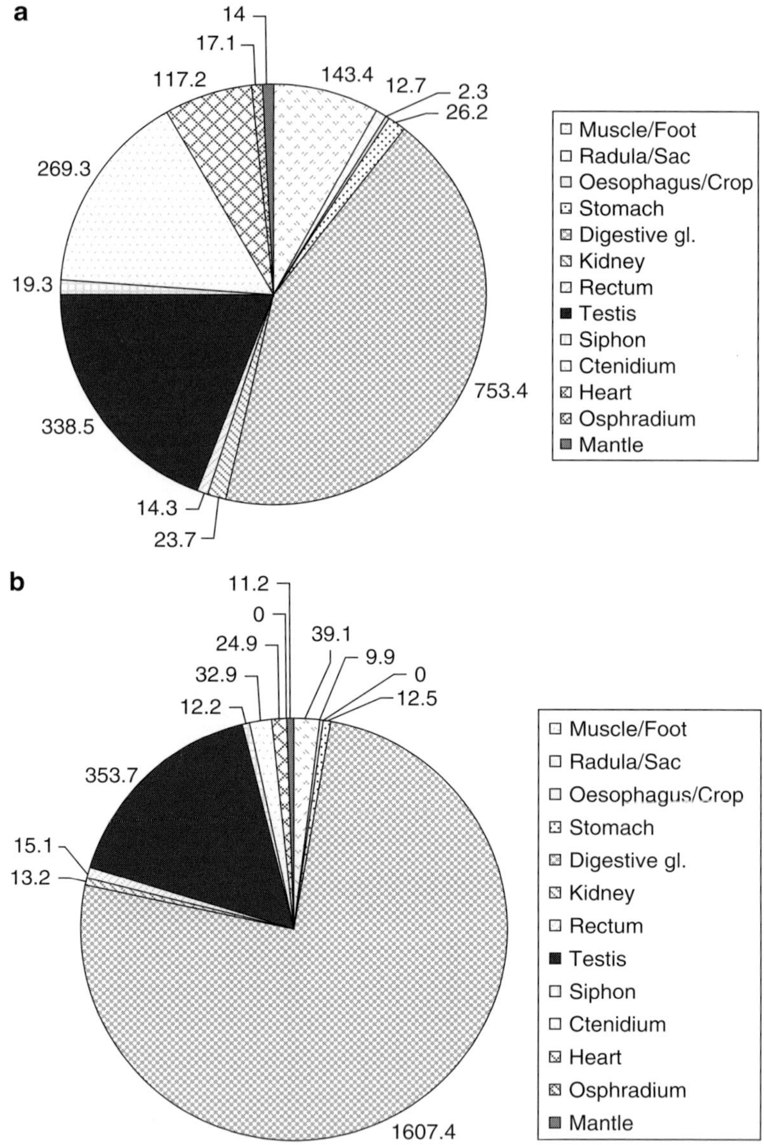

Fig. 8.5 Total body burden of organotin compounds in the ivory shell (*Babylonia japonica*) from Yodoe, Tottori, Japan (June 1991) (unit: ng): (**a**) tributyltin in males; (**b**) triphenyltin in males;

survey of imposex and organotin concentrations in tissues of *T. clavigera* (Horiguchi et al. 1994), contamination with TBT and TPT was relatively high along the coast of Tottori Prefecture, especially in Miho Bay, where the *B. japonica* specimens used in this study were collected.

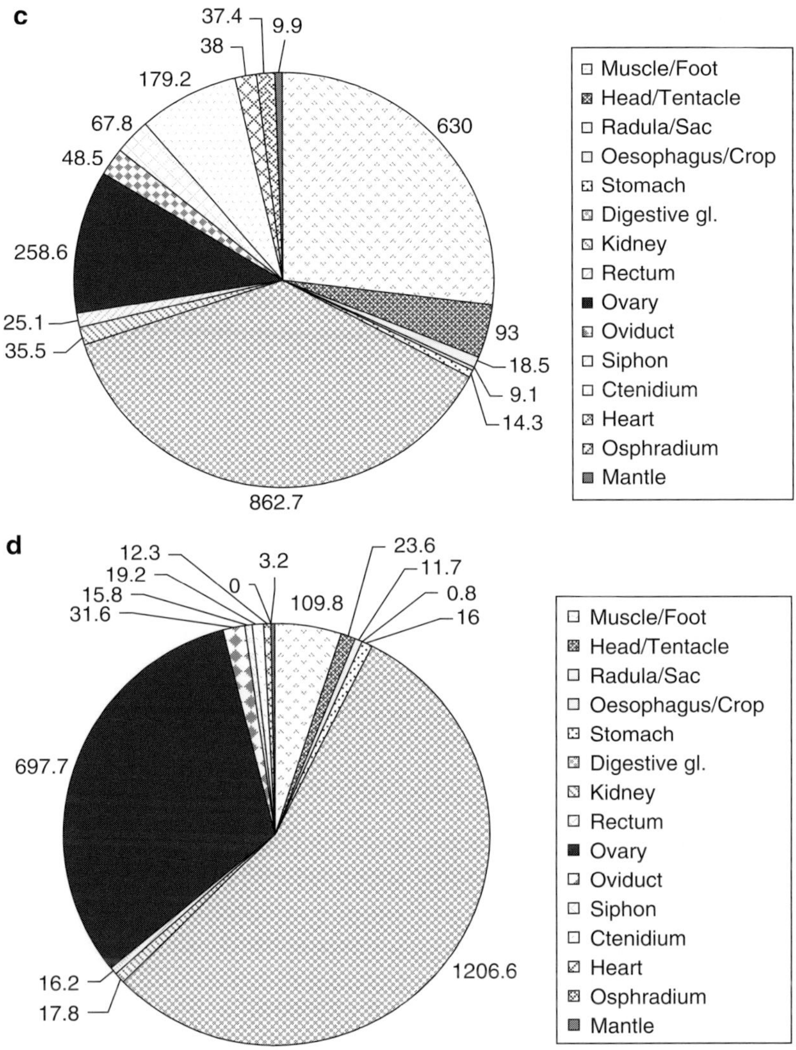

Fig. 8.5 (continued) (**c**) tributyltin in females (including imposex individuals); (**d**) triphenyltin in females (including imposex individuals) (Horiguchi et al. 2006)

Concentrations of TBT and TPT were relatively high in the ovaries of females (Horiguchi et al. 2006; Fig. 8.4). Both TBT and TPT concentrations in gonads were positively correlated with penis length in females (Horiguchi et al. 2006; Fig. 8.6), as was the case with *T. clavigera* (Horiguchi et al. 1994; Shim et al. 2000). Laboratory experiments revealed that both TBT and TPT induced or promoted the development of imposex in *T. clavigera* (Horiguchi et al. 1995, 1997a); therefore, imposex could be caused by TBT or TPT in *B. japonica* as well. However, it is difficult to estimate the threshold concentration of TBT and/or TPT that induces

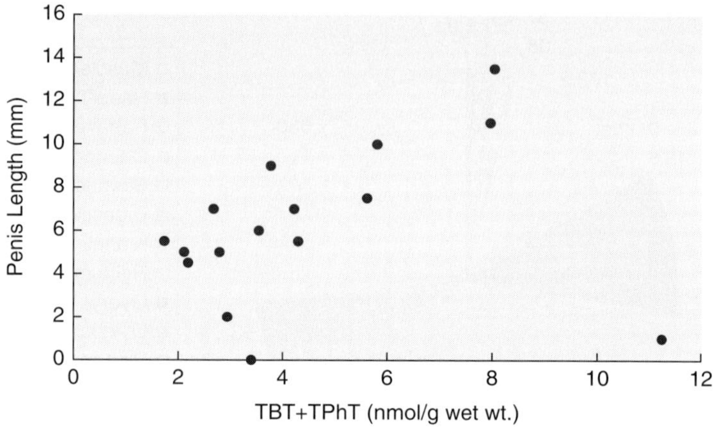

Fig. 8.6 Relationship between triorganotin (the sum of tributyltin and triphenyltin) concentrations in gonads and penis length in female *Babylonia japonica* (Horiguchi et al. 2006)

the development of imposex in *B. japonica*. Laboratory flow-through exposure experiments with *B. japonica*, using TBT and TPT, are needed to estimate the threshold concentration for the development of imposex. The estimated threshold concentration of TBT (in whole body tissues) inducing the development of imposex was reported to be approximately 20 ng Sn/g dry weight (corresponding to approximately 10–12.5 ng TBT/g wet weight, supposing that the concentration on a dry weight basis is 4–5 times that on a wet weight basis) for *N. lapillus* (Gibbs et al. 1987), and to be 10–20 ng/g wet weight for *T. clavigera* (Horiguchi et al. 1994). Because of limited experimental and analytical data for *B. japonica*, however, it is difficult to compare the sensitivity to TBT and TPT between this and other gastropod species, such as *N. lapillus*, *Ocenebra erinacea, Urosalpinx cinerea* and *T. clavigera* (Bryan et al. 1987; Gibbs et al. 1987, 1990, 1991; Horiguchi et al. 1994, 1995).

The planktonic stage of *B. japonica* is estimated to last approximately 4–5 days (Hamada et al. 1988, 1989). This means the recruitment of veliger larvae from other populations inhabiting remote, less contaminated areas is unlikely. Reproductive failure accompanied by imposex in females could result in extirpation of the *B. japonica* population within several years, because the number of offspring produced by adult *B. japonica* in the population is likely to continue to decrease. The existence and duration of a free-swimming phase during larval development is an important factor in determining the linkage between impaired reproductive ability, caused by imposex, to population decline (Bryan et al. 1986; Gibbs and Bryan 1986; Gibbs et al. 1988, 1990, 1991; Horiguchi et al. 2006).

In conclusion, it is suggested that reproductive failure (suppressed ovarian maturation and ovarian spermatogenesis) in adult females with imposex, possibly induced by TBT or TPT from antifouling paints, could have brought about the marked decline in *B. japonica* populations that has been observed.

8.4 Could Ovo-Testis and a Disturbed Reproductive Cycle in the Giant Abalone *Haliotis madaka* be Linked with Organotin Contamination in a Site of Population Decline?

A remarkable population decline has been observed in Japanese abalone since the 1970s (Fig. 8.7), although much effort (e.g. artificial production and release of juvenile abalone into the sea) has been made to enhance stocks (Imai et al. 2006). The proportion of artificially released individuals, which are distinguishable from natural stocks by the green color of the tips of the shells, has exceeded 95% of the total abalone captured in some areas, such as Jogashima (Kanagawa Prefecture) (Imai et al. 2006). This suggests that reproduction in natural abalone stocks is declining.

Reduced abalone recruitment may result from several factors, including mass mortality of larvae and/or juveniles (due to sudden large changes in seawater temperature, food availability, increased predation and/or increased incidence of disease), reduced egg production, low fertilization rate (possibly due to pollutants in the marine environment) and/or overfishing (by commercial fishery). The causal factors for such population declines in abalone have been sought, but are still unknown (Imai et al. 2006).

Imposex, the superimposition of male sexual organs on female gastropod mollusks, bringing about reproductive failure in severely affected individuals, is known to be an endocrine disruption in gastropods, which is typically induced by TBT and TPT from antifouling paints (Smith 1971; Gibbs and Bryan 1986; Gibbs et al. 1987; Horiguchi et al. 1997a).

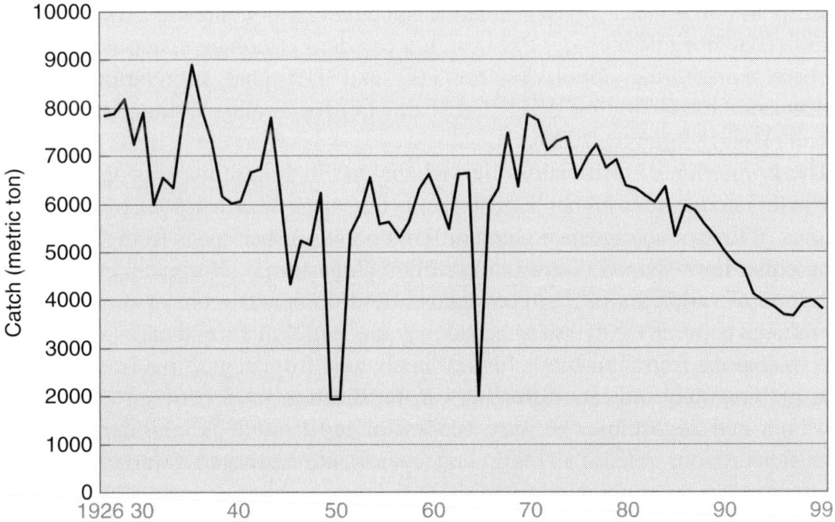

Fig. 8.7 Temporal trend of the total catch of abalone in Japan

The areas where abalone populations have decreased remarkably and the period when this occurred correspond broadly to sites contaminated with organotin compounds and sites with a history of marine pollution by organotins, respectively. Therefore, it is hypothesized that endocrine disruption in abalone has been caused by organotins, and has contributed to population decline (Horiguchi et al. 2000).

A total of 15 *Haliotis madaka* (giant abalone) individuals more than 10 cm in length were collected monthly from two sites between September 1995 and November 1996. The first site, at Tsushima, Nagasaki Prefecture, was a reference site, while the second, at Jogashima, Kanagawa Prefecture, was representative of areas where abalone populations have declined drastically. In Tsushima, abalone stocks are relatively stable and contamination levels of organotin compounds, such as TBT and TPT, are very low (Horiguchi et al. 1997b). Jogashima was known to be heavily contaminated with organotins, following an imposex survey there of the rock shell, *Thais clavigera*, and is one of the most contaminated sites in Japan with regard to TBT and TPT (Horiguchi et al. 1997b). The abalone specimens were divided into two groups: more than half of the gonad samples were used for histological examination and the rest for chemical analysis (Horiguchi et al. 2000).

Gonad samples for histological examination were fixed in Bouin's fluid, embedded in paraffin and stained with Hematoxylin-Eosin. Scores for the development of reproductive cells were applied to quantitatively evaluate the gonadal maturation of abalone, using the developmental stages described in Tomita (1967, 1968). The individual reproductive developmental score was defined as the mean value of a histogram of these scores for the reproductive cells of each abalone. The population reproductive developmental score was defined as the monthly mean value of the individual reproductive developmental scores (Horiguchi et al. 2000), which is the same as the method applied for *B. japonica*.

Chemical analysis of organotin (butyltin and phenyltin) compounds in tissues (muscle and/or gonad) of each abalone specimen was conducted by the method described in Horiguchi et al. (1994) with a certified reference material of Japanese sea bass, *Lateolabrax japonicus*, for TBT and TPT analysis (National Institute for Environmental Studies; NIES CRM no. 11) for quality assurance and quality control (Horiguchi et al. 2000).

The proportion of artificially released abalone in Jogashima was approximately 90% in this study, much higher than that from Tsushima (less than 5%). Morphological features of the gonad/digestive gland differed between specimens from the two sites, being either horn-shaped (Tsushima) or blunt (Jogashima) (Horiguchi et al. 2000).

Temporal variations in the reproductive developmental score of the populations also differed between the two sites: gonad maturation of females and males was synchronous in abalone from Tsushima, but not in abalone from Jogashima ($p < 0.05$; Figs. 8.8a, b). This may indicate differences in fertilization rates between abalone from Tsushima and Jogashima, because successful fertilization is considered to result from synchronous release of eggs and sperm into seawater. Ovarian maturation also seemed to be suppressed in females from Jogashima, compared to Tsushima (Fig. 8.8b) probably due to the presence of immature females in Jogashima throughout the spawning season. Testicular maturation seemed to be more frequently

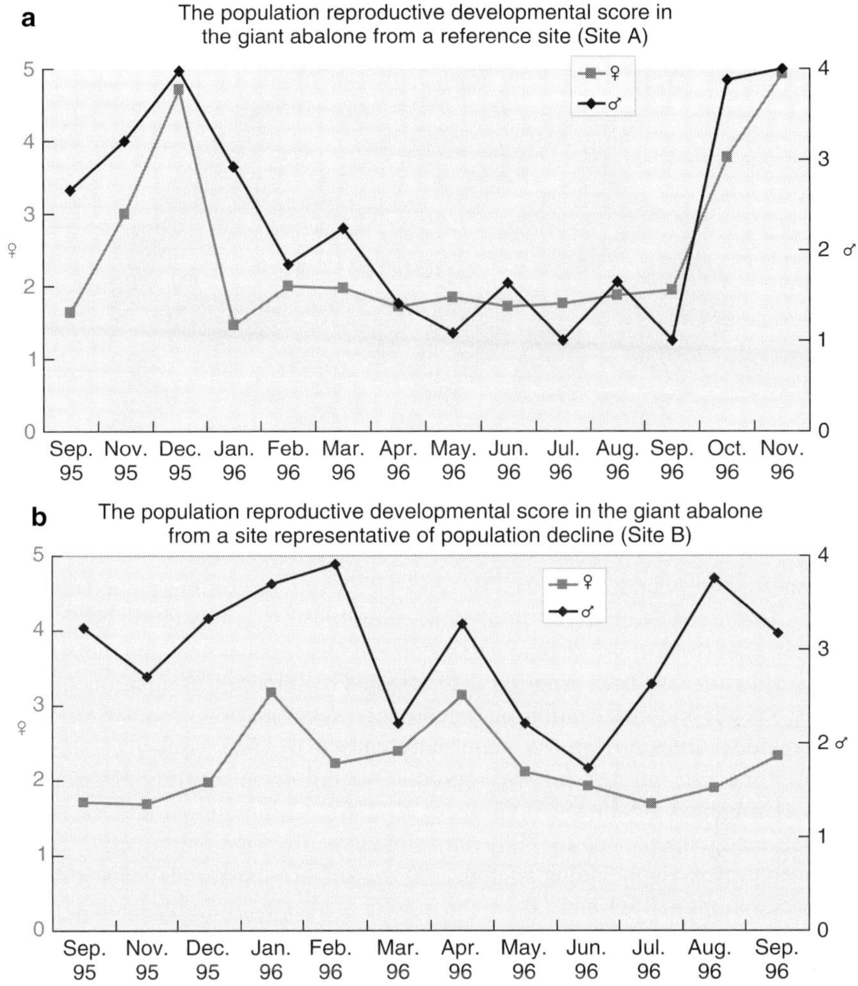

Fig. 8.8 Reproductive cycle of the giant abalone, *Haliotis madaka* (Horiguchi et al. 2000). (a) Tsushima, Nagasaki Prefecture, Japan (a reference site); (b) Jogashima, Kanagawa Prefecture, Japan (a site representative of areas where abalone populations have declined drastically)

observed in male abalone from Jogashima than from Tsushima (Fig. 8.8b). These gonadal features possibly suggest low reproductive success in giant abalone populations around Jogashima (Horiguchi et al. 2000).

Eleven of 54 females (approximately 20%) from Jogashima were observed to be masculinized; most of the gonadal tissues were ovaries with a small amount of testis tissue (i.e. an ovo-testis) (Horiguchi et al. 2000; Fig. 8.9). Either spermatogenesis (13%) or seminiferous tubule-like structure formation (8%) was observed (Horiguchi et al. 2000). This phenomenon of ovo-testis formation is basically similar to imposex in meso- and neogastropods, which is known to be induced

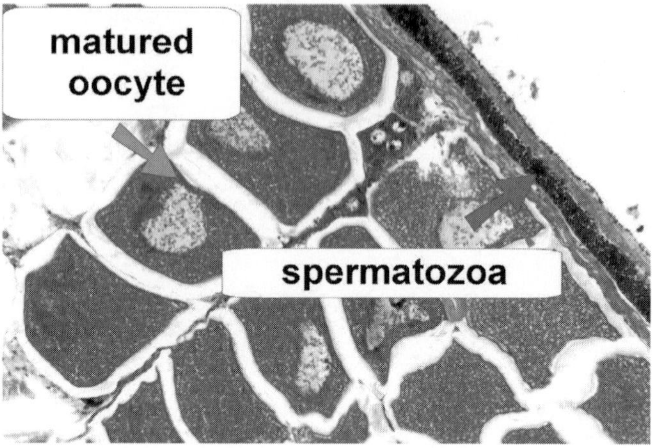

Fig. 8.9 Spermatogenesis in ovary of the giant abalone (*H. madaka*) (Horiguchi et al. 2000). This masculinized female abalone was collected at Jogashima, Japan in April 1996

by organotin compounds, such as TBT and TPT from antifouling paints, although no penis formation is observed in abalone (Smith 1971; Gibbs et al. 1987, 1988; Horiguchi et al. 1997a). More than 150 species of gastropods (neo- and mesogastropods) worldwide have been reported to be affected by imposex, as mentioned above. Intersex, i.e. the masculinization of female accessory sex organs was observed in the periwinkle *Littorina littorea,* reportedly caused by TBT (Bauer et al. 1995). Both imposex and intersex involve reproductive failure in severely affected individuals (Gibbs and Bryan 1986; Gibbs et al. 1988; Oehlmann et al. 1996). Thus, organotin compounds, such as TBT, may similarly affect the reproductive systems in archaeogastropods including abalone.

Concentrations of TBT and TPT in the muscle of abalone from Jogashima ($n = 83$) of 4.9 ± 4.4 and 6.3 ± 6.6 ng/g wet wt, respectively, were significantly higher than those from Tsushima ($n = 125$) ($p < 0.01$) of 0.8 ± 0.8 and 0.6 ± 1.3 ng/g wet wt, respectively. Organotin concentrations in ovary and testis were 10 to 20 times higher than those in muscle (e.g. average concentrations of TBT and TPT in the ovary of abalone from Tsushima were 31.8 and 22.1 ng/g wet wt, while those in the testis were 22.7 and 14.6 ng/g wet wt, respectively.). Therefore, it was suspected that organotin pollution had caused masculinization of female giant abalone from Jogashima (Horiguchi et al. 2000).

In addition, a 7-month *in situ* exposure experiment was conducted, using 40 abalone from Tsushima that were caged near a shipyard in Jogashima, from June 1998 to January 1999 (from the immature to the mature stage). The exposed abalone were fed brown algae, *Ecklonia cava*, once or twice a week during the experimental period. They were collected in January 1999 for histological examination and chemical analysis. This 7-month *in situ* exposure experiment resulted in spermatogenesis in the ovary of approximately 90% of exposed females (Horiguchi et al. 2000; Fig. 8.10). TBT and TPT levels in the muscle of the abalone were from

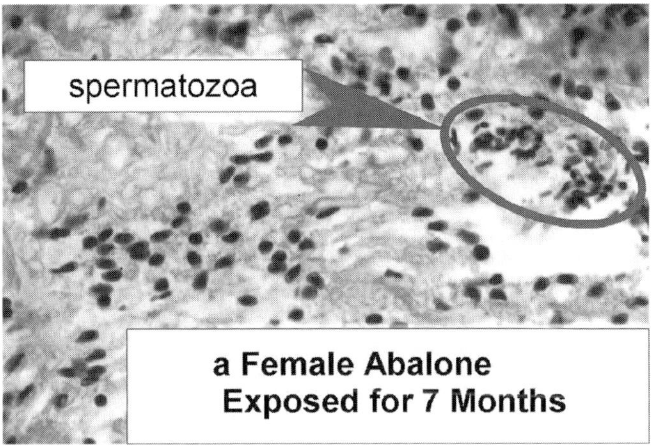

Fig. 8.10 Spermatogenesis in ovary of abalone (*H. gigantea*) exposed *in situ* near a shipyard in Jogashima, Japan (Horiguchi et al. 2000). Ovarian spermatogenesis was observed in approximately 90% of exposed females (collected from reference site, Tsushima, Japan and then used for this *in situ* exposure experiment for 7 months (from June 1998 to January 1999))

0.9 ± 0.4 and 1.3 ± 1.4 ng/g wet wt (n = 15), to 5.0 ± 0.2 and 21.5 ± 2.1 ng/g wet wt (n = 40), respectively ($p < 0.01$) (Horiguchi et al. 2000).

Subsequently, 2-month flow-through exposure experiments of TBT and TPT were conducted with abalone, *Haliotis gigantea,* to examine whether TBT and/or TPT induced spermatogenesis in females. Nominal concentrations of 100 ng/l of TBT and 100 ng/l of TPT caused significant formation of spermatids, spermatozoa and seminiferous tubule-like structures (spermatogenesis) in ovaries of exposed females (Horiguchi et al. 2002; Fig. 8.11). There were also significantly more contracted primary oocytes observed in ovaries of females exposed to either TBT or TPT than in ovaries of controls (Horiguchi et al. 2002). No significant histological changes were observed in testis of exposed males (Horiguchi et al. 2002). This ovarian spermatogenesis caused by TBT and/or TPT exposure seems very similar to the masculinization of mesogastropods and neogastropods, such as imposex. Remarkably high concentrations of TBT and TPT were observed in the head (including ganglia of the central nervous system), compared to concentrations in muscles: 68.32 ± 4.75 ng TBT/g and 1406.39 ± 11.32 ng TPT/g in the head, compared to 2.38 ± 0.81 ng TBT/g and 126.07 ± 68.04 ng TPT/g in muscles (on a wet tissue basis) (Horiguchi et al. 2002). Accumulation of TBT and TPT in the head may disturb reproductive hormonal regulators through neuropeptides released from ganglia. This may be one of the inducers for spermatogenesis in the ovaries of female abalone.

Thus, it was hypothesized that endocrine disruption, resulting in spermatogenesis in the ovary of giant abalone around the shipyard in Jogashima, was caused by TBT and/or TPT, and that organotin compounds from antifouling paints could be one of the causal factors of the observed abalone population decline.

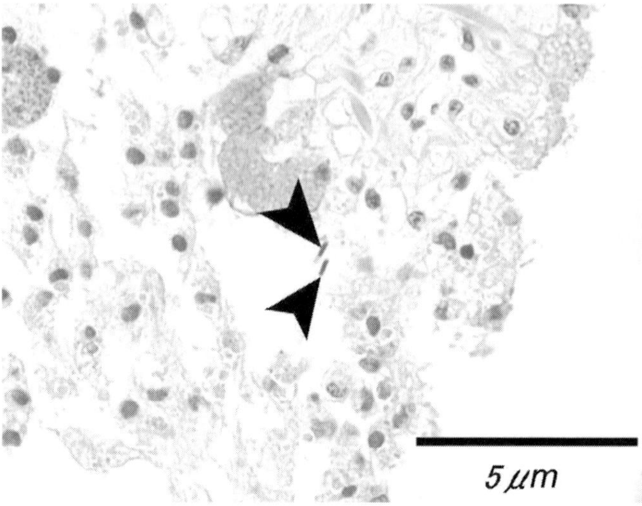

Fig. 8.11 Spermatogenesis in ovary of abalone (*H. gigantea*) exposed to 100 ng/l of TBT in a laboratory flow-through exposure system for 2 months. Ovarian spermatogenesis was observed in female *H. gigantea*, collected at a reference site, Tsushima, Japan and then used for this flow-through exposure experiment

Histological examination of gonads as well as chemical analysis of organotin compounds in tissues of the giant abalone, *Haliotis madaka*, was conducted to evaluate continuing endocrine disruption in abalone populations in Japan (Horiguchi et al. 2005). Abalone specimens were collected from two different areas, Tsushima as a reference site and Jogashima as a site representative of declining abalone populations where serious organotin contamination had been observed, each month from January 1998 to March 1999. Scores were given to the development stages of reproductive cells in the ovary and testis, the same as in Horiguchi et al. (2000), to evaluate the degree of sexual maturation by calculating the mean value of a histogram of these scores for the reproductive cells of each abalone (Horiguchi et al. 2005). The temporal variation in the degree of sexual maturation showed that female and male abalone from Tsushima matured synchronously, while those from Jogashima did not (Horiguchi et al. 2005), which was similar to results of the previous study during September 1995–November 1996 (Horiguchi et al. 2000). Approximately 19% of female abalone from Jogashima were masculinized with an ovo-testis (Horiguchi et al. 2005), which was also similar to the results of Horiguchi et al. (2000). Chemical analyses showed that concentrations of total butyltins (TBT, dibutyltin (DBT) and monobutyltin (MBT): ΣBTs) and total phenyltins (TPT, diphenyltin (DPT) and monophenyltin (MPT): ΣPTs) in the muscle of abalone from Jogashima ($n = 73$) of 7.8 ± 9.0 and 4.5 ± 6.8 ng/g wet wt, respectively, were significantly higher than those from Tsushima ($n = 87$) of 4.7 ± 4.9 and 0.8 ± 1.7 ng/g wet wt, respectively ($p < 0.05$ for ΣBTs; $p < 0.001$ for ΣPTs) (Horiguchi et al. 2005). Concentrations of TBT and TPT in the muscle of abalone from Jogashima ($n = 73$)

of 2.2 ± 2.5 and 5.8 ± 5.1 ng/g wet wt, respectively, were insignificantly and significantly higher than those from Tsushima ($n = 87$) of 0.4 ± 0.6 and 0.5 ± 0.9 ng/g wet wt, respectively ($p > 0.05$ for TBT; $p < 0.001$ for TPT) (Horiguchi et al. 2005). Thus, endocrine disruption as well as contamination by organotins in the giant abalone from Jogashima is still persisting.

8.5 Legislation Affecting Production, Import and Use of Organotin Compounds in Japan, and Future Perspectives on Organotin Pollution and Gastropod Populations

The causative agents of gastropod imposex and intersex, together with masculinization of female abalone, namely TBT and TPT compounds (TBTs and TPTs) have been used worldwide as antifouling agents in paints for ships and fishing nets since the mid-1960s, although TPT use has been much lower compared to TBT (Horiguchi et al. 1994; Goldberg 1986).

In Japan, 14 TBTs and 7 TPTs have been registered as existing chemical substances by the government (Horiguchi et al. 1994). Approximately 70%, 20% and 10% of total amount of TBTs produced and/or imported in Japan had been used in antifouling paints for ships, those for fishing nets and the other purposes (e.g., materials for different TBT species), respectively (Horiguchi et al. 1994). TPTs had also been used for agricultural chemicals (Horiguchi et al. 1994). Production, import and use of TBTs and TPTs have been strictly regulated in Japan since 1990 (Horiguchi et al. 1994). These uses of tri-organotins were reported to have been completely stopped by 1997, although evidence suggests illegal TBT use in antifouling paints in some areas (Horiguchi 2000; Horiguchi et al. 1994; Horiguchi et al., unpublished data, 2001).

A new international treaty, the International Convention on the Control of Harmful Anti-fouling System on Ships (abbreviated to AFS Convention) was adopted at the International Maritime Organization (IMO) in October 2001 for the worldwide ban of TBT and TPT (IMO 2001). The AFS Convention is expected to finally come into effect in September 2008, and field surveys should continue to be conducted to observe reproductive, anatomical, histopathological and ecological effects on affected molluscan populations together with temporal trend monitoring.

References

Bauer B, Fioroni P, Ide I et al. (1995) TBT effects on the female genital system of *Littorina littorea*, a possible indicator of tributyltin pollution. Hydrobiologia 309:15–27.

Bech M (2002a) A survey of imposex in muricids from 1996 to 2000 and identification of optimal indicators of tributyltin contamination along the east coast of Phuket Island, Thailand. Mar Pollut Bull 44:887–896.

Bech M (2002b) Imposex and tributyltin contamination as consequence of the establishment of a marina and increasing yachting activities at Phuket Island, Thailand. Environ Pollut 117: 421–429.

Blaber SJM (1970) The occurrence of a penis-like outgrowth behind the right tentacle in spent females of *Nucella lapillus* (L.). Proc Malacol Soc Lond 39:231–233.

Bryan GW, Gibbs PE, Hummerstone LG et al. (1986) The decline of the gastropod *Nucella lapillus* around south-west England: evidence for the effect of tributyltin from antifouling paints. J Mar Biol Assoc UK 66:611–640.

Bryan GW, Gibbs PE, Burt GR et al. (1987) The effects of tributyltin (TBT) accumulation on adult dog-whelks, *Nucella lapillus*: long-term field and laboratory experiments. J Mar Biol Assoc UK 67:525–544.

Bryan GW, Gibbs PE, Burt GR (1988) A comparison of the effectiveness of tri-*n*-butyltin chloride and five other organotin compounds in promoting the development of imposex in the dog-whelk, *Nucella lapillus*. J Mar Biol Assoc UK 68:733–744.

Coelho MR, Fuentes S, Bebianno MJ (2001) TBT effects on the larvae of *Ruditapes decussates*. J Mar Biol Assoc UK 81:259–265.

Fioroni P, Oehlmann J, Stroben E (1991) The pseudohermaphroditism of prosobranchs; morphological aspects. Zool Anz 226:1–26.

Fretter V (1984) Prosobranchs. In: Wilbur KM (ed) The Mollusca Vol. 7. Reproduction (Tompa AS, Verdonk NH, van den Biggelaar JAM, eds). Academic, Orlando, FL: pp.1–45.

Gibbs PE, Bryan GW (1986) Reproductive failure in populations of the dog-whelk, *Nucella lapillus*, caused by imposex induced by tributyltin from antifouling paints. J Mar Biol Assoc UK 66:767–777.

Gibbs PE, Bryan GW (1996) TBT-induced imposex in neogastropod snails: masculinization to mass extinction. In: de Mora SJ (ed) Tributyltin: Case Study of an Environmental Contaminant. Cambridge University Press, Cambridge: pp.212–236.

Gibbs PE, Bryan GW, Pascoe PL et al. (1987) The use of the dog-whelk, *Nucella lapillus*, as an indicator of tributyltin (TBT) contamination. J Mar Biol Assoc UK 67:507–523.

Gibbs PE, Pascoe PL, Burt GR (1988) Sex change in the female dog-whelk, *Nucella lapillus*, induced by tributyltin from antifouling paints. J Mar Biol Assoc UK 68:715–731.

Gibbs PE, Bryan GW, Pascoe PL et al. (1990) Reproductive abnormalitics in female *Ocenebra erinacea* (Gastropoda) resulting from tributyltin-induced imposex. J Mar Biol Assoc UK 70: 639–656.

Gibbs PE, Spencer BE, Pascoe PL (1991) The American oyster drill, *Urosalpinx cinerea* (Gastropoda): evidence of decline in an imposex affected population (R. Blackwater, Essex). J Mar Biol Assoc UK 71:827–838.

Goldberg ED (1986) TBT: an environmental dilemma. Environment 28:17–20, 42–44.

Hamada M, Inoue M (1993) Seed production of the ivory shell, *Babylonia japonica* Reeve [in Japanese]. Bull Tottori Prefect Fish Exp Station 193–195.

Hamada M, Inoue M (1994) Seed production of the ivory shell, *Babylonia japonica* Reeve [in Japanese]. Bull Tottori Prefect Fish Exp Station 182–184.

Hamada M, Inoue M (1995) Seed production of the ivory shell, *Babylonia japonica* Reeve [in Japanese]. Bull Tottori Prefect Fish Exp Station 204–206.

Hamada F, Kanazawa T, Yamamoto E (1988) Seed production of the ivory shell, *Babylonia japonica* Reeve [in Japanese]. Bull Tottori Prefect Fish Exp Station 6:110–116.

Hamada F, Kanazawa T, Yamamoto E (1989) Seed production of the ivory shell, *Babylonia japonica* Reeve [in Japanese]. Bull Tottori Prefect Fish Exp Station 7:103–109.

Horiguchi T (2000) Molluscs. In: Kawai S, Koyama J (eds) Problems of Endocrine Disruptors in Fisheries Environment. [in Japanese] Koseisha-Koseikaku, Tokyo: pp.54–72.

Horiguchi T (2004) Gastropods. In: Takeuchi I, Tanabe S, Hino A (eds) Biological Monitoring for Anthropogenic Chemicals in Low Doze [in Japanese]. Koseisha-Koseikaku, Tokyo: pp.37–67.

Horiguchi T, Shimizu M (1992) Effects on aquatic organisms, mainly on molluscs. In: Satomi Y, Shimizu M (eds) Organotin Pollution and Its Effects on Aquatic Organisms. [in Japanese] Koseisha-Koseikaku, Tokyo: pp.99–135.

Horiguchi T, Shiraishi H, Shimizu M et al. (1994) Imposex and organotin compounds in *Thais clavigera* and *T. bonni* in Japan. J Mar Biol Assoc UK 74:651–669.

Horiguchi T, Shiraishi H, Shimizu M et al. (1995) Imposex in Japanese gastropods (Neogastropoda and Mesogastropoda): effects of tributyltin and triphenyltin from antifouling paints. Mar Pollut Bull 31:402–405.

Horiguchi T, Shiraishi H, Shimizu M et al. (1997a) Effects of triphenyltin chloride and five other organotin compounds on the development of imposex in the rock shell, *Thais clavigera*. Environ Pollut 95:85–91.

Horiguchi T, Shiraishi H, Shimizu M et al. (1997b) Imposex in sea snails, caused by organotin (tributyltin and triphenyltin) pollution in Japan: a survey. Appl Organomet Chem 11:451–455.

Horiguchi T, Takiguchi N, Cho HS, Kojima M, Kaya M, Shiraishi H, Morita M, Hirose H, Shimizu M (2000) Ovo-testis and disturbed reproductive cycle in the giant abalone, *Haliotis madaka*: possible linkage with organotin contamination in a site of population decline. Mar Environ Res 50:223–229.

Horiguchi T, Kojima M, Kaya M et al. (2002) Tributyltin and triphenyltin induce spermatogenesis in ovary of female abalone, *Haliotis gigantea*. Mar Environ Res 54:679–684.

Horiguchi T, Shiraishi H, Morita M (2003) Specific tissue distributions of organotin compounds in prosobranch gastropods. [Abstract] SETAC 24th Annual Meeting Abstract Book p.290.

Horiguchi T, Kojima M, Takiguchi N et al. (2005) Continuing observation of disturbed reproductive cycle and ovarian spermatogenesis in the giant abalone, *Haliotis madaka,* from an organotin-contaminated site of Japan. Mar Pollut Bull 51:817–822.

Horiguchi T, Kojima M, Hamada F et al. (2006) Impact of tributyltin and triphenyltin on ivory shell (*Babylonia japonica*) populations. Environ. Health Perspectives 114:Supplement 1:13–19.

Imai T, Takiguchi N, Horiguchi T (2006) Estimation of the causal factor for declining abalone populations in Jogashima, Kanagawa, Japan. Bull Kanagawa Prefect Fish Technol Center 1:51–58, 2006 [in Japanese].

Inoue S, Oshima Y, Nagai K et al. (2004) Effect of maternal exposure to tributyltin on reproduction of the pearl oyster (*Pinctada fucata martensii*). Environ Toxicol Chem 23:1276–1281.

International Maritime Organization (IMO) (2001) Adoption of the final act of the conference and any instruments, recommendations and resolutions resulting from the work of the conference. International Convention on the Control of Harmful Anti-fouling Systems on Ships, 2001 (AFS/CONF/26, 18 October 2001). London, UK: International Maritime Organization, 1–26.

Kajikawa A (1984) Sexual characteristics, spawning abilities and their percentage occurrence in populations of *Babylonia japonica* Reeve [in Japanese]. Bull Tottori Prefect Fish Exp Station 2:31–32.

Kajikawa A, Yamamoto E, Masutani R (1983) Sexual characteristics and spawning abilities in the ivory shell, *Babylonia japonica* Reeve [in Japanese]. Bull Tottori Prefect Fish Exp Station 1:16–18.

Lapota D, Rosenberger DE, Platter-Rieger MF et al. (1993) Growth and survival of *Mytilus edulis* larvae exposed to low levels of dibutyltin and tributyltin. Mar Biol 115:413–419.

Li Q, Osada M, Takahashi K et al. (1997). Accumulation and depuration of tributyltin oxide and its effect on the fertilization and embryonic development in the Pacific oyster, *Crassostrea gigas*. Bull Environ Contam Toxicol 58:489–496.

Marshall DJ, Rajkumar A (2003) Imposex in the indigenous *Nassarius kraussianus* (Mollusca: Neogastropoda) from South African harbours. Mar Pollut Bull 46:1150–1155.

Mensink B, Boon JP, ten Hallers-Tjabbes CC et al. (1997) Bioaccumulation of organotin compounds and imposex occurrence in a marine food chain (Eastern Scheldt, The Netherlands). Environ Technol 18:1235–1245.

Nakayama K, Oshima Y, Nagafuchi K et al. (2005) Early-life-stage toxicity in offspring from exposed parent medaka, *Oryzias latipes*, to mixtures of tributyltin and polychlorinated biphenyls. Environ Toxicol Chem 24:591–596.

Oehlmann J, Fioroni P, Stroben E et al. (1996) Tributyltin (TBT) effects on *Ocinebrina aciculata* (Gastropoda: Muricidae): imposex development, sterilization, sex change and population decline. Sci Total Environ 188:205–223.

Ruiz JM, Bryan GW, Gibbs PE (1995) Acute and chronic toxicity of tributyltin (TBT) to pediveliger larvae of the bivalve *Scrobularia plana*. Mar Biol 124:119–126.

Schulte-Oehlmann U, Oehlmann J, Fioroni P et al. (1997) Imposex and reproductive failure in *Hydrobia ulvae* (Gastropoda: Prosobranchia). Mar Biol 128:257–266.

Shim WJ, Kahng SH, Hong SH et al. (2000) Imposex in the rock shell, *Thais clavigera*, as evidence of organotin contamination in the marine environment of Korea. Mar Environ Res 49:435–451.

Smith BS (1971) Sexuality in the American mud snail, *Nassarius obsoletus* Say. Proc Malacol Soc Lond 39:377–378.

Sole M, Morcillo Y, Porte C (1998) Imposex in the commercial snail *Bolinus brandaris* in the northwestern Mediterranean. Environ Pollut 99:241–246.

Takamaru T, Fuji A (1981) Reproductive cycles in populations of *Neptunea arthritica* from the southern coastal waters of Hokkaido, Japan [in Japanese]. Suisan Zoshoku 29:78–87.

ten Hallers-Tjabbes CC, Wegener JW, Van Hattum BA et al. (2003) Imposex and organotin concentrations in *Buccinum undatum* and *Neptunea antiqua* from the North Sea: relationship to shipping density and hydrographical conditions. Mar Environ Res 55:203–233.

Terlizzi A, Delos AL, Garaventa F et al. (2004) Limited effectiveness of marine protected areas: imposex in *Hexaplex trunculus* (Gastropoda: Muricidae) populations from Italian marine reserves. Mar Pollut Bull 48:188–192.

Tomita K (1967) *Annual Report of Hokkaido Prefectural Fisheries Experimental Station* (7):1–7 (in Japanese)

Tomita K (1968) *Annual Report of Hokkaido Prefectural Fisheries Experimental Station* (9): 56–61 (in Japanese)

Uki N (1989) Sexual maturation, development and growth in gastropods, and their control [in Japanese]. In: Takashima F, Hanyu I (eds) Reproductive Biology of Fish and Shellfish. Midori Shobo, Tokyo: pp.367–417.

Section 3
Toxicity of Organotin Compounds in Aquatic Organisms

Chapter 9
Toxicity for Aquatic Organisms

Kazuhiko Mochida and Kazunori Fujii

9.1 Introduction

The potential effects of so-called endocrine disrupting chemicals (EDCs) on fish reproduction have been a growing concern since the early 1990s (Colborn et al. 1993; Jobling et al. 1996). Adverse reproductive health effects, possibly a result of EDCs, have been observed in wild fish populations (Jobling et al. 1996; Hashimoto et al. 2000; Larsson et al. 2000). Although much of the research on EDCs has focused on estrogenic compounds, tributyltin (TBT) is also thought to act as an endocrine disruptor (Matthiessen and Gibbs 1998; McAllister and Kime 2003; Nakanishi et al. 2006). Indeed, masculinization has been induced in the Japanese flounder (*Paralichthys olivaceus*) by TBT administration (Shimasaki et al. 2003). TBT exposure has also induced masculinization, as well as irreversible sperm damage, in the zebrafish (*Danio rerio*), although direct evidence of aromatase inhibition was not shown (McAllister and Kime 2003). Additionally, Shimizu and Kimura (1987) observed that long-term exposure to tributyltin oxide (TBTO) resulted in significant depression of the gonad somatic index in male goby (*Chasmicthys dolichognathus*). Thus, there is evidence for TBT-induced reproductive toxicity in fish. In view of the need to maintain and protect wild populations of fish, it is essential that scientists clarify the adverse effects of TBT, especially with regard to reproduction.

In this chapter, I would like to focus on the toxicity of TBT on fish reproduction and review progress in this research field. Sublethal effects (e.g., growth inhibition, histological changes) caused by acute and chronic toxicity of TBT and other organotin compounds to several marine fish species have been described elsewhere (see the review by Fent (1996)) and are not addressed in this chapter.

K. Mochida and K. Fujii
National Research Institute of Fisheries and Environment of Inland Sea,
Fisheries Research Agency, 2-17-5 Maruishi, Hatsukaichi, Hiroshima 739-0452, Japan

T. Arai et al. (eds.), *Ecotoxicology of Antifouling Biocides*,
DOI: 10.1007/978-4-431-85709-9_9,

9.2 Toxicity to Fish Reproduction

9.2.1 Toxic Effect of TBT During Sex Differentiation

Available data on the toxic effects of TBT on various aspects of reproduction are summarized in Table 9.1.

Induction of masculinization in fish was first reported by Shimasaki et al. (2003). They examined the effect of TBT on the sex differentiation process in fish, using larvae of genetically female Japanese flounder that were obtained by mating normal females with sex-reversed, meiotic gynogenetic males, and found that sex reversal of genetically female flounder into phenotypic males was induced by dietary exposure to TBT (Table 9.1). In the same year, McAllister and Kime (2003) reported that the result of TBT exposure at an environmentally relevant level (0.1 ng/l) on zebrafish from hatching to 70 days was a male-biased population. Santos et al. (2006) also reported a masculinizing effect of TBT on zebrafish (Table 9.1), and we have confirmed that TBT-exposure at 500 ng/l on mummichog (*Fundulus heteroclitus*) from embryo to 70 days post hatch also induced a male-biased population (Mochida et al., unpublished data, 2007). Inhibition of P450 aromatase activity by TBT toxicity has been hypothesized as one of the causes of the masculinizing effect of TBT (Mathiessen and Gibbs 1998). CYP19 is a P450 aromatase and is an enzyme responsible for estrogen biosynthesis from androgens (Simpson et al. 1994). CYP19 gene products have also been identified in several teleost species including goldfish (*Carassius auratus*), zebrafish, and Nile tilapia (*Oreochromis niloticus*) (Kwon et al. 2001; Kishida and Callard 2001; Trant et al. 2001). Teleost fishes have two structurally distinct CYP19 isoforms. One of these forms, CYP19A1, is predominantly expressed in the ovary and plays an important role in sex-differentiation (Kishida and Callard 2001; Kazeto et al. 2004). Aromatase inhibitors may be particularly important endocrine disruptors since estrogens are indispensable in the reproductive development of both male and female fish (Sharpe 2001). In grey mullet (*Mugil cephalus*), high gonadal aromatase activity was observed during sexual differentiation (Chang et al. 1999). Kwon et al. (2000) demonstrated that exposing genetically female Nile tilapia during the period of sexual differentiation to fadrozole, which is a nonsteroidal aromatase inhibitor, via their daily diet from 7 to 37 days post hatch caused masculinization in a dose-dependent manner. In addition, fadrozole injected into male coho salmon (*Oncorhynchus kisutch*) in early life inhibited estradiol production during the period of sexual maturation (Afonso et al. 2000). These results indicate that aromatase activity is a key factor in sexual differentiation in fish. Fent and Stegeman (1993) analyzed liver microsomal protein of a marine fish, scup (*Stenotomus chrysops*), that had been injected with TBT at a dose of 16.3 mg/kg body weight and demonstrated that several kinds of P450 proteins were degraded. A similar result was obtained from TBT-exposed olive flounder (*Paralichthys olivaceus*) with hepatic CYP content decreasing with increasing exposure concentration of TBT (Shim et al. 2003). There have also been several research studies that demonstrated reduction of P450 aromatase gene expression by TBT exposure (Shimasaki et al. 2003;

Table 9.1 Toxic effects of tributyltin (TBT) on various aspects of fish reproduction

Toxic effect	Fish species	Compounds	Exposure	Duration	LOEC (nominal)	Reference
Sex differentiation						
Masculinization						
	Japanese flounder (XX genetically female) (*Paralichthys olivaceus*)	TBTO	Dietary	From 35 to 100-d after hatch	0.1 µg/µg diet	Shimasaki et al. (2003)
	Zebrafish (*Danio rerio*)	TBT	Water	70-d from hatching	0.1 ng/l	McAllister and Kime (2003)
	Zebrafish (*Danio rerio*)	TBTCI	Dietary	120-d from 5-dpf larvae	25 ng/g diet	Santos et al. (2006)
Spermatogenesis						
Depression of gonad somatic index (GSI)	Salt-water goby (*Chasmichthys dolichognathus*)	TBTO	Water	120-d during maturing stage	2.1 µg/l	Shimizu and Kimura (1987)
Depression of GSI and histological damage	Mummichog (*Fundulus heteroclitus*)	TBTO	Water	2-wks during maturing stage	3.9 µg/l[a]	Mochida et al. (2007)
Inhibition of germ cell proliferation	Mummichog (*Fundulus heteroclitus*)	TBTO	Water	2-wks during maturing stage	2.1 µg/l[a]	Mochida et al. (2007)
Decrease in expression of some spermatogenesis-related gene	Mummichog (*Fundulus heteroclitus*)	TBTO	Water	2-wks during maturing stage	4 µg/l	Mochida et al. (2007, 2008)
Decrease in sperm motility	Zebrafish (*Danio rerio*)	TBT	Water	70-d from hatching	10 ng/l	McAllister and Kime (2003)
All sperm lacked flagella Increase in milt volume	Zebrafish (*Danio rerio*)	TBT	Water	70-d from hatching	100 ng/l	McAllister and Kime (2003)
Decrease in sperm counts	Guppy (*Poecilia reticulata*)	TBT	Water	21-d (adult)	11.2 ng/l	Haubruge et al. (2000)

(continued)

Table 9.1 (continued)

Toxic effect	Fish species	Compounds	Exposure	Duration	LOEC (nominal)	Reference
Oogenesis						
Decrease in number of floating egg rate and hatchability	Japanese whiting (*Sillago japonica*)	TBTO	Dietary	30-d (adult reproductive season)	200 μg/g diet	Shimasaki et al. (2006)
Increase in the ratio of testosterone to estradiol-17 β in ovary	Cuvier (*Sebastiscus marmoratus*)	TBT	Water	50-d (maturing stage)	10 ng/l	Zhang et al. (2007)
Increase in apoptotic ovarian follicular cells	Cuvier (*Sebastiscus marmoratus*)	TBT	Water	50-d (maturing stage)	10 ng/l	Zhang et al. (2007)

LOEC, lowest observed effect concentration; TBTO, tributyltin oxide; TBTCI, tributyltin chloride; water, waterborne exposure.
d, days; wks, weeks; dpf, days post-fertilization
Numbers in brackets show reference for this data.
[a]Actual toxicant concentration.

Lyssimachou et al. 2006). P450 aromatase gene expression was reduced in gonads of Japanese flounder that had induced masculinization from the administration of TBT at 1.0 μg/g diet (Shimasaki et al. 2003). Using brain aromatase gene isoforms, Lyssimachou et al. (2006) studied the effect of TBT on the brain steroidogenic pathway of juvenile Atlantic salmon (*Salmo salar*) and revealed that TBT exposure at 250 ng/l for 7 days inhibited gene expression of the aromatase isoforms.

Recent studies have shown that inhibition of sex steroid metabolism plays an important role as a non-genomic pathway in invertebrate endocrine disruption by xenobiotics, such as TBT (Janer et al. 2005; Janer and Porte 2007). Steroid metabolic pathways, such as sulfation by sulfotransferase and glucuronidation by UDP-glucuronosyltransferase, are considered to have an important role in inhibiting biological activity of the steroid and increasing excretion of the steroid (Janer and Porte 2007). The inhibitory effect of TBT on the metabolic activity was also reported for some fish species (Thibaut and Porte 2004; Martin-Skilton et al. 2006).

Pathways for steroid synthesis and metabolism and the pathways that are reported to be inhibited by TBT are illustrated in Fig. 9.1. The precise mechanisms that induce masculinization by TBT exposure have not been clarified, but the weight of evidence indicates that toxic levels of TBT result in the accumulation of testosterone

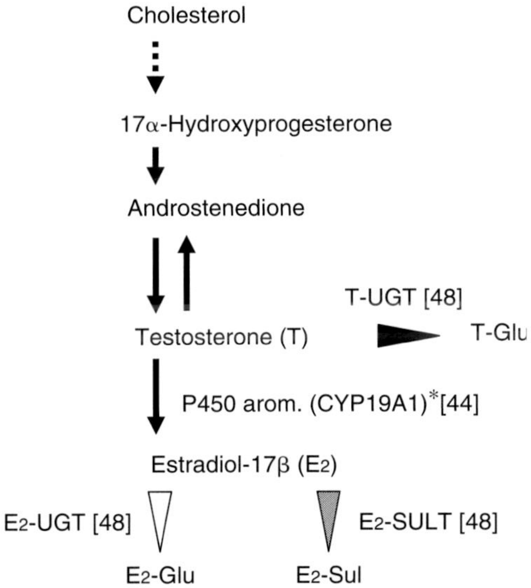

Fig. 9.1 Simplified illustration of the pathways for steroid synthesis and metabolism. Asterisk indicates a pathway that was reported to be inhibited by tributyltin (TBT) only. Black arrow head, shaded arrow head, and open arrow head indicate pathways that were reported to be inhibited by TBT, triphenyltin (TPT), or dibutyltin, by TBT or TPT, and by TPT only, respectively. Numbers in brackets show references for this data. Glu, glucuronide; Sul, sulphate; SULT, sulfotransferase; UGT, UDP-glucuronosyltransferase

by the inhibition of aromatase activity and disruption of the metabolic pathways of steroids that are involved in sex differentiation, and this almost certainly contributes to TBT-induced masculinization in fish.

A recent study of marine gastropods clearly indicated the relationship between the retinoid X receptor (RXR) and induction of development of imposex in female gastropods (Nishikawa et al. 2004). Sequence analysis of the RXR homologue from the rock shell (*Thais claviger*) revealed that the ligand-binding domain of rock shell RXR was homologous to vertebrate RXR and bound to both 9-cis retinoic acid (RA), which is the natural ligand of RXR, and TBT. Injection of RA into female rock shells induced imposex, although the precise mechanism of differentiation of male genital tracts in female gastropods remains unknown (Nishikawa et al. 2004). To date, no studies have indicated that treatment with RA could induce masculinization in any teleost species. RXR homologue has also been cloned from zebrafish (Jones et al. 1995; Waxman and Yelon 2007), and RA signaling is important for multiple aspects of embryonic development and tissue homeostasis in zebrafish and Japanese medaka (*Oryzias latipes*) (Grandel et al. 2002; Hayashida et al. 2004). In addition, recent research has revealed that precise regulation of retinoid levels during fetal gonad development provides the molecular control mechanism for germ cell sex determination in mice (Bowles et al. 2006). Although there is no evidence that RA is involved in sex differentiation in teleosts, it would be worth clarifying whether disruption by TBT of the RA signaling pathway during sex determination occurs and plays a role in masculinization.

9.2.2 Toxic Effects of TBT on Spermatogenesis

In this section, the effect of TBT on gonadal maturation of male fish is reviewed. As shown in Table 9.1, several studies have reported a toxic effect of TBT on spermatogenesis during testicular maturing stages.

Until recently, no studies have addressed the molecular mechanism by which TBT might affect fish gametogenesis. Therefore, we have focused on elucidating the effect of TBT on spermatogenesis (Mochida et al. 2007, 2008). First, we exposed the mummichog (*Fundulus heteroclitus*) to TBTO for 2 or 4 weeks and examined histological damage to the testes. Because depressing the proliferation of spermatogenic cells directly correlates with reduced numbers of spermatozoa, the proliferating activity can be an excellent marker for the evaluation of toxic effects on reproduction. We also examined the effect of TBTO exposure to germ cell proliferation by measuring the bromodeoxyuridine incorporation ratio into the germ cells in testes. We then used complementary DNA (cDNA) subtraction and differential display methods to detect differential gene expression in the testes of control and TBTO-exposed fish in order to elucidate the molecular mechanism by which TBTO may affect fish spermatogenesis (Mochida et al. 2007, 2008).

Histological damage to the testes after exposure of the mummichog to TBTO at 5.8 μg/l for 2 weeks was primarily detected in the epithelial cells of the seminal

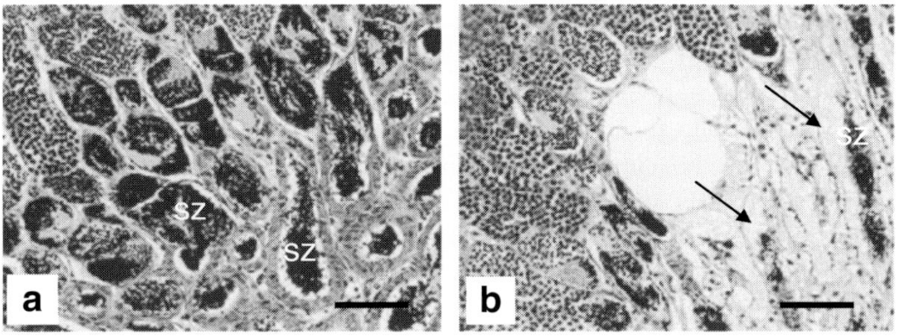

Fig. 9.2 Histopathological changes in testes of mummichog after exposure to TBTO. Normal testis (a) and a typical example of a damaged testis of fish exposed to TBTO at 7.1 μg/l for 4 weeks (b). Note that epithelium formed a net-like structure and that the lumen of the seminal duct lacking spermatozoa was prominent (*arrows* in b). Sz, spermatozoa. Bars, 100 μm

ducts. In a typical testis, the epithelial cells of the seminal ducts shrank and formed a net-like structure (Fig. 9.2) (Mochida et al. 2007), and the number of spermatozoa in the seminal ducts tended to be slightly reduced. The loss of spermatids and spermatozoa probably occurred because of the inability of these epithelial cells to maintain the germ cells. In the research by Haubruge et al. (2000), reduced sperm counts in guppies (*Poecilia reticulata*) following exposure to TBT is suggested as being caused by damage to normal Sertoli cell function, which facilitates the transport of maturing sperm into the testicular deferent duct. This finding is in agreement with our histological observations of testes of mummichog exposed to TBTO. Exposure to TBTO also markedly depressed spermatogenic cell proliferation in a concentration-dependent manner (Mochida et al. 2007), although the mechanism of action of TBTO inhibition of germ cell proliferation remains unknown.

To date, we have identified six differentially expressed genes in the testes of the TBTO-treated fish. Most of the predicted proteins encoded by the differentially expressed genes were present in spermatids and spermatozoa and were thought to play important roles in the sperm maturation process, including sperm motility (Mochida et al. 2008). That most of the predicted proteins were present in spermatids and spermatozoa is in agreement with the abovementioned fact that epithelial cells and germ cells in the vicinity of the seminal ducts were primarily damaged by TBT toxicity. Among these differentially expressed genes, the role of creatine kinase (CK) and dynein heavy chain (DHC) have been well studied (Tombes and Shapiro 1985; Wallimann et al. 1986). Briefly, CK is an enzyme participating in an energy shuttle that utilizes phosphocreatine to transfer the energy from ATP, generated by mitochondria in the sperm midpiece, to axonemal dynein (Tombes and Shapiro 1985), a motor protein involved in axonemal beating (Gibbons 1981). We also analyzed the expression levels of these genes by real-time polymerase chain reaction (PCR) and confirmed that expression of the CK gene and the DHC gene were reduced by exposure to TBTO. Although the toxic effect of TBTO might not be the cause of endocrine-mediated alteration of spermatogenesis, as this is another toxicological pathway of testicular damage caused by TBT, further research needs to be undertaken.

McAllister and Kime (2003) investigated sperm damage in zebrafish that had long-term exposure to environmental levels of TBT. They exposed fish to TBT for 70 days and examined sperm quality in adults 3–5 months after cessation of exposure. In the group receiving TBT at 10 ng/l, sperm motility was significantly lower than that of the control group and all sperm lacked flagella, and in the group exposed to TBT at 100 ng/l, milt volume increased other than the above-mentioned toxic effects (Table 9.1). In TBTO exposure studies using mummichog, a decrease in the expression level of the CK gene and the DHC gene was observed (Mochida et al. 2008). DHC is one of the components of sperm flagella, and also CK is involved in flagellar beating. It is also intriguing to examine expression levels of zebrafish genes by using the exposure-cessation method as carried out by McAllister and Kime (2003). In adult zebrafish with early life exposure to TBT, the toxic effect to sperm continued to be observed even 3–5 months after TBT exposure (McAllister and Kime 2003). As suggested by the authors, this is possibly due to the bioaccumulation properties of TBT, since it can be highly concentrated in fish tissue (Yamada et al. 1994). In some fish species, TBT bioaccumulates to high levels in the serum or plasma and is bound to a TBT-binding protein (Oshima et al. 1997; Shimasaki et al. 2002). Research on the involvement of this protein in the accumulation of TBT in testes will help to elucidate the mechanism of expression of the toxic effect to zebrafish spermatogenesis. Further studies should be conducted to clarify the effect of short-term exposure to environmentally relevant concentrations of TBT in the early-life stages on fish spermatogenesis, along with the effect of long-term exposure to similar concentrations of TBT.

Rurangwa et al. (2002) analyzed sperm motility of African catfish (*Clarias gariepinus*) and common carp (*Cyprinus carpio*) after direct exposure to TBT. A decrease in the duration and intensity of motility was observed in both catfish and common carp spermatozoa after exposure to TBT for 24 h at 0.27 μg/l and 2.7 μg/l, respectively. Decreased motility of TBT-exposed sperm is probably due to inhibition of energy metabolism, because TBT caused an immediate decrease in ATP content in catfish spermatozoa after exposure. Working with herring (*Clupea harengus*), Grzyb et al. (2003) demonstrated that creatine kinase was released due to sperm cell membrane degradation resulting from direct exposure of sperm to TBT.

9.2.3 Toxic Effects on Oogenesis

Few studies have addressed the toxic effect of TBT on fish oogenesis. Shimasaki et al. (2006) examined the effect of TBT especially on the quality of the eggs of Japanese whiting (*Sillago japonica*) that were fed TBT during the active spawning period and observed a decrease in the number of floating egg and hatching viability (Table 9.1). Zhang et al. (2007) exposed female cuvier (*Sebasticus marmoratus*) to TBT at environmentally relevant concentrations and observed an increase in the ratio of testosterone to estradiol-17β and a delay in oogenesis (Table 9.1). Since estradiol-17β plays an important role in oogenesis, such as the production of egg yolk protein in liver (for reviews see Mommsen and Walsh (1988)), these effects are possibly caused by the inhibition of P450 aromatase activity by TBT. In addition, an increase in the occurrence of apoptotic follicular cells in TBT-exposed female cuvier was also observed (2007). Mochida et al. (2008) also found that the frequency of occurrence of apoptotic germ cells in testes increased in mummichog exposed to TBTO. Further studies are needed to clarify the mechanism that induces apoptosis by TBT in fish germ cells.

9.2.4 Toxic Effects of TBT to Other Reproduction-Related Phenomena

In addition to the toxic effect of TBT on reproduction, TBT toxicity also affects sexual behavior. Sexual displays, such as dancing, decreased in TBT-exposed male Japanese medaka (1 μg/g body weight for 3 weeks) (Nakayama et al. 2004). These treated fish also had a low fertility rate and failed to spawn, which was possibly linked to the low fertilization rate that was observed during the study (Nakayama et al. 2004). Given that preferential accumulation of TBT occurs in the nervous system, including the brain (Rouleau et al. 2003; Triebskorn et al. 1994b), it is likely that the brain involvement in spawning behavior is damaged. Alteration of swimming behavior from TBT exposure was also reported in rainbow trout (*Oncorhynchus mykiss*) (Triebskorn et al. 1994a) and carp (Schmidt et al. 2004, 2005). Impairment of sexual behavior during the spawning season could directly influence the conservation of wild fish populations. The toxicity of TBT and its effects on behavior, including brain activity, is a research area that should be studied more actively and thoroughly.

Several studies have revealed genotoxic effects of TBT on fish (Tiano et al. 2001; Ferraro et al. 2004; Micael et al. 2007). Because genetic damage may cause heritable mutations that could influence the adaptability of wild fish populations in the aquatic environment, further research should be conducted on the physiological changes caused by the genotoxicity of TBT.

9.3 Conclusions

Many studies have shown that the toxic effects of TBT are expressed in various aspects of fish reproduction. Environmentally relevant concentrations of TBT could disturb the sex differentiation process in zebrafish (McAllister and Kime 2003), while in Japanese medaka, no masculinization was induced at 60 days post hatch by nanoinjection of TBT into ova (Hano et al. 2007). Further studies should be carried out focusing on the sensitivity of the sex differentiation process and/or different expression profiles of aromatase to TBT toxicity among fish species. Finally, I should also briefly mention the toxic effect of several TBT metabolites on fish reproduction. To date, only one study that I am aware of has examined the toxicity of TBT metabolites on fish reproduction. As shown in Fig. 9.1, triphenyltin and dibutyltin inhibited enzyme activity involved in steroid metabolism in common carp (Thibaut and Porte 2004). Additional investigations that address the toxic effects of long-term exposure of TBT metabolites on fish reproduction are, therefore, also warranted.

Acknowledgements We are grateful to the members of the Ecotoxicology section of the National Research Institute of Fisheries and Environment of Inland Sea, Fisheries Research Agency for their kind assistance. This work was supported, in part, by a Grant-in-aid (Hazardous Chemicals) from the Ministry of Agriculture, Forestry, and Fisheries of Japan (HC-07-2232-2).

References

Afonso LOB, Iwama GK, Smith J et al. (2000) Effects of the aromatase inhibitor Fadrozole on reproductive steroids and spermiation in male coho salmon (*Oncorhynchus kisutch*) during sexual maturation. Aquaculture 188:175–187

Bowles J, Knight D, Smith C et al. (2006) Retinoid signaling determines germ cell fate in mice. Science 312:596–600

Chang C-F, Hung C-Y, Chiang M-C et al. (1999) The concentrations of plasma sex steroids and gonadal aromatase during controlled sex differentiation in grey mullet, *Mugil cephalus*. Aquaculture 177:37–45

Colborn T, vom Saal FS, Soto AM (1993) Developmental effects of endocrine-disrupting chemicals in wildlife and humans. Environ Health Perspect 101:378–384

Fent K (1996) Ecotoxicology of organotin compounds. Crit Rev Toxicol 26:1–117

Fent K, Stegeman JJ (1993) Effects of tributyltin in vivo and hepatic cytochrome P450 forms in marine fish. Aquat Toxicol 24:219–240

Ferraro MVM, Fenocchio AS, Mantovani MS et al. (2004) Mutagenic effects of tributyltin and inorganic lead (Pb II) on the fish *H. malabaricus* as evaluated using the comet assay and the piscine micronucleus and chromosome aberration tests. Genet Mol Biol 27:103–107

Gibbons IR (1981) Cilia and flagella of eukaryotes. J Cell Biol 91:107s–124s

Grandel H, Lun K, Rauch G-J et al. (2002) Retinoic acid signaling in the zebrafish embryo is necessary during pre-segmentation stages to pattern the anterior-posterior axis of the CNS and to induce a pectoral fin bud. Development 129:2851–2865

Grzyb K, Rhychłowski M, Biegniewska A et al. (2003) Quantitative determination of creatine kinase release from herring (*Clupea harengus*) spermatozoa induced by tributyltin. Comp Biochem Physiol C 134:207–213

Hano T, Oshima Y, Kim SG et al. (2007) Tributyltin causes abnormal development in embryos of medaka, *Oryzias latipes*. Chemosphere 69:927–933

Hashimoto S, Bessho H, Hara A et al. (2000) Elevated serum vitellogenin levels and gonadal abnormalities in wild male flounder (*Pleuronectes yokohame*) from Tokyo bay, Japan. Mar Environ Res 49:37–53

Haubruge E, Petit F, Gage MJG (2000) Reduced sperm counts in guppies (*Poecilia reticulata*) following exposure to low levels of tributyltin and bisphenol A. Proc R Soc Lond B 267:2333–2337

Hayashida Y, Kawamura T, Hori-e R et al. (2004) Retinoic acid and its receptors are required for expression of aryl hydrocarbon receptor mRNA and embryonic development of blood vessel and bone in the medaka fish, *Oryzias latipes*. Zool Sci 21:541–551

Janer G, Porte C (2007) Sex steroids and potential mechanisms of non-genomic endocrine disruption in invertebrates. Ecotoxicology 16:145–160

Janer G, Sternberg RM, LeBlanc GA et al. (2005) Testosterone conjugating activities in invertebrates: are they targets for endocrine disruptors? Aquat Toxicol 71:273–282

Jobling S, Sheahan D, Osborne JA et al. (1996) Inhibition of testicular growth in rainbow trout (*Oncorhynchus mykiss*) exposed to estrogeninc alkylphenolic chemicals. Environ Toxicol Chem 15:194–202

Jones BB, Ohno CK, Allenby G et al. (1995) New retinoid X receptor subtypes in zebra fish (*Danio rerio*) differentially modulate transcription and do not bind 9-cis retinoic acid. Mol Cell Biol 15:5226–5234

Kazeto Y, Place AR, Trant JM (2004) Effects of endocrine disrupting chemicals on the expression of CYP19 genes in zebrafish (*Danio rerio*) juveniles. Aquat Toxicol 69:25–34

Kishida M, Callard GV (2001) Distinct cytochrome P450 aromatase isoforms in zebrafish (*Danio rerio*) brain and ovary are differentially programmed and estrogen regulated during early development. Endocrinology 142:740–750

Kwon JY, Haghpanah V, Kogson-Hurtado LM et al. (2000) Masculinization of genetic female Nile tilapia (*Oreochromis niloticus*) by dietary administration of an aromatase inhibitor during sexual differentiation. J Exp Zool 287:46–53

Kwon JY, McAndrew BJ, Penman DJ (2001) Cloning of brain aromatase gene and expression of brain and ovarian aromatase genes during sexual differentiation in genetic male and female Nile tilapia *Oreochromis niloticus*. Mol Reprod Develop 59:359–370

Larsson DGJ, Hallman H, Forlin L (2000) More male fish embryos near a pulp mill. Environ Toxicol Chem 19:2911–2917

Lyssimachou A, Jenssen BM, Arukwe A (2006) Brain cytochrome P450 aromatase gene isoforms and activity levels in Atlantic salmon after water borne exposure to nominal environmental concentrations of the pharmaceutical ethynylestradiol and antifoulant tributyltin. Toxicol Sci 91:82–92

Martin-Skilton R, Coughtrie MWH, Porte C (2006) Sulfotransferase activities towards xenobiotics and estradiol in two marine fish species (*Mullus barbatus* and *Lepidorhombus boscii*): characterization and inhibition by endocrine disrupters. Aquat Toxicol 79:24–30

Matthiessen P, Gibbs PE (1998) Critical appraisal of the evidence for tributyltin-mediated endocrine disruption in mollusks. Environ Toxicol Chem 17:37–43

McAllister BG, Kime DE (2003) Early life exposure to environmental levels of the aromatase inhibitor tributyltin causes masculinisation and irreversible sperm damage in zebrafish (*Danio rerio*). Aquat Toxicol 65:309–316

Micael J, Reis-Henriques MA, Carvalho AP et al. (2007) Genotoxic effects of binary mixtures of xenoandrogens (tributyltin, triphenyltin) and a xenoestrogen (ethinylestradiol) in a partial life-cycle test with zebrafish (*Danio rerio*). Environ Int 33:1035–1039

Mochida K, Ito K, Kono K et al. (2007) Molecular and histological evaluation of tributyltin toxicity on spermatogenesis in a marine fish, the mummichog (*Fundulus heteroclitus*). Aquat Toxicol 83:73–83

Mochida K, Ito, K, Onduka T et al. (2008) Genes differentially expressed in testis of a marine fish, the mummichog (*Fundulus heteroclitus*) after exposure to tributyltin oxide. Coast Mar Sci 32:57–66

Mommsen TP, Walsh PJ (1988) Vitellogenesis and oocyte assembly. In: Hoar WS, Randall DJ (eds) Fish Physiology, Academic, New York, 11A:347–406

Nakanishi T, Nishikawa J, Tanaka K (2006) Molecular targets of organotin compounds in endocrine disruption: do organotin compounds function as aromatase inhibitors in mammals? Environ Sci 13:89–100

Nakayama K, Oshima Y, Yamaguchi T et al. (2004) Fertilization success and sexual behavior in male medaka, *Oryzias latipes*, exposed to tributyltin. Chemosphere 55:1331–1337

Nishikawa J, Mamiya S, Kanayama T et al. (2004) Involvement of the retinoid X receptor in the development of imposex caused by organotins in gastropods. Environ Sci Technol 38:6271–6276

Oshima Y, Nirmala K, Go J et al. (1997) High accumulation of tributyltin in blood among the tissues of fish and applicability to environmental monitoring. Environ Toxicol Chem 16:1515–1517

Rouleau C, Xiong Z-H, Pacepavicius G et al. (2003) Uptake of waterborne tributyltin in the brain of fish: axonal transport as a proposed mechanism. Environ Sci Technol 37:3298–3302

Rurangwa E, Biegniewska A, Slominska E et al. (2002) Effect of tributyltin on adenylate content and enzyme activities of teleost sperm: a biochemical approach to study the mechanisms of toxicant reduced spermatozoa motility. Comp Biochem Physiol C 131:335–344

Santos MM, Micael J, Carvalho AP et al. (2006) Estrogens counteract the masculinizing effect of tributyltin in zebrafish. Comp Biochem Physiol C 142:151–155

Schmidt K, Steinberg CEW, Pflugmacher S et al. (2004) Xenobiotic substances such as PCB mixtures (Aroclor 1254) and TBT can influence swimming behavior and biotransformation activity (GST) of carp (*Cyprinus carpio*). Environ Toxicol 19:460–470

Schmidt K, Staaks GBO, Pflugmacher S et al. (2005) Impact of PCB mixture (Aloclor 1254) and TBT and a mixture of both on swimming behavior, body growth and enzymatic biotransformation activities (GST) of young carp (*Cyprinus carpio*). Aquat Toxicol 71:49–59

Sharpe RM (2001) Hormones and testis development and the possible adverse effects of environmental chemicals. Toxicol Lett 120:221–232

Shim WJ, Jeon JK, Hong SH et al. (2003) Accumulation of tributyltin in olive flounder, *Paralichthys olivaceus*: its effect on hepatic cytochrome P450. Environ Contam Toxicol 44:390–397

Shimasaki Y, Oshima Y, Yokota Y et al. (2002) Purification and identification of a tributyltin-binding protein from serum of Japanese flounder, *Paralichthys olivaceus*. Environ Toxicol Chem 21:1229–1235

Shimasaki Y, Kitano T, Oshima Y et al. (2003) Tributyltin causes masculinization in fish. Environ Toxicol Chem 22:141–144

Shimasaki Y, Oshima Y, Inoue S et al. (2006) Effect of tributyltin on reproduction in Japanese whiting, *Sillago japonica*. Mar Environ Res 62:S245–S248

Shimizu A, Kimura S (1987) Effect of Bis(tribtyltin) oxide on gonadal development of a salt-water goby, *Chasmichtys dolichognathus*: exposure during maturing period (in Japanese with English abstract). Bull Tokai Reg Fish Res Lab 123:45–49

Simpson ER, Mahendroo MS, Means GD et al. (1994) Aromatase cytochrome P450, the enzyme responsible for estrogen biosynthesis. Endocrin Rev 15:342–354

Thibaut R, Porte C (2004) Effects of endocrine disrupters on sex steroid synthesis and metabolism pathways in fish. J Steroid Biochem Mol Biol 92:485–494

Tiano L, Fedeli D, Moretti M et al. (2001) DNA damage induced by organotins on trout-nucleated erythrocytes. Appl Organomet Chem 15:575–580

Tombes RM, Shapiro BM (1985) Metabolite channeling: a phosphorylcreatine shuttle to mediate high energy phosphate transport between sperm mitochondrion and tail. Cell 41:325–334

Trant JM, Gavasso S, Ackers J et al. (2001) Developmental expression of cytochrome P450 aromatase genes (CYP19a and CYP19b) in zebrafish fry (*Danio rerio*). J Exp Zool 290:475–483

Triebskorn R, Köhler H-R, Flemming J et al. (1994a) Evaluation of bis(tri-n-butyltin)oxide (TBTO) neurotoxicity in rainbow trout (*Oncorhynchus mykiss*). I. Behaviour, weight increase, and tin content. Aquat Toxicol 30:189–197

Triebskorn R, Köhler H-R, Körtje K-H et al. (1994b) Evaluation of bis(tri-n-butyltin)oxide (TBTO) neurotoxicity in rainbow trout (*Oncorhynchus mykiss*). II. Ultrastructural diagnosis and tin localization by energy filtering transmission electron microscopy (EFTEM). Aquat Toxicol 30:199–213

Wallimann T, Moser H, Zurbriggen B et al. (1986) Creatine kinase isoenzymes in spermatozoa. J Muscle Res Cell Motil 7:25–34

Waxman JS, Yelon D (2007) Comparison of the expression patterns of newly identified zebrafish retinoic acid and retinoid X receptors. Develop Dyn 236:587–595

Yamada H, Tateishi M, Yakayanagi K (1994) Bioaccumulation of organotin compounds in the red sea bream (*Pagrus major*) by two uptake pathways: dietary uptake and direct uptake from water. Environ Toxicol Chem 13:1415–1422

Zhang J, Zuo Z, Chen Y et al. (2007) Effect of tributyltin on the development of ovary in female cuvier (*Sebastiscus marmoratus*). Aquat Toxicol 83:174–179

Chapter 10
Biological Effects of Tributyltin on the Caprellidea (Crustacea: Amphipoda)

Madoka Ohji

10.1 Introduction

During the past several decades, butyltin compounds (BTs), one of the representative groups of organotin compounds (OTs), have been widely used as an antifouling agent in paints for boats, ships, and aquaculture nets (Fent 1996, Champ and Seligman 1996), thus these compounds have been found in a variety of marine organisms, often at concentrations exceeding acute or chronic toxicity levels (Bryan and Gibbs 1991; Alzieu 1996). The hazardous effects of antifouling paints containing BTs in marine ecosystem have become a significant environmental issue all over the world (Champ and Wade 1996; Bosselmann 1996). To prevent the destruction of marine ecosystems, BT application to small boats and fish farming equipment has been banned or regulated in developed countries since the late 1980s (Champ and Wade 1996; Bosselmann 1996). Nevertheless, significant accumulation of BTs has been noted at various trophic levels in the marine food chain including plankton, algae, crustaceans, fishes and cetaceans, indicating that BTs impact continues to be felt in marine ecosystems.

Tri-organotins, tributyltin (TBT) are reported to be the most toxic compounds, and at nanogram-per-liter levels, TBT has adverse effects on many aquatic organisms, for example, producing retardation of regenerative growth, delayed molting, reduction in burrowing activity and deformities in limbs in the fiddler crab (Weis and Perlmutter 1987; Weis et al. 1987; Weis and Kim 1988), impairment of egg

M. Ohji
International Coastal Research Center, Ocean Research Institute, The University of Tokyo, 2-106-1 Akahama, Otsuchi, Iwate 028-1102, Japan

T. Arai et al. (eds.), *Ecotoxicology of Antifouling Biocides*,
DOI: 10.1007/978-4-431-85709-9_10, © Springer 2009

production in the calanoid copepod (Johansen and Møhlenberg 1987), reduction in larval growth in the silverside (Hall et al. 1988) and avoidance reactions in the Baltic amphipod (Laughlin et al. 1984). Recently, a relationship between metabolic capacity, accumulation and toxicity of BTs in marine organisms has been reported in terms of comparisons of BT residue levels in organisms at various trophic levels in the food chain (Fent 1996; Takahashi et al. 1999; Ohji et al. 2002a). The results indicate that though BTs accumulated in most organisms at levels up to 70,000 times higher than those in seawater, no significant biomagnification was observed in the higher levels of the food chain (Takahashi et al. 1999). High concentrations have, however, been found in lower trophic animals such as caprellids. It seems that TBT accumulates specifically for the caprellids in the marine ecosystem regardless of the trophic level in the food chain, and it can be a break point for the disturbance in the natural food chain structure. It is considered causing them to accumulate BTs at elevated concentrations because of their lower metabolic capacity to degrade TBT (Ohji et al. 2002a). The BTs seem to be accumulated in a species specific manner. Thus, studying the implications of species-specific accumulation and the biological effects of BTs on caprellids may provide some clues to understanding the accumulation mechanisms in the coastal ecosystem as well as the mode of action of BTs in organisms.

Caprellid amphipods are small crustaceans (1–3 cm in body length), and are distributed worldwide, living especially in algae beds, on buoys, and on aquaculture nets of the subtidal zone in temperate regions (McCain and Steinberg 1970). Caprellids are an important trophic link, as one of the dominant secondary producers between unicellular algae and fish in coastal water ecosystem and form an important prey resources for small fish in coastal water ecosystems (Fuse 1962, Caine 1989; Holbrook and Schmitt 1992). The generation length and life span of *Caprella* have been well investigated (Takeuchi and Hirano 1991). *Caprella danilevskii* has a short generation duration of 25.6 days which includes the incubation time of embryos and maturation time of hatched juveniles, and has a shortened life-span of 1–3 months. Therefore, caprellids may prove to be a convenient and important model for the study of the biological effects of TBT in coastal ecosystems. Recently, use of caprellids to monitor small temporal and spatial changes in baseline concentrations of BTs was proposed – '*Caprella* watch' (Ohji et al. 2002a). However, little information is presently available regarding TBT's biological effects on such characteristics as sex ratio, survival, growth and reproduction. The determination of such effects is a prerequisite to reliable biomonitoring of the state of coastal ecosystem using *C. danilevskii* as a model.

TBT has had marked effects on the development of imposex in female dog whelks during exposure experiments after hatching (Gibbs et al. 1988). Several hypotheses have been proposed concerning the imposex induction mechanisms, such as those involving cytochrome P450–mediated aromatase inhibition, testosterone excretion–inhibition, functional disorder of female cerebropleural ganglia, and involvement of a neuropeptide–APGWamide (Bettin et al. 1996, Ronis and Mason 1996, Féral and Le Gall 1983, Oberdörster and McClellan-Green 2002, 2003).

Recently, it has been reported that TBT and TPT are high affinity ligands for the human retinoid X receptors (hRXRs) and that the natural ligand of RXR significantly caused the development of imposex in female rock shells (Nishikawa et al. 2004). It is reported that intersex individuals in the caprellid were observed in coastal waters (Takeuchi 1990). Therefore, sex disturbance might also occur in response to TBT exposure after hatching in the caprellids. However, the action mechanism of TBT may differ among organisms and the effects of TBT exposure might also differ according to developmental stage. Therefore, in the present study, two periods of TBT exposure were investigated, including post-hatching and during the embryonic stage in order to examine the biological effects of TBT on the caprellid.

This chapter summarizes experiments to examine: the sensitivity to TBT, capacity to metabolize TBT, and biological effects, on the caprellid amphipods. Compared to the caprellids, similar acute toxicity experiments were conducted on the gammarids, which have a similar ecological niche, habitat, body size and life history (Fuse 1962; Myers 1971; Dahl 1977; Imada et al. 1981; Hiwatari and Kajihara 1988; Hong 1988; Sedberry 1988; Takeuchi and Hirano 1991; Holbrook and Schmitt 1992; Horinouchi and Sano 2000). Population-level effects of chemical pollutants are evaluated in terms of decrement of mean extinction time of populations based on LC_{50} values, and estimating extinction risk of populations is utilized for the conservation of wildlife (Tanaka and Nakanishi 2000). Furthermore, the biological effects of TBT exposure at ambient water levels on the caprellid amphipod, *Caprella danilevskii* Czerniavski were examined after hatching and during the embryonic stage. The results form the basis of discussions on the biological impact of TBT on caprellids and the fluctuation of abundance of this species in the coastal ecosystems.

10.2 Acute Toxicity

10.2.1 Materials and Methods

10.2.1.1 Specimens

Five species of caprellids, *Caprella equilibra*, *C. penantis* R-type, *C. verrucosa*, *C. subinermis* and *C. danilevskii*, and three species of gammarids, *Jassa slatteryi*, *Cerapus erae* and *Eohaustorioides* sp. were collected by SCUBA and dredging from Otsuchi Bay, northeastern Japan (Fig. 10.1). Specimens were used for the experiment within 2 h after collection.

In order to clarify the metabolic capacity to degrade TBT, BT accumulation and the proportions of TBT and its derivatives (DBT and MBT) were analyzed in each species collected at the same time and from the same location as the samples used in the acute toxicity tests. Immediately after collection, the samples were placed in polyethylene bags and frozen at -80°C until chemical analysis.

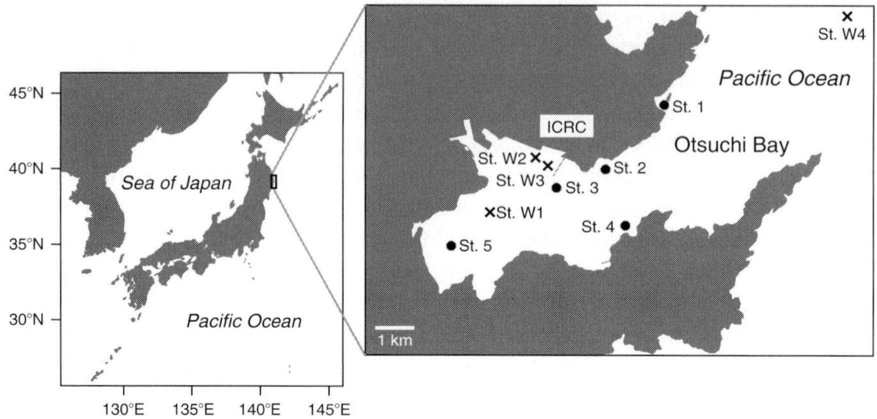

Fig. 10.1 Map showing the sampling locations in Otsuchi Bay, Japan. ● and X indicate the sampling sites of specimens and seawater, respectively. ICRC indicates the location of International Coastal Research Center, Ocean Research Institute, The University of Tokyo

10.2.1.2 Seawater and Tributyltin Solution

The seawater for control and dilution was collected from St. W4 10 m below the surface where TBT was anticipated to be low (Fig. 10.1).

Test solutions of tributyltin chloride (TBTCl) were made as follows. Prior to TBT exposure experiments, the seawater was filtered through a 0.47-μm Millipore filter. A primary solution of 500 μg TBTCl l^{-1} was made by adding 0.5 ml of 2,000 mg TBTCl l^{-1} in acetone solution to 2 l of seawater and was then stirred for 12 h by a magnetic stirrer. A solution of 0.05 ml acetone l^{-1} was used as the control, and dilution was also made by adding 0.1 ml of acetone to 2 l of seawater. After stirring, the bottle was plugged and stored at 4°C. The highest concentration test solution (100 μg TBTCl l^{-1}) was prepared from the primary 500 μg TBTCl l^{-1} solution, which was diluted with filtered seawater. The other five test concentrations (0.001, 0.01, 0.1, 1 and 10 TBTCl μg l^{-1}) were prepared by diluting the 100 μg TBTCl l^{-1} solution with 0.05 ml l^{-1} acetone solution.

To determine the concentrations and proportions of BT residues in the seawater of Otsuchi Bay, seawater samples were collected at a depth of 0.5 m at Sts. W 1–3 with 1 l polycarbonate bottles (Fig. 10.1). The seawater collected was immediately acidified with 1 ml of 12 M HCl and stored at 4°C in the dark until chemical analysis. The seawater for control and dilution collected at a depth of 10 m at St. W4 outside the bay was also analyzed and stored in a 20 l poly tank.

10.2.1.3 Acute Toxicity Experiments

The acute toxicity test was modified from the ecological effect testing method in the risk assessment program of the Organization for Economic Cooperation and Development (OECD) (OECD 1998).

Two deep Petri dishes (9 cm in diameter, 6 cm in height) containing 250 ml of each test solution were prepared 6 h prior to the experiments. Three glass rods (0.1 cm in diameter, 3 cm in length) in each Petri dish were used as substrates. Caprellids and gammarids collected from Otsuchi Bay were immediately brought back to the laboratory, and seven or eight individuals were maintained in each Petri dish at 20°C without food. Survival rates at each test concentration were observed for 48 h. After the experiment, organisms were fixed in 10% formalin. Body lengths were measured from the basal part of antenna I on the head to the posterior end of pereonite VII in caprellids and urosome III in gammarids, respectively.

10.2.1.4 Chemical Analysis

Analysis of BTs was conducted by GC-FPD after derivatization using a Grignard reagent "*n*-propyl magnesium bromide" (Ohji et al. 2002a). This method was slightly modified from previously reported methods (Harino and Fukushima 1992; Iwata et al. 1994; Environment Agency Japan 1990). Briefly, for seawater samples, acidification with HCl and extraction with 0.1% tropolone-benzene was performed. The moisture in the solvent was removed with anhydrous Na_2SO_4. BTs in the extract were then propylated by adding *n*-propyl magnesium bromide as a Grignard reagent. After decomposition of the excess Grignard reagent with 1 M H_2SO_4, the derivatized extract was transferred to 10% benzene-hexane. The extract was then passed through a Florisil packed glass column (eluting with hexane). The final hexane elute from the column was concentrated to 5 ml and subjected to GC quantification. For biological samples, 1–2 g (wet wt) of the whole bodies of crustaceans were homogenized with 0.1% tropolone-acetone and 2 M HCl. The homogenate was centrifuged at 3,000 rpm, and BTs in the supernatant were transferred to 0.1% tropolone-benzene. The other steps were similar to those for seawater.

Sample extracts were analyzed by capillary gas chromatography with flame photometric detection (GC-FPD: Hewlett-Packard 5890 Series II gas chromatograph with a DB-1 capillary column). Tributyltin chloride, dibutyltin dichloride and monobutyltin trichloride of known amounts (0.1 μg each) spiked into uncontaminated seawater and krill containing undetectable levels of BT residues were concurrently run with samples through the whole analytical procedure as external standards for seawater and biological samples, respectively. Procedural blanks were included with every batch of samples to check for interfering compounds. The concentrations refer to TBT, DBT and MBT as corresponding ion. The concentration of TBT^+ corresponds to 0.89 times that of TBTCl.

10.2.2 Results

10.2.2.1 Tributyltin Concentration in the Test Seawater Solution

The average TBT concentration which was produced for 100 μg TBTCl l^{-1} before the experiments was 104 ± 8.7 μg TBTCl l^{-1} (Mean ± SD) (n = 7). This confirmed the accuracy of the test concentration in the medium. In addition, four other test concentrations (0.1, 1, 10 and 100 μg TBTCl l^{-1}) were also analyzed after the experiments, and average TBT concentrations were 0.079 ± 0.01, 0.90 ± 0.10, 8.6 ± 0.78 and 95 ± 4.9 μg TBTCl l^{-1} (n = 2), respectively. This confirmed that the concentrations remained approximately the same even after 48 h. Therefore, the possibility of TBT adsorption on the surface of the glass containers and any evaporation during preparation of solutions could be eliminated.

10.2.2.2 Acute Toxicity in Amphipods

The 48-h LC_{50} values determined for five species of caprellids ranged from 1.2 μg TBTCl l^{-1} in *Caprella penantis* R-type to 6.6 μg TBTCl l^{-1} in *C. equilibra* (n = 5) (Fig. 10.2). These values for caprellids were significantly lower than those for the three species of gammarids which had LC_{50} values ranging from 17.8 μg TBTCl l^{-1} in *Jassa slatteryi* to 23.1 μg TBTCl l^{-1} in *Eohaustorioides* sp. (n = 3) (Mann-Whitney *U*-test, $p < 0.05$). The body lengths of the five species of caprellids were 4.9–8.0 mm, while those of the three species of gammarids were 3.2–6.2 mm.

10.2.2.3 Residue Profile of Butyltins in Seawater

Butyltins were detected in seawater collected from Sts. W1 to 3 (Table 10.1). At St. W3, TBT was the predominant compound at a concentration of 19 ng l^{-1}, accounting for 63.1% of the total BTs (ΣBTs = TBT + DBT + MBT), followed by MBT, 5.8 ng l^{-1}

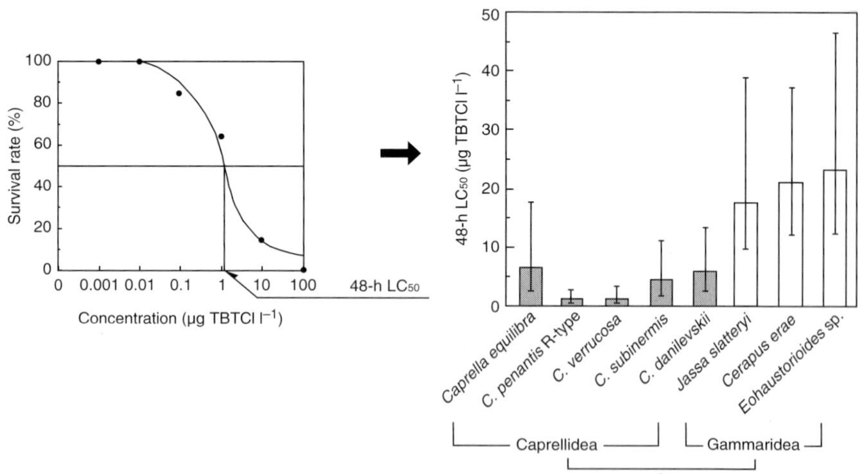

Fig. 10.2 Comparison of 48h-LC_{50} values for TBTCl in caprellid and gammarid amphipods (Crustacea). Bars indicate 95% confidence intervals. The left figure shows the dose-response curve for 48h-LC_{50} of *Caprella verrucosa* as a representative. Mann-Whitney *U*-test, $^{*}p < 0.05$

Table 10.1 Butyltin concentrations of the seawater (ng l^{-1}) and caprellid and gammarid amphipods (Crustacea) (ng g^{-1} wet wt) collected from Otsuchi Bay, Japan

Medium	TBT	DBT	MBT	ΣBTs
Seawater				
St. W1	<2.0	<3.0	6.2	6.2
St. W2	<2.0	<3.0	<5.0	–
St. W3	19	5.3	5.8	24
St. W4	<2.0	<3.0	<5.0	–
Caprellidea				
Caprella equilibra	55	8.6	7.4	64
Caprella penantis R-type	38	<1.0	9.9	38
Caprella verrucosa	81	12	11	93
Caprella subinermis	29	7.2	10	36
Caprella danilevskii	29	<1.0	16	29
Gammaridea				
Jassa slatteryi	6.8	4.9	14	12
Cerapus erae	9	11	29	20
Eohaustorioides sp.	30	24	46	54

ΣBTs = TBT + DBT + MBT

(19.3%) and DBT, 5.3 ng l^{-1} (17.6%). Concentrations of TBT, DBT and MBT in seawater from St. W4 were below the detectable levels.

10.2.2.4 Accumulation Profile of Butyltins in Amphipods

Concentrations of ΣBTs in caprellids collected from Otsuchi Bay were 45–105 ng g^{-1} wet wt (n = 5), which were comparable to those in gammarids (26–100 ng g^{-1} wet wt) (n = 3) (Table 10.1). In caprellids, TBT was the predominant compound and accounted for 72% of the ΣBT concentrations (n = 5) (Fig. 10.3). In contrast, in gammarids, TBT was less than 25% and the breakdown products, DBT and MBT, were the predominant compounds contributing to 75% of the ΣBTs (n = 3).

10.3 Chronic Toxicity – Exposure After Hatching

10.3.1 Materials and Methods

10.3.1.1 Specimens

Caprella danilevskii was collected by SCUBA from the rocky shore in Ostuchi Bay, northeastern Japan, and the specimens were brought back to the laboratory and kept in an aquarium provided with running seawater. Premature females and mature

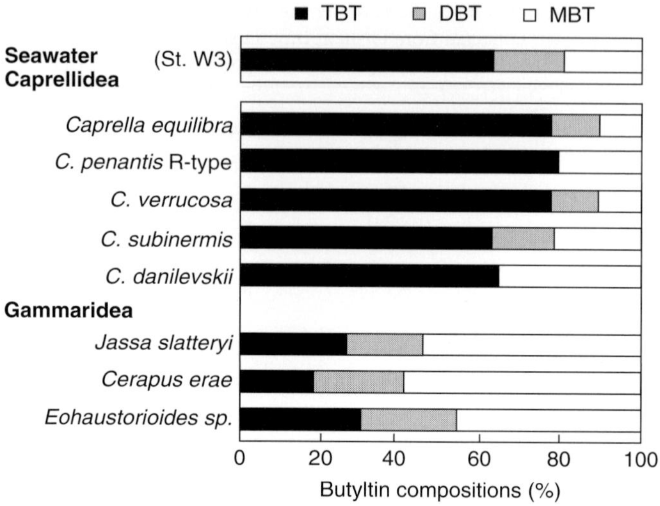

Fig. 10.3 Butyltin speciation in the seawater of St. W3 and the whole body of caprellid and gammarid amphipods (Crustacea) collected from Otsuchi Bay, Japan

males were sorted and provided for the experiments (Fig. 10.4). These specimens were kept in deep Petri dishes (6 cm in diameter, 6 cm in height) which contained filtered seawater with a Teflon mesh piece (2 × 2 cm) as substrate, and maintained at 20°C under a 12:12 h light: dark photoperiod. One ovigerous mature female was allocated per dish, and a total of four females were prepared for an exposure experiment (20 females in five concentration-exposure experiments). Diatom colonies *Chaetoceros calcitrans* (Paulsen) Takano were added once a day to each Petri dish; this amount was more than sufficient to meet the daily dietary demands of the caprellids. The seawater in each dish was changed every day, and Petri dishes and Teflon mesh pieces were replaced every 2 days. After the confirmation that premature females had reached the mature stage, mature females were allowed to copulate with males, thus stimulating the release of eggs in the brood pouch. After releasing eggs into the brood pouch, mature males were transferred to other Petri dishes, and ovigerous mature females were held within the filtered seawater. The condition of mature females such as hatching and the emergence of juveniles was observed under a binocular microscope at 12 h intervals each day.

10.3.1.2 Seawater and Tributyltin Solution

The seawater used for these experiments was collected at 10 m depth outside Otsuchi Bay, where TBT concentrations at 0.5 and 10 m deep were confirmed to be less than the detection limit (Ohji et al. 2002a), and stored in a 20 l polyethylene tank. The tributyltin-seawater solution and the control seawater that contained only

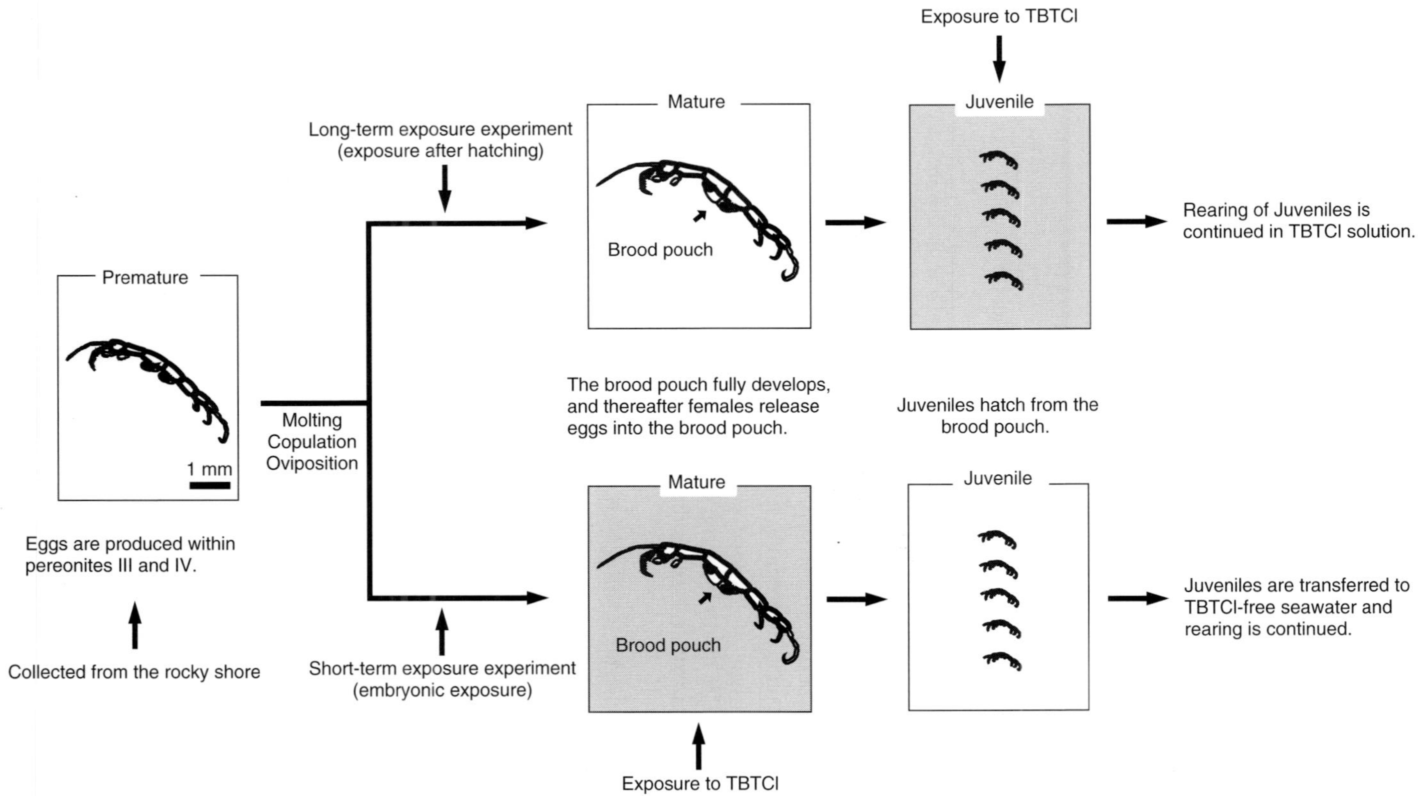

Fig. 10.4 Schematic view of the experimental methodology used to investigate the chronic toxicity of TBTCl

acetone were made in the following procedure. Prior to the TBT exposure experiments, the seawater was filtered through a 0.47 μm Millipore filter. A solution of 10,000 ng TBTCl l^{-1} was made by adding 5 μl of 2,000 mg TBTCl l^{-1} acetone solution to 1 l of seawater, and thereafter the solution was stirred for 12 h. Control and diluent solutions were prepared using 5 μl acetone l^{-1} seawater. In the present study, five test concentrations of TBTCl (0, 10, 100, 1,000 and 10,000 ng l^{-1}) were prepared by dilution of the stock solution. These solutions were freshly prepared each week. The five test concentrations of TBTCl were measured to confirm the accuracy of TBTCl present in the test solutions during the experiment in the previous report (Ohji et al. 2002a). The concentrations remained the same between pre- and post-experiments.

10.3.1.3 Exposure Experiments After Hatching

After hatching from the brood pouch, juveniles were transferred into Petri dishes containing the TBTCl solution at each concentration with Teflon mesh pieces set as a substrate, and were continued to rear. Two juveniles were allocated per dish, and a total of 20–27 specimens were used for the exposure experiment (122 juveniles in five exposure experiments). The juveniles that emerged from the brood pouch were classified as instar I. Their body lengths were measured from the basal part of the antenna I on the head to the posterior end of pereonite VII. The sex was determined from instar II. The maturity of the females was divided into three stages: immature, premature and mature based on the morphology of the oostegites on pereonites III and IV.

Mature females were allowed to copulate with mature males collected from the field and to release eggs in the brood pouch. The number of eggs in the brood pouch was counted at the same time each day. In the present study, oogenesis in the premature stage, and embryo development and new oogenesis in the mature stage were distinguishable under the binocular microscope. Males and females that survived over 50 days were fixed with10% formalin, as were the animals that died during the experiment period. The sex of the hatched juveniles was determined from the presence of oostegites in females and the development of gnathopod II and the presence of abdominal appendages in males.

10.3.2 Results

10.3.2.1 Sex Ratio

Sex ratio of male to female was 55.0% and 45.0%, respectively in the control (0 ng TBTCl l^{-1}) (Fig. 10.5). The ratio was almost constant in spite of increasing of the TBTCl concentrations ranging from 50.0% (1,000 ng TBTCl l^{-1}) to 55.6% (100 ng TBTCl l^{-1}) for males and ranging from 44.4% (100 ng TBTCl l^{-1}) to

50.0% (1,000 ng TBTCl l^{-1}) for females, although all specimens died in the 10,000 ng TBTCl l^{-1} experiment because of acute toxic concentration for this species (Ohji et al. 2002a). No significant differences were found in the sex ratio between the control and other three concentrations of TBTCl (chi-squared test, $p > 0.5$).

10.3.2.2 Survival

As the TBTCl concentrations were increased, survival rates within 50 days after hatching decreased, 25.0% in 10 ng l^{-1}, 11.1% in 100 ng l^{-1} and 8.3% in 1,000 ng l^{-1} (Fig. 10.6). All specimens died in 10,000 ng TBTCl l^{-1} within 4 days after hatching, while all control specimens survived (100%). Significant differences were found in the survival rate between the control and the other four concentrations of TBTCl (log-rank test, $p < 0.0001$).

10.3.2.3 Growth

In each concentration of TBTCl except for 10,000 ng l^{-1}, body length increased as the organism became older (Fig. 10.7). However, significant differences were seen in body length between the control and 100 ng TBTCl l^{-1} and between the control and 1,000 ng TBTCl l^{-1} in each instar after instar II of either males or females (Mann-Whitney *U*-test, $p < 0.05$). No significant difference was found in the body length between the control and 10 ng TBTCl l^{-1} in each instar of either males or females (Mann-Whitney *U*-test, $p > 0.05$). This indicates that a decrease in growth rate results after exposure to 100 and 1,000 ng TBTCl l^{-1} in spite of the organism's sex.

The number of days required between hatching to the instar X, which corresponds to the experimental period in the control, 10 and 100 ng TBTCl l^{-1}, was approximately

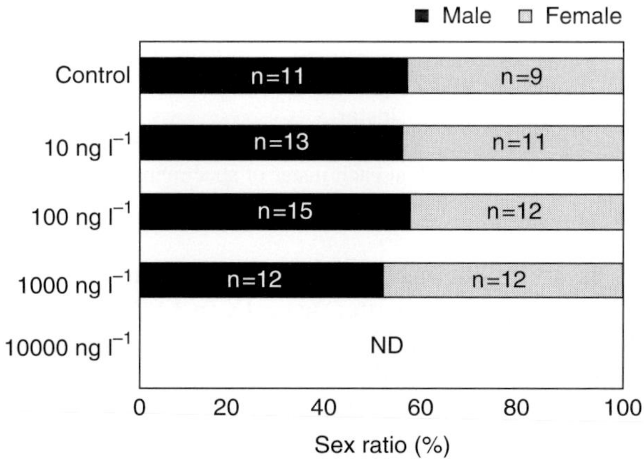

Fig. 10.5 Sex ratio in juveniles exposed to TBTCl after hatching. ND indicates no data because of the death of all specimens

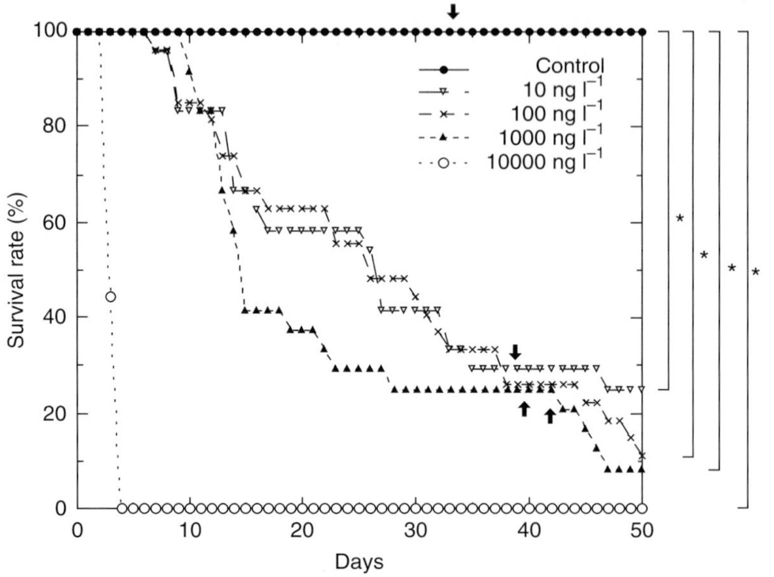

Fig. 10.6 Survival rate of specimens exposed to TBTCl after hatching. The number of hatched juveniles was calculated as 100%. Arrows indicate the days required from hatching to maturation in females. Log-rank test, *p < 0.0001

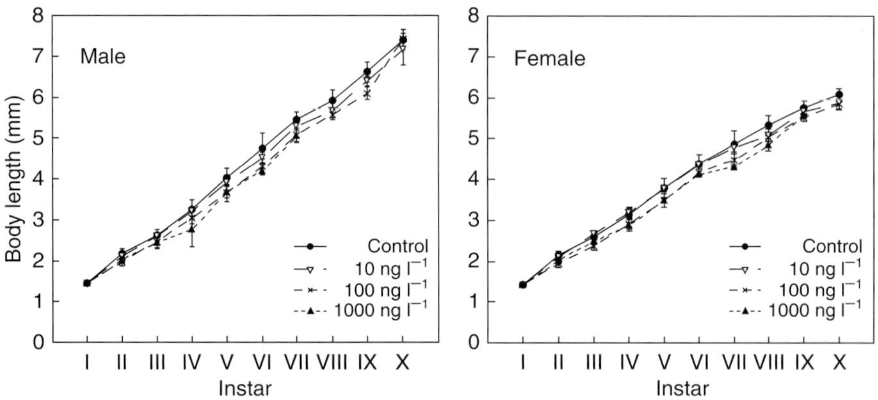

Fig. 10.7 Body length at each instar of specimens exposed to TBTCl after hatching

50 days for both males and females. Although all male specimens died after instar VII and females died after instar IX in 1,000 ng TBTCl l^{-1} (Fig. 10.8), a significant difference was found in the day required from hatching to each instar between the control and 1,000 ng TBTCl l^{-1} in both males and females (Mann-Whitney *U*-test, $p < 0.05$). However, no significant differences were found between other combinations (Mann-Whitney *U*-test, $p > 0.05$).

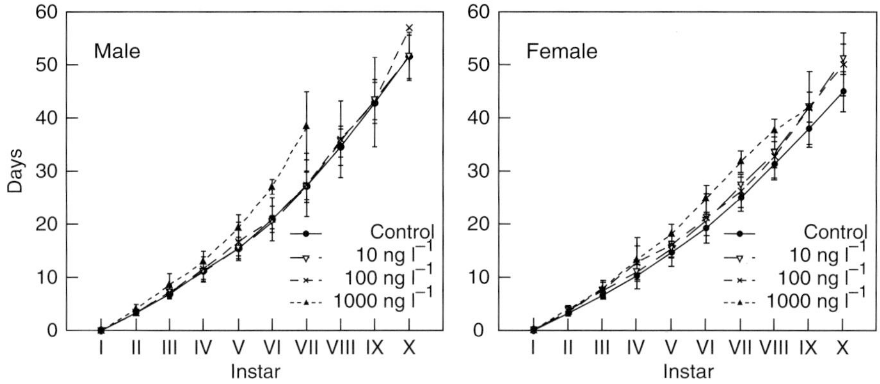

Fig. 10.8 Days required from hatching to each instar of specimens exposed to TBTCl after hatching

Table 10.2 Instar and the days required from hatching to maturation of juveniles exposed to TBTCl after hatching

Concentration (ng TBTCl l^{-1})	Instar	Day
Control	VIII ± 0.5	33 ± 1.6
10	VIII ± 1.0	39 ± 1.4
100	VIII ± 0.6	39 ± 2.3
1,000	IX	42
10,000	ND	ND

Numerical data and ND indicate mean and standard deviation and no data because of death of all specimens, respectively

10.3.2.4 Maturation and Reproduction

The instar and day required from hatching to maturity in the female caprellid ranged from VIII to IX and from 33 to 42 days, respectively (Table 10.2). Significant differences were seen in the day required from hatching to maturity between the control and 10 ng TBTCl l^{-1} and between the control and 100 ng TBTCl l^{-1} (Mann-Whitney *U*-test, $p < 0.05$), while no significant differences were seen in the instar required from hatching to maturity for all other combinations (Mann-Whitney *U*-test, $p > 0.05$). Though all specimens were observed to mature completely during instar VIII and instar IX in the control, several specimens died at premature and immature stages during those instars and instar X in other TBTCl concentrations. This suggests that a delay in the day required from hatching to maturity is caused by exposure in TBTCl.

After maturation, the number of eggs in the brood pouch, the number of juveniles hatched and the period from spawning to juvenile hatching ranged from 2.0 to 2.7, from 0.3 to 2.7 and from 5.0 to 6.0 days in the increasing TBTCl concentration (Table 10.3). A significant difference was found in the number of juveniles hatched between control and 100 ng TBTCl l^{-1} (Mann-Whitney *U*-test, $p < 0.05$).

Table 10.3 Reproductive conditions of mature females exposed to TBTCl after hatching

Concentration (ng TBTCl l^{-1})	Number of embryos spawned	Number of juveniles hatched	Incubation period of embryos	Duration of instar
Control	2.7 ± 1.7	2.7 ± 1.7	5.0 ± 0.0	7.0 ± 0.9
10	2.0 ± 2.8	0.5 ± 0.7	6.0	7.5 ± 2.1
100	2.7 ± 3.1	0.3 ± 0.6	6.0	9.7 ± 2.1
1,000	ND	ND	ND	ND
10,000	ND	ND	ND	ND

Numerical data and ND indicate mean and standard deviation and no data because of death of all specimens, respectively

Molting interval after maturation ranged from 7.0 to 9.7 days in each TBTCl concentration. A significant difference was seen between the control and 100 ng TBTCl l^{-1} (Mann-Whitney *U*-test, $p < 0.05$). This indicates that a delay in the molting interval is caused by exposure to TBTCl.

The decrease in the number of eggs because of eggs dropping from the brood pouch (brood loss) was found in one of two females that reached the mature stage, while the other did not succeed in egg formation in 10 ng TBTCl l^{-1}. In 100 ng TBTCl l^{-1}, brood loss was found in two of three females that reached the mature stage and the other one did not succeed in egg formation. The percentages of hatching success after spawning were decreased as the TBTCl concentration increased except for 1,000 and 10,000 ng l^{-1}, i.e. 100% in the control, 25.0% in 10 ng l^{-1} and 12.5% in 100 ng l^{-1}. In 1,000 and 10,000 ng l^{-1}, all specimens died before or after reaching maturation.

10.3.2.5 Morphological Alterations

During the experiment period, morphological alterations such as loss of gill, contraction of gill, necrosis and loss of pereopod, cramp of pereopod (mis-clinging to the substrate with pereopod), molting disorder, and substances attached on the surface of the body were observed, and their occurrence increased as the TBTCl concentration increased (Fig. 10.9).

10.4 Chronic Toxicity – Embryonic Exposure

10.4.1 Materials and Methods

10.4.1.1 Specimens

Caprella danilevskii was collected by SCUBA from the rocky shore in Uchiura Bay, Japan, after which specimens were immediately brought to the laboratory and kept in an aquarium provided with running seawater. Premature females and mature males were sorted and provided for the experiments (Fig. 10.4).

10.4.1.2 Seawater and Tributyltin Solution

The seawater used for the present experiments was collected from a depth of 10 m outside Otsuchi Bay.

A tributyltin-seawater solution and the control seawater were made according to our previously described method in the section of TBT exposure after hatching. In the present study, five test concentrations of TBTCl (0, 10, 100, 1,000 and 10,000 ng l^{-1}) were prepared using dilute solution. These solutions were made every week. The five test concentrations of TBTCl were measured to confirm the accuracy of TBTCl present in those test solutions during the experiment in the previous report (Ohji et al. 2002a). The concentrations remained the same between pre- and post-experiments.

10.4.1.3 Embryonic Exposure Experiments

After confirmation that premature females had reached the mature stage, these parental females were allowed to copulate with males, and spawning was stimulated

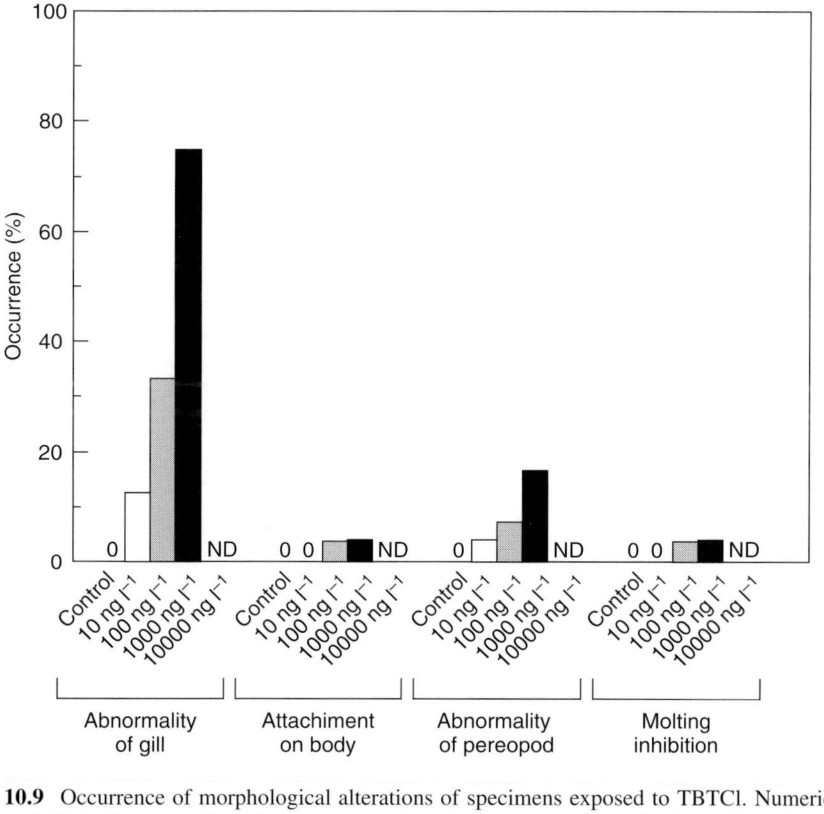

Fig. 10.9 Occurrence of morphological alterations of specimens exposed to TBTCl. Numerical data and ND indicate number of specimens and no data because of the death of all specimens, respectively

(first mature stage in parent) (Fig. 10.4). After spawning in the brood pouch, ovigerous mature females were transferred to Petri dishes (6 cm in diameter, 6 cm in height) containing each concentration of TBTCl, respectively, with a Teflon mesh piece (2 × 2 cm) as a substrate; specimens were then maintained at 20°C and a 12:12 h light: dark photoperiod. One ovigerous mature female was allocated per dish, and a total of 11 females were used for the exposure experiment (55 females in five exposure experiments). Colonies of diatom *Chaetoceros calcitrans* (Paulsen) Takano were added to each Petri dish once a day; this amount was sufficient to supply the daily dietary demands of the caprellids. The seawater in each dish was changed every day, and Petri dishes and Teflon mesh pieces were replaced every 2 days. The condition of ovigerous parental females and egg number in the brood pouch were observed each day at the same time under a binocular microscope.

Specimens were exposed to five concentrations (0, 10, 100, 1,000 and 10,000 ng l^{-1}) of TBTCl for 5 days, which corresponded to the period of embryonic development. After being released from the brood pouch, the juveniles were transferred into the filtered seawater containing neither TBTCl nor acetone. Two juveniles were allocated per dish, and a total of 11–25 specimens were used for the exposure experiment (68 juveniles in five exposure experiments). The juveniles released from the brood pouch were classified as instar I. At each instar, the body length of every juvenile was measured. The sex was determined from instar II.

Furthermore, parental females were also transferred to the filtered seawater. After molting, these females recopulated with a mature male that was collected from the field. After spawning (second mature stage in parent), the eggs were counted at each concentration of TBTCl to examine the effects of TBTCl on the oogenesis stage.

After reaching maturity, female juveniles exposed to TBTCl during the embryonic period were allowed to copulate with mature males collected from the field, and spawning was stimulated (first generation of offspring). The eggs in the brood pouch were counted at the same time each day. After juveniles were released from the brood pouch, these juveniles were continued to rear until instar II, and the sex was determined under the light microscope (second generation of offspring). Males and females that survived over 50 days were fixed with 10% formalin. The animals that died during the experiment period were also fixed with 10% formalin.

10.4.2 Results

10.4.2.1 Condition of Parental Females

Eleven ovigerous females were allocated to each concentration compartment of TBTCl (0, 10, 100, 1,000, and 10,000 ng l^{-1}). The number of eggs per female ranged from 2.3 ± 1.7 (mean ± SD) to 3.5 ± 2.2 in the brood pouch (Table 10.4). No significant differences were found in the number of eggs spawned between the control and the other four concentrations of TBTCl (Mann-Whitney *U*-test, $p>0.1$). A number of deaths of ovigerous females exposed for 5 days were observed at more

Table 10.4 Reproductive conditions of mature female exposed to TBTCl during the 5 days which corresponds to the first mature stage

Concentration (ng TBTCl l^{-1})	Number of embryos spawned	Number of juveniles hatched
First spawning		
Control	2.3 ± 1.7	2.3 ± 1.7
10	2.4 ± 1.3	1.6 ± 1.6
100	3.5 ± 2.2	1.3 ± 1.9
1,000	2.9 ± 2.3	1.0 ± 1.3
10,000	2.7 ± 1.4	0.0
Second spawning		
Control	3.1 ± 1.8	–
10	1.4 ± 1.4	–
100	1.3 ± 1.9	–
1,000	1.0 ± 1.0	–
10,000	ND	–

Numerical data, ND and bar indicate mean and standard deviation, no data because of death of all specimens, and no observation, respectively

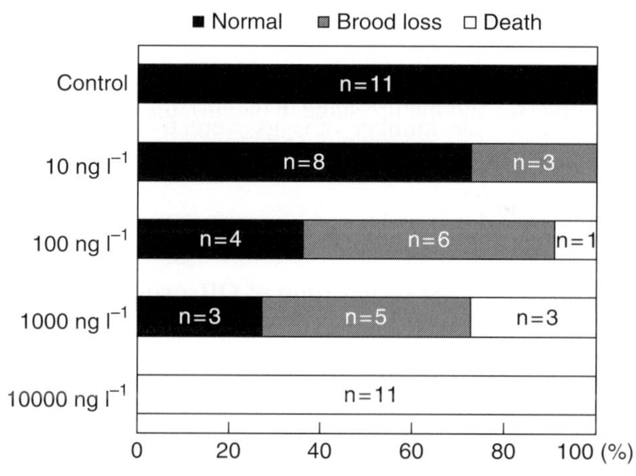

Fig. 10.10 Condition of the parental female in the first mature stage after 5-day exposure to TBTCl

than 100 ng TBTCl l^{-1} (Fig. 10.10) and all specimens died at 10,000 ng TBTCl l^{-1} due to the acute toxic concentration for the species (Ohji et al. 2002a). Brood loss of the females also occurred at concentrations higher than 10 ng TBTCl l^{-1}, ranging from three to six specimens, while no brood loss was observed in the control (0 ng TBTCl l^{-1}) (Fig. 10.10).

The number of eggs per female spawned in the brood pouch in the second mature stage ranged from 1.0 ± 1.0 to 3.1 ± 1.8 (Table 10.4). Significant differences were

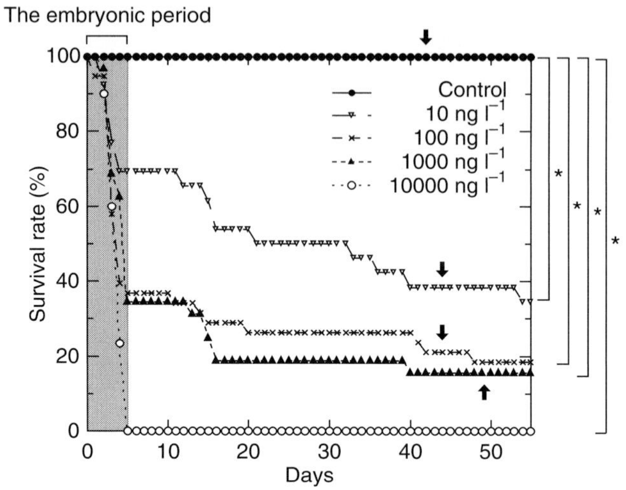

Fig. 10.11 Changes in the survival rate during spawning and sacrifice in offspring exposed to TBTCl during the embryonic stage and thereafter reared in seawater with no TBTCl added. Arrows indicate the days required from hatching to maturation in females. Log-rank test, $^*p < 0.0001$

found in the number of eggs between the control and three concentrations (10, 100 and 1,000 ng l^{-1}) of TBTCl (Mann-Whitney *U*-test, $p<0.05$–0.01). Furthermore, significant differences in the number of eggs were found between the first and second mature stages at 100 and 1,000 ng TBTCl l^{-1} (Wilcoxon's signed-rank test, $p<0.05$) (Table 10.4).

10.4.2.2 Survival in the First Generation of Offspring

The embryo survival rate (estimated from the amount of brood loss, the number of eggs in the brood pouch in dead specimens, and the total number of eggs) during the TBTCl exposure period decreased as the TBTCl concentrations increased, i.e. 69.2% at 10 ng l^{-1}, 36.8% at 100 ng l^{-1}, 34.4% at 1,000 ng l^{-1} and 0% at 10,000 ng l^{-1} (Fig. 10.11). Significant differences were found in the embryo survival rates between the control and the other four concentrations (log-rank test, $p < 0.05$–0.0001).

The number of juveniles hatched per female was 2.3 ± 1.7 in the control. However, it decreased as the TBTCl concentrations increased, ranged from 1.6 ± 1.6 at 10 ng l^{-1} to 0 at10,000 ng l^{-1}. Significant differences were found between control and 1,000 ng TBTCl l^{-1} and between the control and 10,000 ng TBTCl l^{-1} (Mann-Whitney *U*-test, $p < 0.05$–0.0001). Furthermore, significant differences were found between the number of eggs spawned in the brood pouch and the number of juveniles hatched at 100, 1,000 and 10,000 ng TBTCl l^{-1} (Wilcoxon's signed-rank test, $p < 0.05$–0.01) (Table 10.4).

At all concentrations, the survival rate in offspring continued to decrease despite the movement of hatched juveniles into seawater that did not contain both TBTCl

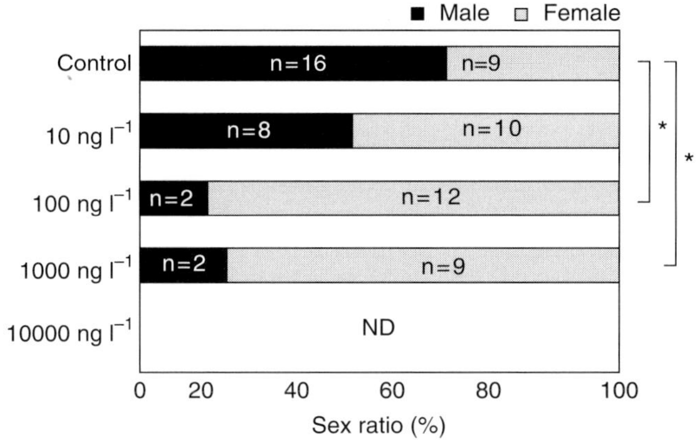

Fig. 10.12 Sex ratio in offspring of the first generation exposed to TBTCl during the embryonic stage. Chi-squared test, $^{*}p < 0.05$. ND indicates no data because of the death of all specimens

and acetone (Fig. 10.11). Significant differences were found in the survival rate between the control and the other four concentrations (log-rank test, $p < 0.0001$). The survival rate of females at maturity decreased to 38.5% at 10 ng TBTCl l^{-1}, 21.1% at 100 ng TBTCl l^{-1}, 15.6% at 1,000 ng TBTCl l^{-1} and 0% at 10,000 ng TBTCl l^{-1}, although the survival rate in the control was 100%. The drastic change in survival rate was observed twice, at 10–15 days and during 35–45 days after spawning.

10.4.2.3 Sex Ratio in the First Generation of Offspring

The female proportions were 36% in the control (Fig. 10.12), corresponding to previous field observations (Takeuchi and Hirano 1991). However, as the TBTCl concentrations increased, the proportion of females increased, i.e. 55.6% at 10 ng l^{-1}, 85.7% at 100 ng l^{-1} and 81.8% at 1,000 ng l^{-1}. Significant differences occurred in the sex proportion between the control and 100 ng TBTCl l^{-1} and between the control and 1,000 ng TBTCl l^{-1} (chi-squared test, $p < 0.01$).

10.4.2.4 Growth, Maturation and Reproduction in the First Generation of Offspring

In the present study, no significant differences were found in the body length in each instar and in the time taken for each instar from hatching between the control and each concentration of TBTCl in either males or females (Mann-Whitney *U*-test $p > 0.05$). These results suggest that no growth or molting inhibition occurs after hatching in response to exposure to TBTCl in the embryonic period.

The instar and day required from hatching to maturity in the female caprellid ranged from VIII to IX and from 37 to 45 days, respectively (Table 10.5). Significant

Table 10.5 First instar and the day required from hatching to maturation of juvenile exposed to TBTCl during the embryonic period

Concentration (ng TBTCl l^{-1})	Instar	Day
Control	VIII ± 0.4	37 ± 2.6
10	IX ± 0.5	39 ± 4.3
100	IX ± 0.0	39 ± 2.5
1,000	IX ± 0.8	45 ± 12.1
10,000	ND	ND

Numerical data and ND indicate mean and standard deviation and no data because of death of all specimens, respectively

differences were seen in the instar required from hatching to maturity between the control and 10 ng TBTCl l^{-1}, between the control and 100 ng TBTCl l^{-1} and between the control and 1,000 ng TBTCl l^{-1} (Mann-Whitney *U*-test, $p < 0.05–0.01$), while no significant differences were seen in the day required from hatching to maturity for all other combinations (Mann-Whitney *U*-test, $p > 0.05$).

In the first mature stage in offspring, oogenesis inhibition and brood loss were observed at 100 and 1,000 ng TBTCl l^{-1}. Three of six mature females exhibited apparent oogenesis inhibition at 100 ng TBTCl l^{-1} and three of five at 1,000 ng TBTCl l^{-1}. Brood loss was apparent in one of six mature females at 100 ng TBTCl l^{-1} and in two of five at 1,000 ng TBTCl l^{-1}. These abnormal ratios during the mature stage increased as the TBTCl concentrations increased, i.e. 0% at the control and at 10 ng TBTCl l^{-1}, 66.7% at 100 ng TBTCl l^{-1} and 100% at 1,000 ng TBTCl l^{-1}.

10.4.2.5 Sex Ratio in the Second Generation of Offspring

The proportion of females in the control and at 10, 100 and 1,000 ng TBTCl l^{-1} were 28.6%, 28.6%, 22.2% and 33.3%, respectively. No significant differences in the sex proportion between control and other concentrations of TBTCl were observed (chi-squared test, $p > 0.5$). These results suggest that TBTCl exposure in the embryonic period does not affect the sex proportion in the second generation.

10.5 Discussion

10.5.1 Differences in Sensitivity and Metabolic Capacity to Degrade Tributyltin Between Caprellid and Gammarid Amphipod

In the present study, the 48-h LC_{50} values in caprellids and gammarids, which belong to the same order, Amphipoda Crustacea, were compared in order to elucidate the acute toxicity of TBT. The 48-h LC_{50} values in caprellids, 1.2–6.6 µg

Table 10.6 Review of the 48-h LC_{50} for TBT in various marine organisms

Organism	Compound	Concentrations ($\mu g\ l^{-1}$)	Temperature (°C)	Reference
Bacillariophyceae				
Skeletonema costatum	TBTO	15.6	ND	Walsh et al. (1985)
Mollusca				
Crassostrea gigas (adults)	TBTO	1874	ND	Thain (1983)
Crassostrea gigas (larvae)	TBTO	1.6	ND	Thain (1983)
Ostrea edulis	TBTO	>312	ND	Thain (1983)
Mytilus edulis (adults)	TBTO	312	ND	Thain (1983)
Mytilus edulis (larvae)	TBTO	2.5	ND	Thain (1983)
Copepoda				
Acartia tonsa	TBTO	1.2	20	Bushong et al. (1987)
Eurytemora affinis	TBT	2.5	20	Hall et al. (1988)
Amphipoda: Caprellidea				
Caprella equilibra	TBTCl	6.6	20	Ohji et al. (2002a)
Caprella penantis R-type	TBTCl	1.2	20	Ohji et al. (2002a)
Caprella verrucosa	TBTCl	1.3	20	Ohji et al. (2002a)
Caprella subinermis	TBTCl	4.6	20	Ohji et al. (2002a)
Caprella danilevskii	TBTCl	5.9	20	Ohji et al. (2002a)
Amphipoda: Gammaridea				
Jassa slatteryi	TBTCl	17.8	20	Ohji et al. (2002a)
Cerapus erae	TBTCl	21.2	20	Ohji et al. (2002a)
Eohaustorioides sp.	TBTCl	23.1	20	Ohji et al. (2002a)
Decapoda				
Crangon crangon (adults)	TBTO	7.4	ND	Thain (1983)
Crangon crangon (larvae)	TBTO	6.9	ND	Thain (1983)
Fish				
Agonus cataphractus	TBTO	27.1	ND	Thain (1983)
Oncorhyunchus mykiss	TBT	23.0	20	Alabaster (1969)
Solea solea (adults)	TBTO	91.5	ND	Thain (1983)
Solea solea (larvae)	TBTO	8.8	ND	Thain (1983)

The concentrations were converted into TBTCl
ND indicates no data

TBTCl l^{-1}, were significantly lower than those in gammarids, 17.8–23.1 μg TBTCl l^{-1}. Moreover, in the comparison of the 48-h LC_{50} values for TBT among various trophic level organisms (Table 10.6), caprellids belong to a sensitive group of organisms. Hayakawa (1976) tested the acute toxicity of the antifouling paint for steel ship's bottoms, which contained TBTO as a dominant component, and reported that *Caprella penantis* was more sensitive than fish, *Atherion elymus* and shrimp, *Leander serrifer*. These facts indicate that caprellids have low resistance to the acute toxicity of TBT. The ecological risk assessment evaluated in terms of LC_{50} values may present a possibility for interpreting the ecological risk of chemical pollutants in the context of population vulnerability (Tanaka and Nakanishi 2000). The concentration at which the intrinsic rate of natural increase corresponds to zero has a highly significant relationship to that of LC_{50} values (Tanaka and Nakanishi 1998). The extinction of a keystone species such as caprellid occupying an influential ecological niche in the food web may induce instability in the coastal ecosystem.

Results for the chemical analysis of field-collected crustacean and seawater samples showed that TBT predominantly accumulated in caprellids and that the proportions of BTs in these organisms were similar to those found in seawater from St. W3 (Fig. 10.3). In contrast to caprellids, TBT's breakdown products, DBT and MBT, were predominant in gammarids (Fig. 10.3). Thus, as with the above acute toxicity, there was a difference in the proportion of TBT among caprellids and gammarids, nevertheless both these groups of amphipods belong to similar trophic levels (Imada et al. 1981; Sedberry 1988; Holbrook and Schmitt 1992; Horinouchi and Sano 2000) and share a similar habitat (Imada et al. 1981; Hong 1988) and are similar body size (Myers 1971; Dahl 1977; Hiwatari and Kajihara 1988; Takeuchi and Hirano 1991) and life history (Myers 1971; Hiwatari and Kajihara 1988; Takeuchi and Hirano 1991; Takahashi et al. (1999) also reported that caprellids accumulated BTs with a significantly high proportion of TBT compared to gammarids. These results suggest that the metabolic capacity of caprellids to degrade TBT is lower than that of gammarids.

It has been known that differences in BT residue levels and the proportion of TBT in organisms are related to environmental and physiological factors. It seems that physiological and ecological characteristics, such as metabolic capacity and trophic levels of each organism, are important factors which influence the pattern of TBT accumulation. Therefore, the results in the present study suggest that the difference in sensitivity to TBT among the amphipods is related to the species-specific capacity to metabolize TBT.

Generally, it is known that several groups of aquatic organisms, e.g. the Annelida, Arthropoda and Mollusca, have the metabolic capacity to degrade TBT (Maguire et al. 1984; Maguire and Tkacz 1985; Lee et al. 1987, 1989; Francois et al. 1989; Thain et al. 1990) and that metabolic capacity varies in different organism groups (Langston 1990; Laughlin et al. 1986; Lee 1986). For example, the crab *Callinectes sapidus*, the fish *Leistomus canthurus*, and the shrimp *Penaeus aztecus* are able to metabolize TBTO, while the oyster *Crassostrea virginica* show only a limited ability to metabolize TBTO (Lee 1986). TBT is known to metabolize by a detoxifying system involving two phases in vivo. The phase-one reactions involve the cytochrome P-450 dependent mixed-function oxygenase (MFO) system which hydroxylates TBT to alpha-, beta-, gamma-, and delta-hydroxydibutyltin derivatives (Fish et al. 1976). The phase-two reactions conjugate sugars or sulfate to hydroxybutyldibutyltin, and these highly polar conjugates are then rapidly eliminated from the organism. The MFO system of vertebrates and invertebrates is associated with the endoplasmic reticulum of the cell and is a multicomponent enzyme system composed of phospholipid, cytochrome p-450, and NADPH cytochrome P-450 reductase (Lu 1976; Lee 1981; Stegeman 1981). Thus, metabolism of a compound generally reduces persistence, increases elimination, and reduces toxicity (Lee 1996). The Mollusca have low cytochrome P-450 content and mixed function oxygenase activity (Lee 1981; Anderson 1985; Livingstone and Farrar 1985). In addition, it is also considered that differences between organisms in terms of metabolic capacity occur due to the inhibition of the cytochrome system by TBT. The binding of TBT to glutathione *S*-transferase and cytochrome P-450 results in the inhibition of these two detoxifying enzyme systems (Henry and Byington 1976; Rosenberg and Drummond 1983). Therefore, it is believed that the cause of the different levels of susceptibility to the acute toxicity of TBT in the two groups of amphipods in the present study (Fig. 10.2) are related to differences in metabolic capacity. Further study is necessary to provide evidence of the linkage of TBT metabolites and TBT metabolizing enzyme systems to the observed effects.

10.5.2 Growth and Morphological Alterations

In the TBT exposure experiments after hatching, the marked delay in growth and molting during the early developmental and mature stages was found to occur regardless of gender in *Caprella danilevskii* (Ohji et al. 2003a) (Fig. 10.13). However, no significant difference was found in the growth and molting inhibition in the embryonic exposure experiments (Ohji et al. 2002b). These considerations suggest that effects on sensitive stages by exposure to TBT may extend over a long period after hatching in the caprellids. Similar growth delays induced by exposure to TBT have also been reported in various other organisms, e.g. the mysid *Acanthomysis sculpta* (Davidson et al. 1986), American oyster *Crassostrea virginica* (Thain 1986), blue mussel *Mytulis edulis* (Strømgren and Bongard 1987) and American lobster *Homarus americanus* (Laughlin and French 1980). Furthermore, several morphological alterations caused by TBT exposure after hatching were observed during the growth of *C. danilevskii*, though no morphological alterations were observed in response to TBT exposure during the embryonic period (Fig. 10.13). The morphological alterations resulting from exposure to TBT such as imposex in the gastropod *Nucella lapillus* (Bryan et al. 1986; Gibbs and Bryan 1986, 1987), shell thickening in the oyster *Crassostrea gigas* (Thain and Waldock 1986) and deformities in regenerated limbs of the crab *Uca Pugilator* (Weis and Kim 1988) have also been reported. Exposure to TBT after hatching may also have induced the cramp of the pereopod observed in this study, since TBT is known to be neurotoxic (Watanabe 1980). Collectively evidence suggests that TBT might act as a developmental toxicant or teratogen, affecting the processes of differentiation and morphogenesis during growth.

10.5.3 Maturation and Reproduction

Conspicuous inhibition of maturation and reproduction occurred in mature females even at nanogram-per-liter levels of TBT exposure (corresponding to present TBT levels in the coastal environment) both after hatching and during

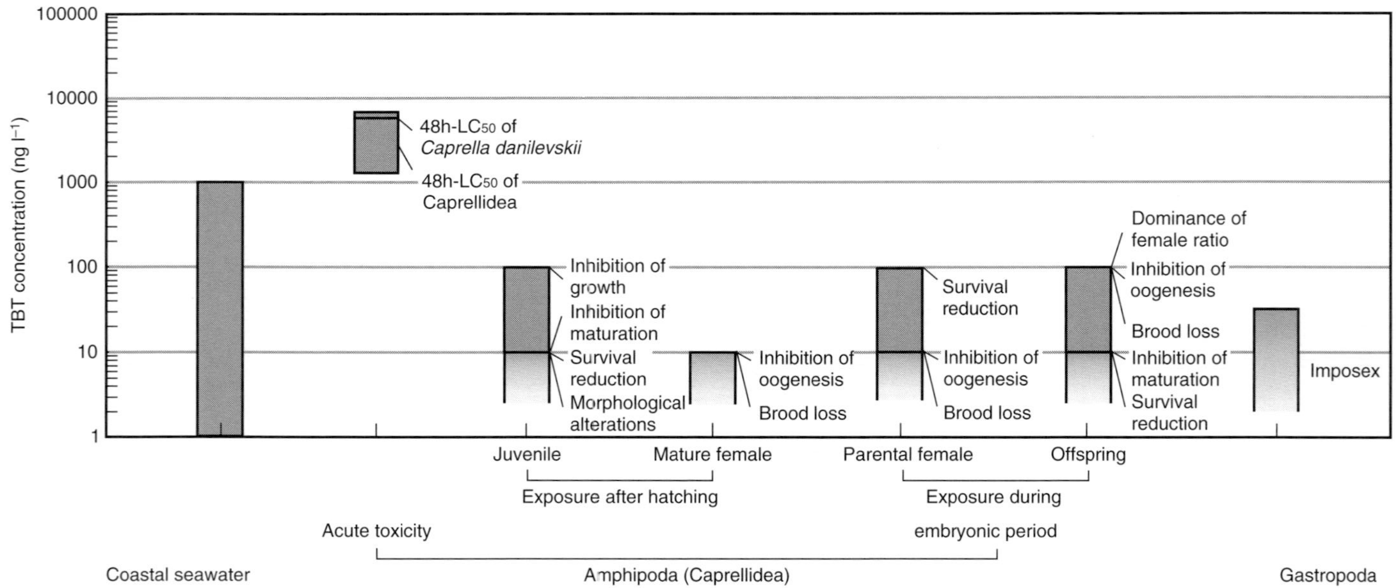

Fig. 10.13 Summary of biological effects of TBT on caprellids in the present studies, and on gastropods, and TBT concentration in coastal seawater. LC_{50} indicates median lethal concentration. LC_{50} values for caprellids are based on Ohji et al. (2002a). Results of experiments of TBT exposure after hatching and during embryonic period based on Ohji et al. (2003a) and Ohji et al. (2002b), respectively. TBT concentrations in coastal seawater are cited from Batley (1996). The occurrence levels of imposex in gastropods are cited from Bryan et al. (1986), Gibbs et al. (1988) and Bettin et al. (1996)

the embryonic stage, although such inhibitions were not apparent in control treatments for *Caprella danilevskii* (Ohji et al. 2002b, 2003a) (Fig. 10.13). In gastropods, masculinization (imposex) by TBT exposure is the superimposition of male sex organs (development of a penis and vas deferens) on female individuals, with this condition leading to reproductive failure and consequently population decline (Bryan et al. 1986, Gibbs and Bryan 1986, 1987; Bettin et al. 1996; Matthiessen and Gibbs 1998) (Fig. 10.13). Therefore, TBT exposure might also affect the maturation and reproduction systems of the caprellids, although no external morphological alterations of reproductive organs were observed in the present study. In caprellids, TBT might not induce morphological alterations in the reproductive organs but cause disruptions in the internal physiological mechanisms concerning maturation and reproduction over the whole life stages. A similar phenomenon of impairment of egg production has been reported in the copepod *Acarita tonsa* (Johansen and Møhlenberg 1987) and in the sea urchin *Paracentrouts lividus* (Girard et al. 1997, 2000) in response to TBT exposure. The cytotoxicity of TBT often results in an arrest of cellular dynamics, leading to apoptosis (Stridh et al. 1999) or a blocking of cell division (Girard et al. 1997) primarily occurring through an alteration of macromolecular syntheses (Snoeij et al. 1988; Girard et al. 1997) or membrane-mediated processes controlling cell signaling. These processes consist primarily of a disruption of calcium homeostasis (Chow et al. 1992; Matsuoka and Igisu 1996) or calcium signaling (Corsini et al. 1997; Girard et al. 1997). Girard et al. (1997, 2000) have found that TBT inhibits sea urchin egg cleavage by altering many of the cellular events related to cell division. Furthermore, Girard et al. (2000) have suggested that the inhibition occurs in response to a few hours of TBT exposure and is sufficient to damage the organism during its embryonic life. A similar inhibition related to egg cleavage might occur in caprellids, resulting in brood loss and oogenesis inhibition in the species. In the present study, impaired reproductive success also occurred in both short- and long-term exposure to TBT. Therefore, our data suggest that nanogram concentrations of TBT similar to those encountered in coastal waters can directly affect reproduction in the caprellids, and that this phenomenon is an environmentally realistic scenario in the coastal ecosystem.

10.5.4 Sex Disturbance by Tributyltin Exposure in Caprellids

It is noteworthy that no significant differences were seen regarding change in the sex ratio by TBT exposure at all levels after hatching in the present study (Ohji et al. 2003a) (Fig. 10.13). However, an increase in the female ratio in hatched juveniles was found in embryonic exposure experiments with *Caprella danilevskii* (Ohji et al. 2002b). As TBT concentrations increase, the proportion of females were found to increase to 55.6% at 10 ng l^{-1}, 85.7% at 100 ng l^{-1} and 81.8% at 1,000 ng l^{-1}. Although the sex ratio in the present study was altered in response to TBT exposure, the number of females was almost constant (9–12) regardless of increases in TBT concentrations. Accordingly, males seem to have a higher sensitivity to TBT than females. However, the survival rate in response to exposure to TBT has been found to be similar regardless of sex in the juvenile stage (Ohji et al. 2004, 2005). These considerations suggest that sex disturbance may be induced during the embryonic stage in caprellids. This phenomenon contrasts with results for gastropod molluscs in which TBT exposure after hatching induced imposex in females (Matthiessen and Gibbs 1998). TBT had marked effects on the development of imposex in the dog whelk in exposure experiments after hatching (Gibbs et al. 1988). Several hypotheses have been proposed about the imposex induction mechanisms, such as those involving cytochrome P450 – mediated aromatase inhibition (Bettin et al. 1996), testosterone excretion – inhibition (Ronis and Mason 1996), functional disorder of female cerebropleural ganglia (Féral and Le Gall 1983), and involvement of a neuropeptide – APGWamide (Oberdörster and McClellan-Green 2002, 2003), although the exact physiological/biochemical pathway was still unclear. Recently, it is reported that TBT and TPT bind the hRXRs with high affinity and that injection of 9-cis retinoic acid (RA), the natural ligand of hRXRs, into females of the rock shell *Thais clavigera* induces the development of imposex (Nishikawa et al. 2004). Cloning of the RXR homologue from *T. clavigera* revealed that the ligand-binding domain of rock shell RXR was very similar to vertebrate RXR and bound to both 9-cis RA and to organotins. These results suggest that RXR plays an important role in the induction/differentiation and growth of male genital tracts in female gastropods. Since this phenomenon differs from our results, it is suggested that the mode of action of TBT may differ among organisms. Furthermore, as the factor of difference of phenomenon among organisms, it is also considered that the effects of TBT exposure might differ according to the developmental stage of the organism. In gastropods, TBT induces sex disturbance after sex determination, while, in contrast, TBT appears to induce sex disturbance prior to sex determination in caprellids. Sex differentiation in crustaceans, i.e. amphipods, isopods and decapods, is known to control by a hormone secreted from the androgenic gland (Charniaux-Cotton 1954; Katakura 1960; Taketomi et al. 1996). It is known that the androgenic gland is produced during the embryonic period. In our previous study, sex disturbance caused by TBT was induced in an earlier embryonic stage in the caprellid (Ohji et al. 2003b). Therefore, it is considered that TBT might affect the production of the androgenic gland or the secretion of androgenic hormone in the caprellid.

Furthermore, as mentioned above, TBT is known to be metabolized by the two phases detoxifying system involving the cytochrome P-450 dependent MFO system. The binding of TBT to glutathione *S*-transferase and cytochrome P-450 results in the inhibition of two detoxifying enzyme systems. Furthermore, cytochrome P-450 systems control the conversion of cholesterol into a variety of hormones. Inhibition or stimulation of cytochrome P-450 systems can result in changes in hormone production or clearance (Levin et al. 1974; Kupfer and Bugler 1976). Therefore, TBT may conceivably affect androgenic hormone production in the caprellids. Further experiments are needed to clarify TBT action in the endocrine systems in this genus.

10.5.5 Survival and Biomass of Caprellids in the Coastal Ecosystem

It is reported here that TBT affects the caprellid community at present. The biomass of the caprellids inhabiting sea grasses communities in the inner of the Otsuchi Bay (49.8–125.0 individuals m^{-2}) were a tenth of that near the mouth of the bay (1112.5 individuals m^{-2}) (Takeuchi and Hino 1997). This significant difference in caprellid biomass between inner and mouth of the Otsuchi Bay might be induced by the difference in TBT concentrations at each site, since TBT concentrations were higher in the inner bay (3.9–19 ng l^{-1}) than at the mouth (less than the detection limit) (Takahashi et al. 1999; Ohji et al. 2002a). Furthermore, prior to 1960, an extremely high biomass of the caprellids was reported from Japan (Fuse 1962). Seasonal fluctuation of the epifaunal animals living in the *Sargassum* zone in Kasaoka Bay, Japan, from 1956 to 1958 were studied, and it was reported that the biomass of the caprellids was 1.3 kg wet wt m^{-2}. In the past decade, such a high biomass and density of caprellid amphipods has not been reported in the coastal waters of Japan or of those of other developed countries. The biomass of caprellids inhabiting the *Sargassum* zone in Otsuchi Bay, Japan, from1993 to 1995 was estimated at 100 g wet wt m^{-2} (Takeuchi 1998). Although a reduction in TBT contamination was recorded after the ban (Environment Agency Japan 1995), TBT concentrations in Japanese coastal waters still persist, ranging from below the detectable level to 160 ng l^{-1} (180 ng l^{-1} as TBTCl) (Takeuchi et al. 2001) and averaging 10 ng l^{-1} (11 ng l^{-1} as TBTCl). Furthermore, the high TBT contamination in marine organisms and seawater continues in the coastal waters of Spain, France and Canada (Chau et al. 1997; Morcillo et al. 1997; Michel and Averty 1999), as well as in Asian and Oceanian countries (Kannan et al. 1995; Kan-antireklap et al. 1997) where no restrictions have been imposed. In small estuaries, marinas and moorings contribute significantly to TBT load, ranging from 24 ng l^{-1} (27 ng l^{-1} as TBTCl) to 2,440 ng l^{-1} (2,740 ng l^{-1} as TBTCl) (Lau 1991; Batley 1996). In the present study, a drastic decrease in the survival rate was observed in response to TBT exposure in both short- and long period even at 10 ng TBTCl l^{-1}, which corresponds to mean concentrations in the coastal waters (Ohji et al. 2002b, 2003a) (Fig. 10.13). Regarding survival rate throughout the whole life history, based on the results of the TBT exposure during the embryonic period (Ohji et al. 2002b) and after hatching (Ohji et al. 2003a), significant differences were also seen in the survival rate between the control and the other four concentrations of TBTCl (Fig. 10.14). The survival rate over 55 days (5 days in the embryonic period and 50 days from juvenile stage to mature stage after hatching) after spawning decreased in a range from 0% (10,000 ng TBTCl l^{-1}) to 17.3% (10 ng TBTCl l^{-1}) (2.9% in 1,000 ng TBTCl l^{-1} and 4.1% in 100 ng TBTCl l^{-1}) except for the control (100%). These considerations all lead to the conclusion that TBT exposure threatens the survival of caprellids through their whole life history, and may have contribute to the decrease in the caprellid biomass in the coastal ecosystems.

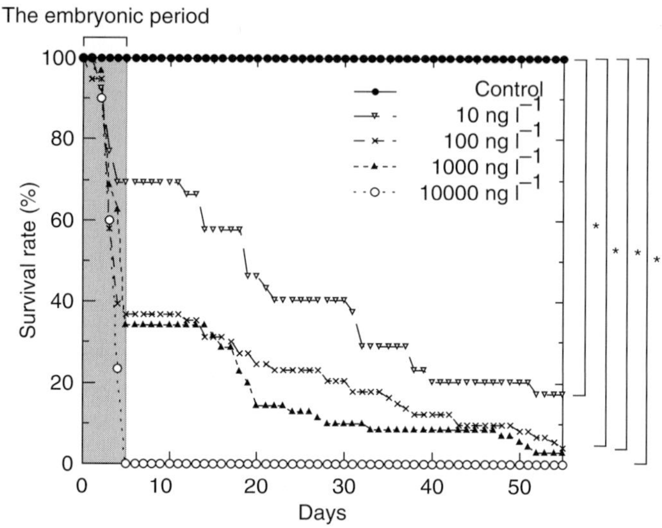

Fig. 10.14 Survival rate throughout whole life history in specimens based on the results of the TBT exposure after hatching (Ohji et al. 2003a) and during the embryonic period (Ohji et al. 2002a). The number of eggs at the beginning of the experiment was calculated as 100%. Log-rank test, $^{*}p < 0.0001$

10.6 Conclusions

In acute toxicity tests, the 48-h LC_{50} values for caprellids were significantly lower than those for gammarids. This suggests that caprellids are more sensitive to TBT than gammarids. Furthermore, in the caprellids, TBT was the predominant compound accumulated, which reflected the BT ratio in seawater, while in the gammarids, TBT's breakdown products (DBT and MBT) predominated. This difference suggests that caprellids have a lower metabolic capacity to degrade TBT than gammarids. Therefore, the difference in sensitivity to TBT among these amphipods could be related to the species-specific capacity to metabolize TBT. Moreover, in the comparison of 48-h LC_{50} values for TBT among various trophic level organisms, the caprellids belong to one of the more sensitive groups of organisms. In chronic toxicity tests, even at ambient water concentrations, exposure to TBT after hatching (50 days) influences survival, growth, maturation, reproduction and morphological alterations in the caprellids. Adverse effects on sex ratio, reproduction, and survival of caprellids have also been observed after TBT exposure only with exposure during the embryonic stage (5 days) at ambient water concentrations. Remarkably, the proportion of females increased dramatically in response to exposure to TBT in the embryonic period, though no significant difference was observed in the sex ratio in response to long-term exposure to TBT at these levels after hatching. These findings suggest that sex disturbance might therefore be induced during the embryonic stage in caprellids. It has been reported that caprellids have a lower metabolic capacity to degrade TBT and therefore accumulate BTs at higher concentrations than other organisms in the coastal ecosystem. Accordingly, TBT exposure, both short- and long-term, in the coastal environment might critically damage the life history characters of caprellids. The impaired reproductive success of a keystone species affects the entire population of species due to drops in the reproductive output below the critical level required for maintaining the population's survival, thus leading to changes in the ecosystem around keystone species. Since caprellids link primary producers to higher consumers in coastal waters, the high ecological risk to caprellids due to their high sensitivity to TBT over their life history may result in a disturbance in the coastal water ecosystem.

References

Alabaster J S (1969) Survival of fish in 164 herbicides, insecticides, fungicides wetting agents and miscellaneous substances. Int Pesticide Control 2:287–297

Alzieu C (1996) Biological effects of tributyltin on marine organisms. In: de Mora S J (ed) Tributyltin: case study of an environmental contaminant. pp 167–211, Cambridge University Press, Cambridge

Anderson R S (1985) Metabolism of a model environmental carcinogen by bivalve molluscs. Mar Environ Res 17:137–140

Batley G (1996) The distribution and fate of tributyltin in the marine environment. In: de Mora S J (ed) Tributyltin: case study of an environmental contaminant. pp 139–166, Cambridge University Press, Cambridge

Bettin C, Oehlmann J, Stroben E (1996) TBT- induced imposex in marine neogastropods is mediated by an increasing androgen level. Helgol Meeresunters 50:299–317

Bosselmann K (1996) Environmental law and tributyltin in the environment. In: de Mora S J (ed) Tributyltin: case study of an environmental contaminant. pp 237–263, Cambridge University Press, Cambridge

Bryan G W, Gibbs P E (1991) Impact of low concentrations of tributyltin (TBT) on marine organisms: a review. In: Newman M C, McIntosh A W (eds) Metal ecotoxicology: concepts and applications. pp 323–361, Lewis Publishers, Michigan

Bryan G W, Gibbs P E, Hummerstone L G et al. (1986) The decline of the gastropod *Nucella lapillus* around south – west England: evidence for the effect of tributyltin from antifouling paints. J Mar Biol Ass UK 66:611–640

Bushong S J, Hall, W S, Johnson W E et al. (1987) Toxicity of tributyltin to selected Chesapeake Bay biota. In: Proceedings of the Organotin Symposium, Oceans '87 Conference, pp 1499–1503, Halifax, Nova Scotia, Canada

Caine E A (1989) Caprellid amphipod behavior and predatory strikes by fish. J Exp Mar Biol Ecol 126:173–180

Champ M A, Seligman P F (1996) An introduction to organotin compounds and their use in antifouling coatings. In: Champ M A, Seligman P F (eds) Organotin: environmental fate and effects. pp 1–25, Chapman & Hall, London

Champ M A, Wade T L (1996) Regulatory policies and strategies for organotin compounds. In: Champ M A, Seligman P F (eds) Organotin: environmental fate and effects. pp 430–458, Chapman & Hall, London

Charniaux-Cotton H (1954) Découverte chez un Crustacé Amphipode (*Orchestia gammarella*) d' une glande endocrine responsible de la différenciation des caractères sexuels primaries et secondaires mâles. CR Hebd Séances, Acad Sci, Paris 239:780–782

Chau Y K, Marguire R J, Brown M et al. (1997) Occurrence of butyltin compounds in mussels in Canada. Appl Organomet Chem11:903–912

Chow S C, Kass G E N, McCave M J et al. (1992) Tributyltin increases cytosolic free Ca^{2+} concentration in thymocytes by mobilizing intracellular Ca^{2+}, activating a Ca^{2+} entry pathway, and inhibiting Ca^{2+} efflux. Arch Biochem Biophys 298:143–149

Corsini E, Viviani B, Marinovich M et al. (1997) Role of mitochondria and calcium ions in trib utyltin-induced gene regulatory pathways. Toxicol Appl Pharmacol 353:136–143

Dahl E (1977) The amphipod functional model and its bearing upon systematics and phylogeny. Zoologica Scripta 6:221–228

Davidson B M, Valkirs A O, Seligman P F (1986) Acute and chronic effects of tributyltin on the mysid *Acanthomysis sculpta*, Crustacea, Mysidacea. In: Proceedings of the Organotin Symposium, Oceans '86 Conference, vol. 4, pp 1219–1225, Washington, DC

Environment Agency Japan (1990) Analytical method for butyltin and phenyltins. Heisei Gan-Nen Do. Technical Report for the Development of Analysis of Artificial Chemicals. Environmental Health and Safety Division, Environmental Health Department, Environment Agency, Tokyo, pp 127–137

Environment Agency Japan (1995) Chemicals in the environment. Environmental Health and Safety Division, Environmental Health Department, Environment Agency, Tokyo

Fent K (1996) Ecotoxicology of organotin compounds. Crit Rev Toxicol 26:1–117

Féral C, Le Gall S (1983) In: Lever J, Boer H H (eds) Molluscan Neuro-endocrinology. pp 173–175, North-Holland Publishing, Amsterdam

Fish R H, Kimmel E C, Casida J E (1976) Bioorganotin chemistry: reactions of tributyltin derivatives with a cytochrome P-450 dependent monooxygenase enzyme system. J Organomet Chem 118:41–54

Francois R, Short F T, Weber J H (1989) Accumulation and persistence of tributyltin in eelgrass (*Zostera marina* L.) tissue. Environ Sci Technol 23:191–196

Fuse S (1962) The animal community in the *Sargassum* belt. Physiol Ecol 11:23–45

Gibbs P E, Bryan G W (1986) Reproductive failure in populations of the dog-whelk, *Nucella lapillus*, caused by imposex induced by tributyltin from antifouling paints. J Mar Biol Ass UK 66:767–777

Gibbs P E, Bryan G W (1987) TBT paints and the demise of the dogwhelk *Nucella lapillus*, Gastropoda. In: Proceedings of the Organotin Symposium, Oceans '87 Conference, vol. 4, pp 1482–1487, Halifax, Nova Scotia

Gibbs P E, Pascoe P L, Burt G R (1988) Sex change in the female dogwhelk, *Nucella lapillus*, induced by tributyltin from antifouling paints. J Mar Biol Ass UK 68:715–731

Girard J P, Ferrua C, Pesando D (1997) Effects of tributyltin on Ca^{2+} homeostasis and mechanisms controlling cell cycling in sea urchin eggs. Aquat Toxicol 38:225–239

Girard J P, Szpunar J, Pedrotti M L et al. (2000) Toxicity of tri-*n*-butyltin to sea urchin eggs and larvae: relation to bioaccumulation at the nanomolar level. Environ Toxicol Chem 19:1272–1277

Hall L W, Bushong S J, Ziegenfuss M C et al. (1988) Acute and chronic effects of tributyltin on a Chesapeake Bay copepod. Environ Toxicol Chem 7:41–46

Harino H, Fukushima M (1992) Simultaneous determination of butyltin and phenyltin compounds in the aquatic environment by gas chromatography. Analytica Chimica Acta 246:91–96

Hayakawa Y (1976) An estimate of the permissionable limits of the number of boats moored in Moroiso Cove. Bull Japan Soc Fish Oceanogr 29:15–29

Henry R A, Byington K H (1976) Inhibition of glutathione *S*-aryltransferase from rat liver by organogermanium, lead and tin compounds. Biochem Pharmacol 25:2291–2295

Hiwatari T, Kajihara T (1988) Experimental studies on the growth and breeding of *Hyale barbicornis* (Amphipoda, Crustacea) at different temperatures. Bull Japan Soc Fish Sci 54:39–43

Holbrook S J, Schmitt R J (1992) Cause and consequences of dietary specialization in surfperches: patch choice and intraspecific competition. Ecology 73:402–412

Hong J S (1988) Amphipod Crustaceans as fouling organisms in Tungnyang Bay, Korea. Mar Fouling 7:1–7

Horinouchi M, Sano M (2000) Food habits of fishes in a *Zostera marina* bed at Aburatsubo, central Japan. Ichthyol Res 47:163–173

Imada K, Hirayama A, Nojima S et al. (1981) The microdistribution of phytal amphipods on *Sargassum* seaweeds. Res Crust 11:124–137

Iwata H, Tanabe S, Miyazaki N et al. (1994) Detection of butyltin compound residues in the blubber of marine mammals. Mar Pollut Bull 28:607–612

Johansen K, Møhlenberg F (1987) Impairment of egg production in *Acartia tonsa* exposed to tributyltin oxide. Ophelia 27:137–141

Kan-antireklap S, Tanabe S, Sanguansin J et al. (1997) Contamination by butyltin compounds and organochlorine residues in green mussel (*Perna viridis*, L.) from Thailand coastal waters. Environ Pollut 94:79–89

Kannan K, Tanabe S, Iwata H et al. (1995) Butyltin in muscle and liver of fish collected from certain Asian and Oceanian countries. Environ Pollut 90:279–290

Katakura Y (1960) Transformation of ovary into testis following implantation of androgenic glands in *Armadillidium vulgare*, an isopod crustacean. Zool Jpn 33:241–244

Kupfer D, Bugler W H (1976) Interactions of chlorinated hydrocarbons with steroid hormones. Federation Proc 35:2603–2608

Langston W J (1990) Bioavailability and effects of TBT in deposit-feeding clams, *Scrobicularia plana*. In: Proceedings of the 3rd International Organotin Symposium, pp 110–113. Vienna, Monaco

Lau M M (1991) Tributyltin antifoulings: a threat to the Hong Kong marine environment. Arch Environ Contam Toxicol 20:299–304

Laughlin R B, French W J (1980) Comparative study of the acute toxicity of a homologous series of trialkyltins to larval shore crabs, *Hemigrapsus nudus*, and lobster, *Homarus americanus*. Bull Environ Contam Toxicol 25:802–809

Laughlin R B, Nordlund K, Linden O (1984) Long-term effects of tributyltin compounds on the Baltic amphipod, *Gammarus oceanicus*. Mar Environ Res 12:243–271

Laughlin R B, French W, Guard H E (1986) Accumulation of bis(tributyltin) oxide by the marine mussel *Mytilus edulis*. Environ Sci Technol 20:884–890

Lee R F (1981) Mixed function oxygenase (MFO) in marine invertebrates. Mar Biol Lett 2:87–105

Lee R F (1986) Metabolism of bis(tributyltin)oxide by estuarine animals. In: Proceedings of the Organotin Symposium, Oceans '86 Conference, pp 1182–1188, Washington, DC

Lee R F (1996) Metabolism of tributyltin by aquatic organisms. In: Champ MA, Seligman P F (eds) Organotin: environmental fate and effects. pp 369–382, Chapman & Hall, London

Lee R F, Valkirs A O, Seligman P F (1987) Fate of tributyltin in estuarine areas. In: Proceedings of the Organotin Symposium, Oceans '87 Conference, pp 1411–1415, Washington, DC

Lee R F, Valkirs A O, Seligman P F (1989) Importance of microalgae in the biodegradation of tributyltin in estuarine waters. Environ Sci Technol 23:1515–1518

Levin W, Welch M W, Conney A H (1974) Increased liver microsomal androgen metabolism by phenobarbital: correlation with decreased androgen action on the seminal vesicles of the rat. J Pharmacol Exp Ther 188:287–292

Livingstone D R, Farrar S V (1985) Responses of the mixed function oxidase system of some bivalve and gastropod molluscs to exposure to polynuclear aromatic and other hydrocarbons. Mar Environ Res 17:101–105

Lu A Y H (1976) Liver microsomal drug metabolizing enzymes: functional components and their properties. Fed Proc 35:2461–2463

Maguire R J, Tkacz R J (1985) Degradation of the tri-*n*-butyltin species in water and sediment from Tronto Harbor. Environ J Agric Food Chem 33:947–953

Maguire R J, Wong P T S, Rhamey J S (1984) Accumulation and metabolism of tri-*n*-butyltin cation by a green alga, *Ankistrodesmus falcatus*. Can J Fish Aquat Sci 41:537–540

Matsuoka M, Igisu H (1996) Induction of c-*fos* expression by tributyltin in PC12 cells: involvement of intracellular Ca^{2+}. Environ Toxicol Pharmacol 2:373–380

Matthiessen P, Gibbs P E (1998) Critical appraisal of the evidence for tributyltin-mediated endocrine disruption in molluscs. Environ Toxicol Chem 17:37–43

McCain J C, Steingerg J E (1970) Amphipoda I. Caprellidea I. Fam. Caprellidae. In: Gruner H E, Holthuis L B (eds) Crustaceorum Catalogus. Pars 2, pp 1–78, SPB Academic Publishing, Amsterdam

Michel P, Averty B (1999) Contamination of French coastal waters by organotin compounds: 1997 update. Mar Pollut Bull 38:268–275

Morcillo Y, Borghi V, Porte C (1997) Survey of organotin compounds in the Western Mediterranean using molluscs and fish as sentinel organisms. Arch Environ Contam Toxicol 32:198–203

Myers A A (1971) Breeding and growth in laboratory-reared *Microdeutopus gryllotalpa* Costa (Amphipoda: Gammaridea). J Nat Hist 5:271–277

Nishikawa J, Mamiya S, Kanayama T et al. (2004) Involvement of the retinoid X receptor in the development of imposex caused by organotins in gastropods. Environ Sci Technol 38: 6271–6276

Oberdörster E, McClellan-Green P (2000) The neuropeptide APGWamide induces imposex induction in the mud snail, *Ilyanassa obsoleta*. Peptides 21:1323–1330

Oberdörster E, McClellan-Green P (2002) Mechanisms of imposex induction in mud snail, *Ilyanassa obsoleta*: TBT as a neurotoxin and aromatase inhibitor. Mar Environ Res 54:715–718

OECD (1998) Fish, short-term toxicity test on embryo and sac-fry stages. OECD guideline for testing of chemicals, proposal for a new guideline 212, pp 1–20

Ohji M, Takeuchi I, Takahashi S et al. (2002a) Differences in the acute toxicities of tributyltin between the Caprellidea and the Gammaridea (Crustacea: Amphipoda). Mar Pollut Bull 44:16–24

Ohji M, Arai T, Miyazaki N (2002b) Effects of tributyltin exposure in the embryonic stage on sex ratio and survival rate in the caprellid amphipod *Caprella danilevskii*. Mar Ecol Prog Ser 235:171–176

Ohji M, Arai T, Miyazaki N (2003a) Chronic effects of tributyltin on the caprellid amphipod *Caprella danilevskii*. Mar Pollut Bull 46:1263–1272

Ohji M, Arai T, Miyazaki N (2003b) Timing of sex disturbance caused by tributyltin exposure during the embryonic stage in the caprellid amphipod, *Caprella danilevskii*. J Mar Biol Ass UK 83:943–944

Ohji M, Arai T, Miyazaki N (2004) Effects of tributyltin on the survival in the caprellid amphipod *Caprella danilevskii*. J Mar Biol Ass UK 84:345–349

Ohji M, Arai T, Miyazaki N (2005) Acute toxicity of tributyltin to the Caprellidea (Crustacea: Amphipoda). Mar Environ Res 59:197–201

Ronis M J J, Mason A Z (1996) The metabolism of testosterone by the periwinkle (Littorina littorea) *in vitro* and *in vivo*: Effects of Tributyl tin. Mar Environ Res 42:161–166

Rosenberg D W, Drummond G S (1983) Direct in vivo effects of bis(tri-*n*-butyltin)oxide on hepatic cytochrome P-450. Biochem Pharmacol 32:3823–3829

Sedberry G R (1988) Food and feeding of black sea bass, *Centropristis striata*, in live bottom habitats in the South Atlantic bight. J Elisha Mitchell Sci Soc 104:35–50

Snoeij N J, Bol-Schoenmakers M, Penninkis A H et al. (1988) Differential effects of tri-*n*-butyltin chloride on macromolecular synthesis and ATP levels of rat thymocyte subpopulations obtained by centrifugal elutriation. Int J Immunopharmacol 10:29–37

Stegeman J J (1981) Polynuclear aromatic hydrocarbons and their metabolism in the marine environment. In Gelboin H V, Ts'o P O P (eds) Polycyclic hydrocarbons and cancer vol. 3, pp 1–60, Academic, New York

Stridh H, Orrenius S, Hampton M B (1999) Caspase involvement in the induction of apoptosis by environmental toxicants tributyltin and triphenyltin. Toxicol Appl Pharmacol 156:141–146

Strømgren T, Bongard T (1987) The effect of tributyltin oxide on growth of *Mytilus edulis*. Mar Pollut Bull 18:30–31

Takahashi S, Tanabe S, Takeuchi I et al. (1999) Distribution and specific bioaccumulation of butyltin compounds in a marine ecosystem. Arch Environ Contam Toxicol 37:50–61

Taketomi Y, Nishikawa S, Koga S (1996) Testis and androgenic gland during development of external sexual characteristics of the crayfish *Procambarus clarkii*. J Crust Biol 16:24–34

Takeuchi I (1990) A preliminary note on a hermaphroditic abnormality of *Caprella danilevskii* Czerniavski (Crustacea: Amphipoda). Ann Rep Mar Ecosystems Res Ctr 10:29–30

Takeuchi I (1998) Dry weight, carbon and nitrogen components of the caprellid amphipods (Crustacea) inhabiting the *Sargassum yezoense* community of Otsuchi Bay, northeastern Japan. Mar Biol 130:417–423

Takeuchi I, Hirano R (1991) Growth and reproduction of *Caprella danilevskii* (Crustacea: Amphipoda) reared in the laboratory. Mar Biol 110:391–397

Takeuchi I, Hino A (1997) Community structure of caprellid amphipods (Crustacea) on seagrasses in Otsuchi Bay, northeastern Japan, with reference to the association of *Caprella japonica* (Schurin) and *Phyllospadix iwatensis* Makino. Fish Sci 63:327–331

Takeuchi I, Takahashi S, Tanabe S et al. (2001) *Caprella* watch; a new approach for monitoring butyltin residues in the ocean. Mar Environ Res 52:97–113

Tanaka Y, Nakanishi J (1998) Ecological risk estimation based on life table evaluation of chronic toxicity. J Japan Soc Water Environ 21:589–595

Tanaka Y, Nakanishi J (2000) Mean extinction time of populations under toxicant stress and ecological risk assessment. Environ Toxicol Chem 19:2856–2862

Thain J E (1983) The acute toxicity of bis (tributyl tin) oxide to the adults and larvae of some marine organisms. ICES, C M E, 10

Thain J E (1986) Toxicity of TBT to bivalves: effects on reproduction, growth and survival. In: Proceedings of the Organotin Symposium, Oceans '86 conference, vol. 4, pp 1306–1313. Washington, DC

Thain J E, Waldock MJ (1986) The impact of tributyltin, TBT antifouling paints on molluscan fisheries. Water Sci Technol 18:193–202

Thain J E, Waldock M J, Finch J et al. (1990) Bioavailability of the residues in sediments to animals. In: Proceedings of the 3rd International Organotin Symposium, pp 84–86, Vienna, Monaco

Walsh G E (1986) Organotin toxicities studies conducted with selected marine organisms at EPAS environmental research laboratory, Gulf Breeze, Florida. In: Proceedings of the Organotin Symposium, Oceans '86 Conference, vol 4, pp 23–25, Washington, DC

Watanabe I (1980) Organotins. In: Spencer P S, Schaumburg H H (eds) Experimental and clinical neurotoxicology. pp. 545–557, Williams & Wilkins, Baltimore, MD

Weis J S, Perlmutter J (1987) Effects of tributyltin on activity and burring behavior of the fiddler crab, *Uca pugirator*. Estuaries 10:342–346

Weis J S, Kim K (1988) Tributyltin is a teratogen in producing deformities in limbs of the fiddler crab, *Uca pugirator*. Arch Environ Contam Toxicol 17:583–587

Weis J S, Gottlieb J, Kwiaikowski J (1987) Tributylyin retards regeneration and produces deformites of limbs in the fiddler crab, *Uca pugirator*. Arch Environ Contam Toxicol 16:321–326

Chapter 11
Toxic Interactions Between Tributyltin and Polychlorinated Biphenyls in Aquatic Organisms

Yuji Oshima, Kei Nakayama, Takeshi Hano, Sang Gyoon Kim, Yohei Shimasaki, Ik Joon Kang, and Tsuneo Honjo

11.1 Introduction

In recent years, discharges of anthropogenic chemicals to the environment have been increasing in association with industrial development. These chemicals and their degradation products are released to the environment, discharged into water, and may ultimately contaminate aquatic organisms. Polychlorinated biphenyls (PCBs) and tributyltin (TBT) are particularly ubiquitous pollutants.

11.1.1 Tributyltin

TBT, which has been widely used as a marine biocide, leaches directly into the aquatic environment and pollutes water, sediments and organisms. TBT is now known to be a typical endocrine disruptor which binds with the retinoid X receptor of gastropods and induces the imposex (Nishikawa et al. 2004). Horiguchi et al. (2006) showed that the TBT and triphenyltin concentrations in the gonads were positively correlated with penis length in females of ivory shell (*Babylonia japonica*). In addition, maternal exposure to TBT clearly resulted in developmental toxicity in the progeny of pearl oysters (*Pinctada fucata martensii*; Inoue et al. 2004) and Manila clams (*Ruditapes philippinarum*; Inoue et al. 2006). Thus, TBT severely affects reproduction in aquatic invertebrates.

Y. Oshima, T. Hano, S.G. Kim, Y. Shimasaki, I.J. Kang, and T. Honjo
Faculty of Agriculture, Graduate School, Kyushu University,
6-10-1 Hakozaki, Higashi-ku, Fukuoka 812-8581, Japan

K. Nakayama
Center for Marine Environmental Studies (CMES),
Ehime University, 2-5 Bunkyo-cho, Matsuyama 790-8577, Japan

T. Arai et al. (eds.), *Ecotoxicology of Antifouling Biocides*,
DOI: 10.1007/978-4-431-85709-9_11, © Springer 2009

Recently, TBT has been reported to induce masculinization – to the extent of complete sex reversal – in genetically female Japanese flounder, *Paralichthys olivaceus* (Shimasaki et al. 2003) and zebrafish (McAllister and Kime 2003; Santos et al. 2006), though not in Japanese medaka, *Oryzias latipes* (Hano et al. 2007).

11.1.2 Polychlorinated Biphenyls

PCBs are globally distributed in sediments, fish, and marine mammals, because of their mass production and environmental persistence (Nakata et al. 1997; Schlenk et al. 2002). They impair reproduction in white perch (*Morone Americana*; Monosson et al. 1994) and zebrafish (*Danio rerio*; Örn et al. 1998). Also, these chemicals are maternally transferred from females to the next generation, and decrease development and hatching success of progeny (Mac et al. 1993; Fisk et al. 1998; Westerlund et al. 2000). Further, PCBs are also neurotoxic and have the potential to alter the behavior of organisms (Damstra et al. 2002). We have previously reported that PCB exposure affects general and schooling behavior of male Japanese medaka *Oryzias latipes* (Nakayama et al. 2004a, 2005a).

11.1.3 TBT and PCB Contamination

We can assume that organisms inhabiting polluted areas are exposed to complex mixtures of chemicals. In coastal areas of the Aireake Sea in Japan, Nakata et al. (2003, 2006) reported contamination of organisms by PCBs, polycyclic aromatic hydrocarbons, TBT and perfluorooctane sulfonate. "Squid Watch" revealed that both TBT and PCBs were detectable in the livers of squid collected from coastal waters and open oceans (Yamada et al. 1997). Hayteas and Duffield (1998) reported the presence of 24.3 μg/g of PCBs and 53.9 μg/g of *p,p*-DDE in blubber of marine mammals. In pelagic fish, PCBs, organochlorine pesticides, TBT and polybrominated diphenyl ether have been detected in skipjack tuna (*Katsuwonus pelamis*) collected from the offshore waters of various regions of the world (Ueno et al. 2003, 2004a, b). Thus, organisms inhabiting polluted areas are likely to be contaminated simultaneously with TBT and PCBs.

11.2 Mixture Toxicity

Aquatic organisms are undoubtedly exposed to many kind chemicals, albeit at low concentrations. Effects of these chemicals is a fundamental issue that cannot be avoided in the field of ecotoxicology. Evaluation of the toxicity of chemical mixtures is a top priority. For example, Toxicity Equivalent Quantity (TEQ) values for dioxins, which include coplanar PCBs, have been calculated from individual components by integration of each equivalent toxicity value, compared with 2,3,7,8-tetrachlorodibenzo-*p*-dioxin (TCDD).

11.2.1 Mixture Toxicity Studies

Mixture toxicities of environmental contaminants have been studied using both *in vitro* and *in vivo* testing systems. *In vitro* yeast estrogen screening assays have demonstrated that mixtures of weak estrogenic chemicals produce significant effects, even when the individual chemicals are combined at below their no-observed-effect concentrations (NOECs; Silva et al. 2002).

Likewise, in many *in vivo* tests, it has been reported that combined exposure of chemicals can exert toxicity, where exposure to individual chemicals have no effect. Mixed exposures to 25 or 50 different organic chemicals, below their NOECs, have been shown to additively affect growth or mobility, respectively, of *Daphnia magna* (Hermens et al. 1985; Deneer et al. 1988). Binary mixture toxicities of antifoulants (TBT, Irgarol, and Sea-Nine) were reported using reproduction of green alga *Scenedesmus vacuolatus* as an endpoint (Arrhenius et al. 2006). Not all interactions are additive: Wang (2006) demonstrated antagonism in the induction of glutathione content of liver sampled form *Sebastiscus marmoratus*, injected intraperitoneally with TBT and benzo(a)pyrene. TBT also antagonized EE2-induced expression of aromatase B mRNA in brain of Atlantic salmon, following waterborne exposure (Lyssimachou et al. 2006). However, concern is greatest where mixtures of chemicals may have more toxic effects compared to the toxicity of individual chemicals. For example, Kim and Cooper (1998) have reported that TCDD and 3,3′,4,4′, 5-pentachlorobiphenyl (PCB 126) additively induce toxic effects in Japanese medaka embryos.

It is well known that coplanar-PCBs induce toxic effects through the aryl hydrocarbon receptor (AhR) and induce cytochrome P4501A (CYP1A). In contrast, TBT is thought to inhibit CYP1A in a dose-dependent manner (Mortensen and Arukwe 2007; Morcillo et al. 2004). However, a low level of TBT has been found to potentiate 3,3′4,4′5-pentachlorobiphenyl (PCB-126)-induced CYP1A *in vitro* (Kannan et al. 1998), and *in vivo* in mice (DeLong and Rice 1997) as well as in channel catfish (*Ictalurus punctatus*; Rice and Roszell 1998). Also, exposure of TBT and/or PCB-126 was reported to disrupt lysozyme levels in catfish (Burton et al. 2002). Therefore, because of these possible interactions, we have studied the combined effect of TBT and PCBs on fish, using Japanese medaka as model animal.

11.2.2 Medaka

Medaka is an extensively studied species, from a perspective of basic biology and also as a model fish for toxicity testing. Much information exists concerning the accumulation, toxicity and endocrine effects of anthropogenic chemicals in this species. Well studied reproductive endpoints, availability of a full life-cycle test, easy handling and small size are among the advantages of using this fish. In our previous studies, mixed administration of TBT and PCBs was shown to affect the development and reproduction of medaka significantly.

11.2.3 Combined Effects of TBT and PCBs on Reproduction and Behaviour of Japanese Medaka

In initial studies, Nirmala and co-workers (1999) fed Japanese medaka daily on each of four diets containing TBT, PCB, combination of TBT and PCB, or no addition (controls). Contaminant doses were 1 μg g^{-1} body weight [b.w.] daily for 3 weeks. The combination of TBT and PCB showed additive effects on the spawning of medaka, resulting in a significant reduction in spawning frequency, and on the number of spawned eggs and fertilization success, when compared with effects in fish fed on either a TBT or PCB diet (Nirmala et al. 1999; Fig. 11.1).

The effects of TBT, PCBs, and a mixture of TBT and PCBs on reproduction, gonadal histology, and sexual behavior in Japanese medaka were subsequently studied further following administration of TBT (1 μg/g b.w. daily), PCBs (1 μg/g b.w. daily), separately and in combination, for 3 weeks (Nakayama et al. 2004b). Reproductive success was determined during week 3 and the sexual behavior of male medaka assessed after the exposure period. The numbers of males that performed "following" and "dancing" were significantly decreased in the TBT + PCBs group. The frequency of "dancing" also decreased after treatment with TBT, whereas sexual behavior in fish that received PCBs only was not affected significantly (Table 11.1).

Combined effects of TBT and PCBs on general behavior of other organisms has been reported. For example, a synergistic effect of TBT and PCBs on the swimming behavior of carp exposed to a mixture of PCB (5 μg/l) and TBT (2 μg/l) has been described (Schmidt et al. 2005a). Schmidt et al. (2005b) have also shown significant effects of the TBT and PCBs in daphnids: these included effects of TBT-PCB mixtures which were approximately additive (swimming behavior) and also synergistic (reproduction).

11.2.4 Transgenerational Effects of Mixtures in Medaka

Nakayama et al. (2005b) examined the fecundity and fertility of the parent fish and assessed the deformity, hatchability, time to hatching and swim-up failure rate of the next generation. Groups of medaka were fed freeze-dried brine shrimp flakes contaminated with a mixture of either 0, 1, 5, or 25 μg TBT/g plus 0 or 25 μg PCB/g, for 21 days. Fertilization success in the third week was significantly decreased by administration of 25 μg TBT/g (77%) compared with the control group (87%). Both TBT and PCBs were maternally transferred into the eggs of the next generation, causing early life stage toxicity. Administration of 1 μg TBT/g was not toxic to embryological development, but abnormal eye development (i.e. small eyes or no eyes) occurred when TBT at the same concentration was mixed with PCBs (6.4%) (Table 11.2). Administration of TBT alone significantly decreased hatchability and increased swim-up failure, and administration of PCBs alone significantly increased time to hatching. Statistical analysis by two-way analysis of variance (ANOVA)

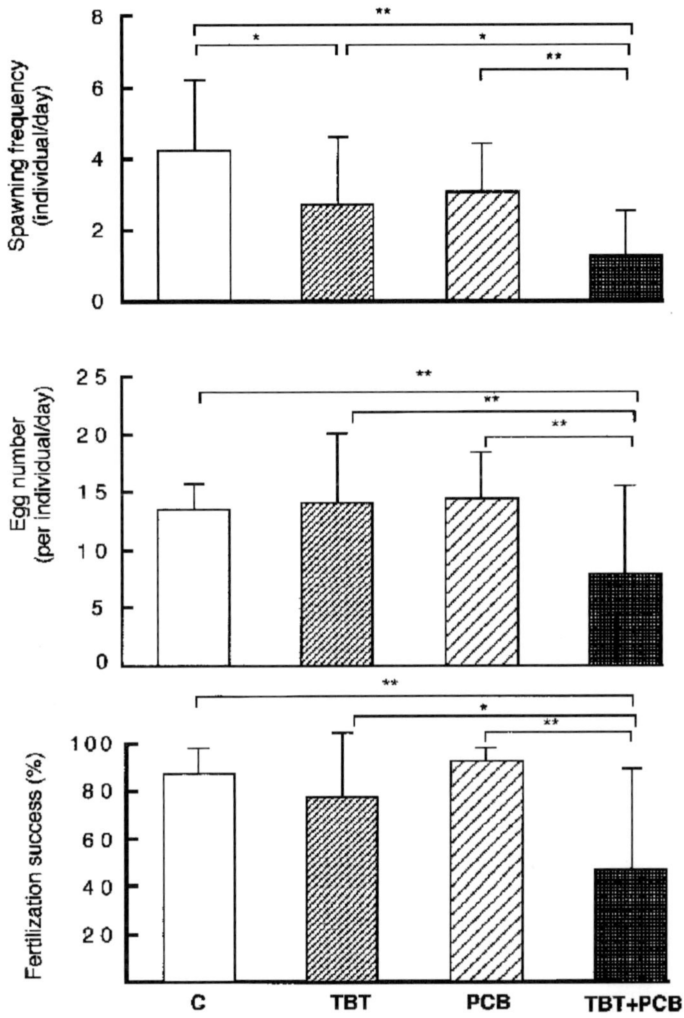

Fig. 11.1 Spawning frequency, egg number, and fertilization success of Japanese medaka *Oryzias latipes* during dietary administration of tributyltin (TBT), PCB, and TBT + PCB (1 mg/g body weight per day) for 14 days after 7-days preadministration of the diets. C = controls. Each group contained 12 females and four males. Error bars represent the SD. $^{*}p < 0.05$, $^{**}p < 0.01$ (Nirmala et al. 1999)

Table 11.1 Sexual behavior of male medaka *Oryzias latipes* treated with tributyltin (TBT) and/or polychlorinated biphenyls (PCBs) (Nakayama and Oshima, 2008)

	Number of male medaka that performed (n = 12)			
Treatment	Following	Dancing	Crossing	Mating
Control	12	9	6	6
TBT	8	5	3	3
PCBs	12	10	7	6
TBT + PCBs	5**	3*	2	2

Asterisks represents significant differences compared to control values (*$p < 0.05$, **$p < 0.01$; Fisher's exact test).

Table 11.2 Early-life effects in embryos spawned by parent medaka *Oryzias latipes* exposed to tributyltin (TBT) with (w/) or without (w/o) polychlorinated biphenyls (PCBs; $n = 6$)

TBT concentration	Swim-up failure (%)		Hatchability (%)		Abnormal eye (%)		Time to hatching (day)	
(µg/g-diet)	w/o PCBs	w/ PCBs[b]	w/o PCBs	w/ PCBs[b]	w/o PCBs	w/ PCBs	w/o PCBs	w/ PCBs[b]
0	2.5 ± 3.0	0.5 ± 0.8	93.3 ± 7.2	90.3 ± 19.7	0	0	6.7 ± 0.2	7.0 ± 0.3[a]
1	1.1 ± 1.7[a]	11.8 ± 11.4[a]	97.1 ± 2.4	79.5 ± 17.8	0	6.4 ± 7.5	6.8 ± 0.3	6.9 ± 0.4[a]
5	10.4 ± 5.8[a]	9.2 ± 4.7[a]	65.4 ± 25.2[a]	85.3 ± 6.7[a]	4.6 ± 3.1	4.4 ± 3.8	6.4 ± 0.4	7.4 ± 0.4[a]
25	5.1 ± 4.0[a]	9.7 ± 4.8[a]	77.0 ± 14.5[a]	86.7 ± 5.1[a]	9.1 ± 14.0	2.3 ± 3.2	6.6 ± 0.3	7.3 ± 0.3[a]

Control fish received 0 µg TBT/g-diet without PCBs.
The concentration of PCBs in the diet was 25 µg/g.
Data are presented as mean ± standard deviation.
[a]Significantly different from control (two-way analysis of variance; $p < 0.05$).
[b]Significant interaction of TBT and PCBs detected (two-way analysis of variance; $p < 0.05$).

detected an interaction between TBT and PCBs in these three parameters. TBT induces abnormal development of the eyes, reduced hatchability, and increased swim-up failure, whereas PCBs delay time to hatching. Administration of mixtures of TBT and PCBs has more adverse effects on the developmental stage of medaka than does that of each chemical alone. Lowest observed effective concentration (LOEC) on swim up success of medaka larvae were suggested to be <20 ng-TBT/g-egg and <31 ng-PCB/g-egg. These transgenerational effects might be attributed to chemicals (TBT and PCBs) transferred from female fish to eggs or transferred effects (e.g. decreased amount of vitamin A in eggs) in females.

11.2.5 Nanoinjection Can Simulate Maternal Transfer of Chemicals

Transgenerational effects of chemicals to the next generation are an important issue. However, methods to study this phenomenon usually require large-scale facilities and long exposure times with parent fish. An alternatively fish embryo-nanoinjection technique has been developed.

Ovo-nanoinjection into fish embryos (Walker et al. 1996) has been shown to be a powerful tool for evaluating the toxicity of individual chemicals (Papoulias et al. 2000a, b; Edmunds et al. 2000; Hano et al. 2005) and also crude extracts of soil (Wilson and Tillitt 1996; Ishaq et al. 1999). In this method, embryos are exposed to a known amount of chemical directly, during early developmental stages. This technique can be performed in a small-scale facility, requires relatively short exposure times, shows good reproducibility, and simulates maternal transfer without the need to chronically expose adult fish.

Embryos of medaka (within 8 h after fertilization) have been exposed to TBT *in ovo,* via nanoinjection, at concentrations of 0 (control), 0.16, 0.80, 3.96, 19.2 and 82.1 ng/egg, using embryonic survival, development and hatching as end points (Hano et al. 2007). Hatched fry were reared until 60 days when they sexually mature, and sexual differentiation was also examined according to genetic and phenotypic sex, based on the existence of DMY (a male determining gene in medaka) and secondary sex characteristics. TBT caused a concentration-dependent mortality and impaired embryonic development (Table 11.3). However, no masculinization was detected in the 60 day post-hatch medaka adults. The lowest observed effective concentration for inducing abnormal embryonic development was estimated to be 0.16 ng/egg (ca. 160 ng/g egg) (Hano et al. 2007).

Nanoinjection of 16 combinations of a binary mixture of TBT (0, 7.5, 15, 30 ng/g-egg) and PCBs (0, 7.5, 15, 30 ng/g-egg) into embryo of medaka has been used to examine the effects on development and swim-up success of embryos, at environmentally relevant levels. The results showed that mixtures of TBT and PCBs induced alterations in the development and swim-up success of embryos, with LOEC estimated to be 7.5 ng-TBT/g-egg + 7.5 ng-PCBs/g-egg (Kim et al., in preparation). These results accord with the LOEC derived from swim-up success

Table 11.3 Toxicity of TBT to the medaka during embryogenesis and around the day of hatching (Hano et al. 2007)

Dose (ng/egg)	Total no. of embryos injected (day 0)	Survivors at 1 dpf	Abnormal development: Abnormal eye development	Abnormal development: Hemorrhage	Abnormal development: Delay in development	No. hatched larvae	Swim up failure
Uninjected	60	60	0	0	2 (3.3)	58 (96.7)	0
0	60	56	0	0	2 (3.6)	54 (96.4)	0
0.16	60	50**	5 (10.0)	0	2 (4.0)	38 (76.0)**	2 (5.2)
0.80	60	47**	6 (12.8)	2 (4.3)	6 (12.8)*	38 (80.1)**	1 (2.6)
3.96	60	38**	11 (28.9)	2 (5.3)	9 (23.7)**	26 (68.4)**	7 (26.9)
19.2	60	12**	3 (25.0)	0	4 (33.3)**	6 (50.0)**	3 (50.0)
82.1	60	2**	1 (50.0)	1 (50)	1 (50.0)**	0**	–

Numbers in parentheses indicate the relative proportion (in %) of survivors at 1 dpf (day post-fertilization).
*$p < 0.01$, **$p < 0.01$.

rates of medaka larvae in reproduction tests (<20 ng-TBT/g-egg with <31 ng-PCBs/g-eggs). The LOEC of TBT and PCBs obtained in these studies are consistent with concentrations reported in fish eggs. These results demonstrate that the toxicity of TBT and PCB mixtures to development of medaka could be attributed to maternal transfer, following exposure of adult female fish to these chemicals.

11.3 Conclusions

Results shown above clearly indicate that combined toxicity of TBT and PCBs is possible in the field. It is therefore important to evaluate the toxicity of mixtures of contaminants at environmentally relevant levels, because organisms inhabiting polluted areas are likely to be affected by chemicals in combination, rather than in isolation.

References

Arrhenius A, Backhaus T, Gronvall F et al. (2006) Effects of three antifouling agents on algal communities and algal reproduction: mixture toxicity studies with TBT, Irgarol, and Sea-Nine. Arch Environ Contam Toxicol 50:335–345

Burton JE, Dorociak IR, Schwedler TE et al. (2002) Circulating lysozyme and hepatic CYP1A activities during a chronic dietary exposure to tributyltin (TBT) and 3,3′,4,4′,5-pentachlorobiphenyl (PCB-126) mixtures in channel catfish, *Ictalurus punctatus*. J Toxicol Environ Health 65:589–602

Damstra T, Barlow S, Bergman A et al. (2002) Global assessment of the state-of-the-science of endocrine disruption. WHO/PCS/EDC/02.2. World Health Organization, Geneva, Switzerland.

DeLong GT, Rice CD (1997) Tributyltin potentiates 3,3′,4,4′,5-pentachlorobiphenyl- induced cytochrome P-4501A-related activity. J Toxicol Environ Health 51:131–148

Deneer JW, Sinnige TL, Seinen W et al. (1988) The joint acute toxicity to *Daphnia magna* of industrial organic chemicals at low concentrations. Aquat Toxicol 12:33–38

Edmunds JSG, McCarthy RA et al. (2000) Permanent and functional male-to-female sex reversal in d-rR strain medaka (*Oryzias latipes*) following embryo microinjection of *o,p*'-DDT. Environ Health Perspect 108:219–224

Fisk AT, Johnston TA (1998) Maternal transfer of organochlorines to eggs of walleye (*Stizostedion vitreum*) in Lake Manitoba and western Lake Superior. J Great Lakes Res 24:917–928

Hano T, Oshima Y, Oe T, Kinoshita M et al. (2005) Quantitative bio-imaging analysis for evaluation of sexual differentiation in germ cells of *olvas-GFP*/ST-II YI medaka (*Oryzias latipes*) nanoinjected in ovo with ethinylestradiol. Environ Toxicol Chem 24:70–77

Hano T, Oshima Y, Kim SG et al. (2007) Tributyltin causes abnormal development in embryos of medaka, *Oryzias latipes*. Chemosphere 69:927–933

Hermens J, Broekhuyzen E, Canton H et al. (1985) Quantitative structure activity relationships and mixture toxicity studies of alcohols and chlorohydrocarbons: effects on growth of *Daphnia magna*. Aquat Toxicol 6:209–217

Horiguchi T, Kojima M, Hamada F et al. (2006) Impact of tributyltin and triphenyltin on ivory shell (*Babylonia japonica*) populations. Environ Health Perspect 114 Suppl 1:13–19.

Ishaq R, Akerman G, Naf C et al. (1999) Organic pollutant characterization and toxicity testing of settling particulate matter by nanoinjection in sea trout (*Salmo trutta*) eggs. Environ Toxicol Chem 18:533–543

Inoue S, Oshima Y, Nagai K et al. (2004) Effect of maternal exposure to tributyltin on reproduction of the pearl oyster (*Pinctada fucata martensii*). Environ Toxicol Chem 23:1276–1281

Inoue S, Oshima Y, Usuki H et al. (2006) Effects of tributyltin maternal and/or waterborne exposure on the embryonic development of the Manila clam, *Ruditapes philippinarum*. Chemosphere 63:881–888

Kannan K, Villeneuve DL, Blankenship AL et al. (1998) Interaction of tributyltin with 3,3′,4,4′, 5-pentachlorobiphenyl-induced ethoxyresorufin O-deethylase activity in rat hepatoma cells. J Toxicol Environ Health 55:373–384

Kim YC, Cooper KR (1998) Interactions of 2,3,7,8-tetrachlorodibenxo-p-dioxin (TCDD) and 3,3′,4,4′,5-pentachlorobiphenyl (PCB 126) for producing lethal and sublethal effects in the Japanese medaka embryos and larvae. Chemosphere 36:409–418

Lyssimachou A, Jenssen BM, Arukwe A (2006) Brain cytochrome P450 aromatase gene isoforms and activity levels in atlantic salmon after waterborne exposure to nominal environmental concentrations of the pharmaceutical ethynylestradiol and antifoulant tributyltin. Toxicol Sci 91:82–92

Mac MJ, Schwartz TR, Edsall CC et al. (1993) Polychlorinated biphenyls in Great Lakes lake trout and their eggs: relations to survival and congener composition 1979–1988. J Great Lakes Res 19:752–765

McAllister BG, Kime DE (2003) Early life exposure to environmental levels of the aromatase inhibitor tributyltin causes masculinisation and irreversible sperm damage in zebrafish (*Danio rerio*). Aquat Toxicol 65:309–316

Monosson E, Fleming WJ, Sullivan CV (1994) Effects of the planar PCB 3,3′, 4,4′-tetrachlorobiphenyl (TCB) on ovarian development, plasma levels of sex steroid hormones and vitellogenin, and progeny survival in the white perch (*Morone americana*). Aquat Toxicol 29:1–19

Morcillo Y, Janer G, O'Hara SC et al. (2004) Interaction of tributyltin with hepatic cytochrome P450 and uridine diphosphate-glucuronosyl transferase systems of fish: in vitro studies. Environ Toxicol Chem 23:990–996

Mortensen AS, Arukwe A (2007) Modulation of xenobiotic biotransformation system and hormonal responses in Atlantic salmon (*Salmo salar*) after exposure to tributyltin (TBT). Comp Biochem Physiol C Toxicol Pharmacol 145:431–441

Nakata H, Tanabe S, Tatsukawa R et al. (1997) Bioaccumulation profiles of polychlorinated biphenyls including coplanar congeners and possible toxicological implications in Baikal seal (*Phoca sibirica*). Environ Pollut 95:57–65

Nakata H, Sakai Y, Miyawaki T et al. (2003) Bioaccumulation and toxic potencies of polychlorinated biphenyls and polycyclic aromatic hydrocarbons in tidal flat and coastal ecosystems of the Ariake Sea, Japan. Environ Sci Technol 37:3513–3521

Nakata H, Kannan K, Nasu T et al. (2006) Perfluorinated contaminants in sediments and aquatic organisms collected from shallow water and tidal flat areas of the Ariake Sea, Japan: environmental fate of perfluorooctane sulfonate in aquatic ecosystems. Environ Sci Technol 40:4916–4921

Nakayama K, Oshima Y, Hiramatsu K et al. (2004a) Alteration of general behavior of male medaka, *Oryzias latipes*, exposed to tributyltin and/or polychlorinated biphenyls. J Fac Agr Kyushu Univ 49:85–92

Nakayama K, Oshima Y, Yamaguchi T et al. (2004b) Fertilization success and sexual behavior in male medaka, *Oryzias latipes*, exposed to tributyltin. Chemosphere 55:1331–1337

Nakayama K, Oshima Y, Hiramatsu K et al. (2005a) Effects of polychlorinated biphenyls on the schooling behavior of Japanese medaka (*Oryzias latipes*). Environ Toxicol Chem 24:2588–2593

Nakayama K, Oshima Y, Nagafuchi K et al. (2005b) Early–life-stage toxicity in offspring from exposed parent medaka, *Oryzias latipes*, to mixtures of tributyltin and polychlorinated biphenyls. Environ Toxicol Chem 24:591–596

Nakayama K, Oshima Y (2008) Adverse effects of tributyltin on reproduction of Japanese medaka, Oryzias latipes. Coast Mar Sci 32:67–76

Nirmala K, Oshima Y, Lee R et al. (1999) Transgenerational toxicity of tributyltin and its combined effects with polychlorinated biphenyls on reproductive processes in Japanese medaka (*Oryzias latipes*). Environ Toxicol Chem 18:717–721

Nishikawa J, Mamiya S, Kanayama T et al. (2004) Involvement of the retinoid X receptor in the development of imposex caused by organotins in gastropods. Environ Sci Technol 38:6271–6276

Örn S, Andersson PL, Förlin L et al. (1998) The impact on reproduction of an orally administered mixture of selected PCBs in zebrafish (*Danio rerio*). Arch Environ Contam Toxicol 35:52–57

Papoulias DM, Noltie DB, Tillitt DE (2000a) An in vivo model fish system to test chemical effects on sexual differentiation and development: exposure to ethinylestradiol. Aquat Toxicol 48:37–50

Papoulias DM, Noltie DB, Tillitt DE (2000b) Effects of methyl testosterone exposure on sexual differentiation in medaka, *Oryzias latipes*. Mar Environ Res 50:181–184

Rice CD, Roszell LE (1998) Tributyltin modulates 3,3′,4,4′,5-pentachlorobiphenyl (PCB-126)-induced hepatic CYP1A activity in channel catfish, *Ictalurus punctatus*. J Toxicol Environ Health 55:197–212

Santos MM, Micael J, Carvalho AP et al. (2006) Estrogens counteract the masculinizing effect of tributyltin in zebrafish. Comp Biochem Physiol C Toxicol Pharmacol 142:151–155

Schlenk D, Sapozhnikova Y, Baquirian JP et al. (2002) Predicting chemical contaminants in freshwater sediments through the use of historical biochemical endpoints in resident fish species. Environ Toxicol Chem 21:2138–2145

Schmidt K, Staaks GB, Pflugmacher S et al. (2005a) Impact of PCB mixture (Aroclor 1254) and TBT and a mixture of both on swimming behavior, body growth and enzymatic biotransformation activities (GST) of young carp (*Cyprinus carpio*). Aquat Toxicol 71:49–59

Schmidt K, Steinberg CEW, Staaks GBO (2005b) Influence of a xenobiotic mixture (PCB and TBT) compared to single substances on swimming behavior or reproduction of *Daphnia magna*. Acta Hydrochim Hydrobiol 33:287–300

Shimasaki Y, Kitano T, Oshima Y et al. (2003) Tributyltin causes masculinization in fish. Environ Toxicol Chem 22:141–144

Silva E, Rajapakse N, Kortenkamp A (2002) Something from "nothing"—eight weak estrogenic chemicals combined at concentrations below NOECs produce significant mixture effects. Environ Sci Technol 36:1751–1756

Ueno D, Takahashi S, Tanaka H et al. (2003) Global pollution monitoring of PCBs and organochlorine pesticides using skipjack tuna as a bioindicator. Arch Environ Contam Toxicol 45:378–389

Ueno D, Inoue S, Takahashi S et al. (2004a) Global pollution monitoring of butyltin compounds using skipjack tuna as a bioindicator. Environ Pollut 127:1–12

Ueno D, Kajiwara N, Tanaka H et al. (2004b) Global pollution monitoring of polybrominated diphenyl ethers using skipjack tuna as a bioindicator. Environ Sci Technol 38:2312–2316

Walker M, Zabel E, Ackerman G et al. (1996) Fish egg injection as an alternative exposure route for early life stage toxicity studies: description of two unique methods. In Ostrander GK, ed, Techniques in Aquatic Toxicology. Lewis, New York, pp 41–72

Wang C, Zhao Y, Zheng R et al. (2006) Effects of tributyltin, benzo[a]pyrene, and their mixture on antioxidant defense systems in *Sebastiscus marmoratus*. Ecotoxicol Environ Saf 65:381–387

Westerlund L, Billsson K, Andersson PL et al. (2000) Early life-stage mortality in zebrafish (*Danio rerio*) following maternal exposure to polychlorinated biphenyls and estrogen. Environ Toxicol Chem 19:1582–1588

Wilson PJ, Tillitt DE (1996) Rainbow trout embryotoxicity of a complex contaminant mixture extracted from Lake Michigan lake trout. Mar Environ Res 42:129–134

Yamada H, Takayanagi K, Tateishi M et al. (1997) Organotin compounds and polychlorinated biphenyls of livers in squid collected from coastal waters and open oceans. Environ Pollut 96:217–226

Chapter 12
Immunotoxic Effects of Organotin Compounds in Teleost Fish

Ayako Nakayama, Helmut Segner, and Shin'ichiro Kawai

12.1 Evidence of Toxic Effects of Environmental Contaminants on the Immune System of Fish

A number of field studies have provided evidence that environmental toxicants can modulate immune parameters of exposed fish (for overviews see Snieszko 1974; Dunier and Siwicki 1993; Zeeman 1994; Zelikoff 1994; Rice 2001). Also, laboratory studies demonstrated that toxicants impact the immune system of fish (e.g., Weeks et al. 1988; Thuvander and Carlstein 1991; Kaattari et al. 1994; Rice and Schlenk 1995; Sanchez-Dardon et al. 1999; Carlson et al. 2004; Quabius et al. 2005). Consequently, it has been suggested to utilize immune parameters of fish as indicators of environmental pollution (Anderson 1990; Weeks et al. 1992; Wester et al. 1994).

In fish immunotoxicological studies, emphasis has been given to the measurement of single endpoints or functions such as depressed phagocytic activity of macrophages or the alteration of oxidative burst activity of immune cells. However, the ultimate concern is that toxic exposure might increase the susceptibility of fish to pathogen infections. Thus, it is important to elucidate the consequences of changes in molecular or cellular immune parameters for the overall immune system function, since, as pointed out by Rice (2001), alterations at the molecular and cellular level do not necessarily translate into immune modulation at the system level. Rather few studies directly correlated alterations of specific molecular and cellular immune parameters with altered immune system function and/or altered susceptibility to pathogens (Palm et al. 2003; Carlson et al. 2002; Burki et al. 2008). However, some more indirect evidence is available that toxic impact on immune

A. Nakayama and H. Segner
Centre for Fish and Wildlife Health, Institute of Animal Pathology, University of Berne, Switzerland

A. Nakayama and S. Kawai
Graduate School of Human Environmental Sciences, Kobe College, Hyogo, Japan

T. Arai et al. (eds.), *Ecotoxicology of Antifouling Biocides*,
DOI: 10.1007/978-4-431-85709-9_12, © Springer 2009

parameters increases the susceptibility of fish to disease. One example comes from the decline of wild Pacific salmon populations in the USA and Canada. When juvenile chinook salmon (*Oncorhynchus tshawytscha*) collected from estuaries in the Puget Sound area and showing suppression of various immune parameters were infected with *Vibrio anguillarum* in the laboratory, they were more susceptible than fish from non- or less-polluted areas (Arkoosh et al. 1998). Significant differences were seen even 2 months after removal from the contaminated areas, suggesting that the chemical exposure had a lasting effect on disease susceptibility. This assumption is supported by the study of Milston et al. (2003) who showed that short-term contaminant exposure of chinook salmon during early life-history stages resulted in long-term impairment of humoral immune competence. Under complex field conditions, as in Puget Sound, it is difficult if not impossible to provide conclusive evidence that the contaminant-induced immunosuppression is causative to the observed decline of chinook salmon populations. However, in a demographic modeling study, Spromberg and Meador (2005) could show that immune suppression acting through reduction of age-specific survival would produce pronounced changes in the population growth rate. This result highlights the potential of immunotoxicants to adversely affect organism health and population growth of aquatic wildlife.

One factor that complicates the functional interpretation of toxicant-induced changes in single immunological endpoints of fish is the still rather limited knowledge of the immune systems of fish. Although often addressed as "primitive" immune system compared to mammals, the immune systems of fish are highly evolved. Importantly, since water is an excellent transmission medium for pathogens, much better than air, the efficiency of piscine immune systems has to be efficient enough to meet the demands of this challenging environment.

In the following section, we will introduce a few basic components and features of the immune system of teleost fish. For a comprehensive overview on the fish immune system, the reader is referred to the reviews of Iwama and Nakanishi (1996), Rice (2001) and Burnett (2005).

12.2 The Immune System of Teleost Fish

12.2.1 Immune Organs and Cells

While several features of the teleostean immune system are similar to those in the immune systems of higher vertebrates, there exist also a number of unique characters and processes such as the prominence of antipathogenic factors in the skin mucus (Suzuki et al. 2003) or the strong dependence on the innate immunity (Plouffe et al. 2005). In contrast to mammals, teleost fish lack bone marrow and lymph nodes. A clear separation of lymphoid and myeloid tissues is not realized but the "lymphopoietic" organs of teleost contain both lymphoid and myeloid cells in different stages of development. These organs include head kidney, excretory kidney,

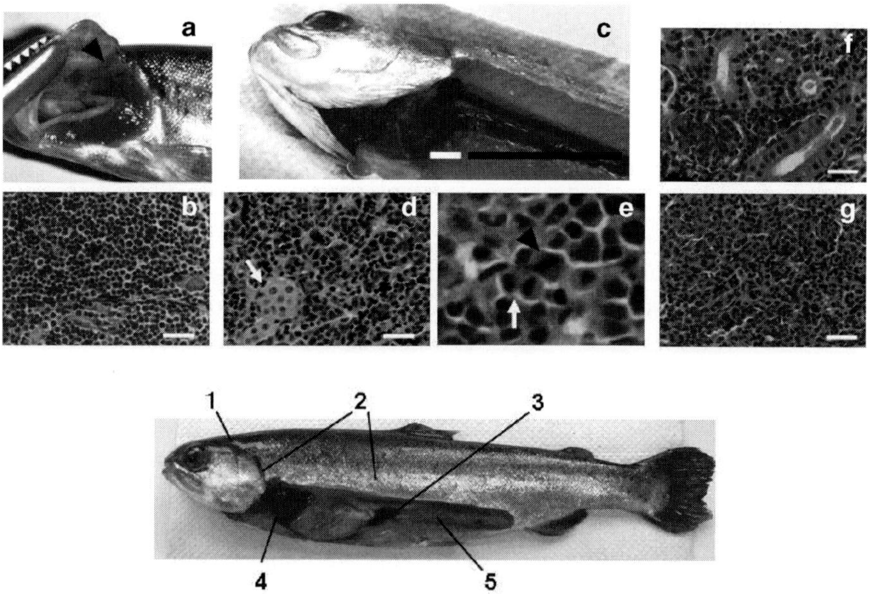

Fig. 12.1 Immune-related organs in rainbow trout (*Oncorhynchus mykiss*). Location of thymus (1 and a; *arrowhead*) and thymus histological appearance (b) occupied with lymphoblasts (thymocytes). Location of kidney (2 and c; the *white* and *black lines* indicate head – *anterior* – kidney and trunk – excretory – kidney. The head kidney (d) histologically consists of hematopoietic tissue with mitotically active (e; *arrowhead*) precursors of erythrocytes and leukocytes (e; *arrow*), lymphoid parts, as well as endocrine (interrenal and chromaffin) cells (d; *arrow*). The trunk kidney consists of hematopoietic tissue, and excretory tissue with renal tubules and glomeruli (f). Spleen (3) serves to remove blood cells from the circulation, as antigen-trapping and presenting organ, as well as hematopoietic organ. Histologically, erythrocyte-rich and lymphocyte-rich fractions are clearly visible (g). (4) Liver, containing resident phagocytes. Gut-associated lymphoid tissue (GALT) is present along the intestine (5). Hematoxylin-Eosin staining, scale bar; 50 μm

thymus and spleen (Fig. 12.1). Further, with pathogens entering the fish usually from the water, intestinal mucosa as well a skin and branchial epithelia function as major immune barriers. Teleosts possess a reticuloendothelial systems (RES) encomprising phagocytes in various locations such as wandering serosal phagocytes in the body cavity, endothelium fixed phagocytes in the liver (although the existence of Kupffer cells homologues is discussed controversially – see Hinton et al. 2001), or microglia in the neuronal system.

Teleost immune cells comprise – according to the present state of knowledge – three types of granulocytes including neutrophils, basophils and eosinophils (the latter are considered to be functionally analogous to the mammalian mast cells), monocytes/macrophages, T- and B-lymphocytes, as well as thrombocytes. In addition, natural cytotoxic cells (NCC) – which are considered to be analogous to natural killer cells in mammals – and cytotoxic T cells requiring MHC class I molecules have been characterized (Fischer et al. 2006). The immune cell composition of lymphoid organs and blood appears to be highly variable both between individual fishes and between fish species.

12.2.2 The Innate, Non-specific Immune System

The innate immune response provides fairly rapid defense mechanisms against pathogens, which are immediately activated after recognition of pathogen associated molecular patterns (PAMP) that are common to many pathogens, for instance, bacterial lipopolysaccharide (LPS). The key role in innate immune system activation is played by pattern recognition receptors (PRR) of the host which recognize either the foreign molecules or endogenous, host-derived alarm molecules (Magnadóttir 2006). Main components of the innate immune system in fish include physical barriers such as skin or endothelia, humoral factors (anti-bacterial peptides, lysozyme, acute phase proteins or complement factors) and phagocytic cells such as granulocytes, monocytes/macrophages and NCC. Humoral factors such as lysozyme are found in plasma, skin mucus as well as in eggs, and are a frequently measured parameter in immunotoxicity studies with fish. Complement factors are synthesized in the liver and probably also in extrahepatic sites (Løvoll et al. 2007; Boshra et al. 2004), and they have a number of functions including opsonization of the pathogen or activation of the acquired immune system. The main functions of the phagocytic cells are to phagocytose tissue debris and microorganisms, to secrete immune response regulating factors and to bridge innate and adaptive immune responses (Secombes and Fletcher 1992). A key feature of phagocytic cells such as macrophages and granulocytes is the respiratory burst activity, i.e. the generation of reactive oxygen intermediates in order to kill pathogenic microorganisms (Sharp and Secombes 1993). Measuring the respiratory burst activity is often used as an indicator of the immunological capacity of the fish, and has been shown to be modulated by a wide variety of environmental toxicants (Rice 2001).

12.2.3 The Acquired, Specific Immune System

The importance of the acquired immune system in bony fish is believed to be secondary to the innate immune system. Cells involved in the specific immune system are T- and B-lymphocytes, which mediate the cellular and humoral response, respectively. Although the characterization of piscine T-lymphocytes is not as far progressed as in mammals, it is clear that fish possess both antigen-presenting T-helper cells (CD4-like) and cytotoxic T-cells (CD8-like) (Fischer et al. 2006). Fish B-lymphocytes produce immunoglobulins which are primarily tetrameric IgMs (Warr 1983), instead of the pentameric immunoglobulins of mammals. The kinetics of antibody production is much slower (weeks to months) in teleosts than in mammals, and the antibody response shows a clear temperature dependency. Toxicant effects on the acquired immune system are often measured as alteration of mitogen-stimulated lymphoproliferation or a change of antibody production.

12.3 Evaluation for Immunotoxic Effects of Tributyltin (TBT) Using the Rainbow Trout Model

Reported immunotoxicies of TBT in fish include thymus atrophy (Wester and Canton 1987; Grinwis et al. 1998), reduction of leukocyte numbers (Grinwis et al. 1998) and impairment of leukocyte functions, for instance, decrease of phagocytic activities (Wishkovsky et al. 1989; Rice and Weeks 1991; Rice et al. 1995; Harford et al. 2005; Nakayama et al. 2005, 2007) and reduced lymphocyte functions (O'Halloran et al. 1998; Regala et al. 2001; Harford et al. 2007). In vitro studies with isolated fish leukocytes indicated that low doses of TBT and short-term exposures to TBT stimulate the production of reactive oxygen species, whereas higher doses or longer exposure period suppress reactive oxygen formation (Rice and Weeks 1991). Possibly, these effects are mediated via an effect of TBT on cellular Ca^{2+} levels (Elferink et al. 1986; Raffray et al. 1993; Rice et al. 1995). Further in vitro immune cell effects reported for TBT include the suppression of mitogen-stimulated lymphocyte proliferation (O'Halloran et al. 1998). A limitation in our current knowledge on TBT-induced immunological alterations in fish is that published studies related observed effects to external TBT concentrations, but not to internal body burdens. Knowledge of the relationship between accumulated TBT doses in the organism and immunotoxic responses would help to compare results between studies and to extrapolate from laboratory-derived effect thresholds to the field situation. In order to better understand the relationship between immunotoxic effects of TBT and accumulated TBT concentrations, we experimentally set out for two types of exposure experiments using the rainbow trout model: an immersion exposure with TBT, and intraperitoneal (ip) injection of TBT. In both approaches, the actual amounts of TBT in target immune tissues, especially in blood were analytically determined in order to link the internal doses to the effects.

12.3.1 Exposure of Rainbow Trout Tributyltin via Water Alters the Number of Neutrophils and Their Respiratory Burst Activity

Our first aim was to determine toxic effects of TBT on phagocyte numbers in rainbow trout. After immersion exposure of trout to 20 μg tributyltin chloride (TBTCl)/l for 5 days, we measured the numbers of head kidney neutrophils and their respiratory burst activities using flow cytometry. The advantage of this method is the discrimination of head kidney leukocyte populations (lymphocytes + thrombocytes, monocytes/macrophages and granulocytes) by size (forward scatter: FSC, X-axis) and granularity (side scatter: SSC, Y-axis). The typical FSC/SSC cytograms of head kidney leukocytes are shown in Fig. 12.2a for the control and Fig. 12.2b for the TBT exposure group. In these cytograms, an increased percentage of neutrophils (marked with dotted line) prepared from the TBT exposure group compared to non-treated group is clearly shown. Moreover, the respiratory burst activity, measured

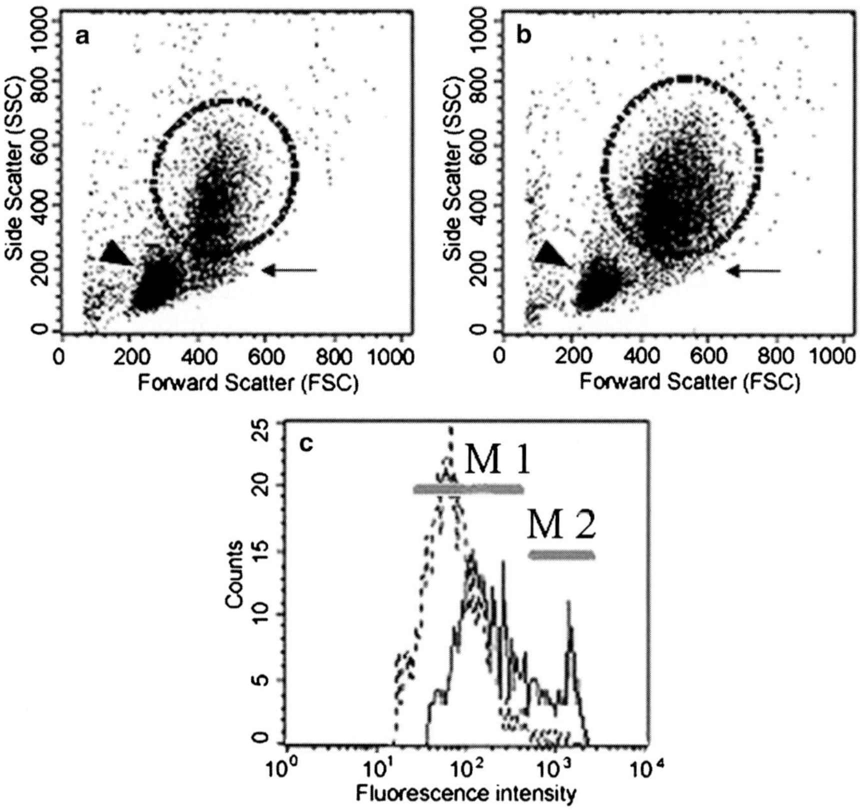

Fig. 12.2 Flow cytometric analysis (a, b) and detection of respiratory burst activity (c) of head kidney leukocytes from rainbow trout exposed to TBT 20 µg/l for 5 days (a, b) Typical FSC/SSC cytogram of one trout from the control group (a) and one from the TBT-treated group (b) shows the distribution of lymphocytes (*arrowheads*), granulocytes (almost neutrophils; *dotted lines*) and monocytes/macrophages (*arrows*). Note the increase of neutrophile granulocytes (b; *dotted line*) in TBT exposed trout. (c) The respiratory burst activity of head kidney leukocytes obtained from trout exposed to TBT 20 µg/l (*dotted line*) decreased, especially the population in control trout (*solid line*) expressing a high activity (10^3 fluorescence intensity; M2-area). Additionally, the mean fluorescence intensity of lower 10^2 fluorescence intensity (M1-area) in TBT exposed trout shifted toward the weaker intensity.

by means of fluorescence intensity, in head kidney leukocytes of TBT exposure group is decreased (Fig. 12.2c). Especially, the number of leukocytes which have a high respiratory burst activity, determined by a high fluorescence intensity around 1,000 (M2), clearly decreased in the exposure group. Probably, the population expressed lower respiratory burst activity, shown as lower fluorescence intensity around 100 (M1), representing a neutrophil population with disturbed functional activities. These results demonstrate firstly, that water-borne exposure of rainbow trout to 20 μg TBTCl/l for 5 days elicited a temporal increase of the neutrophil population together with a concomitant decrease of other leukocyte populations (lymphocytes and monocytes) in the head kidney and secondly, these neutrophils although being increased in number, show reduced respiratory burst activities.

The concentration selected for our experiment – 20 μg TBTCl/l – is rather high and lethal in the long term. Thus, our finding that TBT affected the composition of the leukocyte population in the head kidney and their respiratory burst activity may be questioned. However, results from longer-term exposure experiments with lower TBT concentrations (Oliveira-Ribeiro et al. 2002; Schwaiger et al. 1992) also reported changes in the cellular composition of immune organs such as increased karyorrhexis of lymphoid cells and erythrophagia in spleen. These results support our observations that immune cells are a target of the toxic action of TBT.

The same analyses for composition of head kidney leukocytes were made for trout exposed by immersion to TBTCl at 5 μg/l for 28 days, the results showed no remarkable changes of leukocyte composition, except an increased neutrophil population in the blood after 7 days immersion. To date, TBT accumulation in blood is well studied (Oshima et al. 1997; Shim et al. 2002). In the next section, internal doses of TBT and its effects on a humoral factor in blood are discussed.

12.3.2 Relationship Between TBT Concentration in Blood and Lysozyme Activity in Plasma

The aim of this study was to evaluate the relationship between TBT concentration in the blood and plasma lysozyme activity. TBT was analysed using a gas chromatograph equipped with a mass spectrometer and lysozyme was analysed by decreased turbidity of *Micrococcus* bacterial solution. To this end, rainbow trout were exposed to TBTCl at 0 (control), 5 and 10 μg/l for 5 days. As shown in Fig. 12.3, TBT accumulated in a dose-dependent manner in the blood of exposed fish and the levels of TBT in 5 and 10 μg/l 5day exposure groups were recorded at 5.8 and 10 mg/l blood, respectively. This correlated with a significantly decreased plasma lysozyme activity of fish exposed to 10 μg/l for 5 days, whereas in the 5 μg/l exposure group, lysozyme activity was not affected. The different effect might be due to the border between non-effective and effective internal doses of TBT in blood or due to the short exposure time.

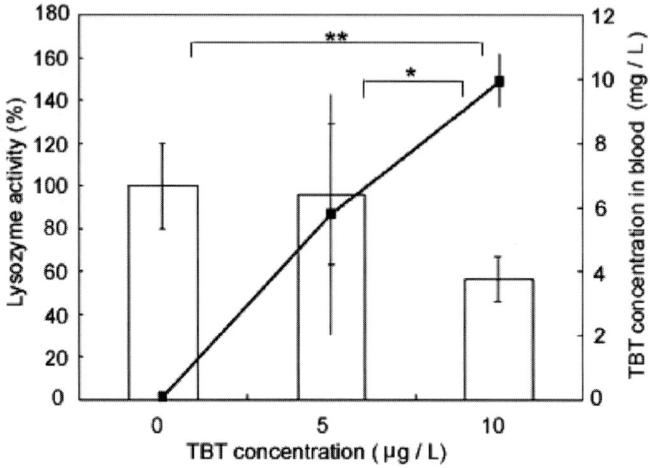

Fig. 12.3 The effects of different concentrations of TBT exposure for 5 days on lysozyme activity and TBT concentration in blood. Lysozyme activities (*columns* □) and TBT concentration in blood (*line* ■) are shown. Lysozyme activity and TBT concentration measurements were made on blood samples from the same individuals. The mean values ± SD are shown the average of fish in each experimental group. The ** and * show significant differences between 0 and 10 μg/l exposure groups, and 5 and 10 μg/l exposure groups of lysozyme activity, for $p < 0.01$ and $p < 0.05$, respectively (Nakayama et al. 2005 with some modifications)

Oshima et al. (1997) reported that a cultured Japanese flounder (*Paralichthys olivaceus*) sold in a market possessed very high accumulated TBT concentrations in the blood. While the experimental levels used in our study were more than 1,000 times higher than TBT concentrations usually detected in an aquatic environment (maximum ppt levels in water), the TBT concentrations recorded in the blood from the experimental fish were only about twofold higher than those detected from the former market fish. Thus, since internal TBT concentrations of the laboratory and the field-caught fish are rather similar, the immunotoxic effects observed in our experimental fish may point to the presence of immunotoxic effects in the cultured flounder. The available evidence from short term laboratory exposures indicate that internal blood TBT concentrations of <~5 mg/l blood do not affect plasma lysozyme activity while fish accumulating 10 mg/l blood or more show significant immunosuppressive changes including modulated leukocyte populations, decreased respiratory burst activities (as shown in Fig. 12.2) as well as down regulated lysozyme activities in plasma.

12.3.3 Enhanced Susceptibility to a Pathogen Correlated with TBT Intraperitoneal Exposure

While the available data clearly indicate that TBT has immunotoxic potency for different fish species, it is less clear whether these TBT-induced changes of immune parameters translate into altered resistance or susceptibility towards pathogenic microorganisms. To address this question, we treated rainbow trout with intraperitoneal (ip) injection of TBT and assessed the consequences on the susceptibility of trout to the bacterial pathogen, *Aeromonas salmonicida*, the causative agent of furunculosis. Rainbow trout reared at 13° were treated with a single ip injection of 2.5 mg TBTCl/kg body weight 7 days prior bacterial challenge. TBT distribution in the tissues was as follows: blood 498 μg/l, Liver 4,037 μg/kg, kidney 2,624 μg/kg and spleen 2,087 μg/kg tissue. Seven days after TBT injection, fish were challenged intraperitoneally with 10^5 colony forming units (CFU) of *A. salmonicida* per fish, and the resulting cumulative mortality was monitored over a 7-day-period. Initial mortalities occurred 4 days after infection with *A. salmonicida* in fish that previously had received TBT treatment. Macroscopic observation in the dead fish revealed an enlarged spleen, liquefied kidney tissue and dilated intestine. Cumulative mortality in the TBT treatment reached 90% by day 7 post-challenge. In contrast, *A. salmonicida* challenge resulted in no mortality in control fish (corn oil injection instead of TBT), although enlarged spleen and hyperemia in the muscle tissue surrounding the injection site were present. These findings suggest that the immunotoxicity of TBT, as shown in the immersion experiment (see 12.3.2.) translates into enhanced pathogen susceptibility of rainbow trout. This observation is particularly remarkable since the tissue TBT levels of rainbow trout from the laboratory challenge experiment were clearly lower than TBT tissue burdens often found wild fish (Oshima et al. 1997; Hassani et al. 2006, summarized by Liu et al. 2006). This observation highlights the risk of TBT accumulation for the disease resistance of wild fish populations (Fig. 12.4).

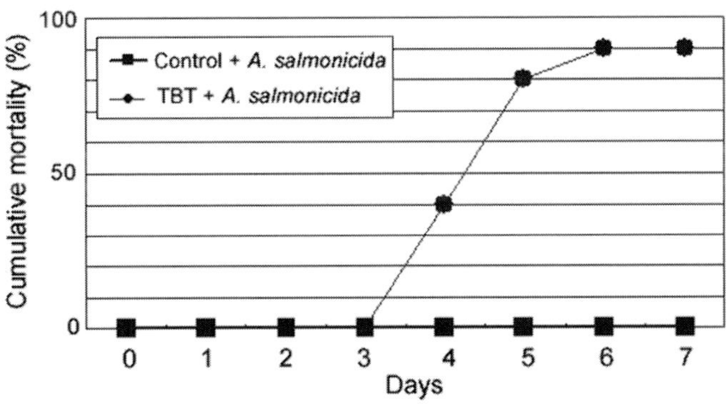

Fig. 12.4 Cumulative mortality. Prior to the bacterial challenge with *Aeromonas salmonicida* (10^5 CFU per fish) at 13°C, treated fish received a single ip injection of TBT 2.5 mg/kg body weight for 7 days (Control fish received corn oil only). Afterwards, fish were challenged with a bacterial pathogen, *Aeromonas salmonicida*, therefore, 0 day indicates 7 days following the TBT administration. After the pathogen challenge, the cumulative mortalities of both control and TBT-injected fish infected with *A. salmonicida* for 7 days are shown (unpublished data)

12.4 Conclusions

Both the literature and our own experimental data provide clear evidence that TBT is immunotoxic to fish, even at low, environmentally relevant concentrations. The available data also provide evidence that this immunotoxic activity can compromise the fish's ability to resist pathogens, thus posing an ecological risk. Future studies on the immunotoxicity of TBT should aim to develop understanding of the immunotoxic mechanisms by which TBT disturbs the resistance to pathogens. Furthermore, greater attention should be given to monitor for indications of compromised immune status in fish from TBT-contaminated areas.

Acknowledgement Thanks are extended to Dr. Hiroya Harino for editorial comments, and Dr. Bernd Köllner for his valuable advices in improving the contents of this chapter.

References

Anderson DP (1990) Immunological indicators: effects of environmental stress on immune protection and disease outbreaks. Am Fish Soc Symp 8: 38–50

Arkoosh MR, Casillas E, Huffman P et al. (1998) Increased susceptibility of juvenile Chinook salmon (*Oncorhynchus tshawytscha*) from a contaminated estuary to the pathogen *Vibrio anguillarum*. Trans Am Fish Soc 127: 360–374

Boshra H, Gelman AE, Sunyer JO et al. (2004) Structural and functional characterization of complement C4 and C1s-like molecules in teleost fish: insights into the evolution of classical and alternative pathways. J Immunol 173: 349–359

Burki R, Krasnov A, Bettge K et al. (2008) Molecular crosstalk between chemical and pathogen stressors and consequences on disease manifestation in rainbow trout. Submitted.

Burnett KG (2005) Impact of environmental toxicants and natural variables on the immune system of fishes. In: Mommsen TP, Moon TW (eds) Biochemistry and Molecular Biology of Fishes, vol. 6, Environmental Toxicology. Elsevier, Amsterdam

Carlson EA, Li Y, Zelikoff JT et al. (2002) Exposure of Japanese Medaka (*Oryzias latipes*) to benzo[a] pyrene suppresses immune function and host resistance against bacterial challenge. Aqua Toxicol 56: 289–301

Carlson EA, Li Y, Zelikoff JT et al. (2004) Suppressive effects of benzo[a]pyrene upon fish immune function: Evolutionarily conserved cellular mechanisms of immunotoxicity. Mar Environ Res 58: 731–734

Dunier M, Siwicki AK (1993) Effects of pesticides and other organic pollutants in the aquatic environment on immunity in fish: a review. Fish Shellfish Immunol 3: 423–438

Elferink JGR, Deierkauf M, Van Stevenick J et al. (1986) Toxicity of organotin compounds for polymorphonuclear leukocytes: The effect on phagocytosis and exocytosis. Biochem Pharmacol 35: 3727–3732

Fischer U, Utke K, Somamoto T et al. (2006) Cytotoxic activities of fish leukocytes. Fish Shellfish Immunol 20: 209–226

Grinwis GCM, Boonstra A, van den Brandhof EJ et al. (1998) Short-term toxicity of bis(tri-*n*-butyltin)oxide in flounder (*Platichthys flesus*): Pathology and immune function. Aquat Toxicol 42: 15–36

Harford AJ, O'Halloran K, Wright PFA et al. (2005) The effects of in vitro pesticide exposures on the phagocytic function of four native Australian freshwater fish. Aquat Toxicol 75: 330–342

Harford AJ, O'Halloran K, Wright PFA et al. (2007) Effect of in vitro and in vivo organotin exposures on the immune functions of Murray cod (*Maccullochella peelii peelii*). Environ Toxicol Chem 26: 1649–1656

Hassani LH, Frenich AG, Benajiba MH et al. (2006) Assessment of butyltin and phenyltin pollutions in the sea bass mullet, *Mugil cephalus*, along the Moroccan and Spanish coasts (Mediterranean Sea). Arch Environ Contam Toxicol 51: 608–614

Hinton DE, Segner H, Braunbeck T et al. (2001) Toxic responses of the liver. In: Schlenk D, Benson WH (eds) Target Organ Toxicity of Marine and Freshwater Teleosts. Taylor & Francis, London/New York

Iwama G, Nakanishi T (1996) The fish immune system. In: Hoar WS, Randall DJ, Farrell AP (eds) Fish Physiology, vol. 15. Academic, San Diego, CA

Kaattari SL, Adkison M, Shapiro D et al. (1994) Mechanisms of immunosuppression by aflatoxin B1. In: Stolen JS, Fletcher TC (eds) Modulators of Fish Immune Responses. SOS Publications, Fair Haven, NJ

Liu SM, Hsia MP, Huang CM et al. (2006) Accumulation of butyltin compounds in cobia Rachycentron canadum raised in offshore aquaculture sites. Sci Total Environ 355: 167–175

Løvoll M, Dalmo RA, Bøgwald J et al. (2007) Extrahepatic synthesis of complement components in the rainbow trout (*Oncorhynchus mykiss*). Fish Shellfish Immunol 23: 721–731

Magnadóttir (2006) Innate immunity of fish (overview). Fish Shellfish Immunol 20: 137–151

Milston RH, Fitzpatrick MS, Vella AT et al. (2003) Short-term exposure of chinook salmon to o,p-DDE or DMSO during early life-history stages causes long-term humoral immunosppression. Environ Health Perspect 111: 1601–1607

Nakayama A, Kurokawa Y, Kawahara E et al. (2005) The immunotoxic effects of tributyltin on non-specific biodefense system in rainbow trout (*Oncorhynchus mykiss*). Jpn J Environ Toxicol 8: 23–35

Nakayama A, Kurokawa Y, Harino H et al. (2007) Effects of tributyltin on the immune system of Japanese flounder (*Paralichthys olivaceus*). Aquat Toxicol 83: 126–133

O'Halloran K, Ahokas JT, Wright PFA et al. (1998) Response of fish immune cells in vitro organotin exposures. Aquat Toxicol 40: 141–156

Oliveira Ribeiro CA, Schatzmann M, Silva de Assis HC et al. (2002) Evaluation of tributyltin subchronic effects in tropical freshwater fish (*Astyanax bimaculatus*, Linnaeus, 1758). Ecotoxicol Environ Saf 51: 161–167

Oshima Y, Nirmala K, Go J et al. (1997) High accumulation of tributyltin in blood among the tissues and applicability to environmental monitoring. Environ Toxicol Chem 16: 1515–1517

Palm RC, Powell DB, Skillman A et al. (2003) Immunocompetence of juvenile chinook salmon against *Listonella anguillarum* following dietary exposure to polycylcic aromatic hydrocarbons. Environ Toxicol Chem 22: 2986–2994

Plouffe DA, Hanington PC, Walsh JG et al. (2005) Comparison of select innate immune mechanisms of fish and mammals. Xenotransplantation 12: 266–277

Quabius ES, Krupp G, Secombes CJ et al. (2005) Polychlorinated biphenyl 126 affects expression of genes involved in stress-immune interaction in primary cultures of rainbow trout anterior kidney cells. Environ Toxicol Chem 24: 3053–3060

Raffray M, McCarthy D, Snowden RT et al. (1993) Apoptosis as a mechanism of tributyltin cytotoxicity to thymocytes: relationship of apoptostic markers to biochemical and cellular effects. Toxicol Appl Pharmacol 119: 122–130

Regala RP, Rice CD, Schwedler TE et al. (2001) The effects of tributyltin (TBT) and 3,3',4,4',5-Pentachlorobiphenyl (PCB-126) mixtures on antibody responses and phagocytes oxidative burst activity in channel catfish, *Ictalurus punctatus*. Arch Environ Contam Toxicol 40: 386–391

Rice CD (2001) Fish immunotoxicology. In: Schlenk D, Benson WH (eds) Target Organ Toxicity in Marine and Freshwater Teleosts, vol. 2. Taylor & Francis, London

Rice CD, Schlenk D (1995) Immune function and cytochrome P4501A activity after acute exposure to 3,3',4,4',5-pentachlorobiphenyl (PCB 126) in channel catfish. J Aqua Anim Health 7: 195–204

Rice CD, Weeks BA (1991) Tributyltin stimulates reactive oxygen formation in toadfish macrophages. Dev Comp Immunol 15: 431–436

Rice CD, Banes MM, Ardelt TC et al. (1995) Immunotoxicity in Channel Catfish, *Ictalurus punctatus*, following acute exposure to tributyltin. Arch Environ Contam Toxicol 28: 464–470

Sanchez-Dardon J, Voccia I, Hontela A et al (1999) Immunomodulation by heavy metals tested individually or in mixtures in rainbow trout exposed in vivo. Environ Toxicol Chem 18: 1492–1497

Schwaiger J, Bucher F, Hermann F et al (1992) A prolonged toxicity study on the effects of sublethal concentrations of bis(tri-*n*-butyltin)oxide (TBTO): histopathological and histochemical findings in rainbow trout (*Oncorhynchus mykiss*). Aquat Toxicol 23: 31–48

Secombes CJ, Fletcher TC (1992) The role of phagocytes in the protective mechanisms of fish. Annu Rev Fish Dis 2: 53–71

Sharp GJE, Secombes CJ (1993) The role of reactive oxygen species in the killing of the bacterial fish pathogen *Aeromonas salmonicida* by rainbow trout macrophages. Fish shellfish immunol 3: 119–129

Shim WJ, Jeon JK, Oh JR, Kim NS et al (2002) Accumulation of tributyltin in the blood of fish: its application for monitoring in the marine environment. Environ Toxicol Chem 21: 1451–1455

Snieszko SF (1974) The effects of environmental stress on outbreaks of infectious diseases of fishes. J Fish Biol 6: 374–383

Spromberg JA, Meador JP (2005) Relating results of chronic toxicity responses to population-level effects: modeling effects on wild Chinook salmon populations. Integr Environ Assess Manag 1: 9–21

Suzuki Y, Tasumi S, Tsutsui S et al (2003) Molecular diversity of skin mucus lectins in fish. Comp Biochem Physiol B 136: 723–730

Thuvander A, Carlstein M (1991) Sublethal exposure of rainbow trout, *Oncorhynchus mykiss*, to polychlorinated biphenyls: effects on the humoral immune response to *Vibrio anguollarum*. Fish Shellfish Immunol 1: 77–86

Warr GW (1983) Immunogloblin of the toadfish, *Spheroides glaber*. Comp Biochem Physiol B 76: 507–514

Weeks BA, Warinner JE, Mathews ES et al (1988) Influence of toxicants on phagocytosis, pinocytosis, and melanin accumulation by fish macrophages. Aqua Toxicol 11: 395–438

Weeks BA, Anderson DP, DuFour AP et al (1992) Immunological biomarkers to assess environmental stress. In: Huggett RJ, Kimerle RA, Mehrle PM, Bergmann HL (eds) Biomarkers: Biochemical, Physiological, and Histological Markers of Anthropogenic Stress. Lewis, Boca Raton, FL

Wester PW, Canton JH (1987) Histopathological study of *Poecilia reticulate* (guppy) after long-term exposure to bis(tri-*n*-butyltin)oxide (TBTO) and di-*n*-butyltindichloride (DBTC). Aquat Toxicol 10: 143–165

Wester PW, Vethaak AD, van Muiswinkel WB et al (1994) Fish as biomarkers in immunotoxicology. Toxicol 86: 213–232

Wishkovsky A, Mathews ES, Weeks BA et al (1989) Effect of tributyltin on the chemiluminescent response of phagocytes from three species of estuarine fish. Arch Environ Contam Toxicol 18: 826–831

Zeeman M (1994) Comparative immunotoxicology and risk assessment. In: Stolen JS (ed) Modulators of Fish Immune Responses. SOS Publications, Fair Haven, NJ

Zelikoff JT (1994) Fish immunotoxicology. In: Dean JH, Luster MI, Munson AE, Kimber I (eds) Immunotoxicology and Immunopharmacology, 2nd edn. Raven Press, New York

Section 4
Genetic and Physiological Impacts of Organotin Compounds

Chapter 13
Genetic Impacts of Organotin Compounds

Tsuyoshi Nakanishi and Jun-ichi Nishikawa

Abbreviations AR: Androgen receptor; CYP: Cytochrome P450; 9cRA: 9 *cis*-retinoic acid; DBD: DNA binding domain; EDC: Endocrine disrupting chemical; ER: Estrogen receptor; hCG: Human chorionic gonadotropin; HSD: Hydroxysteroid dehydrogenase; LBD: Ligand binding domain; NR: Nuclear receptor; PKA: Protein kinase A; PPAR: Peroxisome proliferator-activated receptor; RXR: Retinoid X receptor; TBT: Tributyltin; TPT: Triphenyltin

13.1 Introduction

The concept of endocrine disruption was introduced at the Work Session on "Chemically Induced Alterations in Sexual Development: The Wildlife/Human Connection" in 1991. At this session it was pointed out that a number of environmental chemicals affect hormonal systems and have adverse health effects on wildlife and probably on humans. Such chemicals are referred to as endocrine disrupting chemicals (EDCs), and their effects have emerged as a major environmental issue. The nuclear receptors (NRs) of intrinsic hormone systems are likely to be targets of EDCs, because their intrinsic ligands are fat-soluble and low-molecular-weight agents, as are the environmental pollutants. Examples can be found among persistent organochlorine pollutants (DDT, PCBs), plasticizers (pthalates), detergents (alkylphenols) and birth control pills (ethynylestradiol) (McLachlan 2001). The effects of synthetic chemicals on sex hormone receptors such as the estrogen receptor (ER) and androgen receptor (AR) have attracted much attention, focusing on the reproductive failures observed in wildlife.

T. Nakanishi
Laboratory of Hygienics, Gifu Pharmaceutical University, Gifu, Gifu, Japan

Department of Toxicology, Graduate School of Pharmaceutical Sciences, Osaka University, Suita, Osaka, Japan

J.-ichi Nishikawa
Laboratory of Health Sciences, School of Pharmacy and Pharmaceutical Sciences, Mukogawa Women's University, Nishinomiya, Hyogo, Japan

T. Arai et al. (eds.), *Ecotoxicology of Antifouling Biocides*,
DOI: 10.1007/978-4-431-85709-9_13, © Springer 2009

Among them, the imposex phenomenon in marine gastropods provides one of the clearest examples of endocrine disruption in wildlife. While many field studies have demonstrated the adverse effects of organotins upon female gastropods, the mechanism underlying the imposex phenomenon has not been fully elucidated. Organotin compounds have been widely used as antifouling paints for ships and fishing nets since the 1960s and have thus been released into marine environments. Aquatic invertebrates, particularly marine gastropods, are extremely sensitive to organotin compounds and undergo changes in sexual identity in response to exposure. Most marine gastropods in organotin-polluted areas have shown reproductive failure due to oviduct blockage by vas deferens formation, resulting in population decline or mass extinction (Bryan et al. 1988; ten Hallers-Tjabbes et al. 1994). This phenomenon is called "imposex" as an abbreviation of "imposed sexual organs", because male genital organs, such as the penis and vas deferens, are imposed upon female organs (Smith 1971). Approximately 150 species of imposex-affected gastropods have been found in the world (Fent 1996; Matthiessen et al. 1999). Despite several hypotheses on the cause of imposex induction, such as aromatase inhibition, testosterone excretion-inhibition, functional disorder of the female cerebropleural ganglia, and involvement of neuropeptide APGWamide (Bettin et al. 1996; Ronis and Mason 1996; Oberdörster and McClellan-Green 2000, 2002), the mechanism through which they induce and promote the development of a penis-like structure and a vas deferens in female gastropods remains obscure.

It is well known that steroidal sex hormones such as 17β-estradiol and 5α-dihydrotestosterone exert important roles in physiological processes, including sexual development and reproduction in vertebrates. However, homologues of ER and AR, which mediate the action of sex steroids, have not been found in invertebrates (Escriva et al. 1997). Because gastropods are mollusks, they may not have a functional receptor for testosterone, suggesting that vertebrate-type sex hormones may not be involved in male sexual development in the gastropods. Recently, we and Grün et al. have shown, independently, that tributyltin (TBT) and triphenyltin (TPT) are high-affinity ligands for the human retinoid X receptor (RXR) and the peroxisome proliferator-activated receptor (PPAR) γ, which are members of NR superfamily (Kanayama et al. 2005; Grün et al. 2006). In addition, functional homologues of RXR have been cloned from Japanese and European gastropods (*Thais clavigera* and *Nucella lapillus*). TBT binds to either of them at environmentally relevant levels and the natural ligand of RXR, 9-*cis* retinoic acid, induces imposex in both gastropods (Nishikawa et al. 2004; Castro et al. 2007). The mechanistic impacts of the overall findings are discussed.

13.2 Differences in Nuclear Receptors Between Invertebrates and Vertebrates

Members of the NR transcription factor function as concentration dependent sensors of cognate ligands to coordinate gene regulation of developmental and homeostatic hormone signalling pathways. NRs are structurally related proteins that consist of a highly conserved DNA-binding domain (DBD) and a moderately conserved ligand-binding domain (LBD). The mammalian NRs mediate the actions of small molecular agents such as steroid hormones (e.g., estrogens, androgens, progesterone, glucocorticoids, mineralocorticoids), retinoic acids (all-*trans* and 9-*cis* isomers), thyroid hormone, 1.25 $(OH)_2$ vitamin D_3, and fatty acids. In addition to these receptors, the superfamily also contains a large number of so-called orphan NRs whose ligands do not exist or have not been identified (Giguére 1999). Phylogenetic study and extensive polymerase chain reaction (PCR) surveys have revealed that NR genes appeared very early on during metazoan evolution, but could not be found in fungi, plants, or unicellular eukaryotes (Escriva et al. 1997, 2000). By virtue of genome projects, we now know that *Homo sapiens*, *Ciona intestinalis*, *Drosophila melanogaster*, and *Caenorhabditis elegans,* respectively, have 48, 17, 21, and 284 kinds of NR genes (Maglich et al. 2001; Yagi et al. 2003). There is a striking difference between vertebrates and invertebrates with respect to their NR sets. For instance, receptors for sex and adrenal steroid hormones have not been found in ascidian, fruit fly and round-worm. Although distant ancestors of the ER have been found in octopus, snails, and other mollusks, these ER-like proteins do not bind to estrogens and were constitutive activated transcription factors like the orphan NRs (Thornton et al. 2003; Keay et al. 2006; Bannister et al. 2007). In addition to the absence of sex steroid receptors, cytochrome P450 (CYP) enzymes related to sex steroid biosynthesis have not been identified in invertebrates except for amphioxus *Branchiostoma belcheri*, which is considered to be evolutionarily closer to vertebrates than other invertebrates (Campbell et al. 2004; Rewitz et al. 2006; Mizuta and Kubokawa 2007). The data collected until now suggest that the existence or function of sex steroids is different between vertebrates and invertebrates.

13.3 Molecular Factors Related to Imposex

Imposex is induced by TBT at concentrations as low as 1 ng/l of tin (Sn) (Gibbs et al. 1987; Axiak et al. 1995) and is used extensively all over the world as a biomarker to monitor TBT pollution (ten Hallers-Tjabbes et al. 1994; Horiguchi et al. 1997a; Terlizzi et al. 1998, 2004). Not only TBT, but also TPT, has been shown to have a strong effect on the development of imposex in *Thais clavigera* (Horiguchi et al. 1997b). Historically, several hypotheses have been proposed to explain the chain of events and molecular factors leading to imposex development. The original work of Féral and Le Gall with transplantation experiments suggested that the hierarchic involvement of two illusive factors, termed the retrogressive factor (RF) and the penis morphogenic factor (PMF) (Féral and Le Gall 1983). They also suggested the fundamental role of two anatomical structures, the pedal and cerebropleural ganglia. Subsequently, the neuropeptide APGWamide was proposed as the putative PMF, because injection of APGWamide significantly induces imposex in the mud snail *Ilyanassa obsoleta* (Oberdörster and McClellan-Green 2000, 2002). They proposed that APGWamide is abnormally released by an external stimulus such as TBT exposure and causes the development of male sex characteristics. Despite these observations, APGWamide failed to promote imposex in the prosobranch gastropod *Bolinus brandaries* (Santos et al. 2006).

Inhibition of aromatase, the key enzyme required for conversion of androgens to estrogens, has also been proposed as a potential driver for imposex development. An aromatase enzyme complex consists of the microsomal CYP19 and the flavoprotein nicotinamide adenine dinucleotide phosphate reduced-form reductase. Bettin et al. reported that TBT increases androgen levels through inhibition of aromatase activity in marine neogastropods at relatively high doses (Bettin et al. 1996). The TBT also inhibits the catalytic activity of human aromatase or a granulose cell-like tumor cell line (Cooke 2002; Heidrich et al. 2001; Saitoh et al. 2001). However, the CYP19 gene has not been found outside chordates (Mizuta and Kubokawa 2007). Therefore, it is doubtful whether the inhibitory effect of TBT on aromatase activity is a cause of the imposex in molluscs.

New insight is coming from a quite different direction. We reported that TBT and TPT are high-affinity ligands for RXR and PPARγ by comprehensive ligand screen with human NRs (Kanayama et al. 2005; Grün et al. 2006). We employed an *in vitro* molecular interaction screen between human NRs and coactivators, together with a yeast two-hybrid system, to test suspected EDCs for receptor-mediated activation (Kanayama et al. 2003). Surprisingly, organotins such as TBT and TPT act as potent nanomolar activators of both RXR and PPARγ. In addition, these compounds showed the transactivation function of RXR and PPARγ in mammalian culture cells (Kanayama et al. 2005; Grün et al. 2006). The effectiveness of each organotin compound was comparable to that of the natural ligand of RXR, 9-*cis* retinoic acid (9cRA) or the well-known PPARγ ligand rosiglitazone. The dose ranges of TBT and TPT that induced transactivation were from 10 to 100 nM, which do not cause significant apoptosis or necrosis of mammalian culture cells in general. These results indicate that organotin compounds function as RXR or PPARγ agonists in mammalian cells.

13.4 Existence of the Retinoid X Receptor in Marine Gastropods

Although TBT and TPT apparently activate human RXR and PPARγ, the question is whether these receptors exist in gastropods or not. As described in section 13.2, the composition of members of the NR superfamily is quite different between vertebrates and invertebrates. The subgroup members of the thyroid hormone receptor, retinoic acid receptor, vitamin D receptor and PPAR appear to have been late acquisitions during the evolution of the NR superfamily (Escriva et al. 1997; Laudet 1997). Therefore, PPARγ might not be present in marine gastropods. In contrast, RXR is special among the NR superfamily. It is widely conserved in the evolutionary tree and its homologue, called ultraspiracle (USP), is found even in arthropods (Laudet 1997).

The RXR homologue has been cloned from *Thais clavigera* (Nishikawa et al. 2004) and more recently from *Nucella lapillus* (Castro et al. 2007). Either of these RXRs has a DBD composed of two C_2C_2-type zinc finger motifs and a putative LBD in the C-terminal region. The highest similarity with other species is in the DBD, where 85–90% of the amino acids residues are identical. The LBD of gastropod RXR also shows considerable similarity with that of vertebrate RXRs but has much less similarity with USP, the RXR homologue first found in *Drosophila melanogaster*. Although RXR binds 9cRA in organisms ranging from cnidarians (*Tripedalia cystophora*) to vertebrates, USP from arthropods is unable to do so (Heyman et al. 1992; Mangelsdorf et al. 1992; Henrich and Brown 1995; Kostrouch et al. 1998). As expected from the similarity of gastropod homologues to vertebrate RXR, the binding of gastropod RXR to 9cRA has been confirmed experimentally (Nishikawa et al. 2004; Castro et al. 2007). The dissociation constant in the binding of 9cRA to gastropod RXR is 15.2 nM, which is similar to the values reported for vertebrate RXRs (1–10 nM) (Heyman et al. 1992). Gastropod RXR also binds to organotin compounds, even though the 50% inhibitory concentration (IC_{50}) values are larger than for 9cRA (Nishikawa et al. 2004).

To examine the involvement of RXR in the development of imposex in *T. clavigera*, an *in vivo* injection experiment was carried out. Imposex was significantly induced in female *T. clavigera*, which received the injection of 9cRA, and substantial penis growth was observed in them after 1 month of 9cRA injections (Nishikawa et al. 2004). Through a combination of exposure experiments, Castro et al. (2007) also showed that 9cRA induces imposex in *N. lapillus* to the same degree as the positive control (TBT). Methoprene acid, a selective ligand for RXR, also induces imposex, albeit to lower degree than that observed for 9cRA and TBT (Castro et al. 2007). In this context, RXR plays an important role in the induction/differentiation and growth of male genital tracts in female gastropods. It is possible that sexual differentiation in primitive species is regulated by retinoid signaling instead of steroids. Meanwhile, we do not know whether gastropods inherently possess a pathway for the biosynthesis of 9cRA. Therefore, we do not know whether 9cRA is a real hormone or whether similar derivatives are. We need to identify the active compound from gastropods.

13.5 Possible Human Exposure to Organotin Compounds

The potent biocidal properties of organotins extended their uses to the production of high value food crops and industrial processes, in addition to antifouling biocides for marine vessels. Some organotins are used in food contact packing and drink containers. Human exposure to non-point sources of organotins may occur mainly through contaminated dietary sources, such as seafood, shellfish and food crops. Daily intakes of TBT oxide (TBTO) determined in Japan by the duplicated-position method were 4.7 ± 7.0 μg/day in 1991 (n = 39) and 2.2 ± 2.2 μg/day in 1992 (n = 40). Using the market based method, the daily intake was estimated at

6–9 μg/day in 1991 and 6–7 μg/day in 1992 (Tsuda et al. 1995). In Finland, TPT was detected as the predominant compound at levels up to 1.11 ng/g fresh weight in fish and seafoods (Rantakokko et al. 2006). In addition, a variety of mono- and dialkyltins, and significant contaminating trialkyl species, are also used extensively in the manufacture of polyolefein plastics (PVC) as a heat stabilizer during polymerization, bringing them into closer contact with drinking water and food supplies (Takahashi et al. 1999; Appel 2004).

The information on human exposure to organotin compounds is limited. In a study of eight volunteers from Germany, TPT was detectable in serum in the concentration range 0.17–0.67 mg/l (Lo et al. 2003). In a study of 38 volunteers from the USA, Kannan et al. reported that monobutyltin (MBT), dibutyltin (DBT) and TBT were detected in 53%, 81%, and 70% of the 32 blood samples tested. Blood concentrations of MBT, DBT and TBT were 8.17 ± 8.56, 4.94 ± 3.83, and 8.18 ± 15.4 ng/ml, respectively (Kannan et al. 1999). The toxicological significance of the concentrations of organotins measured in these studies is unknown. However, the potential exposure of humans to organotins has aroused great concern about their potential toxicity. Animal experiments suggested that the spectrum of potential adverse chronic systemic effects of organotins is quite broad and includes primary immunosuppressive, endocrinopathic, and neurotoxic effect, as well as potential ocular, dermal, cardiovascular, upper respiratory, pulmonary, gastrointestinal, blood dyscrasias, reproductive/teratogenic/developmental, liver, kidney, bioaccumulative, and possibly carcinogenic activity. Although many reports have described the potential toxicity of organotins, the critical target molecules for the toxicity of organotin compounds remain unclear.

13.6 Enzyme Inhibition by Organotins

The synthesis of sex steroids from cholesterol requires trafficking process between mitochondria and smooth endoplasmic reticulum, and many enzymatic steps. In *in vitro* experiments, butyltins were shown to exhibit structure-related inhibition of the aromatase activity from human placenta (Heidrich et al. 2001) or transfected cells (Cooke 2002). However, effective inhibition of aromatase by organotins occurs only in the micromolar range. TBT and TPT are generally toxic to mammalian cells and they cause apoptosis or necrosis at micromolar levels (Saitoh et al. 2001; Nakanishi et al. 2002, 2006; Watanabe et al. 2003). In human choriocarcinoma cell lines, JAr, JEG-3, and BeWo, exposure to greater than 300 nM TBT or TPT markedly decreases DNA and protein synthesis (Nakanishi et al. 2002, 2006). In addition, a high concentration of TBT (above 1 μM) inhibits the catalytic activity of human 5α-reductase I and II (Doering et al. 2002), rat 3β-hydroxysteroid dehydrogenase (3β-HSD) (McVey and Cooke 2003) and pig 17β-HSD I (Ohno et al. 2005). At similar concentration ranges, TPT also inhibits the catalytic activity of aromatase, 5α-reductase II, 17β-HSD I and III (Lo et al. 2003). These observations suggest that enzyme inhibition by organotins is not specific to aromatase.

We should take into account the strong cytotoxicity and non-specific effects of organotin compounds, when measured *in vitro*.

13.7 Organotin Compounds Affect Endocrine Functions in Human Placenta and Ovary

In a recent study, we investigated the effects of organotin compounds on aromatase (Nakanishi et al. 2002, 2005) and 17β-HSD I, which converts low-activity estrone to high-activity estradiol (Nakanishi et al. 2006), in human choriocarcinoma cells. Both TBT and TPT increased the catalytic activity of aromatase and 17β-HSD I, along with their mRNA expression, in a dose-dependent fashion, following exposure to non-toxic concentration ranges (3–100 nM). These data indicate that organotins perturb the steroidogenic function through transcriptional regulation in human placental cells, not through direct enzyme inhibition. In addition, these organotin compounds also markedly stimulated hCG production in the same concentration ranges, along with its mRNA expression (Nakanishi et al. 2002, 2005). These results suggest that the placenta represents a potential target organ for organotin compounds in pregnant women and that endocrine-disrupting effects might be the result of local changes in estrogen and hCG concentrations.

In contrast to the above results, however, Saitoh et al. (2001) reported that 20 ng/ml (about 60 nM) TBT and TPT suppressed both the activity and gene expression of aromatase in the human ovarian granulose-like cell line, KGN. This discrepancy in the action of organotins on gene expression is due to the tissue-specific expression of aromatase, which is strictly regulated. Human *CYP19* is a single-copy gene composed of 10 exons; exons II to X encode the aromatase protein, as well as the 3′ untranslated region of mRNA common to all estrogen-producing tissues (Simpson et al. 1994). A number of variations of exon I exist. These encode the 5′ untranslated regions of various *CYP19* mRNAs, which are selectively expressed in some tissues by alternative splicing (Simpson et al. 1994; Sebastian and Bulun 2001; Bulun et al. 2003). The tissue-specific expression of *CYP19* appears to be mediated by tissue-specific promoters lying upstream of the respective exon I sequences, and by transcription factors binding to specific regions of each promoter. In the placenta, *CYP19* is driven by the placental major promoter (I.1), and the transcript contains exon I.1, located approximately 89 kb upstream from exon II. On the other hand, ovarian transcripts contain a sequence at the 5′-end immediately upstream of the translation start site, because gene expression in the ovary uses a proximal promoter (II). In ovarian granulosa cells, the expression of *CYP19* is strongly regulated by the steroidogenic tissue-specific transcriptional factor, Ad4Bp/SF-1, via promoter II. In contrast, Ad4Bp/SF-1 is expressed at very low levels in the human placenta and may not play an important role in activation of the placental major promoter I.1 (Bamberger et al. 1996; Simpson et al. 1997). Saitoh et al. suggest that the effects of organotin compounds in KGN cells are caused partly by association with Ad4Bp/SF-1 (Saitoh et al. 2001). It is therefore likely that the action of organotin compounds in human placental cells is induced by a pathway clearly different from that in ovarian granulosa cells, giving rise to the promotion of aromatase activity and mRNA expression.

In human placental cells, all mRNA expressions of aromatase, 17β-HSD I and hCG are controlled by cAMP-dependent intracellular signal pathways; however, neither TBT nor TPT exerted any effect on intracellular cAMP production (Nakanishi et al. 2002). In addition, there is little possibility that these organotin compounds affect the cAMP-protein kinase A (PKA) pathway in the human ovary, because it stimulates aromatase gene expression through promoter II (Michael et al. 1995). The possible target of these organotin compounds may be a signaling pathway common to the gene expression of aromatase, 17β-HSD I and hCG in the human placenta and ovary.

13.8 Regulation of Aromatase Gene Expression by Organotin Compounds Through RXR or PPARγ Activation in Humans

Gene expression of human aromatase is regulated by the activation of PPARγ and/or RXR. In the human placenta, a selective RXR ligand stimulates aromatase gene expression; however, a selective PPARγ ligand has little or no effect on aromatase gene expression (Sun et al. 1998; Nakanishi et al. 2005). In addition, the PPAR ligand 15-deoxy-$D^{12,14}$-prostaglandin J_2, farnesoid X receptor ligand chenodeoxycholic acid and liver X receptor ligand T0901317, which are agonists of permissive heterodimer partners of RXR, all failed to increase mRNA expression of aromatase in human choriocarcinoma cells (Nakanishi et al. 2005). It is suggested that none of these permissive heterodimers are involved in aromatase expression in the human placenta and that RXR homodimer may be required for the regulation of aromatase expression.

Unlike in the placenta, both RXR- and PPARγ-selective ligands suppress aromatase gene expression in the ovary (Mu et al. 2000, 2001; Fan et al. 2005). However, it was suggested that PPARγ/RXR may inhibit promoter II lying upstream of the ovarian major exon I (PII) by an indirect mechanism because of the absence of a PPARγ/RXR response element in promoter II of aromatase (Mu et al. 2001). A transcriptional factor, nuclear factor-κB, interacts with the ovarian promoter II sequence of aromatase and up-regulates its gene expression in the human ovary. In addition, activation of the PPARγ/RXR heterodimer interferes with the interaction between NF-κB and promoter II sequence of aromatase (Fan et al. 2005). PPARγ/RXR, in the ovary, may regulate aromatase gene expression via the NF-κB signaling pathway.

In light of these findings, human aromatase expression regulated by organotin compounds may involve the activation of PPARγ and/or RXR (Saitoh et al. 2001; Nakanishi et al. 2002, 2005), because the aromatase expression pattern induced in the human placenta and ovary by activation of PPARγ and/or RXR is similar to that induced by organotin compounds. It has already been found, as supporting evidence, that organotin compounds stimulate the expression of a luciferase reporter construct containing the human placental promoter I.1 sequence of aromatase via a ligand-dependent signaling pathway of RXR (Nakanishi et al. 2005).

13.9 Potential Toxicity by Organotin Compounds Through RXR or PPARγ Activation in Mammals

PPARγ is activated by a variety of fatty acids and a class of synthetic antidiabetic agents, thiazolidinediones that are used to treat type II diabetes and reverse insulin resistance in the whole body by sensitizing the muscle and liver tissue to insulin (Lehmann et al. 1995). In addition, PPARγ also serves as an essential regulator of adipocyte differentiation and lipid storage in mature adipocytes (Tontonoz et al. 1994). Unfortunately, the adipogenic activity of PPARγ may result in undesirable effects such as obesity. RXR agonists also activate the PPARγ/RXR heterodimer and act as insulin-sensitizing agonists in rodents (Mukherjee et al. 1997), underscoring the potential effects of both PPARγ and RXR agonists on diabetes and obesity. In light of these previous findings, we evaluated the effects of TPT and TBT on adipogenesis and found that these organotins stimulate the differentiation of preadipocyte 3T3-L1 cells into adipocytes (Kanayama et al. 2005). These results suggested that organotin compounds are a potential obesogen. A recent study from Grün et al. showed that, *in vivo*, acute exposure to TBT in adult mice resulted in coordinate regulation of lipogenic PPARγ/RXR target gene expression in adipose tissue and liver, and modulated adipocyte differentiation factors such as a members of the CCAAT/enhancer binding protein family and sterol regulatory element-binding protein 1c (Grün et al. 2006). Furthermore, developmental exposure *in utero* led to a fatty liver (hepatic steatosis) phenotype and enhanced lipid staining of neonatal fat deposits, and resulting in a significant increase in the epididymal fat pad size of mice later in life (Grün et al. 2006). Whether this occurs through increased lipid storage, an increase in adipocyte number, or a combination of both is currently unresolved. However, activation of PPARγ/RXR induced by organotin compounds represents a compelling mechanistic example of a class of environmental pollutants that have the ability to impact key adipogenic factors, fat deposit size, and function.

Exposure of rats *in utero* to TBT induces a dramatic increase in the incidence of low-birth-weight fetuses because of maternal hypothyroidism (Adeeko et al. 2003). On the other hand, the RXR agonist bexarotene causes clinically significant hypothyroidism in patients with cutaneous T-cell lymphoma (Duvic et al. 2001), and experimental exposure of rats to LG100268 (a selective RXR agonist) induces the acute phase of hypothyroidism (Liu et al. 2002). Similarities between the toxicity of TBT and selective RXR agonists suggest that at least some of the toxic effects of organotin compounds may be mediated by RXR.

Yamabe et al. (2000) reported that TBT and TPT enhance the proliferation of androgen-dependent human prostate cancer cells and the transactivation of AR.

However, the AR antagonist flutamide cannot inhibit organotin-mediated AR transactivation (Yamabe et al. 2000), and these organotin compounds do not function as AR agonists in a yeast two-hybrid system (Nishikawa et al. 2004). Recently, RXR was found to function as a novel co-regulator of AR, and 9cRA inhibits AR activity through the activation of RXR (Chuang et al. 2005). Although it remains unclear whether the co-regulators recruited by organotin-activated RXR are different from those recruited by 9cRA, RXR activation by organotins might be involved in the AR transactivation induced by them.

Taken together, these compounds may cause adverse effects on mammals through the activation of PPARγ and/or RXR because of the above-described toxic effects of organotin compounds in human cells and experimental animals.

13.10 Conclusions

Although organotin compounds inhibit the enzymatic activity of aromatase, their effective concentration is cytotoxic. In this review, we have proposed the activation of RXR and/or PPARγ as a novel mechanism for organotin-induced negative impacts on invertebrates and vertebrates. We reported that RXR plays an important role in the development of gastropod imposex, by showing the cloning of RXR homologues from marine gastropods, binding of organotins to those receptors, and imposex induction by injection of 9cRA (Nishikawa et al. 2004; Castro et al. 2007). These findings indicated that RXR activation is also a critical event for endocrine disruption of organotins in gastropods. However, it is possible that organotin compounds affect target molecules other than PPARγ and RXR. For instance, organotin compounds have been shown to enhance histone acetyltransferase activity (Osada et al. 2005). Further studies are needed to clarify the precise action mechanism of the toxicity of organotin compounds in mammals and gastropods.

References

Adeeko A, Li D, Forsyth DS et al. (2003) Effects of *in utero* tributyltin chloride exposure in the rat on pregnancy outcome. Toxicol Sci 74:407–415

Appel KE (2004) Organotin compounds: toxicokinetic aspects. Drug Metab Rev 36:763–786

Axiak V, Vella AJ, Micaleff D et al. (1995) Imposex in *Hexaplex trunculus* (Gastropoda: Muricidae): first results from biomonitoring of tributyltin contamination in the Mediterranean. Mar Biol 121:685–691

Bamberger AM, Ezzat S, Cao B et al. (1996) Expression of steroidogenic factor-1 (SF-1) mRNA and protein in the human placenta. Mol Hum Reprod 2:457–461

Bannister R, Beresford N, May D et al. (2007) Novel estrogen receptor-related transcripts in Marisa *cornuarietis*; a fresh-water snail with reported sensitivity to estrorogenic chemicals. Environ Sci Technol 41:2643–2650

Bettin C, Oehlmann J, Stroben E (1996) TBT-induced imposex in marine neo-gastropods is mediated by an increasing androgen level. Helgolander Meeresunters 50:299–317

Bryan GW, Gibbs PE, Burt GR (1988) A comparison of the effectiveness of tri-n-butyltin chloride and five other organotin compounds in promoting the development of imposex in the dog-whelk, *Nucella lapillus*. J Mar Biol Assoc UK 68:733–744

Bulun SE, Sebastian S, Takayama K et al. (2003) The human CYP19 (aromatase P450) gene: update on physiologic roles and genomic organization of promoters. J Steroid Biochem Mol Biol 86:219–224

Campbell RK, Satoh N, Degnan BM (2004) Piecing together evolution of the vertebrates endocrine system. Trend Genet 20:359–366

Castro LFC, Lima D, Machado A et al. (2007) Imposex induction is mediated through the retinoid X receptor signaling pathway in the neogastropod Nucella lapillus. Aquat Toxicol 85:57–66

Chuang KH, Lee YF, Lin WJ et al. (2005) 9-cis-retinoic acid inhibits androgen receptor activity through activation of retinoid X receptor. Mol Endocrinol 19:1200–1212

Cooke GM (2002) Effect of organotins on human aromatase activity *in vitro*. Toxicol Lett 126:121–130

Doering DD, Steckelbroeck S, Doering T et al (2002) Effects of butyltins on human 5α-reductase type 1 and type 2 activity. Steroids 67:859–867

Duvic M, Hymes K, Heald P et al. (2001) Bexarotene is effective and safe for treatment of refractory advanced-stage cutaneous T-cell lymphoma: multinational phase II-III trial results. J Clin Oncol 19:2456–2471

Escriva H, Safi R, Hanni C et al. (1997) Ligand binding was acquired during evolution of nuclear receptors. Proc Natl Acad Sci USA 94:6803–6808

Escriva H, Delaunay F, Laudet V (2000) Ligand binding and nuclear receptor evolution. Bioessays 22:717–727

Fan W, Yanase T, Nomura M et al. (2005) Activation of peroxisome proliferator-activated receptor-g and retinoid X receptor inhibits aromatase transcription via nuclear factor-kB. Endocrinology 146:85–92

Fent K (1996) Ecotoxicology of organotin compounds. Crit Rev Toxicol 26:1–117

Féral C, Le Gall S (1983) The influence of a pollutant factor (tributyltin) on the neuroendocrine mechanism responsible for the occurrence of a penis in the females of *Ocenebra erinacea*. In: Lever J, Boer HH (eds) Molluscan neuro-endocrinology, North-Holland Amsterdam, pp. 173–175

Gibbs PE, Bryan GW, Pascoe PL et al. (1987) The use of the dog-whelk, *Nucella lapillus,* as an indicator of tributyltin (TBT) contamination. J Mar Biol Assoc UK 67:507–523

Giguére V (1999) Orphan nuclear receptors: from gene to function. Endocrinol Rev 20:689–725

Grün F, Watanabe H, Zamanian Z et al. (2006) Endocrine-disrupting organotin compounds are potent inducers of adipogenesis in vertebrates. Mol Endocrinol 20:2141–2155

Heidrich DD, Steckelbroeck S, Klingmuller D (2001) Inhibition of human cytochrome P450 aromatase activity by butyltins. Steroids 66:763–769

Henrich VC, Brown NE (1995) Insect nuclear receptors: a developmental and comparative perspective. Insect Biochem Mol Biol 25:881–897

Heyman RA, Mangelsdorf DJ, Dyck JA et al. (1992) 9-cis retinoic acid is a high affinity ligand for the retinoid X receptor. Cell 68:397–406

Horiguchi T, Shiraishi H, Shimizu M et al. (1997a) Imposex in sea snails, caused by organotin (tributyltin and triphenyltin) pollution in Japan: a survey. Appl Organomet Chem 11:452–455

Horiguchi T, Shiraishi H, Shimizu M et al. (1997b) Effects of triphenyltin chloride and five other organotin compounds on the development of imposex in the rock shell, *Thais clavigera*. Environ Pollut 95:452–455

Kanayama T, Mamiya S, Nishihara T et al. (2003) Basis of a high-throughput method for nuclear receptor ligands. J Biochem (Tokyo) 133:791–797

Kanayama T, Kobayashi N, Mamiya S et al. (2005) Organotin compounds promote adipocyte differentiation as agonists of the peroxisome proliferator-activated receptor g/retinoid X receptor pathway. Mol Pharmacol 67:766–774

Kannan K, Senthilkumar K, Elliott JE et al. (1999) Occurrence of butyltin compounds in human blood. Environ Sci Technol 33:1776–1779

Keay J, Bridgham JT, Thornton JW (2006) The *Octopus vulgaris* estrogen receptor is a constitutive transcriptional activator: evolutionary and functional implications. Endocrinology 147:3861–3869

Kostrouch Z, Kostrouchova M, Love W et al. (1998) Retinoic acid X receptor in the diploblast, *Tripedalia cystophora*. Proc Natl Acad Sci USA 95:13442–13447

Laudet V (1997) Evolution of the nuclear receptor superfamily: early diversification from an ancestral orphan receptor. J Mol Endocrinol 19:207–226

Lehmann JM, Moore LB, Smith-Oliver TA et al. (1995) An antidiabetic thiazolidinedione is a high affinity ligand for peroxisome proliferator-activated receptor γ (PPARγ). J Biol Chem 270:12953–12956

Liu S, Ogilvie KM, Klausing K et al. (2002) Mechanism of selective retinoid X receptor agonist-induced hypothyroidism in the rat. Endocrinology 143:2880–2885

Lo S, Allera A, Albers P et al. (2003) Dithioerythritol (DTE) prevents inhibitory effects of triphenyltin (TPT) on the key enzymes of the human sex steroid hormone metabolism. J Steroid Biochem Mol Biol 84:569–576

Maglich JM, Sluder A, Guan X et al. (2001) Comparison of complete nuclear receptor sets from the human, *Caenorhabditis elegans* and *Drosophila* genomes. Genome Biol 2:1–7

Mangelsdorf DJ, Borgemeyer U, Heyman RA et al. (1992) Characterization of three RXR genes that mediate the action of 9-*cis* retinoic acid. Genes Dev 6:329–344

Matthiessen P, Reynoldson T, Billinghurst Z et al. (1999) Field assessment for endocrine disruption in invertebrates. In: deFur P, Crane M, Ingersoll C, Tattersfield L (eds) Endocrine disruption in invertebrates: endocrinology, testing and assessment, SETAC Press, Pensacola, FL, pp. 199–270.

McLachlan JA (2001) Environmental signaling: what embryos and evolution teach us about endocrine disrupting chemicals. Endocrinol Rev 22:319–341

McVey MJ, Cooke GM (2003) Inhibition of rat testis microsomal 3β-hydroxysteroid dehydrogenase activity by tributyltin. J Steroid Biochem Mol Biol 86:99–105

Michael MD, Kilgore MW, Morohashi K et al. (1995) Ad4BP/SF-1 regulates cyclic AMP-induced transcription from the proximal promoter (PII) of the human aromatase P450 (CYP19) gene in the ovary. J Biol Chem 270:13561–13566

Mizuta T, Kubokawa K (2007) Presence of sex steroids and cytochrome P450 genes in amphioxus. Endocrinology 148:3554–3565

Mu YM, Yanase T, Nishi Y et al. (2000) Insulin sensitizer, troglitazone, directly inhibits aromatase activity in human ovarian granulosa cells. Biochem Biophys Res Commun 271:710–713

Mu YM, Yanase T, Nishi Y et al. (2001) Combined treatment with specific ligands for PPARg:RXR nuclear receptor system markedly inhibits the expression of cytochrome P450arom in human granulosa cancer cells. Mol Cell Endocrinol 181:239–248

Mukherjee R, Davies PJA, Crombie DL et al. (1997) Sensitization of diabetic and obese mice to insulin by retinoid X receptor agonists. Nature 386:407–410

Nakanishi T, Kohroki J, Suzuki S et al. (2002) Trialkyltin compounds enhance human CG secretion and aromatase activity in human placental choriocarcinoma cells. J Clin Endocrinol Metab 87:2830–2837

Nakanishi T, Nishikawa J, Hiromori Y et al. (2005) Trialkyltin compounds bind retinoid X receptor to alter human placental endocrine functions. Mol Endocrinol 19:2502–2516

Nakanishi T, Hiromori Y, Yokoyama H et al. (2006) Organotin compounds enhance 17β-hydroxysteroid dehydrogenase type I activity in human choriocarcinoma JAr cells: potential promotion of 17β-estradiol biosynthesis in human placenta. Biochem Pharmacol 71:1349–1357

Nishikawa J, Mamiya S, Kanayama T et al. (2004) Involvement of the retinoid X receptor in the development of imposex caused by organotins in gastropods. Environ Sci Technol 38:6271–6276

Oberdörster E, McClellan-Green P (2000) The neuropeptide APGWamide induces imposex in the mud snail, *Ilyanassa obsoleta*. Peptides 21:1323–1330

Oberdörster E, McClellan-Green P (2002) Mechanisms of imposex induction in the mud snail, *Ilyanassa obsoleta*: TBT as a neurotoxin and aromatase inhibitor. Mar Environ Res 54:715–718

Ohno S, Nakajima, Nakajin S (2005) Triphenyltin and Tributyltin inhibit pig testicular 17b-hydroxysteroid dehydrogenase activity and suppress testicular testosterone biosynthesis. Steroids 70:645–651

Osada S, Nishikawa J, Nakanishi T et al. (2005) Some organotin compounds enhance histone acetyltransferase activity. Toxicol Lett 155:329–335

Rantakokko P, Kuningas T, Saastamoinen K et al. (2006) Dietary intake of organotin compounds in Finland: a market-basket study. Food Addit Contam 23:749–756

Rewitz KF, Styrishave B, Løbner-Olsen A et al. (2006) Marine invertebrate cytochrome P450: emerging insights from vertebrate and insect analogies. Comp Biochem Phyysiol 143:363–381

Ronis MJJ, Mason AZ (1996) The metabolism of testosterone by the periwinkle (*Littorina littorea*) *in vitro* and *in vivo*: effects of tributyltin. Mar Environ Res 42:161–166

Saitoh M, Yanase T, Morinaga H et al. (2001) Tributyltin or triphenyltin inhibits aromatase activity in the human granulosa-like tumor cell line KGN. Biochem Biophys Res Commun 289:198–204

Santos MM, Reis-Henriques MA, Vieira MN et al. (2006) Triphenyltin and tributyltin, single and in combination, promote imposex in the gastropod *Bolinus brandaris*. Ecotox Environ Safe 64:155–162

Sebastian S, Bulun SE (2001) A highly complex organization of the regulatory region of the human CYP19 (aromatase) gene revealed by the Human Genome Project. J Clin Endocrinol Metab 86:4600–4602

Simpson ER, Mahendroo MS, Means GD et al. (1994) Aromatase cytochrome P450, the enzyme responsible for estrogen biosynthesis. Endocrinol Rev 15:342–355

Simpson ER, Michael MD, Agarwal VR et al. (1997) Cytochromes P450 11: expression of the CYP19 (aromatase) gene: an unusual case of alternative promoter usage. FASEB J 11:29–36

Smith BS (1971) Sexuality in the American mud snail, *Nassarius obsoletus* Say. Proc Malac Soc Lond 39:377–378

Sun T, Zhao Y, Mangelsdorf DJ et al. (1998) Characterization of a region upstream of exon I.1 of the human CYP19 (aromatase) gene that mediates regulation by retinoids in human choriocarcinoma cells. Endocrinology 139:1684–1691

Takahashi S, Mukai H, Tanabe S et al. (1999) Butyltin residues in livers of humans and wild terrestrial mammals and in plastic products. Environ Pollut 106:213–218

ten Hallers-Tjabbes CC, Kemp JF, Boon JP (1994) Imposex in whelks *Buccinum undatum* from the open North Sea: relation to shipping traffic intensities. Mar Pollut Bull 28:311–313

Terlizzi A, Geraci S, Minganti V (1998) Tributyltin (TBT) pollution in the coastal waters of Italy as indicated by imposex in *Hexaplex trunculus* (Gastropoda, Muricidae). Mar Pollut Bull 36:749–752

Terlizzi A, Delos, AL, Garaventa F et al. (2004) Limited effectiveness of marine protected areas: imposex in *Hexaplex trunculus* (Gastropoda, Muricidae) populations from Italian marine reserves. Mar Pollut Bull 48:188–192

Thornton JW, Need E, Crews D (2003) Resurrecting the ancestral steroid receptor: ancient origin of estrogen signaling. Science 301:1714–1717

Tontonoz P, Hu E, Spiegelman BM (1994) Stimulation of adipogenesis in fibroblasts by PPARγ2, a lipid-activated transcription factor. Cell 79:1147–1156

Tsuda T, Inoue T, Kojima M et al. (1995) Daily intakes of tributyltin and triphenyltin compounds from meals. J AOAC Int 78:941–943

Watanabe H, Adachi R, Hirayama A et al. (2003) Triphenyltin enhances the neutrophilic differentiation of promyelocytic HL-60 cells. Biochem Biophys Res Commun 306:26–31

Yagi K, Satou Y, Mazet F et al. (2003) A genomewide survey of developmentally relevant genes in *Ciona intestinals*. Dev Genes Evol 213:235–244

Yamabe Y, Hoshino A, Imura N et al. (2000) Enhancement of androgen-dependent transcription and cell proliferation by tributyltin and triphenyltin in human prostate cancer cells. Toxicol Appl Pharmacol 169:177–184

Chapter 14
Effects of Organotins on the Drug Metabolizing Enzymes in Fish

Shin'ichiro Kawai and Ayako Nakayama

14.1 Introduction

Drug metabolizing enzymes play important roles in terrestrial and aquatic organisms for the synthesis, metabolism and excretion of various kinds of chemicals, especially lipophilic compounds including natural and synthetic chemicals. Drug metabolizing systems consist of two phases. In the first phase reaction, oxidative, reductive and hydrolytic reactions are dominant, and water solubility of chemicals increase through these processes. The toxicity of chemicals, however, are not necessarily decreased in the first phase and may be toxified in some cases through metabolic activation. In the second phase, conjugate reactions are dominant, and chemicals modified in the first phase are biotransformed to more water soluble compounds, and are readily excreted. These two-phase biotransformation processes are therefore important in the metabolism of lipophilic compounds such as endogenous steroids in the bodies of all animals, and also in the fates and effects of medicines in the medical and pharmaceutical fields. Since the 1970s, drug metabolizing enzymes have been found to be important in evaluating the toxicities of various lipophilic environmental pollutants, including pesticides and industrial chemicals. Although much knowledge has been accumulated on xenobiotic metabolism using experimental and wild animals, that obtained from aquatic organisms is insufficient for the complete understanding of fates of environmental chemicals incorporated in the body compared with mammals such as rats and mice. Many polycyclic aromatic hydrocarbons (PAHs) are known to induce cytochrome P450(CYP) and the metabolites act as potent carcinogens and/or mutagens and are, therefore, considered as important risk factors in epidemiological and epizootiological cancer. The P450 system is universally distributed in all organisms and plays a key role in the metabolism of xenobiotic compounds such as PAHs,

S. Kawai and A. Nakayama
Department of Biosphere Sciences, Kobe College, 4-1 Okadayama,
Nishinomiya, Hyogo 662-8505, Japan

T. Arai et al. (eds.), *Ecotoxicology of Antifouling Biocides*,
DOI: 10.1007/978-4-431-85709-9_14, © Springer 2009

dioxins, pesticides etc., leading to their detoxification or bioactivation. However, the system is intrinsically important in the metabolism of endogenous substrates including steroids, arachidonic acid, prostaglandins and others.

Thus, drug metabolizing enzyme systems are important for the understanding of both detoxification and bioactivation.

In this chapter, recent reports concerning the relationship between organotins exposure and xenobiotics metabolism by aquatic organisms, mainly fish, are reviewed and discussed.

14.2 Drug Metabolizing Enzymes in Fish

The ability to metabolize xenobiotics such as lipophilic persistent organic pollutants (POPs) is generally considered to be weak in fish compared with mammals. Therefore, various fat-soluble man-made organics, including DDTs and PCBs, readily accumulate in fish tissues. However, there are many reports describing drug metabolizing enzyme activities that are induced by exposure to, or injection of, various kinds of environmental chemicals as shown in Table 14.1 (Ishizuka 1999).

Table 14.1 Environmental relevant inducers and inhibitors of xenobiotic metabolisms

Induction
Aromatic hydrocarbons
PAHs (benzo[a]pyrene, crysene, benzo[k]fluoranthene, benzo[j]fluoranthene), benzen, xylene, buthylated hydroxytoluene, tert-butyl hydroxyquinone, ethoxyquin
Halogenated hydrocarbons
Polychlorinated naphthalene, trichloroethylene, PCBs, PCDDs, PCDFs, hexabromobenzene, buthylated monochlorodiphenylethers
Ethers
Diethyether
Pesticides
Hlogenated pesticides
DDT, alpha-hexachlorocyclohexane, hhlordane, TCPOBOP, mirex, endrin, lindane, dieldrin, toxaphene, trichloro-237, heptachlor, endsulphan
Organophospate pesticides
Malathion
Carbamate pesticides
Carbaryl
Herbicides
CNP, monuron 1967, diuron, tridiphane
Inhibition
Metals
Cadmium, cobalt, manganese, nickel, copper, mercury, organotin
Others
Acrylamide, 3-amino-1.2.3-triazol, paraquat, naphthalene

CYP comprises a family of haemo-proteins that are the dominant catalysts for mono-oxygenase reactions with a large number of drugs, carcinogens and pollutants in terrestrial and aquatic environments.

14.2.1 Cytochrome P450 in Fish

Various kinds of CYPs including CYP1, CYP2, CYP3, CYP4, CYP11 and CYP19-aromatase are known to be distributed in liver, kidney, reproductive gland and brain of fish. As to CYP1 families in fish: CYP1A subfamily (CYP1A1, 1A2, 1A3) CYP1B family and CYP1C family (CYP1C1, 1C2) have been reported. Among these CYP1 families CYP1A are especially important, and DNA responsible for CYP1A has been cloned in several fish species. CYP1A families in fish are induced by the exposure to several PAHs and dioxins. Wild fish have also been investigated regarding the relationship between the levels of several kinds of CYP, including CYP1A, distributing in more than ten species of fish, and water pollution.

β –Naphthoflavone (β –NF) is known to bind to arylhydrocarbon receptor (AhR), and to induce CYP1A activity, and β –NF is popularly used as a CYP1A inducer. Hepatic CYP1A (Ethoxyresorfin-*O*-deethylase:EROD) activity in rainbow trout (*Onchorynchus mykiss*, mean body weight, 80 g) was clearly induced by β –NF exposure at 0.5 µg/l after 24 h, and increased concentration-dependently (Fig. 14.1). EROD activities were induced after 6 h of exposure to β –NF at 1 mg/l, and increased time-dependently (Fig. 14.2) (Nakayama and Kawai 2006).

14.2.2 CYP1A As an Indicator of PAH Pollution in the Aquatic Environments

Aquatic environments including lakes, rivers, estuaries, and coastal areas receive various kinds of anthropogenic inputs, and aquatic organisms are exposed to pollutants such as metals, man-made organic surfactants etc. Several biomarkers have been used to determine sublethal effects of pollutants in fish, for example, metallothionein for metals, plasma vitellogenin for estrogenic substances, brain acetylcholine esterase activities for organophosphorous pesticides and CYP1A1 activities for PAHs and some PCBs. CYP1A1 is one component of the mixed function oxidases (MFO) system, and is important in the detoxification and/or toxification of many PAHs and some PCBs. EROD activity is CYP1A1 dependent and is generally used as a marker of MFO induction. Hepatic EROD activity in flounder (*Platichtthys flesus*) has been measured as an indicator of contaminant exposure in seven English estuaries. Significant induction of EROD activity was clearly observed in flounder from several industrialized estuaries compared to a relatively unpolluted site (Fig. 14.3) (Kirby et al. 1999). Furthermore, close links between EROD induction and hepatic PAH/PCBs concentrations was noticed, as shown in Table 14.2.

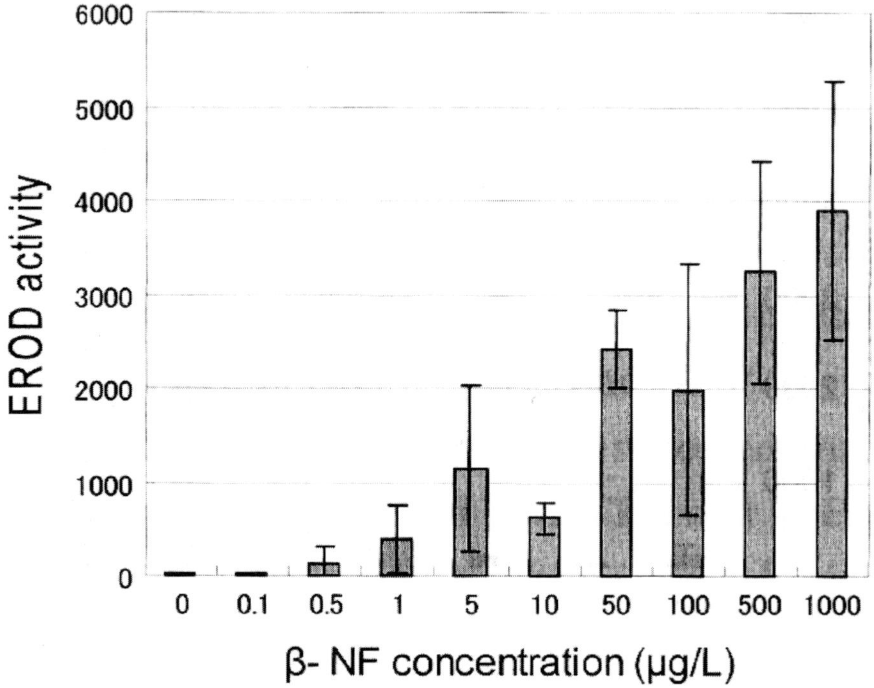

Fig. 14.1 Ethoxyresorufin *O*-deethylase (EROD) activity induced by different concentration of β-naphthoflavone (BNF) ranging from 0.1 to 1,000 μg/l in rainbow trout. EROD activities were expressed as unit/min/individual and means ± standard error (n = 5 for each group, except to control and 0.1 μg/l exposure group (n = 1))

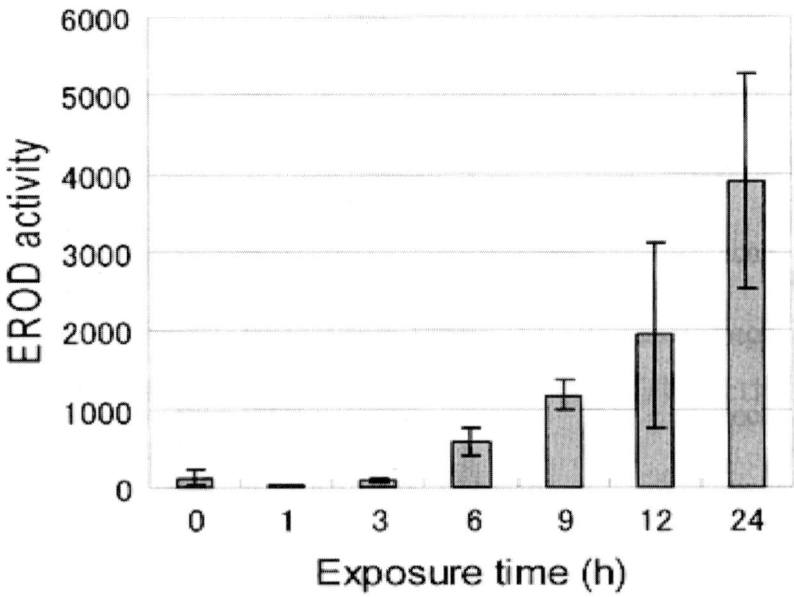

Fig. 14.2 Ethoxyresorufin *O*-deethylase (EROD) activity induced by different exposure time ranging from 0 to 24 h using 1 mg/l of β-naphthoflavone (BNF) in rainbow trout. EROD activities were expressed as unit/min/individual and means ± standard error (n = 5 for each group)

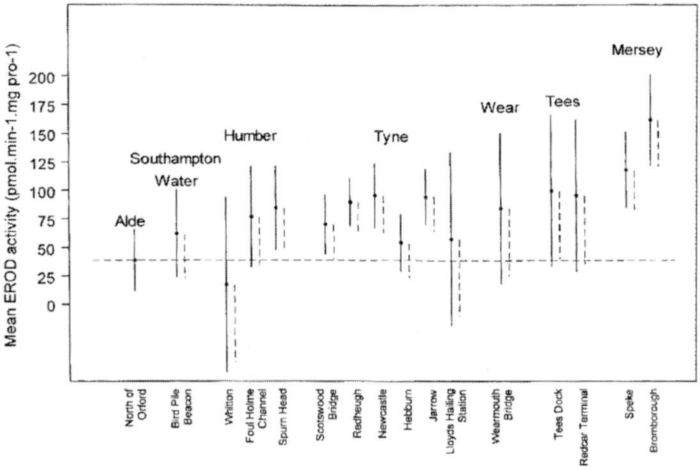

Fig. 14.3 Mean hepatic EROD activity in flounder form English estuarine sites (● is the mean EROD level; – the 95% confidence interval for mean; *horizontal* --- the mean EROD level at reference (Alde) site; *vertical* --- the graphical representation on one-sided *t*-test against reference site)

Table 14.2 Mean total PAH and PCB in bulked samples of flounder liver from selected estuaries

Location	PAH (μg kg^{-1} wet wt)	PCB (μg kg^{-1} wet wt)
River Mersey	363	1,082
River Tees	365	n/a
River Humber	71.2–100.6	424
River alde	109.7	32

n/a – not analysed.

Fuel oil spilled in 1969 and distributing in sediment of Buzzards Bay, Massachusetts was monitored by the induction of EROD activity in marsh fish (*Fundulus heteroclitus*) in 1989. Although residual concentrations of oil or biodegraded metabolites were in trace concentrations, these hydrocarbons appear to induce EROD activity as shown in Fig. 14.4. The fish collected from the spilled area were held in clean water to test whether the induced EROD activity decreased, or not. EROD activity fell to 10% of initial activity after 13 weeks, indicating that initial activity was not fixed and represented recent induction. (Teal et al. 1992).

Rainbow trout injected intraperitoneally with PCBs and polybrominated biphenyls (PBBs) showed a ten-fold increase in aryl hydrocarbon hydoxylase (AHH) activity within 7 days compared with a control group, and the high activity remained constant after 2–3 weeks (Elcombe and Lech 1978).

CYP1A (EROD) activity of gobioid fish (Gobio gobio) living close to a sewage treatment plant (STP) was significantly higher than fish living upstream of the STP (Faller et al., 2003). This suggests that the treated effluent still contained chemicals inducing CYP1A activity. From these findings CYP1A (EROD) activity is an excellent biomarker of pollution by some PAHs or organochlorines.

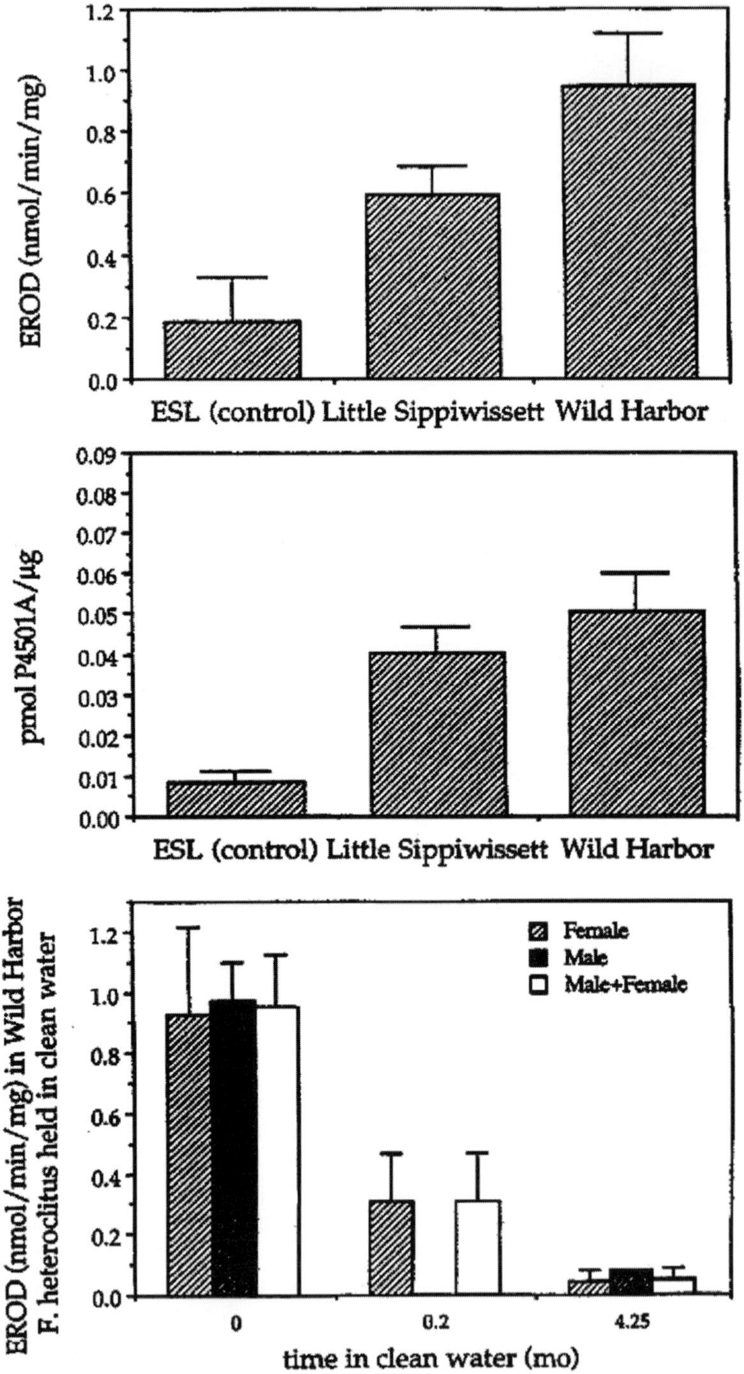

Fig. 14.4 Top – Erod (ethoxy resorufin O-deethylase) activity in *Fundulus heteroclitus* from the oiled Wild Harbor marsh, a reference marsh in Buzzards Bay, Little Sippewissett, and fish held under clean conditions at ESL. Middle – concentrations of P4501A in the same fish. Bottom – the time course of decrease in EROD activity in fish held in clean water

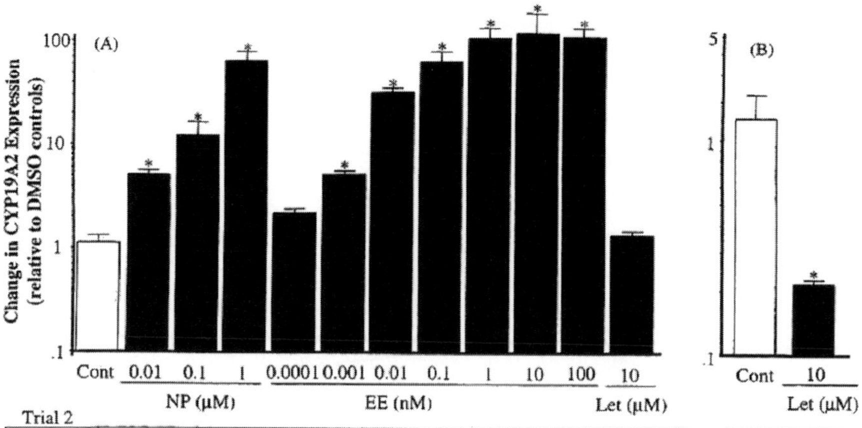

Trial 2

Mean fold	1.14	1.87	8.94	56.26	1.47	2.40	16.42	41.39	64.90	70.80	66.22	0.96
P value to control	-	< 0.05*	< 0.05*	< 0.05*	0.23	< 0.05*	< 0.05*	< 0.05*	< 0.05*	< 0.05*	< 0.05*	0.44

1.24	0.25
-	< 0.05*

Fig. 14.5 Response of CYP19A2 transcription to the estrogenic chemicals, NP and EE, and an aromatase inhibitor, letrazole. CYP19A2 transcript abundance in zebrafish juveniles exposed to xenobiotics for 3 days (a) and 30 days (b). Transcript abundance is expressed relative to that of the vehicle control group. The results represent the mean ± SEM on six samples. The asterisk indicates statistically significant differences ($P < 0.05$)

Cytochrome P450 aromatase (CYP19) is the key steroidgenic enzyme responsible for conversion of androgens to estrogens, which play a critical role in developmental sex differentiation and the adult reproductive cycle in vertebrates. The influence of multiple classes of endocrine disrupting chemicals on the transcript abundance of two CYP19A1 and A2 were investigated in juvenile zebrafish. Estrogenic compounds such as nonylphenol and a pharmaceutical estrogen (ethinylestradiol) strongly enhanced the expression of the CYP19A2 gene in a dose-dependent manner. Exposure to benzo[a]pyrene (BaP) also significantly increased CYP19A2 transcript abundance (Fig. 14.5) (Kazeto et al. 2004)

14.3 Inhibition of CYP Activities by Organotins

Organotins such as TBT and triphenyltin (TPT) have been widely used as antifouling biocides where it is highly toxic to aquatic organisms. It has also been reported to inhibit cytochrome P450.

14.3.1 CYP1A Activity and Organotins

Hepatic microsomes were incubated in vitro with TBT and TPT, and various components analyzed. EROD activity was strongly inhibited by TBT and TPT in a

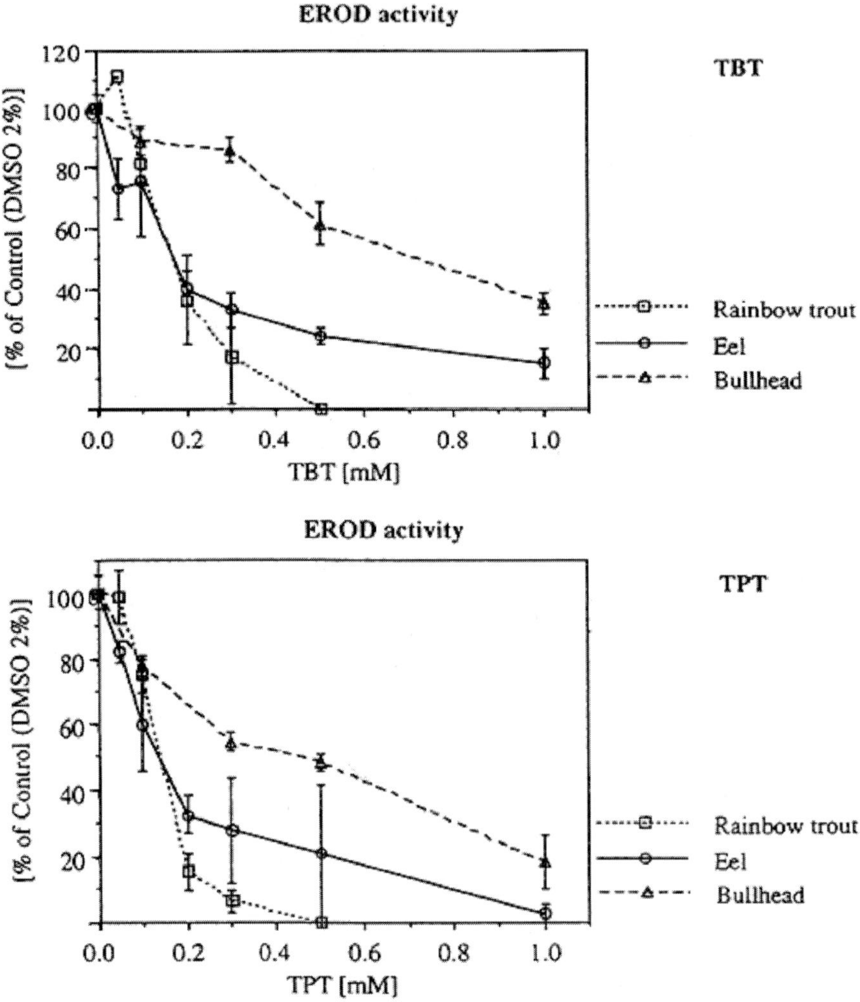

Fig. 14.6 Concentration dependence of the inhibition of ethoxyresorufin *O*-deethylase (EROD) activity after 5 min incubation at 30°C in presence of TBT (*top*) and TPT (*below*) in DMSO in rainbow trout, eel, and bullhead. Averages ± SEM of at least three separate determinations

concentration-dependent manner in fish, including rainbow trout (*Oncorhynchus mykiss*), European eel (*Anguilla anguilla*) and bullhead (*Cottus gobio*) (Fent and Bucheli 1994). Rainbow trout microsomes were more sensitive than were eel or bullhead microsomes (Fig. 14.6). In all fish, both organotins led to a time- and concentration-dependent decrease in spectral total microsomal P-450 content, together with formation of cytochrome p420. As shown in Fig. 14.7, the TPT-induced decrease in absorbance at 450 nm was accompanied by an increase of absorbance at 420 nm. TPT led to a greater inactivation of P450 enzyme than TBT, and caused

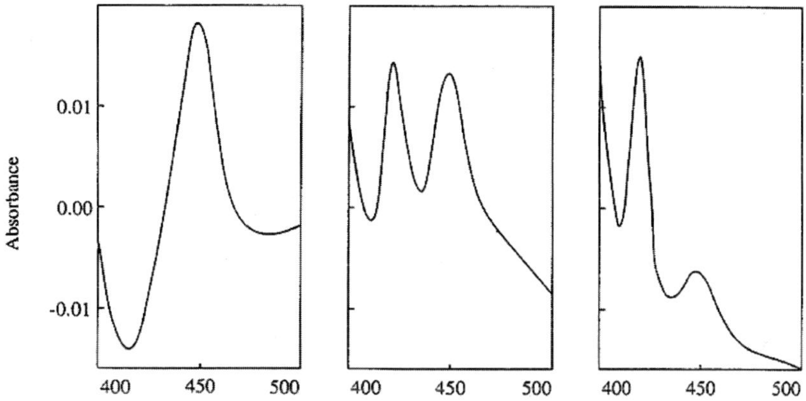

Fig. 14.7 Decrease of absorbance at 450 nm (cytochrome P450) and increase of absorbance at 420 nm (formation of cytochrome P420) in rainbow trout microsomes. The difference spectra were recorded after incubation of microsomal suspension with 2% DMSO (*left*), 0.2 mM TPT (*middle*), and 0.5 mM TPT (*right*) in DMSO for 5 min at 30°C. With increasing concentrations of TPT, suspensions became turbid and the baseline shifted

a 50% loss in all fish at 0.08 mM TPT, whereas in the case of TBT, a 50% loss occurred at 0.18 mM in rainbow trout, 0.30 mM in bullhead, and 0.83 mM in eel.

After hepatic EROD activity in rainbow trout was induced by 0.5 mg/l of β–NF exposure for 24 h, fish were exposed to 10 µg/l of TBT for 5 days. EROD activity was clearly inhibited compared with the group exposed to β–NF only (Fig. 14.8) (Nakayama and Kawai 2006).

EROD activity and the content of microsomal cytochrome P-450 in liver of flounder (*Platichthys flesus*) from Langesudfjord, Norway, were reported to be positively correlated with the field pollution gradient (Stegeman et al. 1988).

Different doses of TBT (0, 3.3, 8.1 and 16.3 mg/kg body weight) were injected intraperitoneally to scup (*Stenotomus chrysops*), and P450 and cytochrome b_5 content in the microsome fraction of liver homogenate were measured after 24 h (Fent and Stegeman 1993). Cytochrome b_5 content was not changed, however, P450 content was markedly decreased in the fish administered 16.3 mg/kg of TBT and moreover, P420 content clearly increased in all TBT injected groups (Table 14.3). This demonstrates that TBT induces the inactivation of P450 and inhibits CYP1A activity (Fent and Stegeman 1993).

Clear relationships between TBT concentration in the liver of reared olive flounder (*Paralichthys olivaceus*) and P450 content has been recorded (Fig. 14.9), and very low levels of TBT, as low as 3.65 ng/l, being the actual TBT level in the coastal area, affected the P450 content (Shim et al. 2003). In vitro CYP1A-EROD activity of Mullet (*Mullus barbatus*) and flounder (*Platichthys flesus*) was reduced at the TBT level of 0.1 mM (30 mg/l) (Morcillo et al. 2004).

Multiple biological effects of TBT on juvenile salmon (*Salmo salar*) exposed for 7 days to TBT at 50 and 250 µg/l in water were investigated. Hepatic samples were

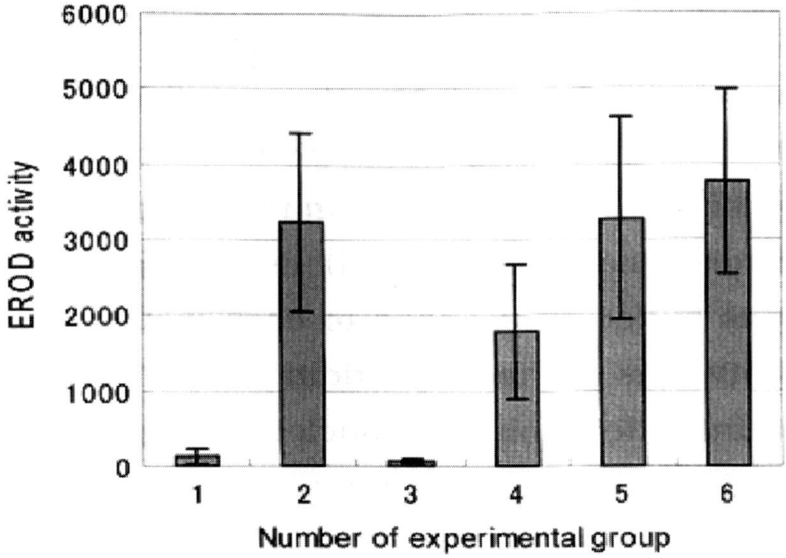

Fig. 14.8 Ethoxyresorufin *O*-deethylase (EROD) activity induced by 0.5 mg/l of β-Naphthoflavone (BNF) and/or 10 μg/l of tributyltin (TBT) under different experimental groups in rainbow trout. The number in this figure shows the experimental groups; (1) control, (2) BNF 0.5 mg/l for 24 h exposure, (3) TBT 10 μg/l for 5 days exposure, (4) BNF 0.5 mg/l for 24 h exposure after fish were exposed to 10 μg/l of TBT for 5 days, (5) BNF 0.5 mg/l for 5 days exposure, and (6) TBT 10 μg/l and BNF 0.5 mg/l for 5 days at the same time exposure. EROD activities were expressed by unit/min/individual and means ± standard error (n = 5 for each group)

Table 14.3 Effect of TBT in vivo on total cytochrome P450 and cytochrome b5

TBT dose (mg/kg)	Animals (n)	Total cytochrome P450 (pmol mg^{-1} protein)[a]	Cytochrome P420 (pmol mg^{-1} protein)[a]	Cytochrome b5 (pmol mg^{-1} protein)[a]
0 (1% ethanol)	5	83.4 ± 10.5	0	38.5 ± 5.0
3.3	4	112.3 ± 11.6	5.1 ± 5.9	49.8 ± 7.8
8.1	4	91.9 ± 11.6	3.5 ± 3.4	48.9 ± 7.6
16.3	5	61.9 ± 11.6	113.5 ± 38.1	44.2 ± 6.0

All values represent averages of separate determinations ± SEM.
[a]Average of 2–3 determinations per animal.

analyzed for gene expression patterns in the hormonal and xenobiotic biotransformation pathways using real-time PCR methods. TBT produced concentration-specific decreases in estrogen receptor-α (ER-α), vitellogenin (Vtg), zona radiate protein (Zr-protein) and increase of ER-β and androgen receptor-β (AR-β) in the hormonal pathway (Fig. 14.10). In the xenobiotic biotransformation pathway, TBT produced apparent increases and decreases at respective low and high concentration, on AhR α, AhR nuclear translocator (ARNT) and AhR repressor (AhRR) mRNA. The expression of CYP1A1 and glutathione *S*-transferase (GST) which

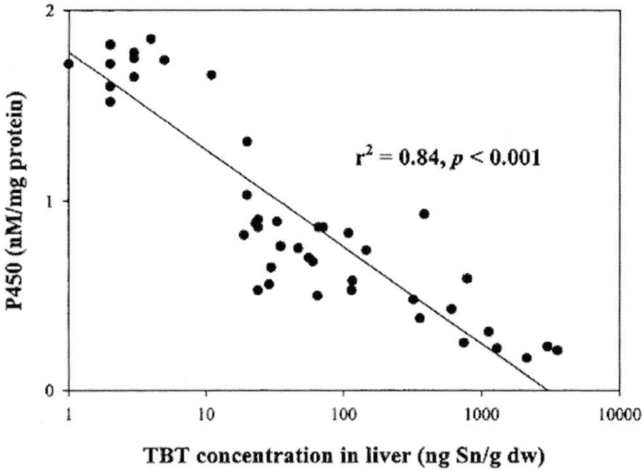

Fig. 14.9 Relationship between hepatic cytochrome P450 content and tributyltin (TBT) concentrations in liver of olive flounder exposed to TBT chloride, including control and five different exposure groups at all the sampling periods

plays an important role in the conjugation process of chemicals, showed a TBT concentration-dependent decrease. Immunochemical analysis of CYP1A1 protein levels confirmed the TBT effects observed at the transcriptional levels. These findings suggest endocrine effects of TBT, in addition to effects on hepatic CYP1A isoenzymes at the transcriptional level that causes the increase of protein and enzymatic levels (Mortensen and Arukwe 2007).

Decreases in P450 content and the activities of EROD and penthoxyresorufin *O*-depentylase (PROD, catalyzed by CYP2B subfamily) by TBT were measured in vitro using hepatic microsomes of a Steller sea lion (*Eumetopias jubatus*) and a Dall's porpoise (*Phocoenoides dalli*). EROD activity (Fig. 14.11) was more sensitive to TBT than P450 content and PROD activity in both species. TBT concentrations that affected P450 content and activity were over ten times higher than the values found in the liver of various marine mammals (Kim et al. 1998).

14.3.2 Interaction of Organotins and Benzo[a]pyrene on CYP1A Activity

Benzo[a]pyrene (BaP), a widespread carcinogenic polycyclic aromatic hydrocarbon, is metabolized and bioactivated to carcinogenic BaP diol-epoxide metabolite primarily by hepatic CYP1A. Interaction between BaP and TBT was investigated in male brook trout (*Salvelinus fontinalis*). Short-term (48 h), single exposure to a high dose (10 mg/kg) of TBT inhibited both the in vivo metabolism and bioactivation of BaP at least by inhibiting the BaP-mediated induction of CYP1A-mediated EROD

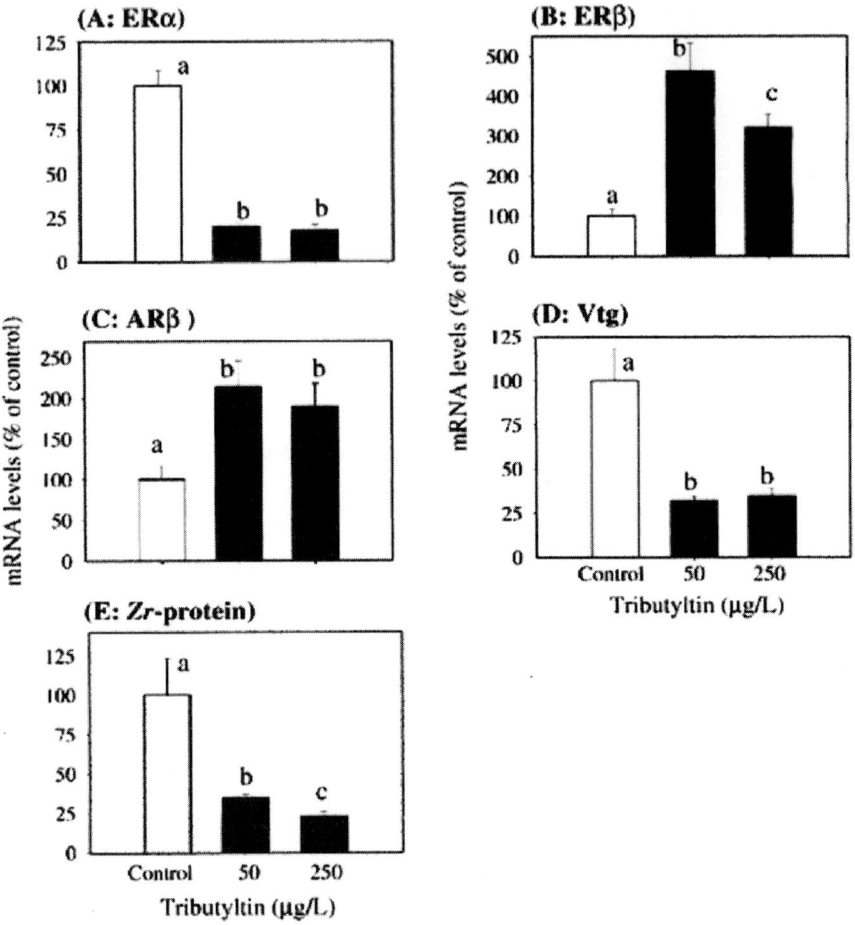

Fig. 14.10 Modulation of hepatic estrogen receptors (ERα and ERβ: a and b, respectively), androgen receptor-β (ARβ: c), vitellogenin (Vtg: d) and eggshell *zona radiata* protein (*Zr*-protein: e) mRNA levels of juvenile Atlantic salmon after exposure to nominal waterborne tributyltin (TBT) concentrations. Real-time PCR of mRNA expression levels with gene-sequence primer pairs of control, 50 and 250 µg TBT/l after 7 days exposure. All values represent the mean ($n = 6$) ± standard error of the mean (SEM). Different letters denote exposure groups that are significantly different ($p < 0.05$), analyzed using multi-parametric analysis of variance (ANOVA)

activity (Padrós et al. 2000). Further mechanistic evidence of mutual metabolic interactions between BaP and TBT in response to long term (56 days), repeated exposures (every 6 days) to low doses (BaP 3 mg/kg, TBT 0.3 mg/kg) was investigated in juvenile arctic charr (*Salvelinus alpinus*). Blood, bile and liver were collected and analyzed for biomarkers associated with P450 activity, BaP metabolism and bioactivation, and TBT metabolism (Padrós et al. 2003). TBT significantly inhibited the induction of hepatic CYP1A-mediated EROD activity at 8 days after

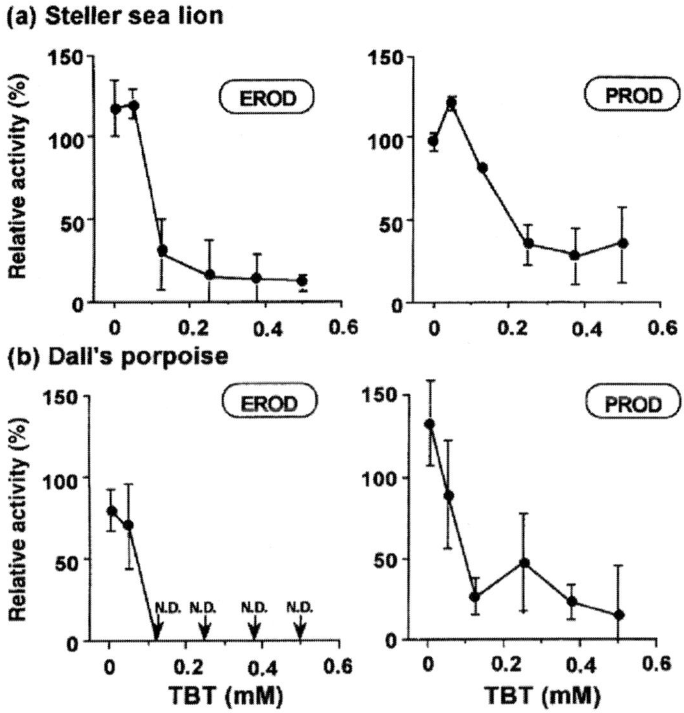

Fig. 14.11 Variations in EROD and PROD activities with different TBT concentrations. Average ± standard deviation of three separate determinations are given. Y-axis revealed a relative activity (%) to control (spiked with ethanol alone). (N.D.: not detected)

two intraperitoneal injections (Fig. 14.12). The formation of biliary BaP metabolites was also inhibited by exposure to TBT. A single high dose of TBT might antagonize the metabolism and bioactivation of BaP at least by inhibiting the induction of CYP1A. Histopathological examinations of liver, kidney and pseudobranch tissue samples originating from these same fish revealed higher lesion incidences at all sampling time points (days 8, 32 and 56) among BaP-exposed fish compared with fish exposed to either TBT alone or combined with BaP. The severity of lesions, like necrosis was also higher in BaP-exposed fish. These results suggest that TBT can antagonize BaP toxicity in fish exposed to both pollutants under controlled laboratory conditions. In contrast, BaP does not appear to provide protection against TBT toxicity (Ribeiro et al. 2007).

Metabolic interaction between TBT and benzo[a]pyrene (BaP) was also reported in scorpion fish (*Sebasticus marmoratus*) which were given a single intraperitoneal injection of TBT (0.5, 1, 5, and 10 mg/kg), BaP (0.5, 1, 5 and 10 mg/kg), or both in combination (0.5, 1, 5, and 10 mg/kg) (Wang et al. 2006). Samples were collected for biochemical analysis 7 days after injection. Co-treatment with TBT and BaP at the highest dose (10 mg/kg) resulted in inhibition of the glutathione S-transferase activity.

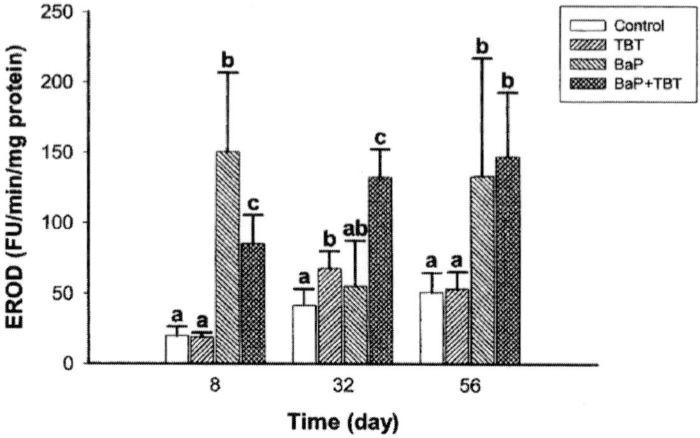

Fig. 14.12 Interactive effects of BaP and TBT on hepatic microsomal EROD activity (FU/min/mg protein). Fish were repeatedly exposed to BaP (3 mg/kg), TBT (0.3 mg/kg), or both in combination; control fish received corn oil vehicle alone. Data are presented as means ± SEM (n = 6 individual fish per treatment per time point). Treatments at a given time point not sharing a common letter are significantly different at $p < 0.005$ as assessed by one-way ANOVA with Student–Newman–Keuls test

14.4 CYP19 Aromatase Activity and Organotins

The effects of organotin compounds on human placental aromatase activity has been examined, and TBT was found to be a partial competitive inhibitor of aromatase to androstenedione, but did not affect electron transfer from NADPH to aromatase by inhibiting NADPH reductase (Heidrich et al. 2001). Imposex in neogastropods is well known to be caused by organotins, and organotins inhibit aromatase activity. However, whether aromatase inhibition is directly responsible for imposex or not is now uncertain, and other mechanism such as the involvement of the retinoid X receptor have recently been suggested (Nishikawa et al. 2004).

14.5 Conclusions

Among CYP families, CYP1A is an excellent biomarker for PAH pollution in aquatic environments, and is also responsible for the fate and behavior of various lipophilic chemicals. CYP1A-mediated EROD activity is inhibited by organotins such as TBT and TPT in vivo and in vitro, suggesting that detoxification and bioactivation are affected by the exposure of fish to organotins. Interaction between TBT and PAHs such as BaP, being a strong inducer of CYP1A, are also important for the evaluation of toxicities of both chemicals.

References

Elcombe CR, Lech JJ (1978) Induction if monooxygenation in rainbow trout by polybrominated biphenyls: a comparative study. Environ. Health Perspect. 23: 309–314.

Faller P, Kobler B, Peter A et al. (2003) Stress status of gudgeon (*Gobio gobio*) from rivers in Switzerland with and without input of sewage treatment plant effluent. Environ. Toxicol. Chem. 22:2063–2072.

Fent K, Bucheli TD (1994) Inhibition of hepatic microsomal monooxygenase system by organotins in vitro in freshwater fish. Aquat. Toxicol. 28: 107–126.

Fent K, Stegeman JJ (1993) Effects of tributyltin in vivo on hepatic cytochrome P450 forms in marine fish. Aquat. Toxicol. 24:.219–240.

Heidrich DD, Steckelbroeck S, Klingmuller D (2001) Inhibition of human cytochrome P450 aromatase activity by butyltins. Steroids 66: 763–769.

Ishizuka M (1999) Xenobiotic metabolizing enzymes in aquatic animals as biomarkers for levels of environmental pollution. Jpn. J. Environ. Toxicol. 2: 3–17.

Kazeto Y, Place AR, Trant JM (2004) Effects of endocrine disrupting chemicals on the expression of CYP19 genes in zebrafish (*Danio reio*) juveniles. Aquat. Toxicol. 69: 25–34.

Kim GB, Nakata H, Tanabe S (1998) In vitro inhibition of hepatic cytochrome P450 and enzyme activity by butyltin compounds in marine mammals. Environ. Pollut. 99: 225–261.

Kirby MF, Matthiessen P, Neall P et al. (1999) Hepatic EROD activity in flounder (*Platichthys flesus*) as an indicator of contaminant exposure in English estuaries. Mar. Pollut. Bull. 38: 676–686.

Morcillo Y, Janer G, O'Hara S CM et al. (2004) Interaction of tributyltin with hepatic cytochrome P450 and uridine diphosphate-glucuronosyl transferase systems of fish: in vitro studies. Environ. Toxicol. Chem. 23: 990–996.

Mortensen A S, Arukwe A (2007) Modulation of xenobiotic biotransformation system and hormonal responses in Atlantic salmon (*Salmo salar*) after exposure to tributyltin (TBT). Comp. Biochem. Physiol. Part C 145: 431–441.

Nakayama A, Kawai S (2006) Metabolic Interaction between β-naphthoflavone as an inducer of hepatic cytochrome P4501A and tributyltin in rainbow trout (*Oncoryynchus mykiss*). Hum. Sci. 9: 47–52.

Nishikawa J, Mamiya S, Kanayama T et al. (2004) Involvement of theretinoid X receptor in the development of imposex caused by organotins in gastropod. Environ. Sci. Technol. 38: 6271–6276.

Padrós J, Pelletier É, Reader S et al. (2000) Mitual in vivo interaction between benzo[a]pyrene and tributyltin in brook trout (*Salvelinus fontinalis*). Environ. Toxicol. Chem. 19: 1019–1027.

Padrós J, Pelletier É, Ribeiro CO (2003) Metabolic interactions between low doses of benzo (a) pyrene and tributyltin in arctic charr (*Salvelinus alpinus*): a long-term in vivo study. Toxicol. Appl. Pharmacol. 192: 45–55.

Ribeiro CA O, Padrós J, Domingos FX V et al. (2007) Histopathological evidence of antagonistic effects of tributyltin on benzo (a) pyrene toxicity in the Arctic charr (*Salvelinus alpinus*). Sci. Total Environ. 372: 549–553.

Shim WJ, Jeon JK, Hong SH et al. (2003) Accumulation of tributyltin in Olive Flounder, *Paralichthys olivaceus*: its effects on hepatic cytochrome P450. Arch. Environ. Contam. Toxicol. 44: 390–397.

Stegeman JJ, Woodin BR, Goksoyr A (1988) Apparent cytochrome P-450 induction as an indication of exposure to environmental chemicals in the flounder *Platichthys flesus*. Marine Ecol. Progress series 46: 55–60.

Teal JM, Farrington JW, Burns KA et al. (1992) The West Falmouth oil spill after 20 years: fate of fuel oil compounds and effects on animals. Mar. Pollut. Bull. 24: 607–614.

Wang CG, Zheng RH, Ding X et al. (2006) Effect of tributyltin, Benzo[a]pyrene, and their mixture on the hepatic monooxygenase system in *Sebastiscus marmoratus*. Bull. Environ. Contam. Toxicol. 75: 1214–1219.

Section 5
Bioaccumulation of Organotin Compounds in Aquatic Organisms

Chapter 15
Plankton

Inneke F. M. Rumengan and Madoka Ohji

15.1 Introduction

Plankton constitutes the largest component of the world's biomass, exerting a vital influence on aquatic life as well as forming the basis of aquatic food webs. Plankton are potentially suitable indicators of any type of contamination in seawater, because of their quick responses to toxicants and other chemicals. Assessing their level of contamination can provide a strong explanation for level of contamination in higher tropic levels of the food chain. Even though many experts question the significance of tributyltin (TBT) biomagnification from plankton, as the primary assemblage in the food chain, plankton would be the first target for organotin compounds released into the water column and a source for assimilation in higher organisms.

Research over the last 25 years has highlighted that organotin accumulates in a variety of marine organisms, from plankton (Harino et al. 1998, 1999; Takahashi et al. 1999) to high-level predators (Tanabe et al. 1998). Organotin compounds have been responsible for many deleterious effects on non-target aquatic life (Fent and Meier 1994; Ohji et al. 2002a, b, 2003a, b, 2004, 2005; Grzyb et al. 2003). Many studies have been conducted regarding organotin impacts on plankton including phytoplankton (Laughlin et al. 1986a; Maguire et al. 1984; Maeda et al. 1990; Reader and Pelletier 1992; Beaumont et al. 1987; Avery et al. 1993; Mooney and Patching 1995; St-Louis et al. 1997; Tsang et al. 1999; Rumampuk et al. 2004); holozooplankton species, such as copepods (Linden et al. 1979; U'ren 1983; Bushong et al. 1987;

I.F.M. Rumengan
Faculty of Fisheries and Marine Science, Sam Ratulangi University,
Kampus Unsrat Bahu, Manado 95115, Indonesia

M. Ohji
International Coastal Research Center, Ocean Research Institute, The University of Tokyo,
2-106-1 Akahama, Otsuchi, Iwate 028-1102, Japan

T. Arai et al. (eds.), *Ecotoxicology of Antifouling Biocides*,
DOI: 10.1007/978-4-431-85709-9_15, © Springer 2009

Hall et al. 1987), rotifers (Cochrane et al. 1991; Snell et al. 1991 a, b; DelValls et al. 1997; Sun et al. 2001; Jeon et al. 2003; van den Brink and Kater 2006), daphnia (Steinhauser et al. 1985; Kline et al. 1989), mysids (Davidson et al. 1986a, b), and meroplankton such as larvae of amphipods and mussels (Laughlin et al. 1986b; Beaumont and Budd 1984), and early life stages of fish (Seinen et al. 1981; Pinkney et al. 1990). Compared to laboratory data, field surveys on accumulation of organotin compounds in plankton are very limited (Harino et al. 1999; Takahashi et al. 1999). Mostly, assessment of organotin impacts on plankton are based on laboratory toxicity tests. There has also been an attempt to use zooplankton as a tool to assess biological impacts of contaminated sediments (van den Brink and Kater 2006). Nevertheless, whilst biomagnification of organotin compounds in aquatic ecosystem has been considered (Cooney 1988; Guruge et al. 1996; Hu et al. 2006), there are very limited data on the transfer of organotin compounds from plankton to higher taxa, either in field surveys (Takahashi et al. 1999), or in experimental studies (Sun et al. 2001).

This chapter addresses the bioaccumulation of organotin compounds and their toxicity effects on plankton biology. This includes discussion of the provision of a realistic sampling strategy for assessing pollution impact on plankton in general. Factual information relating to the ability of plankton to accumulate organotin compounds is drawn largely from laboratory studies, since very few data are available from field surveys. Some toxicological effects of organotin compounds on phytoplankton and zooplankton are described based on observations made in controlled laboratory experiments. The concluding section sets out recommendations for a long term monitoring program on plankton dynamics, to assess the effectiveness of the ban on organotin compounds from 2008.

15.2 Sampling Strategy

Many errors and omissions in assessing the impact of organotin compounds on plankton can occur as a result of poor understanding of plankton characteristics. A sampling strategy for plankton should be designed by considering some important biological properties of plankton, especially their microscopic features and tempo-spatial distribution.

15.2.1 Plankton Net

Phytoplankton range in size from 1 to 1,000 μm in diameter (Omori and Ikeda 1984; Parsons et al. 1990). Phytoplankton have high growth rates, and populations may double in size within a few hours to a few days. Thus, phytoplankton population sizes and species composition are extremely variable over short periods of time. Zooplankton are typically larger than phytoplankton and range in size from 200 μm in diameter to more than 2,000 μm. Zooplankton also have high growth rates, and population sizes may double in size within a few days to a few weeks (Parsons et al. 1990).

Harino et al. (1999), for example, collected plankton samples with a net of 71 μm mesh. The sample collected was assumed to be zooplankton, because this mesh-size was too large to collect a representative phytoplankton sample. A rigorous sampling protocol for phytoplankton would employ water bottles to collect representative samples and use the Utermohl technique to count the collected species (Parsons et al. 1984). It is possible to filter a certain amount of this sea water through an appropriate net to separate phytoplankton from suspended solids and other particles, including zooplankton. If only zooplankton are required, a net of less than 70 μm mesh is considered too small to collect a representative zooplankton sample. Such a hand plankton net is problematic, because it may result in the loss of organisms, by extrusion through the mesh when larger organisms are compressed against the net. In addition, a fine net often clogs, leading to deterioration of its collecting capability. This will cause a big error in the calculation of the volume of water filtered by the net (DeBernadi 1984; Omori and Ikeda 1984). There are a variety of nets for sampling of zooplankton on a course spatial scale in both horizontal and vertical planes. Nets with a larger aperture (>100 μm in the open sea and 200 μm in coastal waters) are recommended. For sampling of the most abundant zooplankton, copepods, a net of 300 μm is normally used. In practice, for usual towing speeds (0.7–1.0 m s^{-1}), it is advisable to use a net with a mesh size of about 75% of the width of the smallest organisms to be sampled (Omori and Ikeda 1984). For better collection of a large spectrum of zooplankton, it would be better to use simultaneously two nets with different mesh sizes.

15.2.2 Sampling Depth and Time

For assessing organotin impact, it is not recommended to conduct "surface sampling," whereby the nets are dragged across the surface of the water from motor boats. It is well-accepted that sampling at the ocean surface does not yield representative marine plankton population samples, since the surface layer often contains freshwater lenses as a result of precipitation and land run-off. This is especially true in a semiclosed areas, such as in bays which are fed by rivers that provide a significant freshwater source. Moreover, regarding the zooplankton collections, sampling time (day or night time) would be important to consider, since much of the zooplankton population migrates out of the euphotic zone during the day and is only present near the surface at night (Parsons et al. 1990). Therefore, zooplankton samples are best collected during night time. It is better to tow the net vertically at a speed of 1 m s^{-1} (Omori and Ikeda 1984) from the bottom to the surface in coastal areas. Thus, the data reported will be representative of the zooplankton populations at the sampling sites. In addition, the various samples should be collected during the same time of day or tidal period, to avoid any substantial variation in the samples. The boats used to collect the samples should travel at consistent speeds during all sample runs, and perform collections over identical tow lengths. Moreover, it is important to avoid sample spillage during transfer from the cod-end of the collection net to the sample bottle, to make samples useful for comparative purposes.

In summary, the sample collection protocol employed for the assessment of organotin impact in plankton should follow standard scientific methods and consider the basic bioecological factors that control plankton diversity and population dynamics. In order to assess environmental condition in terms of the plankton community, it is also necessary to implement a long term monitoring strategy, preferably over at least 1 year to cover all seasonal fluctuations.

15.2.3 Sampling Frequency

From an ecological standpoint, plankton examination has to be done on community composition, rather than on some particular species, over a length of time encompassing different seasons. A significant feature to consider for proper assessment is the horizontal patchiness of phytoplankton and zooplankton biomass, caused by the wind-produced Langmuir vortices within the water column (Parsons et al. 1990). Due to this patchiness, it is only by taking many repetitive samples over time that one can obtain a statistical average of species that accurately reflects the plankton community. The ability to detect differences in plankton species only a few meters apart may be an important requirement when assessing the impact of any anthropogenic contaminant, including organotin compounds. As is well known, the effects of tempo-spatial scales will influence the choice of single or multiple sampling designs; continuous, discrete or grid survey sampling at fixed stations; and the need to account for diurnal/nocturnal, monthly/seasonal, and inter-annual variations. Any sampling protocol that is intended to collect a representative sample of the plankton population in a marine environment must therefore be designed to compensate for the patchy distributions of plankton. Equally, an appropriate approach to assess the impact of organotin compounds on plankton dynamics is likely to involve time-series monitoring of plankton communities with an emphasis on morphological examination, species diversity, and spatial and temporal distribution. This could also provide more insight on the occurrence of indicator species that have the natural capability of binding certain organotin compounds.

15.3 Ecological Effects

15.3.1 Accumulation

Organotin compounds represent ubiquitous contaminants in aqueous ecosystems, and are most often present at lower nanomolar concentrations in water but then accumulate in aquatic organisms (Atanasov et al. 2005; Ohji et al. 2002a, 2007). Accumulation is here defined as the net process by which the chemical concentration in an organism achieves a level exceeding that in organic solids (biota-organic solids accumulation factor, BSAF) or water (bioconcentration factor, BCF), as a result of chemical uptake through all possible routes of exposure (water and food) and elimination from all possible routes (Veltman et al. 2006).

The pattern and extent of organotin accumulation is markedly dependent upon concentration in water. Those in the low nanogram per liter range will probably result in little significant bioaccumulation. Accumulation of organotin compounds by microorganisms including plankton warrants special consideration since they are the basis of many food chains (Laughlin et al. 1986a). Although little attention has been paid to the biota at the lower end of the marine food chain, there are a few reports that have demonstrated very high organotin concentrations in plankton from contaminated areas. For example, accumulations of organotin compounds in plankton from the Port of Osaka have been found to be 150,000 times higher than the concentration in seawater (Harino et al. 1999). Similarly, Takahashi et al. (1999) have estimated BCF of up to 70,000 in plankton and other organisms.

Among butyltin species, TBT is generally most persistent at the lower trophic levels (see, for example, Harino et al. 1999). Certain organisms in the food chain may have a low capacity to degrade TBT, and therefore may accumulate butyltins at elevated levels; those with a strong ability to metabolize TBT are less likely to accumulate organotin compounds (Sun et al. 2001). The half-life of TBT ranges from 4 to 19 days in seawater (Hall et al. 2000), but is typically only 2–6 days in phytoplankton (Lee et al. 1989; Reader and Pelletier 1992). Experimental studies by Sun et al. (2001) have indicated that the half-life of butyltins in algae (*Platymonas* sp.), rotifers (*Brachionus plicatilis*), and mysids (*Neomysis awatschensis Brandt*) are ~1, 2 and 5 days, respectively. Tsang et al. (1999) observed that two *Chlorella* species (*C. vulgaris* and *Chlorella* sp) can metabolize TBT to the less toxic dibutyltin (DBT) and play an important role in TBT biodegradation in aquatic systems. At steady-state, bioconcentration factors for butyltins in alga (*Platymonas* sp.), rotifers (*B. plicatilis*), and mysids (*Neomysis awatschensis Brandt*) were 3,300, 17,300 and 10,900, respectively, and even higher for TBT (Sun et al. 2001). The capacity of rotifers to accumulate butyltins could be associated with their filter-feeding behaviour. Other micro-algae such as *Isochrysis galbana* have a reported BCF of 5,500 for TBT, a value similar to that of several aquatic macro-organisms (Laughlin et al. 1986b); however the BCF for the green alga *Ankistrodesmus falcatus* may be as high as 30,000 (Maguire et al. 1984) implying substantial inter-species variation. Few data are available for triphenyltin (TPT) in planktonic organisms; the BCFs in *Daphnia* is as low as 190 at pH8 (Looser et al. 1998).

Accumulation of TBT in *Daphnia magna* is reported to be significantly higher at pH 8 than at pH 6 (Fent and Looser 1995), attributable to the occurrence of TBT as a neutral hydroxide at the higher pH. This is consistent with results of Avery et al. (1993) who found that pH and salinity could influence the accumulation of organotin compounds in microalga *Chlorella emersonii*, and cyanobacteria (*Synechocystis* and *Plectonema boryanum*). TBT uptake appears to be greater in freshwater ecosystems; furthermore resulting development of resistance to TBT is seen in microorganisms from TBT-polluted freshwater sites, but may not be apparent in estuarine ecosystems.

The only comprehensive data on organotin accumulation in plankton are those reported by Harino et al. (1999), based on their long term field surveys of plankton

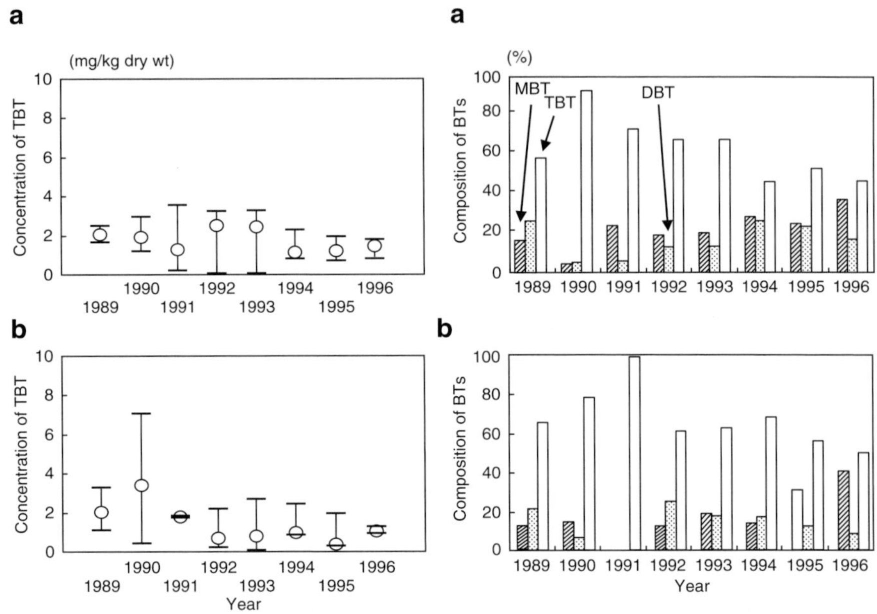

Fig. 15.1 Representative changes of TBT concentrations (*left*) and total butyltins compositions (*right*) in plankton from the Port of Osaka, Japan. TBT; tributyltin, DBT; dibutyltin, MBT; monobutyltin. a: Station A4, b: Station A5. The vessels moored near Station A4 are mainly trading boats. Large-hull vessels weighing 10,000–20,000 t sail and are moored at Station A5. The composition of butyltins was calculated from the median of butyltin concentrations in each year (After Harino et al. 1999. With permission)

in the Port of Osaka, Japan, sampled using a 71 μm-plankton net. Figure 15.1 shows their results over 7 years, based on collection made four times a year (February, June, August and November) from 1989 to 1996. It is clear that, although the concentrations of TBT and total butyltins have varied, they were generally uniform during 1992–1996 and TBT was the dominant butyltin compound. The concentrations of TBT ranged from undetectable values (ND) to ~7 mg kg^{-1} dry wt – much higher than corresponding burdens of TPT (from ND to 1.93 mg kg^{-1} dry wt). The concentrations of diphenyltin (DPT) and monophenyltin (MPT) were in the range ND–2.64 and ND–2.39 mg kg^{-1} dry wt, respectively. The levels of total phenyltins (PTs) in plankton from each station during 1989–1990 were higher than those during 1991–1996, presumably reflecting the effectiveness of local antifouling legislation.

15.3.2 Transfer of Organotin Compounds Through Food Chains

Plankton could provide a valuable means to assess the threat of biomagnification to public health, especially from fish and other seafood of that might consume contaminated plankton. A positive relationship has been found between organisms at different trophic levels and concentrations of TPT, indicating trophic magnification of TPT is possible in aquatic food chains; however, no significant evidence has been found for other organotin compounds (Hu et al. 2006). Assessing TBT levels in the environment, and in a variety of marine organisms within it, might provide better evidence for such a phenomenon. For example, sampling in Taiwanese Harbors between 2001 to 2004, Lee et al. (2006) found no significant correlations between TBT concentrations in fish and those in water or sediment, suggesting that TBT accumulations in fish might arise principally from their food. Similarly whilst triphenyltins were detected in most fish, they were not detectable in seawater and found in only a few sediment samples, indicating, circumstantially, that there has been biomagnification along the aquatic food chain. Lee et al. (2006) also calculated values for the hazard index (estimated chronic daily intake/chronic reference dose or estimated no effect daily exposure) and found mean hazard indices of TBT and TPT >1, which suggests that consumption of fish from Taiwanese harbor areas might pose a potential risk to human health.

Other evidence exists which implies that organotin compounds can be transferred through the food chain, notably that based on filter feeders and predatory species (Cooney 1988). *Mytilus edulis,* for example, accumulates TBT absorbed to phytoplankton, *Isochrysis galbana*, more rapidly than direct uptake from water (Laughlin et al. 1986b). Hu et al. (2006) also demonstrated the possibility of food chain transfer of organotins by determining levels of TBT, TPT and their metabolites in plankton, five benthic invertebrate species, and six fish species collected from Bohai Bay, North China. The concentrations of TPT in marine fish were unexpectedly higher than those of TBT. A positive relationship was found between trophic levels and concentrations of TPT, indicating trophic magnification of TPT in this food web. On the other hand, concentrations of TBT, DBT and monobutyltin (MBT) did not exhibit statistically significant trends with trophic levels. Analysis of organotin compounds in the water and surface sediment from Bohai Bay revealed low inputs of TPT to the environment, which indicated that the high concentrations of TPT found in fish from Bohai Bay were due to the food web magnification of TPT.

A study of food chain interactions among three relatively low trophic levels, e.g. alga (*Platymonas* sp.), rotifers (*B. plicatilis*), and mysids (*Neomysis awatschensis*), has been conducted by Sun et al. (2001). The food chain transfer number *f* was calculated from the difference between the bioaccumulation factor (BAF)[*] and the bioconcentration factor (BCF)[†] divided by the ratio of the concentration of contaminant in food and the concentration of contaminant in ambient water. If values of *f* are >1.0 biomagnification is inferred, whereas an *f* value <1.0 indicates that biomagnification is not taking place. For BTs, the food chain transfer number *f* was 1.44 between algae (*Platymonas* sp.) and rotifers (*B. plicatilis*), but only 0.59 between

---

* Bioaccumulation factor (BAF) is the ratio of the concentration of the contaminant in an organism derived from both water and food, and the concentration of the contaminant in ambient water.

† BCF is the ratio of the contaminant concentration in organisms which is derived from water only, and concentration of the contaminant in ambient water.

rotifers and mysids (*Neomysis awatschensis* Brandt). The likelihood of butyltin biomagnification clearly depends on the organism involved. Butyltin is a relatively moderate lipophilic compound, and the possibility of biomagnification will depend on the biochemical characteristics of individual food chain constituents.

15.4 Toxicological Effects

15.4.1 Plankton-Organotin Interactions

The general order of toxicity of organotin compounds to microorganisms, including plankton, increases with the number and chain length of organic groups bonded to the tin atom. The organotin compounds exert their toxic effects primarily through interactions with membrane lipids (Eng et al. 1988; Avery et al. 1993), because of their lipophilicity. White et al. (1999) proved that the site of action of organotin compounds is both at the cytoplasmic membrane and intracellular level. Tributyltin may cross biomembranes more easily than the hydrophilic cations (Fent and Looser 1995; Veltman et al. 2006), resulting in higher accumulation levels. Biosorption studies on a fungus, cyanobacteria, and microalgae indicate that cell surface binding alone occurred in these organisms, while studies on the effects of TBT on certain microbial enzymes indicated that in some bacteria, TBT can interact with cytosolic enzymes.

Interactions between plankton and organotin compounds are influenced by environmental conditions. In aquatic systems, both pH and salinity can determine the speciation of organotin compounds and therefore reactivity. These environmental factors may also alter selectivity for resistant microorganisms in polluted systems. Tin-resistant microorganisms have been identified, and resistance can be either plasmid or chromosomally mediated. In one TBT-resistant organism, an *Altermonas* sp., an efficient efflux system was suggested as the resistance mechanism. Biotransformation of organotin compounds by debutylation or methylation has been observed by White et al. (1999). Studies on the toxicity of organotin compounds with plankton have to be undertaken with great care, however, as some of these compounds can bind to container walls and any plastic apparatus. Furthermore, organotin toxicity in the field may be substantially underestimated by laboratory simulations (USEPA 2003). There are numerous reports that TBT binds to the sides of aquaria and to the organisms assayed, which contributes to an underestimation of potential toxicity.

15.4.2 Acute Toxicity

Freshwater species tend to accumulate organotin compounds in higher concentrations than marine organisms (Extoxnet 1996). The accumulation of organotin compounds in phytoplankton has been determined in relatively few species. Generally, the toxicity of organotin compounds is influenced more by the alkyl substituents than the ionic substituent which may form the rest of the molecule (Hall and Pinkney 1985; Extoxnet 1996). Such trends have been reported by Fargasová and Kizlink (1996) who studied the inhibitory effect of various organotin compounds on growth of the freshwater alga *Scenedesmus quadricauda* under standardized conditions. The most toxic compounds were tributyltin oxide (TBTO) and triphenyltin chloride (TPTCl), while dibutyl tin bis-N, N-diethyldithiocarbamate and dimethyltin bis(N, N-diethyldithiocarbamate) were found have least inhibitive effects on algal growth. Maguire et al. (1984) exposed the freshwater green alga, *Ankistrodesmus falcatus,* to high concentrations of TBT (20–40 μg l^{-1}) for 4 weeks, implying a high degree of tolerance. However, TBTO has been shown to inhibit cell survival of marine unicellular algae at very low concentrations; the 72-h EC_{50} ranges from 0.33 to 1.03 μg l^{-1} (USEPA 2003). High sensitivity to TBTO is also shown by the marine diatom, *Skeletonema costatum.* Beaumont and Newman (1986) found that cells died within 2 days when exposed to TBTO at 5.0 μg l^{-1}. Studies by Avery et al. (1993) on the biosorption of a range of organotin compounds by cyanobacteria *Synechocystis* PCC 6803 and *Plectonema boryanum* and the microalga *Chlorella emersonii*, have shown that, despite the lower toxicity of dibutyltin chloride (DBTCl) compared to TBTCl, biosorption of the former compound was greater in the three phytoplankton tested. Effects of pH were least evident in *C. emersonii*, but an increase in salinity resulted in reduction in TBTCl uptake (Avery et al. 1993).

Rumampuk et al. (2004) have conducted acute toxicity tests with TBT on marine phytoplankton, *Tetraselmis tetrathele, Nannochloropsis oculata* and *Dunaliella* spp. by examining the chlorophyll a and b contents. After 6 h-exposure to three concentrations of TBT (0.1, 0.5 and 1 μg l^{-1}), *N. oculata* and *Dunaliella* had slightly higher chlorophyll a and b contents in the lowest TBT concentration tested (0.1 μg l^{-1}) than those in controls, but as the TBT concentration increased their chlorophyll contents decreased. The three levels of TBT tested were below the no observable effect concentration (NOEC) for *T. tetrathele*, while the lowest observable effect concentrations (LOEC) for *N. oculata* and *Dunaliella* were between 0.1–0.5 μg l^{-1}. Among the three species, *N. oculata* has the highest sensitivity towards TBT. An LC_{50} value of 8.5 μg l^{-1} TBT has been determined in another algal species, *Isochrysis galbana*, whilst whole phytoplankton assemblages from the York River (USA) appear to be more sensitive, with an LC_{50} value of 5 μg l^{-1} TBT (Hall et al. 2000).

Algal responses to the other organotin compounds were evaluated by Mooney and Patching (1995). They conducted short term acute toxicity tests with TPT and DPT on the respiration and photosynthesis of two marine microoalgal species, a diatom, *Skeletonema costatum* (Grev.) Cleve and a chlorophyte, *Dunaliella tertiolecta* (Butcher). Triphenyltin (TPT) at concentrations in the nanomolar – micromolar range inhibited both photosynthesis and respiration after 30 min exposure. Triphenyltin concentrations inhibiting gross photosynthesis and respiration were two to three orders of magnitude higher for *D. tertiolecta* than for *S. costatum.* The latter alga did not grow over a period of 10 days in the presence of 3.3 nM TBTO, whereas growth of *D. tertiolecta* and *Pavlova lutheri* was only slightly reduced at this concentration. The freshwater green alga, *Ankistrodesmus falcatus* exhibits similar sensitivity to TPT to that shown by *S. costatum,* with an LC_{50} value of 0.005 µM TPT (Wong et al. 1982). As previously mentioned, toxicity can be significantly influenced by physical environmental factors such as temperature, pH and salinity.

The phenomenon of 'hormesis', defined as a stimulatory effect of sub-inhibitory concentrations of any toxic substance, on any organism (Stebbing 1997; Calabrese and Baldwin 1998) was reported to occur in phytoplankton by Mooney and Patching (1995). Photosynthesis of *D. tertiolecta* was subject to slight but significant stimulation immediately after exposed to 2.1 µM TPT. Hormesis has also been reported in *Pavlova lutheri* exposed to low levels of TBTO (Beaumont and Newman 1986; Mooney and Patching 1995). Rumampuk et al. (2004) have also detected such a phenomenon for both *N. oculata* and *Dunaliella* spp. when exposed for 6 h at 0.1 µg l^{-1}, and for *T. tetrathele* at 0.1–1 µg l^{-1} TBT. The chlorophyll a and b contents of this alga were, almost doubled by exposure to 1 µg l^{-1} TBT. Hormesis in zooplankton, for example in rotifers, *B. plicatilis,* has also been reported. Cochrane et al. (1991) demonstrated that, when exposed to sub-lethal concentration of some toxicants, including TBT, the rotifers were stimulated to produce more stress proteins, including a four to five-fold increase in heat shock proteins. More studies on such phenomena are obviously required in other plankton species to examine on how they react toward stressors under different environmental conditions.

Acute toxicity data for zooplankton are limited to a few species. The 48 h-LC_{50} value for *A. tonsa* of 1.1 µg l^{-1} TBT (or 1.16 µg l^{-1} TBTO), determined by Bushong et al. (1987), was similar to the 96 h-LC_{50} value of 1.0 µg l^{-1} TBTO recorded by U'ren (1983). U'ren suggested that the threshold concentration of acute toxicity (i.e. the concentration at which toxicity starts) for this zooplankton is below 0.3 µg l^{-1} TBTO. The 96 h-LC_{50} of another copepod, the harpacticoid, *Nitocra spinipes* was slightly higher at 2 µg l^{-1} TBTO (Linden et al. 1979), whilst the estuarine copepod *Eurytemora affinis* may be more sensitive with a 72 h-LC_{50} value of 0.6 µg l^{-1} TBT (Bushong et al. 1987; Hall et al. 1987).

The acute toxicity of TBT in rotifers has been evaluated mainly on *Brachionus* spp. The freshwater rotifer *B. calyciflorus* is much more sensitive than *B. plicatilis,* with 24 h-LC_{50} values of 190 and 300 µg l^{-1}, respectively (Snell et al. 1991a, b). Life table analysis of rotifer survival in the controls revealed that 72 h is the longest acute test possible for this animal without feeding. However, Jeon et al. (2003) determined much lower 96 h-LC_{50} values, ranging from 0.5 to 8 µg l^{-1} depending on the compound: For tributyltins the order of toxicity was tributyltin acetate (1.1 µg l^{-1}) > tributyltin chloride (2.0 µg l^{-1}) > tributyltin benzoate (3.3 µg l^{-1}) ppb) > tributyltin oxide (5.6 µg l^{-1}), whilst among triphenyltins the sequence was triphenyltin fluoride (1.0 µg l^{-1}) > triphenyltin chloride (1.1 µg l^{-1}) > triphenyltin hydroxide (1.6 µg l^{-1}). For rotifers, TPT was more toxic than TBT, with toxicity dependent to some extent on alkyl or aryl group substitution.

The water flea *Daphnia magna* is used frequently in acute toxicity tests. The 48-h LC_{50} values for this organism were 0.03 and 0.46 mg Sn l^{-1} for dimethyltin and monomethyltin species, respectively (Steinhauser et al. 1985). For TBTO, the 48 h-LC_{50} value *for D. magna* was 3 µg l^{-1} TBTO (Polster and Halacka 1971, cited in

U'ren 1983). Although *D. pulex* and *Ceriodaphnia dubia* are much smaller than *D. magna*, and therefore, more difficult to count, Kline et al. (1989) have attempted to assay them in 48 h exposures to triphenyltin hydroxide (TPTH) in static acute toxicity tests. All cladocerans survived at the lowest TPTH concentrations (~0.6 μg l^{-1}). There was one surviving *C. dubia* in the high concentration, 42.9 μg l^{-1}, although all died at 21.4 μg l^{-1}. At TPTH concentrations of 9.8 –12.2 μg l^{-1}, *D. pulex* and *D. magna* were not acutely affected (immobilized) although *C. dubia* was slightly affected in this range. Effects were exhibited in all species at TPTH concentrations of 20 μg l^{-1} and above.

USEPA (2003) has summarized useable zooplankton acute toxicity data on TBT. LC_{50} and EC_{50} values, calculated or interpolated graphically from original data, indicated the following order of sensitivity to tributyltin oxide: calanoid *A. tonsa* (0.24 μg l^{-1}), mysid *A. sculpta* (0.42–1.68 μg l^{-1}), harpacticoid *N. spinipes* (1.9 μg l^{-1}), *Metamysiddopsis elongate* (<0.97–6.8 μg l^{-1}), mysid *Mysidopsis bahia* (1.1–2.2 μg l^{-1}), and *D. magna* (1.58–66.3 μg l^{-1}). Data of Snell et al. 1991a, b) indicating much higher resistance in *B. calyciflorus* and *B. plicatilis* (up to 190 and 300 μg l^{-1} TBT, respectively) were not included in the USEPA summary.

Generally, meroplankton stages of any tested species are more sensitive to organotin compounds than the adults. Beaumont and Budd (1984) exposed veliger larvae of the mussel *Mytilus edulis* to TBTO for 15 days and found that no larvae survived longer than 5 days in 10 μg l^{-1} TBTO, or longer than 10 days in 1 μg l^{-1} TBTO. Lobster larvae *Homarus americanus* did not survive more than 6 days when exposed to 5 μg l^{-1} TBTO (Laughlin and French 1980, cited in U'ren 1883).

15.4.3 Chronic Toxicity

Determination of the LOEC is a prerequisite for any chronic toxicity test. Most antifouling compounds have LOEC values at microgram per liter levels, which represent the initial toxicity threshold of a chemical (Fernandez-Alba et al. 2002; USEPA 2003). Because plankton are more sensitive, generally, than the higher taxa, most chronic toxicity tests for plankton are conducted at 0.001–0.1 μg l^{-1} levels. Generally, the no observable effects level (NOEL) for phytoplankton and zooplankton is ~1 ng l^{-1} (Linley-Adam 1999; Extoxnet 1993). Beaumont and Newman (1986) found that, at levels in the range 0.001–1.092 μg l^{-1}, TBT restricted the growth of microalgae, implying that the LOEC must be ~0.001 μg l^{-1}. Walsh et al. (1985) observed growth rate inhibition of *S. costatum* when exposed to 0.33 μg l^{-1} of tributyltin oxide. However, *S. costatum* appears to be particularly efficient at degrading TBT, a process which takes place in the cytosol rather than the external surface of the diatom (Lee et al. 1989; Reader and Pelletier 1992). Cyanobacteria and microalgae appear equally sensitive to inhibition by TBTCl (Wong et al. 1982).

Chronic toxicity tests for zooplankton have involved copepods and daphnids, mainly. The available data on chronic values for TBT, including those for some copepod species, are all below 0.01 μg l^{-1} (USEPA 2003). *A. tonsa* showed inhibition of development at 0.003 μg l^{-1}. Many reports demonstrate that reductions in growth occur in commercially or ecologically important marine species at concentrations of TBT less than the final chronic value of 0.0658 μg l^{-1}. Hall et al. (1987) performed chronic tests on *E. affinis*, and found that the mean brood sizes of egg-carrying females exposed to 0.5 μg l^{-1} TBT were significantly reduced, down to 1.3% of the size in control, after 3 days, and no neonates survived after 6 days. Tributyltin had a significant effect at 0.2 μg l^{-1} TBT, whilst the threshold in these experiments was apparently 0.1 μg l^{-1} TBT – much higher than that in *A. tonsa*. U'ren (1983) found that some copepods exposed to 0.3 μg l^{-1} TBTO (equivalent to 0.285 μg l^{-1} TBT) became moribund at 96 h and effective survival was only 63% after 6 days. Organotin compounds may also inhibit the developmental rate of larvae of marine copepods at 1 ng l^{-1}, whereas the survival of larvae was affected at 15–20 ng l^{-1}: *A. tonsa* is particularly sensitive to TBT (Extoxnet 1996). In general, most other marine fauna, including fish, are more resistant to TBT than copepods (Hall et al. 1987).

Many toxicity data provided by USEPA (2003) for plankton, including a freshwater cladoceran, were those of Brooke et al. (1986), including results showing 40% survival of adult *D. magna* at a TBT concentration of 0.5 μg l^{-1}, and 100% survival at 0.2 μg l^{-1}. The chronic limits are 0.1 and 0.2 μg l^{-1}, based upon reproductive effects on adult daphnids. The overall chronic value for *D. magna* is 0.1414 μg l^{-1} (geometric mean of the chronic limits), and the acute-chronic ratio of 30.41 is calculated using the acute value of 4.3 μg l^{-1} from the same study. Two partial life-cycle toxicity tests have been conducted using the copepod, *E. affinis* (Hall et al. 1987). The chronic value was <0.088 μg l^{-1} in this test. The percentage survival of neonates was significantly reduced (73% lower than control survival) at 0.22 μg l^{-1}. The chronic value in this test was 0.14 μg l^{-1}, while the geometric means of the NOEC and LOEC were 0.094 and 0.224 μg l^{-1}, respectively.

Compared to copepods and daphnids, the saltwater mysid, *Acanthomysis sculpta* is more resistant to TBT. The effects of TBT on survival, growth, and reproduction of *A. sculpta* were determined by Davidson et al. (1986a, b) in five separate tests lasting from 28 to 63 days. The number of juveniles released per female at a TBT concentration of 0.19 μg l^{-1} was 50% of the number released in the control treatment. Numbers of females releasing viable juveniles was reduced in 0.19 and 0.33 μg l^{-1} TBT. At concentrations of 0.38 μg l^{-1} and above, survival and weight of female mysids was reduced; all mysids in 0.48 μg l^{-1} died. Based on reproductive effects, the geometric mean of chronic values was 0.13 μg l^{-1}. Freshwater rotifers *B. calyciflorus* were found to be even more resistant to TBT than mysids, because their hatching rate was not reduced until the exposure concentration reached 72 μg l^{-1} (USEPA 2003). However, this rotifer is reported more sensitive than estuarine species *B. plicatilis* (Snell et al. 1991a, b).

Comprehensive studies on the effects of the organotin-containing acaricide azocyclotin, on freshwater micro-organisms, have been conducted by Fliedner et al. (1997). Plankton, as well as physico-chemical water parameters, were analyzed regularly for effects over a period of 7 months. Zooplankton was affected at nominal concentrations >45 μg l^{-1}. Phyllopoda and nauplii of copepods reacted sensitively, whereas direct effects on rotifers and ostracods were not observed.

Within the phytoplankton, picoplankton <2 μm and algae of 2–10 μm were inhibited at nominal concentrations >135 μg l^{-1} and blooms of less sensitive species were observed.

Chronic toxicity effects on meroplankton are comparable to those for holozooplankton. For example, the studies of Brooke et al. (1986), as cited in USEPA (2003), on an early life-stage of the fathead minnow *Pimephales promelas,* showed that all fish exposed to the highest TBT exposure concentration of 2.20 μg l^{-1} died during a 32 days test. Survival was not reduced at 0.92 μg l^{-1} or any of the lower TBT concentrations, although the mean weight of the surviving fish was reduced by up to 48% (at 0.92) when compared to the control fish. Mean length of fry at the end of the test was significantly reduced at concentrations >0.45 μg l^{-1}. Thus the mean chronic value was estimated as 0.26 μg l^{-1} and the acute-chronic ratio was10.01, calculated using an acute value of 2.6 μg l^{-1} derived from the same study. Beaumont and Budd (1984) have reported ~50% mortality in the larvae of mussels *Mytilus edulis* exposed to 0.1 μg l^{-1} TBTO for 15 days (i.e., 15-days LC_{50} approximately 0.1 μg l^{-1} tributyltin oxide), and most surviving larvae were moribund and had grown significantly more slowly than controls. It is clear that TBT causes adverse reproductive and developmental effects in aquatic organisms at very low concentrations (USEPA 2003).

Chronic effects of TBT on plankton are clearly highly varied amongst plankton species. Nevertheless, from the early 1980s to the present day, laboratory studies have continued to show that, at low levels, the chronic effects organotin compounds may have considerable impacts on plankton dynamics and morphology in culture.

15.5 Plankton As Bioindicators for Organotin Compounds

Even though toxicity data for plankton are limited to a small number of species,, some zooplankton in particular are good indicators for the assessment of TBT contamination in aquatic environments. McNaught (1992), for example, recommends certain cladocerans such as *Daphnia, Ceriodaphnia,* and potentially others, for use in environmental bioassays, due to their life-history characteristics, including early reproduction, high net reproductive rates, and the potential for many parthenogenetic generations with constant genotypes and low mutation rates. In addition, Snell and Janssen (1995) consider that rotifers are useful in ecotoxicological studies due to the ease and speed of making quantitative measurements of mortality and reproduction, their sensitivity to common pollutants, and the commercial availability of cysts. As an example, the rotifers *Brachionus* spp. have proved to be a better indicator of environmental gradients than whole zooplankton assemblages. Hence, this taxon can be considered a potentially valuable group for more intensive monitoring and conservation planning (Austoni et al. 2006). However, McNaught (1992) argues that using rotifers such as *Keratella cochlearis* (Gosse) is not a good choice, because of their high mutation rates. He also warns that genetic adaptation to common contaminants is usual among animals with short generation times and high net reproduction rates. In maintaining zooplankton culture stocks in the laboratory, it is vital to check physiological acclimation. As for clones used in bioassays, sensitivity tests should be conducted alongside toxicity testing, at least one a month. Further, cultures must be kept distant from areas used for testing. A shift in sensitivity could take place with generations, perhaps due to latent toxic effects, wherein parthenogenetic females pass the toxicant to offspring, *via* the large amount of lipid contained within their asexual eggs.

There may be other confounding features to consider. The term 'no effect' usually refers to the inability to demonstrate a reduction or inhibition in a biological process. However, some toxic compounds may increase or stimulate a biological process (hormesis), at least initially, in low level exposures. Hormesis could be interpreted as indicative of a particular response of a physiological system to a toxin but may complicate overall interpretation of data-sets.

As a key component of aquatic ecosystems, monitoring of plankton communities would seem to be an important ecological requirement in evaluating any disturbance, including impacts arising from organotin inputs. Intuitively, the number of species would be expected to decrease in polluted circumstances – and the more severe the level of contamination, the greater the reduction. Contamination simplifies the assemblages by reduction of certain taxa and prompts the numerical dominance of a handful of more tolerant species. Consequently, the latter are relatively more abundant in polluted, as opposed to unpolluted, environments – a feature which promotes their value as bioindicator species. In the field however, the complexity of aquatic systems, makes it difficult to assess any species as a specific indicator of organotin-related effects. Disturbances in phytoplankton and zooplankton, either at the population or community level, are usually the result of a combination of factors which are often hard to distinguish.

Monitoring of plankton should therefore involve assessment of spatial and temporal variation at both suspected 'hot-spot' and in uncontaminated areas. There have been few attempts to correlate the changes in zooplankton dynamics with water quality in natural waters. The long term monitoring of butyltin residues in (total) zooplankton from the Port of Osaka provided by Harino et al. (1999) appears to be a unique data set. However, responses of community assemblages, as reported for some zooplankton species that appear typical of eutrophic waters (Gulati 1983; Austoni et al. 2006) have scarcely been addressed. Whilst, Javed (2006) has proposed certain species of phytoplankton and zooplankton as suitable indicators for metal pollution in some rivers in Pakistan, such information has never been generated for organotin pollution in any aquatic system.

References

Atanasov AG, Nashev LG, Tam S et al. (2005) Organotins disrupt the 11(beta)-hydroxysteroid dehydrogenase type 2-dependent local inactivation of glucocorticoids. Environ Health Perspect 113:1600–1606

Austoni M, Cardoso AC, Deviller G et al. (2006) Indicators for the assessment of thresholds and points of non-return. A Thresholds Project Report. European Commission. Dir. General Joint Research Centre. EUR 22130EN. European Community, pp54

Avery SV, Codd GA, Gadd GM (1993) Biosorption of tributyltin and other organotin compounds by cyanobacteria and microalgae. Appl Microbiol Biotechnol 39:812–817

Beaumont AR, Budd MD (1984) High mortality of the larvae of the common mussel at low concentrations of triutyltin. Mar Pollut Bull 15:402–405

Beaumont AR, Newman PB (1986) Low levels of tributyltin reduce growth of marine micro-algae. Mar Pollut Bull 17:457–461

Beaumont AR, Mills DK, Newman PB (1987) Some effects of tributyltin (TBT) on marine algae. CH2498-4/87/0000-148&$1.00.IEEEE. pp1488–1493

Brooke LT, Call DJ, Poirier SH et al. (1986) Acute toxicity and chronic effects of bis (tri-n-butyltin) oxide to several species of freshwater organisms. Center for Lake Superior Environmental Studies, University of Wisconsin-Superior, Superior, WI, 20 pp

Bushong SJ, Hall S, Johnson WE et al. (1987) Toxicity of tributyltin to selected Chesapeake Bay Biota. IEEE 19:1499–1503

Calabrese EJ, Baldwin LA (1998) Hormesis is as a biological hypothesis. Environ Health Perspect 106:357–362

Cochrane BJ, Irby RB, Snell TW (1991) Effects of copper and tributyltin on stress protein abundance in the rotifer *Brachionus plicatilis*. Comp Biochem Physiol 92C:385–390

Cooney JJ (1988) Microbial transformations of tin and tin compounds. J Ind Microbial 3:195–204

Davidson BM, Valkirs AO, Seligman PF (1986a) Acute and chronic effects of tributyltin on the mysid *Acanthomysis sculpta* (Crustacea, Mysidacea). NOSC-TR-1116 or AD-A175-294-8. National Technical Information Service, Springfield, VA

Davidson BM, Valkirs AO, Seligman PF (1986b) Acute and chronic effects of tributyltin on the mysid *Acanthomysis sculpta* (Crustacea, Mysidacea). In: Proceedings of the International Organotin Symposium, Oceans '86, vol 4, pp1219–1225, Washington, DC

DeBernadi R (1984) Methods for the Estimation of Zooplankton Abundance. In: Downing J A, Rigler FH (eds) A manual on methods for the assessment of secondary productivity in fresh waters, pp59–86. IBP Handbook 17. Blackwell Science, Oxford

DelValls TA, Lubián LM, Forja JM et al. (1997) Comparative Ecotoxicology of interstitial waters in littoral ecosystems using Microtox® and the rotifers *Brachionus plicatilis*. Environ Toxicol Chem 16:2323–2332

Eng G, Tierney EJ, Bellama JM et al. (1988) Correlation of molecular total surface area with organotin toxicity for biological and physicochemical applications. Appl Organomet Chem 2:171–175

Extoxnet (Extention Toxicology Network) (1996) Tributyltin. Oregon State University. www.extoxnet.orst.edu/pips/ghindex.html. assessed October 1, 2007

Fargasová A, Kizlink J (1996) Effect of organotin compounds on the growth of the freshwater alga *Scenedesmus quadricauda*. Ecotoxicol Environ Safety 34:156–159

Fent K, Meier W (1994) Effects of triphenyltin on fish early life stages. Arch Environ Contam Toxicol 27: 224–231

Fent K, Looser PW (1995) Bioaccumulation and bioavailability of tributyltin chloride: influence of pH and humic acid. Water Res 29:1631–1637

Fernandez-Alba AR, Hernando MD, Piedra L et al. (2002) Toxicity evaluation of single and mixed antifouling biocides measured with acute toxicity bioassays. Anal Chim Acta 456:303–312

Fliedner A, Remde R, Schafers C (1997) Effects of the organotin pesticide azocyclotin in aquatic microcosms. Chemosphere 35:209–222

Grzyb K, Rychlowski M, Biegniewska A et al. (2003) Quantitative determination of creatine kinase release from herring (*Clupea harengus*) spermatozoa induced by tribu-tyltin. Comp Biochem Physiol 134C:207–213

Gulati RD (1983) Zooplankton and its grazing as indicators of trophic status in Dutch lakes. Environ Monitor Assess 3:343–354

Guruge KS, Tanabe S, Iwata H et al. (1996) Distribution, biomagnification and elimination of butyltin compound residues in common cormorants (*Phalacrocorax carbo*) from Lake Biwa, Japan. Arch Environ Contam Toxicol 31:210–217

Hall LW Jr, Pinkney AE (1985) Acute and sublethal effects of organotin compounds on aquatic biota: an interpretive literature evaluation. CRC Crit Rev Toxicol 14:159–209

Hall LW, Bushong SJ, Ziegenfuss MC et al. (1987) Chronic toxicity of tributyltin to Chesapeake Bay Biota. Water Air Soil Pollut 39:365–376

Hall LW Jr, Scott MC, Killen WD et al. (2000) A probabilistic ecological risk assessment of tributyltin in surface waters of the Chesapeake Bay Watershed. Hum Ecol Risk Assess 6:141–179

Harino H, Fukushima M, Yamamoto Y et al. (1998) Contamination of butyltin and phenyltin compounds in the marine environment of Otsuchi Bay, Japan. Environ Pollut 101:209–214

Harino H, Fukushima M, Kawai S (1999) Temporal trends of organotin compounds in the aquatic environment of the Port of Osaka, Japan. Environ Pollut 105:1–7

Hu J, Zhen H, Wan Y et al. (2006) Trophic magnification of triphenyltin in a marine food web of Bohai Bay, North China: comparison to tributyltin. Environ Sci Technol 40:3142–3147

Javed M (2006) Studies on metal contamination levels in plankton and their role as biological indicator of water pollution in the river Ravi. Pakistan J Biol Sci 313–317

Jeon JK, Lee MH, Lee JS et al. (2003) Toxicity of TBT and TPT compounds on the survival of rotifer (*Brachionus plicatilis*). Korean J Environ Bull 21:158–163

Kline ER, Jarvinen AW, Knuth ML (1989) Acute toxicity of triphenyltin hydroxide to three cladoceran species. Environ Pollut 56:11–17

Laughlin RB Jr, Guard HE, Coleman WM (1986a) Tributyltin in seawater: speciation and octanol-water partition coefficient. Environ Sci Technol 20:201–204

Laughlin RB Jr, French W, Guard HE (1986b) Accumulation of Bis (tributyltin) Oxide by the Marine Mussel *Mytilus edulis*. Environ Sci Technol 20:884–890

Lee CC, Hsieh CY, Tien CJ (2006) Factors influencing organotin distribution in different marine environmental compartments, and their potential health risk. Chemosphere 65:547–559

Lee RF, Valkir AO, Seligmad PF (1989) Importance of microalgae in the biodegradation of tributyltin in estuarine waters. Environ Sci Technol 23:1515–1518

Linden E, Bengtsson B, Svanberg O et al. (1979) The acute toxicity of 78 chemicals and pesticide formulations against two brackish water organisms, the bleak (*Alburnus alburnus*) and the harpacticoid *Nitocra spinipes*. Chemosphere 8:843–851

Linley-Adam G (1999) Harmful effects of the use of anti-fouling paints for ships. The accumulation and impact of organotin compounds on the marine biota within the Mediterranean Region. International Maritime Organization Report. MEPC43/INF14.WPD. pp15

Looser PW, Bertschi S, Fent K (1998) Bioconcentration and bioavailability of organotin compounds: Influence of pH and humic substances. Appl Organomet Chem 12:601–611

Maeda S, Mizoguchi M, Ohki A et al. (1990) Bioaccumulation of zinc and cadmium in freshwater alga, *Chlorella vulgaris*. Chemosphere 21:965–973

Maguire RJ, Wong PTS, Rhamey JS (1984) Accumulation and metabolism of tri-*n*-butyltin cation by a green alga, Ankistrodesmus falcatus. Can J Fish Aquat Sci 41:537–540

McNaught DC (1992) Zooplankters as indicators of ecosystem health: past findings and future directions. J Aquatic Ecosystem Health 1:271–281

Mooney HM, Patching JW (1995) Triphenyltin inhibit photosynthesis and respiration in marine microalgae. J Ind Microbiol 14:265–270

Ohji M, Takeuchi I, Takahashi S et al. (2002a) Differences in the acute toxicities of tributyltin between the Caprellidea and the Gammaridea (Crustacea:Amphipoda). Mar Pollut Bull 44:16–24

Ohji M, Arai T, Miyazaki N (2002b) Effects of tributyltin exposure in the embryonic stage on sex ratio and survival rate in the caprellid amphipod *Caprella danilevskii*. Mar Ecol Prog Ser 235:171–176

Ohji M, Arai T, Miyazaki N (2003a) Chronic effects of tributyltin on the caprellid amphipod *Caprella danilevskii*. Mar Pollut Bull 46:1263–1272

Ohji M, Arai T, Miyazaki N (2003b) Timing of sex disturbance caused by tributyltin exposure during the embryonic stage in the caprellid amphipod, *Caprella danilevskii*. J Mar Biol Ass U K 83:943–944

Ohji M, Arai T, Miyazaki N (2004) Effects of tributyltin on the survival in the caprellid amphipod *Caprella danilevskii*. J Mar Biol Ass UK 84:345–349

Ohji M, Arai T, Miyazaki N (2005) Acute toxicity of tributyltin to the Caprellidea (Crustacea: Amphipoda). Mar Environ Res 59:197–201

Ohji M, Arai T, Midorikawa S et al. (2007) Distribution and fate of organotin compounds in Japanese coastal waters. Water Air Soil Pollut 178:255–265

Omori M, Ikeda T (1984) Methods in marine zooplankton ecology. Wiley-Interscience, New York, pp331

Parsons TR, Maita Y, Lalli CM (1984) A manual of chemical and biological methods for seawater analysis. Pergamon, Oxford, 173pp

Parsons TR, Takahashi M, Hargrave B (1990) Biological oceanographic processes. Pergamon, Oxford, 330pp

Pinkney AE, Matteson LL, Wright DA (1990) Effects of tributyltin on survival, growth, morphometry, and RNA-DNA ratio of larval striped bass, *Morone saxatilis*. Arch Environ Contam Toxicol 19:235–240

Reader S, Pelletier E (1992) Biosorption and degradation of butyltin compounds by the marine diatom *Skeletonema costatum* and the associated bacterial community at low temperature. Bull Environ Contam Toxicol 48:599–607

Rumampuk NDC, Rumengan IFM, Ohji M et al. (2004) Effects of tributyltin on the chlorophyll contents of marine microalga *Tetraselmis tetrathele, Nannochloropsis oculata and Dunaliella* sp. Coast Mar Sci 29:40–44

Seinen W, Helder T, Vernij H et al. (1981) Short term toxicity of tributyltin chloride in rainbow trout (*Salmo gairdneri* - Richardson) yolk sac fry. Sci Total Environ19:155–166

Snell TW, Janssen CR (1995) Rotifers in ecotoxicology: a review. Hydrobiologia 313/314: 231–247

Snell TW, Moffat BD, Janssen C et al. (1991a) Acute toxicity tests using rotifers IV. Effects of cyst age, temperature and salinity on the sensitivity of *Brachionus calyciflorus*. Ecotox Environ Safe 21:308–317

Snell TW, Moffat BD, Janssen C et al. (1991b) Acute toxicity tests using rotifers III. Effects of temperature, strain and exposure time on the sensitivity of *Brachionus plicatilis*. Environ Toxicol Water Qual 6:63–75

Stebbing ARD (1997) A theory for growth hormesis. Belle Newsletter 6 No. 1. Sept. 1997

Steinhauser KG, Amann W, Spath A et al. (1985) Investigations on the aquatic toxicity of organotin compounds. Vom Wasser 65:203–214

St-Louis R, Pelletier E, Marsot P (1997) A mechanistic approach to tributyltin (TBT) sorption by marine microflagellated alga *Pavlova lutheri*. Appl Ognaomet Chem 11:543–550

Sun HW, Dai SG, Huang GL (2001) Bioaccumulation of butyltins via an estuarine food chain. Water Air Soil Pollut 125:55–68

Takahashi S, Tanabe S, Takeuchi S et al. (1999) Distribution and specific bioaccumulation of butyltin compounds in a marine ecosystem. Arch Environ Contam Toxicol 37:50–61

Tanabe S, Prudente M, Mizuno T et al. (1998) Butyltin contamination in marine mammals from North Pacific and Asian coastal waters. Environ Sci Technol 32:193–198

Tsang CK, Lau PS, Tam NFY et al. (1999) Biogradation capacity of tributyltin by two species. Environ Pollut 105:289–297

U'ren SC (1983) Acute toxicity of bis(tributyltin) oxide to a marine copepod. Mar Pollut Bull 14:303–306

USEPA (U.S. Environmental Protection Agency) (2003) Ambient Aquatic Life Water Quality Criteria for Tributyltin. USEPA Office of Research and Development, Environmental. Office of Water 4304T. US EPA 822-R-03-031

van den Brink PJ, Kater BJ (2006) Chemical and biological evaluation of sediments from the Wadden Sea, The Netherlands. Ecotoxicology 15:451–460

Veltman K, Huijbregts MAJ, van den Heuvel-Greve MJ et al. (2006) Organotin accumulation in an estuarine food chain: Comparing field measurements with model estimations. Mar Environ Res 61:511–530

Walsh GE, McLaughlin LL, Lores EM et al. (1985) Effects of organotin compounds on growth and survival of two marine diatoms, *Skeletonema costatum* and *Thalassiosira pseudonana*. Chemosphere 14:383–392

White JS, Tobin JM, Cooney JJ (1999) Organotin compounds and their interactions with microorganisms. Can J Microbiol 45:541–554

World Health Organization (ICS/EHC/89.29), Geneva, Switzerland, 229 pp

Wong PTS, Chau YK, Kramar O et al. (1982) Structure-toxicity relationship of tin compounds on algae. Can J Fish Aquat Sci 39:483–488

Chapter 16
Molluscs

William John Langston and Nicholas Dingle Pope

16.1 Introduction

Concern over organotin (OT) bioaccumulation has focused on molluscs due to the effects of TBT and TPT on reproduction and recruitment in this phylum at extremely low concentrations. There have also been concerns because molluscs form an important component of food chains involving humans. Molluscs represent a significant, if variable, concentration step in the transfer of OTs from water (Bioconcentration Factor-BCF ~10^2–10^5) and sediment (BCF_{sed} up to 10^2) – an attribute which has been harnessed in biomonitoring programmes. Few phyla display comparable abilities for bioconcentration of OTs, which accounts for their sensitivity. However, bioaccumulation is not always simply a function of adsorption of dissolved forms (except at the lowest trophic levels), but may also involve uptake from dietary sources including sediments, and modification by metabolism and excretion, giving rise to much variability. In this chapter we review the pathways and potential for bioaccumulation of OTs in three major classes of the Mollusca, namely gastropods, bivalves and cephalopods. Much of the knowledge gained from studies on OTs will have broader implications – in terms of understanding the processes and timescales of impacts of future persistent contaminants, and, hopefully, in the design of future risk assessment protocols.

W.J. Langston and N.D. Pope
Marine Biological Association, Hill, Plymouth PL1 2PB, UK

T. Arai et al. (eds.), *Ecotoxicology of Antifouling Biocides*,
DOI: 10.1007/978-4-431-85709-9_16, © Springer 2009

16.2 Background – The 'Special Relationship' Between TBT and Molluscs

In the late 1970s and early 1980s, stocks of oysters *Crassostrea gigas* near marinas in France and the UK failed, or were unmarketable due to deformation of the shells. TBT was subsequently linked with reduced spatfall and unnatural shell thickening (Waldock and Thain 1983; Alzieu et al. 1986). Shortly after this, TBT was also shown to cause masculinisation (imposex) in female stenoglossan gastropods *Nucella lapillus,* with widespread population decline observed near to marinas and ports (Bryan et al. 1986). The sensitivity of both shell thickening and imposex responses led to their use as indicators of OT pollution world-wide. Recognition of effects in molluscs resulted in many nations enforcing partial legislation by the early 1990s. Since then it has been shown that numerous non-target species may be adversely affected by OTs, though it is usually mollusc species which are among the most sensitive.

Prior to legislation, elevated concentrations of TBT in spring and summer (linked to increased boating activity) coincided with the onset of the breeding season of many molluscs, exacerbating the threat. However it was not only pulses of TBT in water that posed hazards. TBT preferentially adsorbs on to phytoplankton and suspended particulates ($K_d \sim 10^4$ in fine silts) which constitute the diet of filter-feeders. If deposited in benthic sediments, these particulate loadings can persist, particularly in anoxic conditions (Chapter 5). Organotin-laden paint flakes may also be entrained in sediments near dockyards and marinas. If only a fraction of this particulate material is biologically available, consequences for sediment-dwelling molluscs become important and can be severe. Additionally, the carnivorous diet of cephalopods and some gastropods adds the potential for food chain transfer.

In the following sections, evidence showing how assimilation and degradation pathways influence bioaccumulation is presented. Molluscs clearly exhibit a large range in behaviour with respect to OT. Triorganotins display maximum bioaccumulation within any homologous series: hence accumulation (and toxicity) of TBT/TPT far exceeds that of DBT/DPT and MBT/MPT. However, few studies in the west report significant data for TPT, reflecting much lower usage compared with Japan (where it is now banned), Korea and some other eastern countries. Consequently, much of this review is based on observations on TBT.

16.3 Gastropods

TBT (and TPT) are most harmful to stenoglossan gastropods, by virtue of effects on reproduction, and at least 150 species are now thought to be susceptible to imposex. Body burdens of only 0.01 μg TBT g^{-1} may initiate imposex in the most sensitive species and both dietary and aqueous sources appear equally important for bioaccumulation.

Bioconcentration factors in gastropods exposed to TBT in the laboratory, via water, vary between 10^3 and 10^5. In *Nucella lima* the BCF was ~4×10^3 (at 64 ng TBT l^{-1}) and,

Table 16.1 Summary of TBT bioconcentration parameters in molluscs from field studies and laboratory investigations

Mollusc	Exposure[a]	BCF[b] (water)	BCF[b] (sediment)	Steady state (days)	Half time (days)	Notes[c]	Reference
Gastropods							
Nucella lapillus	L	~30,000 at 8 ng l^{-1}		50–100[d]	50–60	52–80% TBT	Bryan et al. (1986, 1987)
		~16,000 at 50 ng l^{-1}					
		~10,000 at 267 ng l^{-1}				ww (x3 for dw)	
	F(transplant)	~15,000 (~67 ng l^{-1})			~100		
Nucella lima	L	4,390 at 64 ng l^{-1}		~60		22–60% TBT (4 months); ww	Stickle et al. (1990)
		2,050 at 914 ng l^{-1}					
Littorina littorea	F	4,400	3.9			35% TBT; ww	Langston et al. (1987)
Littorina littorea	L	5,100–7,100 (2–400 ng l^{-1})	0.01–0.33	6 m		dw	Bauer et al. (1997)
	F	1,170–34,600				dw	
Littorina littorea	F		0.97			25% TBT	Harino et al. (2005b)
Littorina littorea	F	500–10,000				20–40% TBT winter	Kure and Depledge (1994)
						60–70% TBT spring	
Hinia reticulata	L	11,500 at 9 ng l^{-1}	0.95	70[d]	18–26	dw	Pope (1998)
Bivalves							
Crassostrea gigas	F	25,000				Estimated	Shim et al. (1998)
Crassostrea gigas	L	2,000 at 1,250 ng l^{-1}		14	~10–23	ww	Waldock et al. (1983)
		6,000 at 150 ng l^{-1}					
	F	10,000					

(continued)

Table 16.1 (continued)

Mollusc	Exposure[a]	BCF[b] (water)	BCF[b] (sediment)	Steady state (days)	Half time (days)	Notes[c]	Reference
Ostrea edulis	L	1,000 at 1,250 ng l^{-1}		14	~10–23	ww	Waldock et al. (1983)
		1,500 at 150 ng^{-1}					
	F	2,000					
Aequipecten irradians	L	2,000–10,000		60[d]		ww	Guolan and Yong (1995)
Mytilus edulis	L	1,813 at 800 ng l^{-1}–9,000 at 50 ng l^{-1}		28[d]	36–21	>74% TBT; dw	Yang et al. (2006)
Mytilus edulis	L	5,000 at 500 ng l^{-1}		47[d]	14	ww	Laughlin et al. (1986)
Mytilus edulis	L	7,700–11,000		60[d]		ww	Guolan and Yong (1995)
Mytilus edulis	F(transplant)	5,000–60,000		51[d]	40	ww (x~8 for dw)	Zuolian and Jensen (1989)
Mytilus edulis	F		0.55			28% TBT	Harino et al. (2005b)
Dreissena polymorpha	F(transplant)	900,000				>90% TBT; dw	Van Slooten and Tarradellas (1994)
Mya arenaria	F	133,000	77.9			83% TBT; ww	Langston et al. (1987)
Mya arenaria	F	57,000–220,000	25–384			80–90% TBT; dw	Kure and Depledge (1994)
Mya arenaria	L	91,800 at 50 ng l^{-1}		28[d]	71	>89% TBT; dw	Yang et al. (2006)
		44,800 at 200 ng l^{-1}		28[d]	87	>89% TBT; dw	
		15,538 at 800 ng l^{-1}		28[d]	94	>89% TBT; dw	
Ruditapes decussatus	L	9,000 at 100 ng l^{-1}	<1	20 water 40 sed		dw	Coelho et al. (2002a, c)
Nuculana pernula	F		67–461			77–88% TBT	Strand et al. (2003)
Elliptio complanata	F(transplant)	4,800–18,500	0.1–18	140[d]		Mainly TBT; ww	Chau et al. (1989)

Scrobicularia plana	F	96,000	9	40	50	dw	Langston and Burt (1991)
	L	13,000 at 100 ng l^{-1}	11		28		
Scrobicularia plana	F		3.1			20% TBT	Harino et al. (2005b)
Cerastoderma edule	F		2.7			38% TBT	Harino et al. (2005b)
Macoma balthica	F		3.9			25% TBT	Harino et al. (2005b)
Macoma nasuta	L	10,400 at 200 ng l^{-1}		60			Salazar et al. (1987)
Cephalopods							
Todarodes pacificus	F	18,000–101,000 (digestive gland)				ww	Yamada et al. (1997)

[a]L = laboratory, F = field.

[b]BCF – tissue bioconcentration factor relative to water or sediment.

[c]Notes: BCF_{water} results expressed on wet weight (ww) or dry weight (dw) basis; percentages refer to proportion of extractable BT present as TBT.

[d]Approaching, but not at, steady state.

although not particularly high in comparison with other molluscs (Table 16.1), resultant body burdens were sufficient to cause imposex (Stickle et al. 1990). TBT was also bioaccumulated from food (naturally contaminated mussels) in comparable amounts, though biomagnification of TBT in *N. lima*, relative to food, did not occur.

BCFs in the dog whelk *Nucella lapillus* decrease with increasing contamination, implying dose-dependent tissue saturation; nevertheless at ~50–70 ng TBT l^{-1} the BCF was estimated to be ~16,000 (Bryan et al. 1986, 1987) – higher than in *N. lima* and perhaps explaining the greater sensitivity of *N. lapillus*. Steady state is approached after 50 days aqueous exposure in *N. lapillus* but further accumulation can take place in females, notably during egg development. It is possible that TBT metabolism is more efficient in *N. lima*. However, degradation/elimination of TBT in *N. lapillus* increases (and BCFs decline) with increasing contamination, such that only half the total BT is present as TBT at high doses. Elimination half-times for TBT in *N. lapillus* range from 50 to 60 days, but may vary depending on physiological condition and may be as long as 100 days in the field, perhaps signifying the presence of pools with different rates of exchange (Bryan et al. 1987).

Nucella lapillus fed on ^{14}C-TBT labelled *Mytilus edulis* displayed assimilation efficiencies approaching 100%. When body burdens were compared with dog-whelks exposed via water only, it was estimated that up to half the body burden of TBT could originate from the diet, except perhaps in winter, when feeding rate declines (Bryan et al. 1989). Nevertheless, food chain biomagnification of TBT does not occur because metabolism of TBT appears to be more efficient in dog-whelks than in mussels – only 65% of measurable butyltins was present as the parent compound in *Nucella*, compared with 90% TBT in *Mytilus* (Bryan et al. 1989, 1993a). TBT tissue distributions were indicative of uptake routes and *N. lapillus* fed on TBT-labelled mussels absorbed dietary TBT most efficiently via the digestive gland (where it was also degraded); in contrast absorption of dissolved TBT occurred primarily in the mantle and gills – tissues in direct contact with the water. Kidney was also important in uptake and degradation in both scenarios. Bryan et al. (1993a) provide a useful conceptual model showing relationships between uptake route, degradation and tissue distribution in these carnivorous gastropods.

In a number of neogastropod species in Japan, degrees of imposex have been shown to reflect pollution levels, not only of TBT, but also TPT (Horiguchi et al. 1994). Estimated body burdens of TBT responsible for inducing imposex were 10–20 ng g^{-1}(wet wt) for rock shells, *Thais clavigera* and *T. bronni* – similar to 'effect levels' for the initiation of imposex in *Nucella* spp.

Littorina littorea, a mesogastropod species, exhibits a form of imposex (intersex) but is much less sensitive to TBT than neogastropods – body burdens required to initiate the condition are at least an order of magnitude higher. Nevertheless, these concentrations, and accompanying symptoms, occur near marinas and harbours. The severity of intersex increases as a function of body burden which, in turn, increases in relation to both water and sediment contamination (Bauer et al. 1997). *L. littorea* exhibit rather low BCFs, especially where contamination is high (Table 16.1) and this could contribute to their lower sensitivity. Possible routes of organotin uptake in *Littorina*, besides water, could include the diet of macroalgae (generally low in TBT) and sediment (unlikely, since BCF_{sed} are usually <1), though neither pathway has been quantified. Also, degradation of TBT appears to be rapid in this species, as reflected in seasonal changes in BTs: prior to legislation, approximately 60–70% of the total organotin was present as TBT in spring (presumably because of new inputs from paint) but this proportion fell to 20–40% during autumn and winter, implying effective metabolism.

In the burrowing neogastropod *Hinia reticulata*, BCF factors of 28,000–74,500 have been reported for TBT at seawater concentrations in the range 2.5–47 ng l^{-1} (Stroben et al. 1992; Bryan et al. 1993), though both field studies acknowledge that these BCF values may have been overestimates, due to contributions from other sources (diet). Experiments with ^{14}C-TBT have confirmed *H. reticulata* to be an effective accumulator of TBT from seawater with concentration factors of 10^3–10^4 over a 50 day exposure period (Pope 1998; Table 16.1). However, these studies also revealed that much of the TBT body burden was accumulated from sediments (in equilibrated systems) and that this sediment-derived proportion increased as the sediment TBT concentration increased (31% at 67 ng g^{-1} TBT, 87% at 620 ng g^{-1} TBT). The mechanism whereby *H. reticulata* accumulates TBT from sediment is primarily by direct contact, involving adsorption across the respiratory surfaces and also the exposed tissue of its head/foot – a large flat area of tissue which is in direct contact with the sediment and pore water as the animal burrows. Ingestion of sediment is not considered a major uptake route, since bioaccumulation in individuals with a ligatured proboscis is comparable to that in ‘normal’ snails. In the wild, however, incidental ingestion of some sediment during feeding (on dead biota) cannot be discounted (Pope 1998).

Metabolism of TBT in *Hinia* occurs predominantly in the digestive gland and the more polar debutylated products are excreted via the kidney. Half-times for butyltin depuration, determined following injection of ^{14}C-TBT, were fairly short – between 26 and 18 days (Pope 1998). Within this depuration period, butyltin speciation changed significantly – from >70% as TBT during the first week of depuration to <29% TBT after 90 days. The corresponding DBT proportion increased from <10% to >50% over the same period, while the fraction as MBT remained relatively constant at around 20%. Following exposure via sediment and water, half-times were somewhat longer, ranging from 30 to 66 days.

16.4 Bivalves

This class of mollusc encompasses a variety of filter-feeding types including suspension feeders (oysters, mussels and some clams) and more specialized deposit-feeding clams, giving rise to diverse OT behaviour.

Oysters have been the subject of a number of bioaccumulative and monitoring studies in view of the deleterious effects of OT demonstrated near marinas and harbours (Waldock and Thain 1983; Alzieu et al. 1986; Shim et al. 1998). Uptake of TBT is fairly rapid in both *Crassostrea gigas* and *Ostrea edulis* which reach steady-state within 14 days during laboratory exposures – at BCFs in the range 10^3–10^4 (decreasing with increasing contamination). Bioaccumulation capacity is higher in *C. gigas* than in *O. edulis* (by up to four-fold and nine-fold in laboratory and field, respectively, Table 16.1). Despite initial rapid losses in laboratory-exposed individuals, depuration in *C. gigas* tends to be prolonged in nature, even when sources are reduced, probably due to a slowly exchanging residual pool (Waldock et al. 1983). Loss rates under field conditions are illustrated by the reduction in tin levels in *C. gigas* from Arcachon Bay – from 3.40 $\mu g\ g^{-1}$ dry wt in 1982 (the year the French banned the use of TBT-based paints on all vessels under 25 m length) to 0.50 $\mu g\ g^{-1}$ 2 years later (Alzieu et al. 1986). Fairly slow metabolism of TBT is indicated in laboratory experiments with *Crassostrea virginica,* attributable to low levels of cytochrome P450-mediated transformations (Lee 1986).

Mussels are relatively ubiquitous and TBT accumulation has been shown to be positively related to seawater TBT concentrations, providing time-integrated information on OT contamination. Consequently they too have been used widely as monitoring organisms (Salazar and Salazar 1996; Harino et al. 1998, 1999). Relatively poor correlations between TBT concentrations in mussels (and oysters) and benthic sediments appears to confirm that the water column, including phytoplankton, is the more important source in these suspension feeders (Wade et al. 1990).

As in other molluscs, BCFs in *Mytilus edulis* exposed experimentally to aqueous TBT exhibit an inverse relationship with dose (Table 16.1): for example, at ~50 $ng\ l^{-1}$ the BCF was of the order of 9,000 (Yang et al. 2006) compared with 5,000 at 500 $ng\ l^{-1}$ (Laughlin et al. 1986). Field studies, including transplants, also indicate an inverse (exponential) correlation between BCFs and water concentration (BCF range 5,000–>100,000, but mostly 20–40,000: Zuolian and Jensen 1989; Salazar and Salazar 1996). Extremely high BCFs (900,000) were recorded for the freshwater mussel *Dreissena polymorpha* following transplantation to a TBT-contaminated marina (Van Slooten and Tarradellas 1994).

Loss of OTs from *Mytilus edulis* is fairly rapid during depuration experiments (half life 14–40 days), and is consistent with a first-order kinetic model. However, differences in elimination rate constants have been observed during uptake and release phases, which suggests behaviour in the laboratory and field could be different, contributing to variations in estimated BCF (Laughlin et al. 1986). In nature, higher BCFs could also signify an additional contribution from food. Mussels, like oysters, are epibenthic filter feeders capable of filtration rates of 60 $l\ h^{-1}$ and selective removal of particles down to a few micrometers diameter (principally phytoplankton, together with microorganisms adhering to suspended particulates). Phytoplankton have been confirmed as an important vector for TBT assimilation in mussels, although there is no indication of TBT biomagnification (transfer factor <2) (Laughlin et al. 1986; Guolan and Yong 1995).

Viscera and gills of mussels are dominant sites for TBT bioaccumulation from diet and water, respectively. Despite a general correlation between TBT accumulation and lipid content of mussel tissues (in terms of ranking), TBT concentrations in mussels are an order of magnitude higher than predicted from solubility models, implying that bioaccumulation is not governed by lipid partitioning behaviour alone, but may be modified by physiology – for example, binding to proteins, including enzymes (Laughlin et al. 1986; Guolan and Yong 1995). Degradation does not appear to differ greatly between different mussel tissues and in the exam-

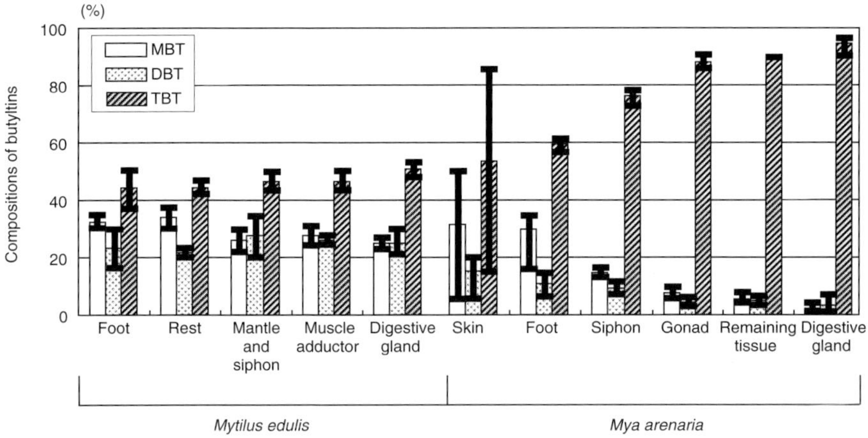

Fig. 16.1 Composition of butyltins in tissues of *Mytilus edulis* and *Mya arenaria*, Mersey Estuary (Reprinted from Harino et al. 2005a. With permission from Elsevier)

ple for *Mytilus edulis* from the Mersey Estuary shown in Fig. 16.1, TBT typically constituted about half the total BT content in most tissues, with the remainder made up equally by DBT and MBT (Harino et al. 2005a).

Infaunal clams in sediments may, in principle, be exposed to additional OT contamination associated with the particulate phase, in addition to that in the water column. The disparity in BCFs between field and laboratory studies often indicates the presence of alternative sources and has prompted several studies on the relative importance of diet (including sediment) on organotin bioaccumulation in clams.

The affinity of TBT for particulates, in particular organic-rich fines (Kd ~10^4), would be anticipated to be particularly important for deposit-feeding clams. Thus, TBT burdens in the infaunal clam *Scrobicularia plana* collected from estuaries around the UK were found to be directly related to sediment loadings and, in experiments, *S. plana* was shown to accumulate >90% of its TBT burden from this source (Langston and Burt 1991). BCF_{sed} were similar in the field and laboratory, at comparable exposures (Table 16.1), and decreased as sediment TBT loadings increased. The kinetics of TBT uptake from sediment indicated the approach of steady-state in *S. plana* after 40 days. Initial loss of TBT from clams during depuration was fairly rapid (half-time 15–30 days depending on dose), but consisted of two compartments, one of which appeared very persistent. In the field the half-time for loss was longer – of the order of 50 days. Given that TBT is likely to persist for some time in sheltered sediments close to harbours and dockyards, the impact on populations of deposit feeders such as *S. plana* may continue for some time compared with other types of mollusc.

Suspension feeding clams *Ruditapes decussatus* appear less vulnerable to TBT in benthic sediments than *S. plana* since, in equilibrated systems, uptake is predominantly (>90%) from water (Coelho et al. 2002a). Although some uptake from sediment does occur, the processing of large amounts of water needed to sustain the suspension-feeding habit of this species is the primary route of TBT assimilation (accounting for body burdens 30 times higher than from sediment). Field surveillance of TBT contamination using *R. decussatus* indicates significant correlation with water, but not sediment, consistent with laboratory results (Coelho et al. 2002b). Suspended microalgae in the water column are a potential source of OTs in clams such as *R. decussatus,* though in experiments using moderate cell densities of ^{14}C TBT – labelled *Isochrysis galbana*, this route appeared to be overshadowed (100-fold) by direct uptake from the aqueous phase (Coelho et al. 2002c). At steady state (40 days) the transfer factor – TBT *R. decussatus*/TBT *I. galbana* – was 0.3, indicating that food chain biomagnification from phytoplankton to clams was not occurring. However, the relative contribution towards body burdens in nature is somewhat uncertain and will vary according to the quantity and quality of micro-algal cells. Where phytoplankton productivity is high, more of the TBT in the water column will be bound to algal cells, representing a larger vector for assimilation. Interestingly, similar BCFs were obtained in *R. decussatus* exposed to either sediments or *Isochrysis,* implying that TBT was assimilated to an equivalent degree from both particulate forms.

Tissue TBT distributions in *Ruditapes decussatus* reflected the major source of uptake-with gills representing the most important site for bioaccumulation from water (Coelho et al. 2002a). The digestive gland accumulated TBT preferentially from food (*Isochrysis galbana*), initially, though after a few weeks of exposure, internal remobilization resulted in a more widespread partitioning of TBT amongst tissues (Coelho et al. 2002c).

The deep-burrowing soft-shell clam *Mya arenaria* is an extraordinary accumulator of BTs, particularly TBT. Butyltins represent a dominant proportion of the total tin in *Mya* (~97%). In contrast, BTs represent ~1% of the total tin in sediment, demonstrating that BTs are considerably more bioavailable than inorganic Sn (the principal particulate form), and are effectively and selectively assimilated. At the height of the TBT problem in the 1980s, BCF in *M. arenaria* from Poole Harbour, UK, were of the order of 133×10^3 on a wet weight basis – or $>500 \times 10^3$ on a dry weight basis – some ten times higher than other bivalves and 30-fold higher than the gastropod *Littorina littorea* (Langston et al. 1987, 1990). Similar conclusions were drawn by Kure and Depledge (1994) who studied *Mya* in Fyn, Denmark. Here, as in the UK, the extent of TBT uptake reflected proximity to marinas and commercial shipping, and varied seasonally according to boating intensity (BCF $57–220 \times 10^3$). Recent studies have established that this distinctive behaviour persists extensively; body burdens in *M. arenaria* exceeded those in a range of other bivalve and gastropod species sampled in the Bohai Sea, and in wider Chinese seafood markets, usually by at least an order of magnitude (Zhou et al. 2003; Yang et al. 2006).

Bioaccumulation of TBT in *Mya* (occasionally $>50\,\mu g\ g^{-1}$ [dry wt] near sources) is enhanced by a relatively low rate of metabolism and excretion – with the bulk (80–90%) of total BT in tissues being present as the parent compound (Langston et al. 1987; Kure and Depledge 1994; see also Fig. 16.1). Kinetics of TBT uptake and loss in *Mya* were compared recently alongside those in mussels *Mytilus edulis,* during aqueous exposures (Yang et al. 2006). Accumulation rate constants for TBT in *Mya* (0.54–2.97) were an order of magnitude higher than those in *Mytilus* (0.062–0.3), driven partly by slower losses: during depuration (28 days) the biological half-life for TBT in clams was approximately three-fold longer than in mussels. Furthermore the percentage of TBT never fell below 89% of the total BT in clams, consistent with a low rate of degradation.

Viscera and gills of *M. arenaria* were shown to be primary sites for accumulation of TBT during aqueous uptake experiments (Kure and Depledge 1994; Yang et al. 2006), and in the wild (Harino et al. 2005a). *Mya* is an infaunal suspension feeder which ingests small detrital particles and microorganisms which are suspended in the water column just above the sediment, and may also absorb and utilize dissolved organic matter (Stewart 1978). Given the high affinity of TBT for such material, coupled with the high filtration rate of *M. arenaria* (up to 541 day^{-1}), it is easy to see why both digestive gland and gills contribute significantly to the high loadings in these clams. Unlike mussels, however, the composition of BTs in different tissues of *Mya* is variable (Fig. 16.1): the low percentage of degradation products MBT and DBT in the digestive gland of *M. arenaria* suggests that metabolism of TBT may be particularly slow here, contributing to the overall 'efficiency' of accumulation in this tissue.

The relative importance of TBT assimilation routes has not be quantified in *M. arenaria,* although the fact that BCFs measured in experimental (aqueous) exposures are much lower than in the field implies that sediment/diet are important TBT vectors in nature. As a general rule, TBT bioconcentration factors, relative to sediment (BCF_{sed}), are elevated in such deep-burrowing clams, compared with surficial and epibenthic molluscs (Table 16.1). BCF_{sed} in the latter are usually below unity, indicating that body burdens in types such as mussels, winkles and some surface-dwelling bivalves are unlikely to be magnified above those in sediment. In contrast *Mya* and *Scrobicularia* exhibit some of the highest BCF_{sed} in estuarine systems, whilst comparable values have been described in clams *Elliptio complanata* exposed to contaminated sediments from freshwater harbours (Chau et al. 1989; Table 16.1). Offshore, deposit-feeding clams such as the protobranch *Nuculana pernula* have particularly high bioaccumulation potential for TBT (an order of magnitude higher than other filter feeding bivalves and gastropods from the same environment) with BCF_{sed} ranging from ~70 to >400 – highly useful in a monitoring context (Strand et al. 2003). Body burdens in *N. pernula* have been shown to reflect TBT gradients in sediments along shipping lanes – for example in the straits between Denmark and Sweden. This relationship was particularly significant following normalization of sediment TBT concentrations to organic content, implying that it is the TBT adsorbed to the organic fraction which is most bioavailable (Strand et al. 2003).

On the available evidence, both feeding style and habitat are important determinants of bioaccumulation potential in bivalves. Coupled with this, variation in the ability to degrade and eliminate TBT results in significant variation in BCFs (Table 16.1). Notable bioaccumulators of TBT, including *Mya arenaria*, *Nuculana pernula* and the horse clam *Tresus capax,* appear to have reduced capabilities for metabolism of TBT, compared with other bivalves from the same environments (Strand et al. 2003; Horiguchi et al. 2003). Such features contribute to a surprisingly wide variation in TBT accumulation patterns.

16.5 Cephalopods

Data on cephalopod body burdens are not extensive but allow an insight into the scale of organotin bioaccumulation in the deep ocean. For example, Takahashi and co-workers (1997) detected organotin (total BT) contamination up to ~400 ng g^{-1} wet wt in cephalopods sampled (along with fish, crustaceans, echinoderms and gastropods) in the aphotic bathyal zone (135–980 m) of the continental slope in Suruga Bay, Japan (Fig. 16.2). BT levels were generally lower than those in shallow-water organisms from the same bay (up to 4,000 ng g^{-1} wet wt in benthic fish), but comparable to those reported in industrialised areas like Tokyo Bay, confirming transport of butyltin pollution to deep-sea ecosystems. Organotin concentrations in cephalopods from Suruga Bay were, in fact, above 'threshold effects levels' in the most sensitive molluscs (~20–100 ng g^{-1} in dog-whelks and oysters), and in excess of levels causing cytotoxicity in cultured fish cell lines. However, effects in cephalopods are largely unstudied. In Nuevo Gulf, off Argentina, one sample of the octopus *Enteroctopus megalocyathus* has been recorded with signs of pseudohermaphroditism (Ortiz and Ré 2006) and though the authors indicated that TBT should not be discarded as an explanation for the observed malformation, the absence of consistent observations linking OTs to reproductive impairment in cephalopods suggests they are less sensitive than gastropods and bivalves.

Compared with other deep sea organisms from Suruga Bay, BTs in cephalopods were often comprised of relatively more DBT and MBT than TBT, particularly in the digestive gland (Takahashi et al. 1997), which may indicate an efficient detoxification

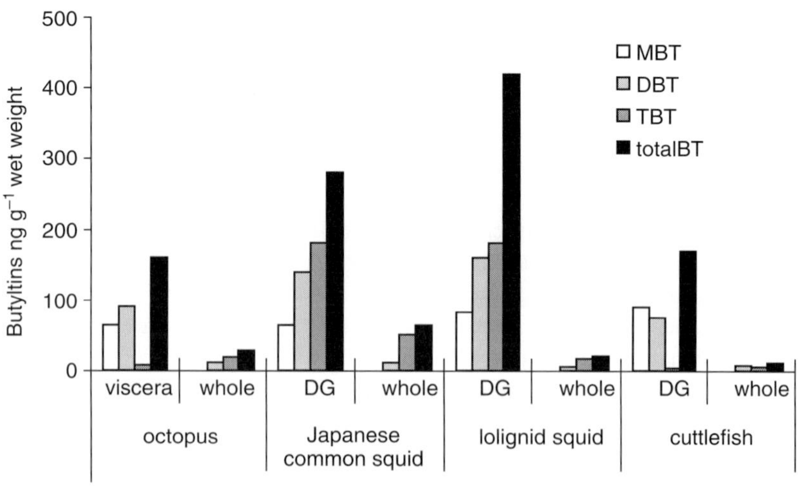

Fig. 16.2 Butyltins in cephalopods from Suruga Bay, Japan (Plotted from data in Takahashi et al. 1997)

system. Nevertheless BTs tend to be concentrated in digestive gland/viscera of squid and octopus, relative to other tissues (Fig. 16.2), implying that the diet is an important uptake pathway.

In more open oceans of the Western Atlantic, deep-sea cephalopods (nine species) have also been reported to contain BTs, but at much lower concentrations (max 2 ng g^{-1} TBT wet wt) than Suruga Bay (Unger et al. 2006). On a larger scale, attempts have been made by Yamada et al. (1997) to use squid as an indicator of OT distributions in the global ocean. The greatest number of samples were from the Sea of Japan and N Pacific, where TBT and TPT concentrations in squid livers increased from ~6 and 8 ng g^{-1}, respectively, in the open ocean, to a maximum of ~300 and 500 ng g^{-1}, respectively, in coastal waters off Japan (closer to sources and concentrated shipping routes). Estimated BCFs in the digestive gland of the squid *Todarodes pacificus* at seven of these sites ranged between 18,000 and 101,000 (mean 48,000), whilst those of TPT appeared to be even higher (~500,000), based on an anticipated, rather than measured values in sea water (Yamada et al. 1997). TPT burdens in squid in this region were generally comparable to, or even higher than, those of TBT, reflecting the history of widespread use of TPT antifouling here. World-wide, TBT concentrations in cephalopods tended to be higher in the northern hemisphere, especially near coasts (up to two orders of magnitude higher), whilst TPT was not detected at all in squid livers collected in the southern hemisphere. These general conclusions must bear the caveats that sampling was spread over 4 years (1989–1993) and that 13 different species were used in different parts of the world. Given the degree of variability seen in other molluscs there is, therefore, some uncertainty over interpretation of trends. Nevertheless the data for squid do present some valuable insights into global patterns in OT distributions.

16.6 Food Chains

Despite the fact that molluscs represent the pinnacle of bioconcentration for OTs, the threat to human consumers appears to be low. As indicated, the ability of molluscs to concentrate OTs is related, partly, to slow rates of metabolism and excretion, whereas most higher organisms possess highly efficient detoxification systems: consequently biomagnification seldom occurs. Only by eating exceptional accumulators of TBT from the most contaminated sites (*Mya* occasionally contains >20 μg TBT g^{-1} dry wt), on a daily basis, would there be a risk of exceeding the tolerable daily intake of 0.25 μg TBT kg^{-1} set by WHO. In essence, this gives the 'all clear' for most types of seafood, since tissue burdens are unlikely to surpass those found in *Mya*. Similar conclusions were drawn by Keithly et al. (1999), based on national *per capita* consumption figures and TBT values in marketable seafoods across four continents (which deviated little from an average of 0.185 μg TBT g^{-1} dry wt). Where evidence for magnification of OT residues along marine food chains has been presented (e.g. Takahashi et al. 1997) – and this seems rare – the concentrations involved remain below extreme values seen in *Mya*. It seems it is molluscs themselves, rather than their consumers, that face the biggest risk of harm from tri-substituted OTs.

16.7 Molluscs As Bioindicators of Trends in OT Contamination

The relatively territorial/sedentary habit of most gastropods and bivalves, coupled with other criteria such as longevity, abundance, ease of collection and size, has led to their widespread use in biomonitoring, though usually on regional scales (in contrast to squid whose distribution and mobility may be more valuable in a global context, as described above). Characteristically high BCFs and biological sensitivity to OT, in bivalves and gastropods, has added to their value as indicators. Furthermore, because the kinetics of uptake and loss have been well studied, we have a reasonable insight as to the timescales over which the responses to changing environmental levels can be monitored. The long half-lives of TBT in *Mya,* for example, suggest suitability for measuring longer-term trends, whereas mussels, which have a rapid turnover time, are more amenable to monitoring short-term change.

Monitoring of spatial patterns in OT contamination, based on body burden surveys of molluscs, has been the subject of a large number of regional and national studies – too numerous to describe in detail here. These affirm the significance of expected sources such as marinas, ports and shipping channels (e.g. Shim et al. 1998; Coelho et al. 2002b; Harino et al. 2003; Rato et al. 2006). Of increasing interest in recent years has been the collection of evidence on temporal trends in molluscs – in particular, the contribution of these data towards the arguments for, or against, a global ban on the use of OTs on the commercial fleet, following the earlier restrictions on leisure craft. Often, results from these surveillance programmes indicate that, where partial bans were imposed in the 1980s–1990s, TBT pollution has declined, particularly at open coastal sites. In the UK this is reflected in the recovery of dogwhelk populations in many areas where, previously, pollution was severe (Birchenough et al. 2002). Many oyster fisheries badly affected by TBT have also returned to normal. Nevertheless, TBT pollution has remained a concern in some coastal and estuarine locations where impacted organisms have been slow to recover and levels of TBT in water and sediment can still exceed environmental quality standards and guidelines. Locally, illegal use of TBT paint or disposal of TBT washings may contribute to these apprehensions.

TBT (and TPT) 'hot-spots' are usually associated with commercial ports and dockyards, where dredging and disposal of contaminated sediments continues to be problematic for environmental managers because of the threat of remobilisation of OTs. Deposit feeding clams such as *Scrobicularia plana* are ideal candidates for directly assessing the bioavailability of these sediment-bound sources. Even at offshore sites, particularly those close to TBT-affected merchant shipping routes and anchorages, recovery can be extremely slow due to persistent residues in sediment.

Combined tissue burden data and imposex measurements in gastropod species such as *Buccinum undatum*, *Neptunea antiqua* and *Nassarius (=Hinia) reticulata*, and bioaccumulation assessment in clams such as *Nuculana pernula* are valuable indicators of trends in these offshore locations (Ten Hallers et al. 2003; Strand et al. 2003; Rato et al. 2006).

Partly because of such mollusc-based evidence, regulation of antifouling paints (with specific reference to triorganotins) was recommended by IMO's Environmental Protection Committee in 1999, with a total phase out of organotin antifouling coatings scheduled for January 2008. Ratification of this legislation was announced in 2007, though there are a number of countries which still do not have restrictions on the use of organotin. Even where legislation is in place, there may still be considerable reservoirs of OT in sediments and the potential for remobilisation and impacts on molluscs and other sensitive species may be locally significant during dredging and disposal operations (Chapter 5). Clearly, there may be some way to go before protection of the environment from OTs, and in particular sensitive molluscan components, is complete. Continued long-term monitoring near major ports, dockyards and shipping-routes, using selected mollusc species, would have obvious advantages in determining the consequences of these actions and charting progress towards recovery.

16.8 Conclusions: Lessons Learned from Organotin Bioaccumulation in Molluscs

Despite the negative effects of TBT, there have at least been some useful lessons for ecotoxicologists and regulators, which stem from observations in molluscs. These include the precautionary warning to 'expect the unexpected' in view of the variability in bioaccumulation, persistence and susceptibility between taxonomic groups. Simple models are unlikely to predict risk, universally. The concept that sediments represent an important and long term source of TBT (for some species) has also been established unequivocally, and should provide important indications of the behaviour of other contaminants, under similar scenarios.

Paradoxically, although antifouling paints were developed to combat barnacle settlement, these crustaceans are among the organisms least affected by TBT (Goldberg 1986). The sensitivity of molluscs to the effects of TBT and TPT lies, partly, with their bioaccumulation potential, which itself is a function of two processes – namely uptake and elimination. Uptake of organic contaminants is often considered a reflection of their lipophilicity and transfer across membranes, however there is little evidence that accumulation is purely lipid-dependent in molluscs (Langston et al. 1987; Takahashi et al. 1997) and internal partitioning may be dictated more by an affinity for proteins. Therefore, unlike most lipophilic contaminants, it is not possible to predict taxonomic variability in bioaccumulation of OTs based on simple partition coefficients (e.g. K_{ow}) alone. Bioaccumulation of OT in molluscs is modified by feeding strategy, and will vary between carnivores, suspension feeders and deposit feeders, according to the bioavailability and concentration in food and sediments, as well as in water. Furthermore, loss rates and depuration times for organotins will vary according to the ability to metabolise and eliminate the parent compound: these rates are likely to depend on a whole gamut of biotic factors, including enzyme activities. Metabolism of TBT is mediated by the cytochrome P450 system, however, the sensitivity of different components of the P450 system to inhibition by OTs appears to be highly variable among molluscs and is likely to result in subtly different modes of action (see review in Langston 1996). The most pertinent example of this involves imposex development in gastropods *Nucella lapillus* which arises as a result of the inhibition, by TBT, of the P450-dependant aromatase responsible for conversion of testosterone to oestrogen (Spooner et al. 1991).

Because of the diversity in uptake pathways and loss processes, BCFs reported for molluscs are extensive in range (Table 16.1). Expressed as biological partition coefficients K_b (=log BCF), values range over almost three orders of magnitude – from 2.6 – 3.9 in herbivorous gastropods like *Littorina littorea* to 5.6–6 in filter feeders such as *Mya arenaria* and *Dreissena polymorpha*. Given the large diversity of molluscs, our understanding of the causes of variability in organotin uptake and elimination processes is probably still far from complete. This unpredictability highlights a more general issue in ecotoxicology, in that many taxonomic groups are likely to be under-represented (and, possibly, under-protected) by current knowledge levels and research effort.

It can be argued that the effects of TBT on molluscs and other non-target organisms were not predicted because there is a narrow approach to risk assessment, based on a standardized suite of toxicity and bioaccumulation tests, with a limited range of organisms. This is, understandably, due to regulatory and financial constraints. Nevertheless, given the serious nature of TBT contamination, and associated costs in resolving the problem, it seems crucial to broaden the ecological relevance of risk assessment, by expanding the mechanisms and end points examined. Molluscs are an obvious key group to consider for greater inclusion in future assessment programmes. Without this broader approach, perception of risk will become dominated by a rather narrow doctrine, focusing on a few end points and mechanisms which may not be sufficiently protective of the more sensitive ecosystem components.

References

Alzieu CL, Sanjuan J, Deltriel JP et al. (1986) Tin contamination in Arachon Bay: effects on oyster shell anomalies. Mar Pollut Bull 17:494–498

Bauer B, Fioroni P, Schulte-Oehlmann U et al. (1997) The use of *Littorina littorea* for tributyltin (TBT) effect monitoring – results from the German TBT survey 1994/1995 and laboratory experiments. Environ Pollut 96:299–309

Birchenough AC, Evans SM, Moss C et al. (2002) Re-colonisation and recovery of populations of dogwhelks *Nucella lapillus* (L.) on shores formerly subject to severe TBT contamination. Mar Pollut Bull 44:652–659

Bryan GW, Gibbs PE, Hummerstone LG et al. (1986) The decline of the gastropod *Nucella lapillus* around south west England: evidence for the effect of tributyltin from antifouling paints. J Mar Biol Assoc UK 66:611–640

Bryan GW, Gibbs PE, Burt GR et al. (1987) The effects of tributyltin (TBT) accumulation on adult dog-whelks, *Nucella lapillus*: long-term field and laboratory experiments. J Mar Biol Assoc UK 67:525–544

Bryan GW, Gibbs PE, Hummerstone LG et al. (1989) Uptake and transformation of ^{14}C-labelled tributyltin chloride by the dog-welk, *Nucella lapillus*: importance of absorption from diet. Mar Environ Res 28:241–245

Bryan GW, Bright DA, Hummerstone LG et al. (1993a) Uptake, tissue distribution and metabolism of ^{14}C-labelled tributyltin (TBT) in the dog-whelk, *Nucella lapillus*. J Mar Biol Assoc UK 73:889–912

Bryan, GW, Burt GR, Gibbs PE et al. (1993b) *Nassarius reticulatus* (Nassariidae, Gastropoda) as an indicator of tributyltin pollution before and after TBT restrictions. J Mar Biol Assoc UK 73: 913–929

Chau YK, Wong PTS, Bengert GA et al. (1989) Bioaccumulation of butyltin compounds by mussels in harbours. Chem Speciat Bioavailab 1:151–156

Coelho MR, Bebianno MJ, Langston WJ (2002a) Routes of TBT uptake in the clam, *Ruditapes decussatus*. I. Water and sediments as vectors of TBT uptake. Mar Environ Res 54:179–192

Coelho MR, Bebianno MJ, Langston WJ (2002b) Organotin levels in the Ria Formosa lagoon. Appl Organomet Chem 16:384–390

Coelho MR, Bebianno MJ, Langston WJ (2002c) Routes of TBT uptake in the clam *Ruditapes decussatus*. II. Food as vector of TBT uptake. Mar Environ Res 54:193–207

Goldberg ED (1986) TBT. An environmental dilemma. Environment 28:17–21

Guolan H, Yong W (1995) Effects of tributyltin chloride on marine bivalve mussels. Water Res 29:1877–1884

Hallers-Tjabbes CC Ten, Wegener J-W, Van Hattum AGM et al. (2003) Organotin levels and imposex development in the benthic gastropods *Buccinum undatum, Neptunea antiqua* from the North Sea. Relations to shipping density and hydrography. Mar Environ Res 55:203–233

Harino H, Fukushima M, Yamamoto Y et al. (1998) Organotin compounds in water, sediment, and biological samples from the Port of Osaka, Japan. Arch Environ Contam Toxicol 35:558–564

Harino H, Fukushima M, Kawai S et al. (1999) Temporal trends of organotin compounds in the aquatic environment of the Port of Osaka, Japan. Environ Pollut 105:1–7

Harino H, Langston WJ, Burt GR et al. (2003) Occurrence of organotin compounds in Mersey and Thames Estuaries a decade after UK TBT legislation. J Mar Biol Assoc UK 83:11–22

Harino H, O'Hara SCM, Burt GR et al. (2005a) Distribution of organotin compounds in tissues of mussels *Mytilus edulis* and clams *Mya arenaria*. Chemosphere 58:877–881

Harino H, O'Hara SCM, Burt GR et al. (2005b) Accumulation of butyltin compounds in benthic biota of the Mersey Estuary. Mar Pollut Bull 50:223–226

Horiguchi T, Shiraishi H, Shimizu M et al. (1994) Imposex and organotin compounds in *Thais clavigera* and *T.bronni* in Japan. J Mar Biol Assoc UK 74:651–669

Horiguchi T, Li Z, Uno S et al. (2003) Contamination of organotin compounds and imposex in molluscs from,Vancouver, Canada. Mar Environ Res 57:75–88

Keithly JC, Cardwell RD, Henderson DG (1999) Tributyltin in seafood from Asia, Australia, Europe, and North America: assessment of human health risks. Hum Ecol Risk Assess 5: 337–354

Kure LK, Depledge MH (1994) Accumulation of organotin in *Littorina littorea* and *Mya arenaria* from Danish coastal waters. Environ Pollut 84:149–157

Langston WJ (1996) Recent developments in TBT ecotoxicology. Toxic Ecotoxic News 3: 179–187

Langston WJ, Burt GR (1991) Bioavailability and the effects of sediment bound TBT in deposit feeding clams *Scrobicularia plana*. Mar Environ Res 32:61–77

Langston WJ, Burt GR, Zhou M (1987) Tin and organotin in water, sediments, and benthic organisms of Poole Harbour. Mar Pollut Bull 18:634–639

Langston WJ, Bryan GW, Burt GR et al. (1990) Assessing the impact of tin and TBT in estuaries and coastal regions. Funct Ecol 4:433–443

Laughlin RB, French W, Guard HE (1986) Accumulation of bis(tributyltin)oxide by the marine mussel *Mytilus edulis*. Environ Sci Technol 20:884–890

Lee RF (1986) Metabolism of Bis (tributyltin) oxide by estuarine organisms. In: Oceans 86 Conference Record Vol 4 Organotin Symposium. New York Institute of Electrical and Electronics Engineers: 1182–1188.

Ortiz N, Ré ME (2006) First report of pseudohermaphroditism in cephalopods. J Mollus Stud 72:321–323

Pope ND (1998) The bioavailability of sediment-bound tributyltin (TBT). Ph.D. thesis, University of Plymouth, Plymouth: 258

Rato M, Sousa M, Quinta R et al. (2006) Imposex and organotin body burden of *Nassarius reticulatus* (L.) in the NW Portuguese continental shelf. Environ Toxicol Chem 25:3213–3220

Salazar MH, Salazar SM (1996) Mussels as bioindicators: effects of TBT on survival, bioaccumulation, and growth under natural conditions. In: Champ MA, Seligman PF (eds) Organotin. Chapman & Hall, London: 305–330

Salazar SM, Davidson BM, Salazar MH et al. (1987) Effects of TBT on marine organisms: Field assessment of a new site-specific bioassay system. In: Proceedings, Oceans 87. Marine Technology Society and IEEE Ocean Engineering Society, Halifax, Canada: 1461–1470

Shim WJ, Oh JR, Kahing S-H et al. (1998). Tributyltin and triphenyltin residues in Pacific oyster (*Crassostrea gigas*) and rock shell (*Thais clavigera*) from the Chinhae Bay system, Korea. J Korean Soc Oceanogr 33:90–99

Spooner N, Gibbs PE, Bryan GW et al. (1991) The effect of tributyltin upon steroid titres in the female dogwhelk, *Nucella lapillus*, and the development of imposex. Mar Environ Res 32:37–49

Stewart MG (1978) Kinetics of neutral amino-acid transport by isolated gill tissue of the bivalve *Mya arenaria* (L.). J Exp Mar Biol Ecol 32:39–52

Stickle WB, Sharp-Dahl JL, Rice SD et al. (1990). Imposex induction in *Nucella lima* (Gmelin) via mode of exposure to tributyltin. J Exp Mar Biol Ecol 143:165–180

Strand J, Jacobsen JA, Pederesen B et al. (2003) Butyltin compounds in sediment and molluscs from the shipping strait between Denmark and Sweden. Environ Pollut 124:7–15

Stroben E, Oehlmann J, Fioroni P et al. (1992). *Hinia reticulata* and *Nucella lapillus* - comparison of two gastropod tributyltin bioindicators. Mar Biol 114:289–296

Takahashi S, Tanabe S, Kubodera T et al. (1997) Butyltin residues in deep sea organisms collected from Suruga Bay, Japan. Environ Sci Technol 31:3103–3109

Unger MA, Vadas GG, Harvey E et al. (2006) Accumulation of persistent organic pollutants and tributyltin in nine species of Atlantic deep-sea cephalopods. Abstract in 11th International Deep-Sea Biology Symposium, National Oceanography Centre, Southampton. http://www.noc.soton.ac.uk/obe/PROJECTS/DEEPSEAS/pdf_files/11DSBS_abstracts.pdf

Van Slooten KB, Tarradellas J (1994) Accumulation, depuration and growth effects of tributyltin in the freshwater bivalve *Dreissena polymorpha* under field conditions. Environ Toxicol Chem 13: 755–762

Wade LW, Garcia-Romero B, Brooks J (1990) Butyltins in sediments and bivalves from U.S. coastal areas. Chemosphere 20:647–662

Waldock MJ, Thain JE (1983) Shell thickening in *Crassostrea gigas*: organotin antifouling or sediment induced? Mar Pollut Bull 14:411–415

Waldock MJ, Thain JE, Miller D et al. (1983) The accumulation and depuration of bis(tributyltin) oxide in oyster: a comparison between the Pacific oyster (*Crassostrea gigas*) and the European flat oyster (*Ostrea edulis*). 1983/E, ICES CM 52: 9pp.

Yamada H, Takayanagi K, Tateishi M et al. (1997) Organotin compounds and polychlorinated biphenyls of livers in squid collected from coastal waters and open oceans. Environ Pollut 96:217–226

Yang RQ, Zhou QF, Jiang GB et al. (2006) Butyltin accumulation in the marine clam *Mya arenaria*: an evaluation of its suitability for monitoring butyltin pollution. Chemosphere 63:1–8

Zhou OF, Li ZY, Jiang GB et al. (2003) Preliminary investigation of a sensitive biomarker of organotin pollution in Chinese coastal aquatic environment and marine organisms. Environ Pollut 125:301–304

Zuolian C, Jensen A (1989) Accumulation of organic and inorganic tin in blue mussel, *Mytilus edulis*, under natural conditions. Mar Pollut Bull 20:281–286

Chapter 17
Fish

Takaomi Arai

17.1 Introduction

Organotins (OTs) are used in a variety of consumer and industrial products such as marine antifouling paints, agricultural pesticides, preservatives, and plastic stabilizers. In particular, butyltins (BTs) and phenyltins (PTs) have been extensively used in boat paints because of their excellent and long-lasting antifouling properties. However, it is well known that BTs and PTs leaching from boats can accumulate in tissues of aquatic organisms causing various deleterious effects.

To understand the contamination status of OTs from fresh water to deep sea ecosystems, various fish species are used as bioindicators. Furthermore, many fish species are economically important as food, thus, to examine the pollution level is mandatory to evaluate risk assessment for human consumption as well as understanding aquatic contamination levels and bioaccumulation.

In this chapter, the status of OT contamination in fish is reviewed, with consideration to their life history.

17.2 Diadromous Fish

17.2.1 Catadromous Fish

The catadromous eel of the genus *Anguilla* is distributed worldwide (Tesch 1977) and performs a spectacular migration of thousands of kilometers between its freshwater and estuarine habitats, and offshore spawning area. Recently, the migratory

T. Arai
International Coastal Research Center, Ocean Research Institute, The University of Tokyo, 2-106-1 Akahama, Otsuchi, Iwate 028-1102, Japan

T. Arai et al. (eds.), *Ecotoxicology of Antifouling Biocides*,
DOI: 10.1007/978-4-431-85709-9_17,

history of several species of anguillid eels has been studied using microchemical analytical techniques to determine the ratios of strontium to calcium (Sr:Ca) in the otoliths of fishes. Tsukamoto and Arai (2001) used these Sr:Ca ratios to classify the migratory histories of Japanese eels collected in Japanese coastal waters into three migratory types: (1) 'sea eels' (spent most of their life in the sea and did not enter freshwater), (2) 'estuarine eels' (inhabited estuaries or switched between different habitats), and (3) 'river eels' (entered and remained in freshwater river habitats after arrival in the estuary). Therefore, it could be considered that there were different ecological risks for pollutants including organotin compounds (OTs) among the three migration types of the species.

A few studies are presently available on TBT concentrations in eels for an approximate comparison of the scale of impact. TBT and TPT concentrations were found to range from 30.5 to 265.9 and from 110.9 to 430.1 ng g^{-1} dry wt, respectively, in the livers of eels from marinas on Lake Grote Poel, the Netherlands (Stäb et al. 1996) (Table 17.1). The concentrations of TBT in the Thames Estuary and Weston Canal in the United Kingdom ranged from 13.9 to 93.4 and from 20.3 to 21.0 ng g^{-1} wet wt, respectively, and TPT concentrations in the same locations ranged from under the detection limit (<2.6) to 42.3 and from 65.6 to 96.6 ng g^{-1} wet wt (Harino et al. 2002) (Table 17.1).

Ohji et al. (2006) found that there were generally no significant correlations between TBT and TPT accumulation and various biological characteristics such as total length (TL), body weight (BW), age and sex in *A. japonica*. The concentrations of TBT and TPT in silver eels (mature eels) were significantly higher than those in yellow eels (immature eels), and the percentages of TBT and TPT were also higher in silver eels than in yellow eels (Fig. 17.1). A positive correlation was found between TBT concentration and the gonad-somatic index (GSI). It is thus considered that silver eels have a higher risk of contamination by TBT than yellow eels. TBT and TPT concentrations in sea eels were significantly higher than those in river eels. In contrast, no significant differences were observed in TBT and TPT concentrations in estuarine eels compared to sea and river eels. These results suggest that sea eels have a higher ecological risk of OTs contamination than river eels during their life history, and the risk of OTs in estuarine eels is considered to be intermediate between that of sea and river eels. Positive linear relationships were found between Sr:Ca ratios and the concentrations of TBT and TPT (Fig. 17.2). These results suggest that the ecological risk of OTs increases, as the sea residence period in the eel become longer. Even at the same maturation stage, TBT and TPT concentrations in sea eels were significantly higher than those in river eels. Thus, it is clear that migratory type is a more important factor for OT accumulation than maturation stage.

17.2.2 Anadromous Fish

Salmonid fish have an anadromous life history pattern where the fish spawn, hatch and spend a period of time in the freshwater environment prior to seaward

Table 17.1 Organotin residues in fish (ng g^{-1} wet wt)

Fish	Country/location	Tissue	MBT	DBT	TBT	MPT	DPT	TPT	References
Diadromous fish									
1. Catadromous fish									
Freshwater eel *Anguilla anguilla*	Netherlands	Muscle	13–63	9–40	50–390	3–29	11–210	93–640	Stab et al. (1996)[a]
Freshwater eel *Anguilla anguilla*	UK	Muscle	19–66	12–24	31–60	<20	<10	<10–31	Harino et al. (2002)[a]
		Liver	29–133	33–79	77–139	<20	<10–21	<10–367	
Freshwater eel *Anguilla japonica*	Japan	Liver	<1–147	3.4–475	1.1–254	<1–122	<1–138	<1–110	Ohji et al. (2006)
2. Anadromous fish									
Three-spined stickleback *Gasterosteus aculeatus*	Poland	Whole body	190–500	500–800	800–1,800				Falandysz et al. (2002)
Masu salmon *Oncorhynchus masou*	Japan	Muscle	7.2–12.7	4.3–6.0	ND–7.2	ND	ND		Ohji et al. (2007)
		Liver	3.1–12.1	2.0–12.5	2.5–13.6	0.7–4.5	1.3–1.1		
Freshwater fish									
Bream *Abramis brama*	Germany	Muscle	<3–18	<3–57	4–481		<3–29	<5–253	Rudel et al. (2007)
Japanese barbell *Hemibarbus barbus*	Japan	Muscle	25	2	20	<1	12	<1	Harino et al. (2000)
Blue gill *Lepomis macrochirus*	Japan	Muscle	37	3	22	<1	12	<1	Harino et al. (2000)
Brackish water fish									
Gray mullet *Mugil cephalus*	Morocco	Muscle		ND–90	ND–8,110		ND–4,000	ND–760	Hassani et al. (2006)
		Liver		ND–6,540	10–18,000		ND–340	ND–2,820	
Japanese sea bass *Lateolabrax japonicus*	Japan	Muscle	83	5	120	9	20	125	Harino et al. (2001)
Marine fish									
1. Coastal fish									
Spanish mackerel *Scomberomorus commerson*	Taiwan	Muscle	25–142	25–105	100–1,234	94–244	161–443	253–556	Lee et al. (2005)

(continued)

Table 17.1 (continued)

Fish	Country/location	Tissue	MBT	DBT	TBT	MPT	DPT	TPT	References
Japanese seabream *Pagrus major*	Taiwan	Muscle	25–146	5.9–143	51–508	110–322	111–467	392–655	Lee et al. (2005)
Ponyfish *Leiogenathus splendens*	Taiwan	Whole body	48–177	69–490	59–1,874				Dong et al. (2004)
Lizardfish *Trachinocephalus myops*	Taiwan	Muscle	5–14	8–35	23–110				Dong et al. (2004)
		Liver	802–2,117	992–7,797	1,233–1,559				
2. Open cean fish									
Bluefun tuna *Thunnus thynnus*	Italy	Muscle	5.9–29	2.2–32	7.3–170				Kannan et al. (1996)
		Liver	21–53	9–440	18–150				
Skipjack tuna *Katuwonus pelamis*	Worldwide	Liver	<1.8–38	<2.4–170	3.8–220				Ueno et al. (2004)
Blue shark *Prionace glauca*	Italy	Liver	4–13	4.3–6	11–23				Kannan et al. (1996)
3. Deep sea fish									
Bone fish *Pterothrissus gissu*	Suruga Bay, Japan	Muscle	<15	<4.0–25	2.3–8.9				Takahashi et al. (1997)
		Liver	<15–240	8.2–690	<2.0–49				
Grenadier *Coelorinchus* sp.	Suruga Bay, Japan	Muscle	<15	5.5–7.7	26–31				Takahashi et al. (1997)
		Liver	19–29	55–67	300–680				
Mesopelagic myctophids *Ceratoscopelus warmingi, Diphus theta, Lampanyctus jordani, L. regalis, Stenobrachius leucopsarus, S. nannochir*	Western North Pacific	Whole body	<5.0–12	<2.0–11	<5.0–35				Takahashi et al. (2000)
Common mora *Mora moro*	Northwestern Mediterranean	Muscle	<1.0	<1.0	<1.0	<1.0	<1.0	3.5	Borghi and Porte (2002)
		Liver	54.5	67	52.1	152	85.2	1,430	

[a]ng g^{-1} dry wt

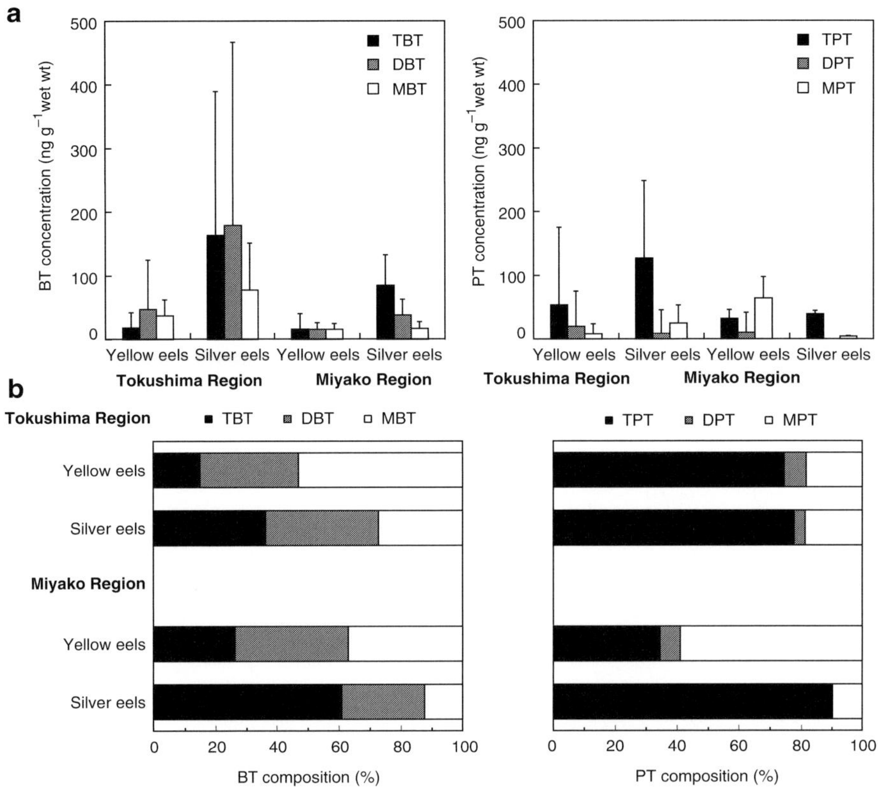

Fig. 17.1 Butyltin (*left*) and phenyltin (*right*) concentrations (a) and compositions (b) in the livers of yellow and silver eels from each region in Japanese eel *Anguilla japonica*

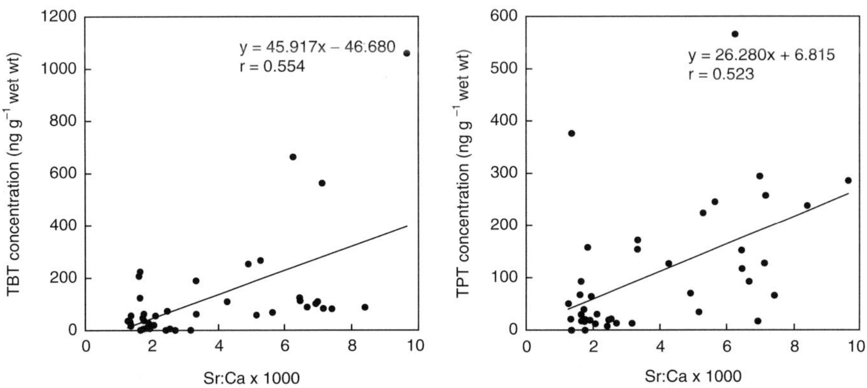

Fig. 17.2 Relationship between tributyltin (*left*) and triphenyltin (*right*) concentrations in the livers and otolith Sr:Ca ratios in Japanese eel *Anguilla japonica*. The higher Sr:Ca ratios indicate the eel depends sea water environment

migration. Some of the fish have life history polymorphism with both sea-run and freshwater-resident migratory histories within a species. Therefore, it could be considered that differences in OTs accumulation may exist between these two migratory types, even within the same species, as well as in the catadromous eel *Anguilla*.

Ohji et al. (2007) demonstrated that the TBT and TPT concentrations in sea-run masu salmon *Oncorhynchus masou* were significantly higher than those in freshwater-resident fish, in spite of their intraspecies relationship (Fig. 17.3). The metabolic capacity of TBT and TPT is considered to be the same between sea-run and freshwater-resident masu salmon, therefore, the proportion of OTs in each fish might be affected by the present status of OTs in the fish's environment (freshwater or sea). This suggests that the difference in the migratory history between sea-run and freshwater-resident fish was reflected in the accumulation profile of TBT and TPT of each migration type. Since TBT and TPT have been used mainly in marine environments, it is considered that the sea-run individuals experience greater exposure

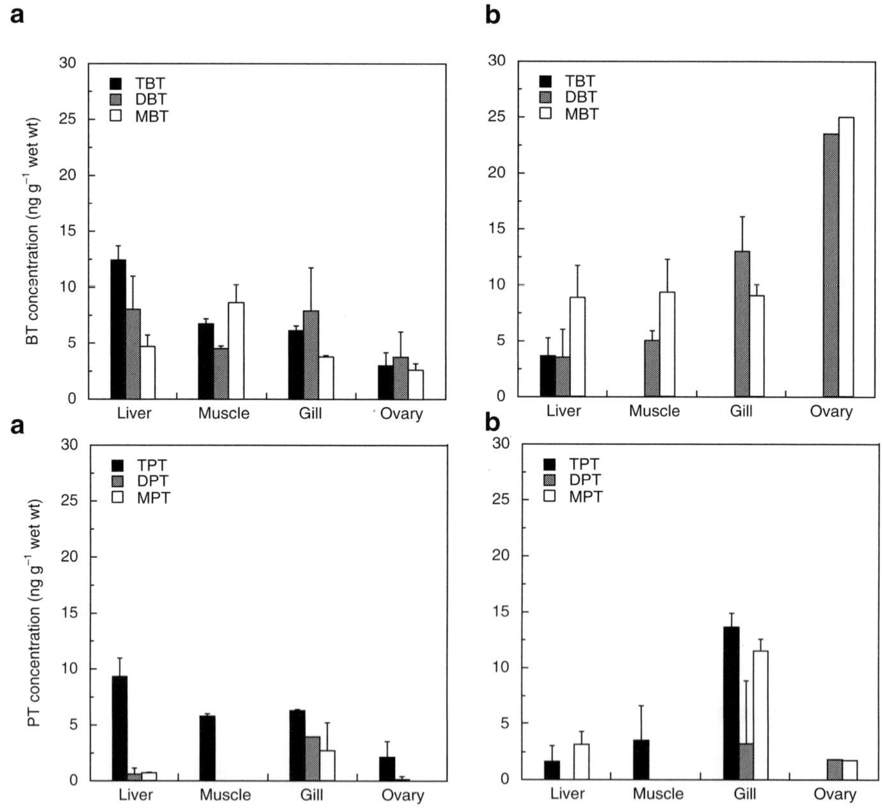

Fig. 17.3 Butyltin (*upper*) and phenyltin (*bottom*) concentrations in liver, muscle, gill, and ovary of (a) sea-run and (b) freshwater-resident masu salmon *Oncorhunchus masou*

to TBT and TPT during the period of migration in coastal and ocean waters. In contrast, since freshwater-resident individuals remain in a freshwater environment throughout their life history, they were only exposed to TBT and TPT to a lesser extent. Furthermore, the relationship between otolith Sr:Ca ratio and TBT concentration clearly showed that TBT concentrations increased significantly with increasing Sr:Ca ratio. Although no correlations were observed in Sr:Ca ratio and TPT concentration, the TPT concentrations in sea-run individuals were significantly higher than those in freshwater-resident fish. These results suggest that the sea-run *O. masou* has a higher ecological risk of TBT and TPT exposure that the freshwater-residents during their life history.

The three-spined stickleback *Gasterosteus aculeatus* is widely distributed in various aquatic habitats such as sea, brackish and freshwaters and has been classified into an anadromous type and a freshwater type, based on its life-style (Wootton 1984). Furthermore, Arai et al. (2003a, b) have recently found populations which have never migrated into freshwater and have spent their entire life history in the sea or estuaries (estuarine resident type). Falandysz et al. (2002) examined three-spined sticklebacks collected from four marinas located along the south-western coast of the Gulf of Gdask, Baltic Sea, Poland. The mean BT concentrations in these fish were between 1,500 and 3,100 ng g^{-1} wet wt and the values followed mean concentrations noted in sediments from the corresponding marina (Table 17.1). Pleasure yachts were identified as the main source of BT pollution of sediment and fish. TBT was the major form of BT present in sediments (between 24% and 43%) and in three-spined sticklebacks (between 54% and 70%), suggesting recent use of marine paints containing TBT on pleasure yachts.

17.3 Non-diadromous Fish

17.3.1 Freshwater Fish

Stäb et al. (1996) examined an extensive study on the presence of organotin compounds in the foodweb of a freshwater lake in the Netherlands. The organotin levels deriving from antifoulants (mainly TBT) and pesticidal usages (TPT) were different for the fish species. The concentration ranges for the different compounds in muscle tissue from individual fish were: <120–6,100 ng g^{-1} TBT, 10–410 ng g^{-1} DBT, 3–93 ng g^{-1} MBT, 110–8,200 ng g^{-1} TPT, 11–890 ng g^{-1} DPT, and <3–48 ng g^{-1} MPT.

Steffen et al. (2003) examined TPT in muscle tissue of roach (*Rutilus rutilus*) from freshwaters in Northern Germany. TPT levels of fish caught in 2001/2002 were in the range <0.3–557 ng g^{-1} (wet wt). The higher levels were predominantly found in regions were potatoes were grown and, probably, TPT containing pesticides were applied. In muscles of bream (*Abramis brama*) sampled in the period 1993–2003 from the rivers Rhine, Elbe, Saale, Mulde, Saar, and from Lake Belau, Germany, TBT was detected in almost all samples, although a decrease in levels has been observed at all sampling sites (Rudel et al. 2007) (Table 17.1). At most sites, the reduction seemed to be a result of the ban on the use of TBT-based antifoulants on small boats, which became effective in Germany in 1989. Highest TBT levels were found in fish from the Elbe near Blankenese (470 ng g^{-1} wet wt; in 1995) and lowest in bream from Lake Belau (<1 ng g^{-1} wet wt; in 2001 and 2003). Highest TPT levels (253 ng g^{-1} wet wt in 1993) were also found in bream caught near Blankenese where the occurrence seemed to be correlated to the former use of TPT as co-toxicant in antifoulants.

Harino et al. (2000) reported that the ratios of TBT to the total BTs and TPT to the total were lower in freshwater fish such as Japanese barbel *Hemibarbus barbus* (20 ng g^{-1} in TBT, 2 ng g^{-1} in DBT, and 25 ng g^{-1} in MBT, <1 ng g^{-1} in TPT, 12 ng g^{-1} in DPT, and <1 ng g^{-1} in MPT) and bluegill *Lepomis macrochirus* (22 ng g^{-1} in TBT, 3 ng g^{-1} in DBT, and 37 ng g^{-1} in MBT, <1 ng g^{-1} in TPT, 12 ng g^{-1} in DPT, and <1 ng g^{-1} in MPT) than those of marine fish collected from the Osaka region of Japan (Table 17.1).

These results suggest that BTs and PTs concentrations in freshwater fish are highly variable among sites and countries either fish collect nearby emission sources such as antifoulants (TBT) and pesticidal usages (TPT) or not.

17.3.2 Brackish Water Fish

Grey mullet *Mugil cephalus* occurs worldwide, where it inhabits estuarine intertidal, freshwater and coastal marine habitats. Hassani et al. (2006) examined the concentrations of BT and PT compounds in fish collected along the northern Mediterranean coast of Morocco, and the south Mediterranean coast of Spain. TBT and TPT were the predominant compounds, and TBT concentrations were higher in liver than in muscle. The total content of BTs in these samples was higher than PT levels. In the Moroccan coast, the highest value of TBT (18,000 ng g^{-1} wet wt) was found in the west harbour of the Tangier site, while the lowest concentration (10 ng g^{-1} wet wt) was detected in the Oued Laoue site (Table 17.1). Concerning TPT, the highest value (4,000 ng g^{-1} wet wt) was found in the harbour of the M'diq Site and the lowest value (10 ng g^{-1} wet wt) was detected in the Oued Laoue site.

Sea bass *Lateolabrax japonicus* also has a similar life history and migration pattern to that of the grey mullet. The concentrations of MBT, DBT and TBT in muscle of the sea bass collected in Osaka Bay, Japan were 83, 5, 120 ng g^{-1} wet wt, respectively. Those of MPT, DPT and TPT were 9, 20, 125 ng g^{-1} wet wt, respectively (Harino et al. 2000) (Table 17.1). In Japanese sea bass *L. japonicus*, there were no significant sex differences in TBT and TBT levels. Furthermore, no correlation between the total length and TBT or TPT concentration was reported in the fish. Nor was any significant correlation of TBT or TPT concentrations to lipid content reported (Harino et al. 2000). Thus, accumulation of TBT and TPT in these fish appears not to be due to biological parameters such as sex and total length.

Brackish water fish spend their life history almost entirely within estuarine regions. Thus, their BTs and PTs levels could be due to greater exposure to heavy boat traffic, shipyard and agriculture activities.

17.4 Marine Fish

17.4.1 Coastal Fish

Total BT concentrations (ΣBTs = TBT + DBT + MBT) in six fish species collected from coastal areas of Taiwan ranged between 259 and 1,364, 82–769, and 169–208 ng g^{-1} dry wt in pelagic, demersal, and cultured species, respectively (Lee et al. 2005) (Table 17.1). Although the life history patterns were different between pelagic and demersal species, they showed no significant differences in total BT concentrations (Lee et al. 2005). Similarly no significant variation in BT concentrations were found between other pelagic and demersal fish species (Kannan et al. 1995). The highest BT concentrations were found locally in fish species collected from fishing harbors (Lee et al. 2005).

Concentrations of total PT compounds (ΣPTs = TPT + DPT + MPT) in muscle tissue of pelagic, demersal and cultured fishes ranged between 534 and 1,120, 532–1,142 and 297–757 ng g^{-1} dry wt, respectively (Lee et al. 2005). Relatively lower PT concentrations were found in muscle tissues of fish from the northwestern Mediterranean (<1–3.5 ng g^{-1} wet wt, Borghi and Porte 2002), from Kaohsiung coastal area, Taiwan (<15.7 ng g^{-1} dry wt, Hung et al. 1998), from Otsuchi Bay, Japan (ND–20 ng g^{-1} wet wt, Harino et al. 1998; 12–152 ng g^{-1} wet wt, Harino et al. 2000), and from coastal locations in the Gulf and Gulf of Oman (<1.5–15.9 ng g^{-1} dry wt, De Mora et al. 2003). These results suggested high PT contamination had been widespread all over the Taiwan coastal areas.

Lee et al. (2005) found that the distribution of BT and PT compounds in coastal fishes was clearly different. It suggests that TPT might not originate from usage of marine antifouling paints like TBT, but from agricultural biocidal usage for protection of crops and vegetables. TPT in fish from the Mediterranean (Morcillo and Porte 2000) and Baltic (Albalat et al. 2002) Seas was also attributed to fungicide usage. However, TBT and TPT in mussels from Otsuchi Bay, Japan showed positive correlation suggesting both were derived from antifoulants (Harino et al. 1998). TPT in fish from Masnou, Spain (Morcillo et al. 1997) was also suggested to be derived from marine antifoulants.

A seasonal variation in the composition and concentration of BTs in coastal fish using benthic ponyfish *Leiogenathus splendens* and lizardfish *Trachinocephalus myops* inhabiting the west coast of Taiwan has been reported (Dong et al. 2004). In the whole body samples of the ponyfish, BT concentrations ranged from 236 to 2,501 ng g^{-1} wet wt, with those in winter considerably higher than in the other seasons (Dong et al. 2004) (Table 17.1). Similarly, BTs composition differed depending upon the season, with TBT (75% and 50%) dominant in winter and spring and DBT (37% and 57%) and MBT (42% and 24%) dominant in summer and autumn, respectively (Dong et al. 2004). In the lizardfish, the concentrations of BTs were one to two orders of magnitude higher in the liver than in the muscle, i.e. 3,058–11,473 vs 36–159 ng g^{-1} wet wt, respectively seasons (Dong et al. 2004). Concentrations of MBT, DBT and TBT in the muscle ranged, respectively, from 5 to 14, 8–35 and 23–110 ng g^{-1} wet wt, with the major compound being TBT (57–69%) in all seasons (Dong et al. 2004). However, in the liver, DBT concentrations, ranging from 992 to 7,797 ng g^{-1} wet wt, differed seasonally with a descending order of autumn > summer > spring seasons (Dong et al. 2004). Meanwhile, TBT (41%) was predominant in spring, whereas DBT (50% and 68%) was most heavily concentrated in summer and autumn seasons (Dong et al. 2004). Thus, seasonally mediated physiological changes, such as dilution due to growth and metabolic compensation, may play important roles in forming different BT accumulation patterns among seasons and organisms.

Furthermore, it is suggested that species-specific differences in TBT metabolism, rather than sediment characteristics or diet, are responsible for differences in proportions of BTs among organisms (Krone et al. 1996; Lee 1991; Takahashi et al. 1999). In exposure experiments with TBT, English sole (*Pleuronectes vetulus*) showed more rapid biotransformation of TBT to DBT than starry flounder (*Platichthys stellatus*) (Krone and Stein 1999). This supports the existence of species-specific differences in TBT metabolism.

These results suggest that seasonally mediated physiological changes, such as dilution due to growth metabolic compensation and species-specific metabolism strongly influence the accumulation pattern of BTs among fish species.

17.4.2 Open Ocean Fish

Pacific bluefin tuna (*Thunnus orientalis*), a highly migratory species, is mainly distributed in the temperate zone of the northern Pacific Ocean (Yamanaka 1982; Bayliff 1994) in contrast to *T. thynnus*, which inhabits the Atlantic Ocean (Collette 1999). Kannan et al. (1996) reported concentrations of BTs in *T. thynnus* collected from the Italian coast of the Mediterranean Sea ranging from 67 to 540 ng g^{-1} wet wt in liver and 16–230 ng g^{-1} wet wt in muscle (Table 17.1). The BTs concentrations in liver of tuna were about eight-fold lower than those of dolphins stranded along the Italian coast of the Adriatic Sea (Kannan et al. 1996). Regarding BT speciation in tuna, DBT was the highest in liver, while TBT and MBT were dominant in the muscle (Kannan et al. 1996).

Skipjack tuna *Katsuwonus pelamis* is a highly migratory pelagic species occurring in all tropical and subtropical waters of the world. The tuna spawns throughout the year and migrates globally after hatching, from the equatorial spawning areas to the northern and southern temperate regions (Collette and Nauen 1983). BTs and total tin (ΣSn), were comprehensively determined in the liver of *K. pelamis* collected from Asian offshore waters, i.e., off-Japan, the Japan Sea, off-Taiwan, the East China Sea, the South China Sea, off-Philippines, off-Indonesia, the Bay of Bengal, off-Seychelles, off-Brazil and open seas of the North Pacific (Ueno et al. 2004). BTs were detected in all the skipjack tuna collected, suggesting widespread contamination of BTs even in offshore waters and open seas on a global scale. Concentrations of ΣBTs (TBT + DBT + MBT) in the tissues of skipjack tuna ranged from 3.8 to 400 ng g^{-1} wet wt and concentrations of MBT, DBT and TBT ranged from 3.8 to 22, <2.4–700 and <1.8–38 ng g^{-1} wet wt, respectively (Ueno et al. 2004) (Table 17.1). Among BTs, TBT was detected at relatively higher concentrations in all locations, whereas the concentrations of DBT and MBT were lower (Ueno et al. 2004). This suggests fresh inputs of TBT into the aquatic environment and the presence of recent sources in the Asian offshore waters, off-Seychelles, off-Brazil and in open seas. The organ and tissue distribution of BTs was examined in *K. pelamis* (Ueno et al. 2004). ΣBTs residue levels were in the order: liver kidney = pyloric coecum > gonad = spleen = gill = muscle (red) > muscle (back) = muscle (stomach) (Ueno et al. 2004). No gender-related difference in concentrations was observed in the order of body tissue distribution. It was noticeable that the concentrations of BTs in the pyloric coecum were comparable to those in the liver. The pyloric coecum is an organ involved in digestion, absorption and storage, therefore, levels in pyloric coecum could reflect BT concentrations in stomach contents. Considering specific accumulation, sex-, body-length differences and migration of skipjack tuna did not seem to affect BT concentrations, indicating rapid reflection of the pollution levels in seawater where and when they were collected (Ueno et al. 2004). High concentrations of BTs were observed in skipjack tuna from offshore waters around Japan, a highly developed and industrialized region (up to 400 ng g^{-1} wet wt) (Ueno et al. 2004). The fish collected from offshore waters around Asian developing countries also revealed levels comparable to those in Japan (up to 270 ng g^{-1} wet wt). High percentages (almost 90%) of BTs in total tin (ΣSn: sum of inorganic tin + organic tin) were found in the liver of skipjack tuna from offshore waters around Asian developing countries (Ueno et al. 2004). The result suggests that the anthropogenic BTs represent the major source of Sn accumulation in skipjack tuna from these regions.

The blue shark, *Prionace glauca*, is found in the world's temperate and tropical oceans. Blue sharks are known to migrate long distances, from New England to South America for example. Kannan et al. (1996) examined BTs concentrations in sharks collected from Italian coast of the Mediterranean Sea. Concentrations of BTs ranged from 1 to 9 ng g^{-1} wet wt in the liver and 75–220 ng g^{-1} in the kidney (Kannan et al. 1996) (Table 17.1), suggesting higher accumulation in kidney than in liver. The higher concentrations found in kidney could be associated with the modified physiological function of this organ in sharks and it has been suggested that shark's kidney might contain elevated levels of glutathione, which plays a role in conjugating metals and/or metalloids (Kannan et al. 1996). The specific accumulation did not find significant correlations between BTs concentrations and body weight or total length (Kannan et al. 1996). In blue shark, TBT was the major BT compound, forming more than 60% of the total BTs concentrations (Kannan et al. 1996).

Although sharks are apex predators in marine ecosystems, the accumulation of BTs in this example seems to be comparable to those in prey organisms such as fish and squid. This indicates that BTs do not accumulate along food chain.

17.4.3 Deep Sea Fish

The family Myctophidae (including about 250 species) are distributed widely in the world oceans and show species-specific diel vertical migration patterns in the water column (Gjosaeter and Kawaguchi 1980). Takahashi et al. (2000) examined BTs concentrations of six species of mesopelagic myctophids collected from 50 to 100 m (night time) and 200–700 m depth (daytime) in the western North Pacific. The highest concentrations of total BTs (MBT + DBT + TBT) were up to 46 ng g^{-1} wet wt. This result suggests the expansion of BT contamination on a global scale. The concentrations of BTs in myctophids were comparable to those in squids from open waters of the western North Pacific (Yamada et al. 1997). However, these concentrations were lower than those in shallow-water fish from Japanese coastal waters (Guruge et al. 1996; Takahashi et al. 1997) and from various coastal regions of the United States (Krone et al. 1996; Kannan et al. 1997) and European countries (Stab et al. 1996; Morcillo et al. 1997). Takahashi et al. (1997) studied BTs concentrations in deep sea fish such as dory *Zenopsis nebulosa*, scorpionfish *Helicolenus hilgendorfi*, bonefish *Pterothrissus gissu*, grenadier *Coelorinchus* sp., *cusk eel Hoplobbrotula armata, greeneye Chlorophthalmus albatrossis and argentine Glossanodon semifasciatus* collected from Suruga Bay, Japan. ΣBTs levels were up to 980 ng g^{-1} wet wt. MBT, DBT and TBT concentrations ranged from <15 to 240, <4.0–690 and <2.0–680 ng g^{-1} wet wt, respectively (Takahashi et al. 1997) (Table 17.1). Borghi and Porte (2002) examined BTs concentrations in five species of deep-sea fish, i.e. common mora *Mora moro*, Mediterranean coding *Lepidion lepidion*, Gunther's grenadier *Coryphaenoides guentheri*, Risso's smooth head *Alepocephalus rostratus* and spiderfish *Bathypterois mediterraneus* from the Gulf of Lions in the northwestern Mediterranean at a depth between 1,000 and 1,800 m. ΣBTs levels ranged from 5 ng g^{-1} wet wt in the liver of *A. rostratus* to 175 ng g^{-1} wet wt in *M. moro* (Borghi and Porte 2002). MBT, DBT and TBT concentrations ranged from <1.0 to 54.5, <1.0–67.0 and <1.0–52.1 ng g^{-1} wet wt, respectively (Borghi and Porte 2002) (Table 17.1). The concentrations of BTs in those fishes collected from North Pacific, Suruga Bay and Gulf of Lions were variable among sites. It might be due to the fresh and continuous input of BTs into the coastal-surface regions through human activities and the lower mobility of BTs in each region.

BTs concentrations have been compared among the various migratory types of mesopelagic myctophid fish (Takahashi et al. 2000). The accumulation patterns of BTs showed a specific trend in accordance with the migration types (Takahashi et al. 2000). The higher concentrations of BTs were found in vertical migratory species that migrate to shallower waters, compared to non-migratory fishes. This suggests that continuing input of BTs to surface waters (e.g., via antifoulings) could be responsible for the higher levels seen in species that exhibit diel vertical migration.

Among BTs, TBT was more dominant than its metabolites (MBT and DBT) in deep sea fish species (Takahashi et al. 1997). This suggests a fresh input of TBT into the deep sea environment, possibly by vertical transport and/or a lower capacity to metabolise TBT. The environment of the deep sea and the lower metabolic capacity of such deep sea fish are like to cause elevated TBT ratios. In the deep water environment, lower temperature, and less sunlight penetration and lower dissolved oxygen could act as a suppressing factor for TBT degradation. These factors suggest that deep sea fish could have an enhanced risk to exposure of TBT than other BT derivatives.

17.5 Conclusions

BTs and PTs compounds occur widely in fish species from freshwater, shallow open sea, and even in deep sea environments. This suggests that BTs and PTs contamination is widespread in all fish species via both terrestrial and marine environmental ecosystems. In diadromous fish species, it is clear that life history is a more important factor for BT accumulation than biological characteristics such as age, growth and maturation, even within the same species. However, there are correlations between BTs and PTs accumulation and those various biological characteristics in other fish species. Further, despite the fish in higher trophic level and predatory behaviour, accumulation of lower levels in those chemicals may be due to differences of metabolic capacity among fish species. Thus, it is mandatory to consider such biological factors when we use fish as a biological indicator for those chemical pollutions.

References

Albalat A, Potrykus J, Pempkowiak J et al. (2002) Assessment of organotin pollution along the Polish coast (Baltic Sea) by using mussels and fish as sentinel organisms. Chemosphere 47: 165–171

Arai T, Goto A, Miyazaki N (2003a) Use of otolith microchemistry to estimate the migratory history of the threespine stickleback, *Gasterosteus aculeatus*. J Mar Biol Assoc UK 83: 223–230

Arai T, Goto A, Miyazaki N (2003b) Migratory history of the threespine stickleback *Gasterosteus aculeatus*. Ichthyol Res 50: 9–14

Bayliff WH (1994) A review of the biology and fisheries for northern bluefin tuna, *Thunnus thynnus*, in the Pacific Ocean. FAO Fish. Tech. Pap. 336/2: 44–295

Borghi V, Porte C (2002) Organotin pollution in deep-sea fish from the northwestern Mediterranean. Environ Sci Technol 36: 4224–4228

Collette BB (1999) Mackerels, molecules, and morphology. Soc Fr Ichthyol 25: 149–164

Collette BB, Nauen CE (1983) FAO Fisheries Synopsis No. 125. FAO species catalogue Vol. 2 scobrids of the world. An annotated and illustrated catalogue of tunas, mackerels, bonitos

and related species known to date. Food and Agriculture Organization of the United Nations, Rome
De Mora SJ, Fowler SW, Cassi R et al. (2003) Assessment of organotin contamination in marine sediments and biota from the Gulf and adjacent region. Mar Pollut Bull 46: 401–409
Dong CD, Chen CW, Liu LL (2004) Seasonal variation in the composition and concentration of butyltin compounds in marine fish of Taiwan. Environ Pollut 131: 509–514
Falandysz J, Brzostowski A, Szpunar J et al. (2002) Butyltins in sediments and three-spined stickleback (*Gasterosteus aculleatus*) from the marinas of the Gulf of Gdansk, Baltic Sea. J Environ Sci Health A 37: 353–363
Gjosaeter J, Kawaguchi K (1980) A review of the worlds resources of mesopelagic fish. FAO Fish Tech Pap 193: 1–151
Guruge KS, Tanabe S, Iwata H et al. (1996) Distribution, biomagnification, and elimination of butyltin compound residues in common cormorants (*Phalacrocorax carbo*) from Lake Biwa, Japan. Arch Environ Contam Toxicol 31: 210–217
Harino H, Fukushima M, Yamamoto Y et al. (1998) Organotin compounds in water, sediment, and biological samples from the Port of Osaka, Japan. Arch Environ Contam Toxicol 35: 558–564
Harino H, Fukusjima M, Kawai S (2000) Accumulation of butyltin and phenyltin compounds in various fish species. Arch Environ Contam Toxicol 39: 13–16
Harino H, O'hara SCM, Burt GR et al. (2002) Butyltin and phenyltin compounds in eels (*Anguilla anguilla*). J Mar Biol Assoc UK 82: 893–901
Hassani LH, Frenich AG, Benajiba MH et al. (2006) Assessment of butyltin and phenyltin pollution in the sea mullet, Mugil cephalus, along the Moroccan and Spanish coasts (Mediterranean Sea). Arch Environ Contam Toxicol 51: 608–614
Hung TC, Lee TY, Liao TF (1998) Determination of butyltins and phenyltins in oysters and fishes from Taiwan coastal waters. Environ Pollut 102: 197–203
Kannan K, Tanabe S, Iwata H et al. (1995) Butyltins in muscle and liver of fish collected from certain Asian and Oceanian countries. Environ Pollut 90: 279–290
Kannan K, Corsolini S, Focardi S et al. (1996) Accumulation pattern of butyltin compounds in dolphin, tuna, and shark collected from Italian coastal waters. Arch Environ Contam Toxicol 31: 19–23
Kannan K, Senthilkumar K, Loganathan BG et al. (1997) Elevated accumulation of tributyltin and its breakdown products in bottlenose dolphins (*Tursiops truncatus*) found stranded along the US Atlantic and Gulf coasts. Environ Sci Technol 31: 296–301
Krone CA, Stein JE (1999) Species dependent biotransformation and tissue distribution of tributyltin in two marine teleosts. Aquat Toxicol 45: 209–222
Krone CA, Stein JE, Varanasi U (1996) Butyltin contamination of sediments and benthic fish from the East, Gulf and Pacific coasts of the United States. Environ Monit Assess 40: 75–89
Lee CC, Wang T, Hsieh CY et al. (2005) Organotin contamination in fishes with different living patterns and its implications for human health risk in Taiwan. Environ Pollut 137: 198–208
Lee RF (1991) Metabolism of tributyltin by marine animals and possible linkages to effects Mar Environ Res 32: 29–35
Morcillo Y, Borghi V, Porte C (1997) Survey of organotin compounds in the western Mediterranean using molluscs and fish as sentinel organisms. Arch Environ Contam Toxicol 32: 198–203
Ohji M, Harino H, Arai T (2006) Differences in organotin accumulation among ecological migratory types of the Japanese eel *Anguilla japonica*. Estuar Coast Shelf Sci 69: 270–290
Ohji M, Arai T, Miyazaki N (2007) Comparison of organotin accumulation in the masu salmon *Oncorhynchus masou* accompanying migratory histories. Estuar Coast Shelf Sci 72: 721–731
Rudel H, Muller J, Steinhanses J et al. (2007) Retrospective monitoring of organotin compounds in freshwater fish from 1988 to 2003: results from the German environmental specimen bank. Chemosphere 66: 1884–1894
Stäb JA, Traas TP, Stroomberg G et al. (1996) Determination of organotin compounds in the foodweb of a shallow freshwater lake in the Netherlands. Arch Environ Contam Toxicol 31: 319–328

Steffen D, Wunsch H, Kämmereit M et al. (2003) Flächendeckendes Biomonitoring zur Triphenylzinnproblematik, Niedersächsisches Landesamt für Ökologie, Hildesheim, Germany

Takahashi S, Tanabe S, Kubodera T (1997) Butyltin residues in deep sea organisms collected from Surnga Bay, Japan. Environ Sci Technol 31: 3103–3109

Takahashi S, Tanabe S, Takeuchi I et al. (1999) Distribution and specific bioaccumulation of butyltin compounds in a marine ecosystem. Arch Environ Contam Toxicol 37: 50–61

Takahashi S, Tanabe S, Kawaguchi K (2000) Organochlorine and butyltin residues in mesopelagic myctophid fishes from the western North Pacific. Environ Sci Technol 34: 5129–5136

Tesch FW (1977) The eel. Biology and management of anguillid eels, Chapman & Hall, London

Tsukamoto K, Arai T (2001) Facultative catadromy of the eel, *Anguilla japonica*, between freshwater and seawater habitats. Mar Ecol Prog Ser 220: 365–376

Ueno D, Inoue S, Takahashi S et al. (2004) Global pollution monitoring of butyltin compounds using skipjack tuna as a bioindicator. Environ Pollut 127: 1–12

Wootton RJ (1984) A functional biology of sticklebacks, Croom Helm, London

Yamada H, Takayanagi K, Tateishi M et al. (1997) Organotin compounds and polychlorinated biphenyls of livers in squid collected from coastal waters and open oceans. Environ Pollut 96: 217–226

Yamanaka H (1982) Fishery biology of the bluefin tuna resource in the Pacific Ocean, 140 p. Japan Fisheries Resources Conservation Association, Tokyo, Japan (In Japanese)

Part III
Environmental Chemistry of Alternative Biocides

Section 6
Analytical Methods

Chapter 18
The Analysis of Antifouling Paint Biocides in Water, Sediment and Biota*

Kevin V. Thomas and Katherine H. Langford

18.1 Introduction

Alternative antifouling biocides to TBT were first detected in environmental surface waters in the early 1990s (Readman et al. 1993). Irgarol 1051 was first detected in the surface waters of marinas on the Côte d'Azur, France at concentrations of up to 1,700 ng l^{-1} (Readman et al. 1993) and in subsequent years the occurrence of Irgarol 1051 was reported in both fresh and marine waters (Scarlett et al. 1999; Thomas et al. 2000; Martinez et al. 2001; Lamoree et al. 2002) These reports established that the alternative antifouling biocides being used to replace the restricted TBT could also be accumulating in the environment and possibly posing a risk to aquatic habitats. Following Irgarol 1051, a number of other compounds were also used as biocidal additives to antifouling paints and methods have been developed to determine their occurrence in environmental waters (Thomas 1998; Piedra et al. 2000; Thomas et al. 2001). The early studies used GC-MS analysis of water extracts to analyse Irgarol 1051 alone; however, as the field developed, multi-residue LC-MS or LC-tandem MS techniques followed that allowed for the simultaneous analysis of the most commonly used biocides and their metabolites (Thomas 1998). However, for certain biocides (e.g. zinc pyrithione) specific methods are predominantly used due to the intrinsic physico-chemical properties that make it a difficult compound to quantitatively analyse (Thomas 1999).

K.V. Thomas and K.H. Langford
Norwegian Institute for Water Research (NIVA), Gaustadalléen 21, NO-0349 Oslo, Norway

* Environmental chemistry of alternative biocides
Part 1 Analytical method

T. Arai et al. (eds.), *Ecotoxicology of Antifouling Biocides*,
DOI: 10.1007/978-4-431-85709-9_18, © Springer 2009

18.2 Irgarol 1051

Irgarol 1051 is a triazine herbicide amenable to GC analysis (Readman et al. 1993). Early methods for the analysis of Irgarol 1051 in seawater used liquid-liquid extraction (LLE) using dichloromethane to extract Irgarol 1051 from water at neutral conditions (Gough et al. 1994). The extracts were then dried using anhydrous sodium sulphate and reduced in volume under nitrogen prior to analysis. DCM extraction typically yields recoveries of >90%. Solid phase extraction (SPE) has also been used to extract Irgarol 1051 from water (Ferrer et al. 1999; Biselli et al. 2000; Bowman et al. 2003; Gatidou et al. 2005). Typically, octadecylsilane (C18) columns have been commonly used, however, not exclusively. Polymeric SPE column materials such as ENV + (Isolute, UK), Oasis HLP (Waters, USA) and StrataX (Phenomenex, USA) have also been successfully used (Ferrer et al. 1999; Pocurull et al. 2000; Hamwijk et al. 2005; Cai et al. 2006) (Table 18.1).

Table 18.1 Solid phase extraction of water samples for biocide analysis

	Extraction phase	Recovery (%)	LOD (ng/l)	Reference
Irgarol	C18 Sep-Pak	81–107	11	Gatidou et al. (2005)
	C18	95		Ferrer et al. (1999)
	C18	93	4	Biselli et al. (2000)
	C18	80–100	1	Bowman et al. (2003)
	ENV+	93–96	11	Gatidou et al. (2005)
	Eni-Chrom P	71–85	11	Gatidou et al. (2005)
	Polymeric	95	2	Ferrer et al. (2001)
	Polymeric	84	10–20	Pocurull et al. (2000)
	Polymeric	106	5	Ferrer et al. (1999)
	Polymeric	89–93		Cai et al. (2006)
	LiChrolute EN	91–95	5	Gimeno et al. (2001)
	SPME–polydimethylsiloxane		50	Lam et al. (2005)
M1	C18 Sep-Pak	79–101	26	Gatidou et al. (2005)
	ENV+	77–110	26	Gatidou et al. (2005)
	Polymeric	87–90		Cai et al. (2006)
	Eni-Chrom P	71–110	26	Gatidou et al. (2005)
	SPME–polydimethylsiloxane		100	Lam et al. (2005)
Diuron	C18 Sep-Pak	98–111	7	Gatidou et al. (2005)
	C18	89		Ferrer et al. (1999)
	ENV+	89–100	7	Gatidou et al. (2005)
	Eni-Chrom P	88–107	7	Gatidou et al. (2005)
	Polymeric	99	10	Ferrer et al. (1999)
	LiChrolute EN	97–99	5	Gimeno et al. (2001)
Dichlofluanid	LiChrolute EN	87–89	5	Gimeno et al. (2001)
	Polymeric	67	10–20	Pocurull et al. (2000)
	Polymeric	<10		Ferrer et al. (1999)

(continued)

Table 18.1 (continued)

	Extraction phase	Recovery (%)	LOD (ng/l)	Reference
	Divinylbenzene disc	72–87	3	Hamwijk et al. (2005)
	C18	<10		Ferrer et al. (1999)
	SPME–PDMS	103–118	2	Lambropoulou et al. (2000)
	SPME–polyacrylate		200	Penalva et al. (1999)
Chlorothalonil	Polymeric	96	2	Ferrer et al. (1999)
	SPME–PDMS	103–124	5	Lambropoulou et al. (2000)
	C18	63		Ferrer et al. (1999)
TCMTB	Polymeric	111	8	Ferrer et al. (1999)
	C18	98		Ferrer et al. (1999)

SPE extraction typically yields recoveries of >90% depending on the choice of material (Table 18.1). A comparison of SPE and DCM liquid-liquid extraction by Liu et al. (1999), reports that both methods work equally well for Irgarol 1051. Other aqueous extraction techniques, such as solid-phase micro extraction (SPME) of Irgarol 1051 is also possible with good recoveries (ca. 90%) (Lam et al. 2005).

It is possible to analyse Irgarol 1051 by gas chromatography-nitrogen selective detection (GC-NPD), gas chromatography-mass spectrometry (GC-MS), or liquid chromatography-mass spectrometry (LC-MS) (Tolosa et al. 1996; Thomas 1998, 2001; Scarlett et al. 1999; Voulvoulis et al. 1999a, b; Biselli et al. 2000; Thomas et al. 2001; Konstantinou et al. 2002; Lamoree et al. 2002; Sakkas et al. 2002; Gatidou et al. 2005; Lambert et al. 2006). Readman et al. (1993) used GC-NPD and GC-MS to quantify Irgarol 1051 alongside other herbicides in C18 SPE extracts of seawater. GC-NPD and GC-MS offer good resolution and sensitive (pg on column) detection. GC-NPD analysis of Irgarol 1051 has a reported limit-of-detection (LOD) of 1 ng l^{-1}, whilst GC-MS also has a reported LOD of around 1 ng l^{-1}.

The typical electron impact (EI) full scan mass spectrum of Irgarol 1051 is shown in Fig. 18.1 and clearly shows the Irgarol 1051 base peak with *m/z* 182 (ion $[M\text{-}NC(CH_3)_3]^+$) and molecular ion at *m/z* 253 (M^+). Typically, Irgarol 1051 in full scan chromatograms is quantified by using *m/z* 253 with *m/z* 182 and *m/z* 238 as additional quantifier ions. GC-MS analysis by selective ion monitoring uses *m/z* 238 and 253 (Connelly et al. 2001). The use of tandem MS can increase the sensitivity of the analyses down to 0.1 ng l^{-1}, through the detection of the product ion corresponding to *m/z* 196 (Steen et al. 1997). For the quantitative analysis of Irgarol 1051 an internal standard is often used, for example atrazine-d_5 (Thomas et al. 2001). More recently, Irgarol 1051's primary metabolite, 2-methylthio-4-*tert*-butylamino-6-amino-s-triazine (M1) has been simultaneously analysed using *m/z* 198 and 157 (Gatidou et al. 2004a). A typical EI mass spectrum is shown in Fig. 18.2.

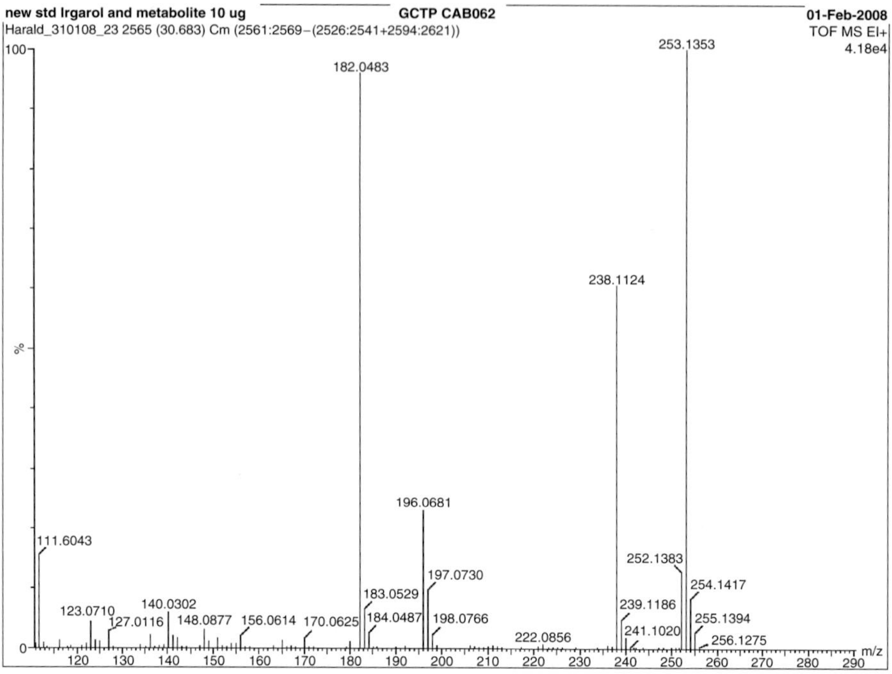

Fig. 18.1 Irgarol GC-TOF-MS spectrum (EI⁺)

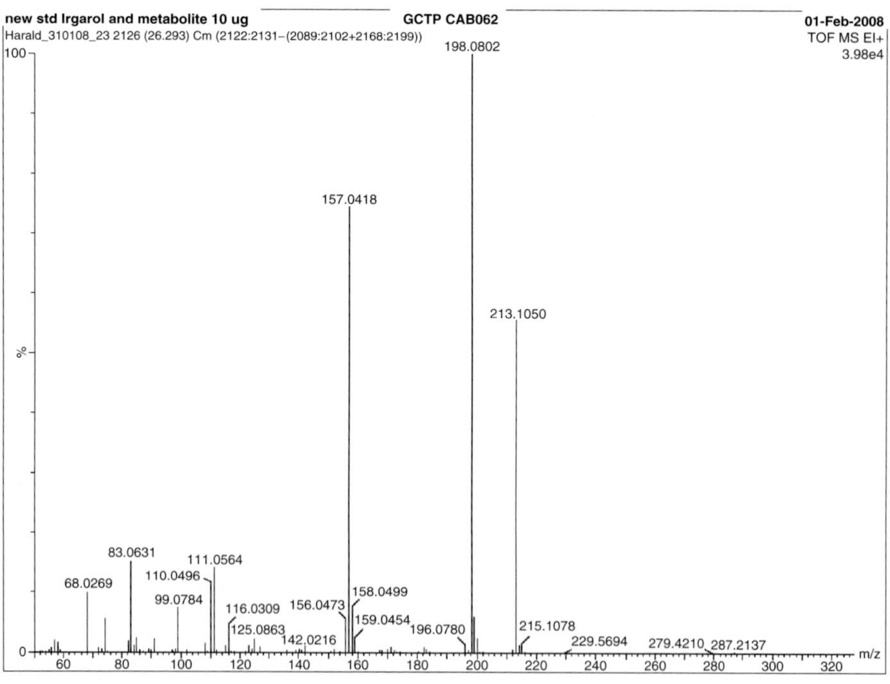

Fig. 18.2 Irgarol metabolite (M1) GC-TOF-MS spectrum (EI⁺)

A need for multi-residue analysis techniques to simultaneously analyse samples for many antifouling biocides led to the development of liquid chromatography-mass spectrometry methods (Thomas 1998; Ferrer et al. 1999). The analysis of Irgarol 1051 by LC-MS also uses SPE extracts which are typically analysed by reverse phase HPLC followed by MS detection (Thomas 1998; Ferrer et al. 1999). Early LC-MS analyses used quadrupole MS operated using SIM in the positive ion mode using the pseudo molecular ion $[M + 1]^+$ *m/z* 254. Similarly to GC-MS, M1 can also be analysed by LC-MS. Advancements in LC-MS technology mean that now it is likely that detection will use a tandem mass spectrometer providing greater specificity than SIM (Lamoree et al. 2002). Tandem MS in the positive mode, using multiple reaction monitoring (MRM), allows the detector to detect the product ions produced from the molecular ion; 254/198 for Irgarol 1051. Lamoree et al. (2002) also used electrospray ionization (ESI) in place of APCI showing that it is possible to use either.

An immunosensor has also been developed for Irgarol 1051 based upon a heterogenous competitive enzyme immunoassay (González-Martínez et al. 1998; Penalva et al. 1999). The method has an LOD of 10 ng l^{-1} and the results are reported to correlate well with those from instrumental analyses.

18.3 Diuron

Diuron has been a very popular antifouling biocide which also has other applications as a non-specific herbicide (Thomas 2001). Analysis of diuron is typically performed by liquid chromatography due to diuron being thermally labile and not directly amenable to GC (Thomas 1998). Derivatisation to compounds more thermally stable is possible, however, only HPLC analysis will be discussed here. Although diuron has been routinely monitored alongside other phenyl urea herbicides, the first antifouling biocide specific method used HPLC-APCIMS, as with Irgarol 1051 (Thomas 1998). Positive APCI (or ESI) of diuron produces an intense pseudomolecular ion ($[M + H]^+$) *m/z* 233 which can be detected by SIM or MRM depending on the type of MS detector being used (Thomas 1998; Lamoree et al. 2002; Lambert et al. 2006). In SIM, diuron is identified by its retention time and monitoring of *m/z* 233. In MS-MS systems two MRM transitions for diuron can be monitored; 233→72 and 233→46. Diuron is considered to be very persistent in seawater, although, aerobic degradation results in the formation of 1-(3,4-dichlorophenyl)-3-methylurea (DCMPU) and 1-(3,4-dichlorophenyl)urea (DCPU), whilst anaerobic degradation in sediments results in the formation of 1-(3-chlorophenyl)-3,1-dimethylurea (CPDU). Analysis of these metabolites in water samples is also possible by LC-MS (Thomas et al. 2002). All ionize to give pseudomolecular ions and characteristic fragment ions which can be used for both detection and quantification (Table 18.2). Recoveries by SPE are typically >80% (Table 18.1) depending on the type of column used.

Table 18.2 *M/z* ratios for biocide analysis

Compound	Instrument	Mode	Molecular ion	Fragment ions	Daughter ion
Irgarol 1052	GC/MS	EI +ve	253	238/182	196
	LC/MS	ESI/APCI +ve	254		198
M1	GC/MS	EI +ve	213	198/157	
Diuron	LC/MS	ESI/APCI +ve	233		72/46
DCOIT	GC/MS	EI−ve/NCI	246	182/169	
	GC/MS	EI−ve/NCI MS/MS	282		170
	LC/MS	APCI +ve	284/282		
Chlorothalonil	GC/MS	EI +ve/NCI		266/264/268/270	
	LC/MS	APCI −ve	264	245	
Dichlofluanid	GC/MS	EI +ve	224	167/123	
	LC/MS	APCI −ve	199	155	
DMSA	GC/MS	EI +ve	200	92/108	
TCMTB	GC/MS	EI +ve	238	180/136/108	
	GC/MS	CI −ve	166	58	
	GC/MS	CI +ve	222	182/210/136	
	LC/MS	APCI −ve	238	166	
TCMS-pyridine	LC/MS	APCI −ve		230/232/234/236	
ZPT	LC/MS	APCI	317	221/319	
TPBP	HPLC	UV	220 nm		

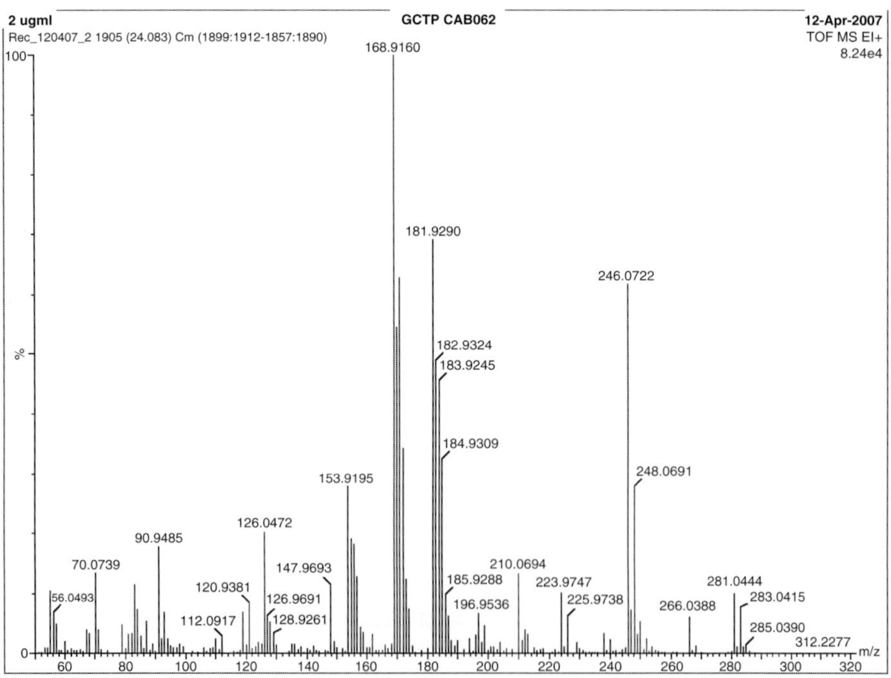

Fig. 18.3 DCOIT GC-TOF-MS spectrum (EI⁺)

18.4 DCOIT (SeaNine 211)

DCOIT, like Irgarol 1051, can be analysed by either GC-MS or LC-MS (Thomas 1998; Steen et al. 2004) following extraction from water. DCOIT is easily extracted from water by C18 or polymeric SPE with good recoveries and reproducibility (Table 18.1). The EI MS spectrum of DCOIT shows a base peak at *m/z* 169 corresponding to the $[C_8H_{17}]^-$ ion, with other ions at *m/z* 246 and 182 (Fig. 18.3). Typically, *m/z* 246 and 169 are used for detection and quantification, with *m/z* 169 being used as additional confirmation. It is possible to achieve LODs as low as 2 ng l^{-1}. Improved sensitivity can be obtained through the use of GC-ECD or – NCI and in particular NCI-MS/MS (Steen et al. 2004) resulting in an LOD of 0.05 ng l^{-1}.

For the determination of DCOIT by LC-MS, PI APCI is often chosen. The APCI or ESI MS of DCOIT shows an intense base peak corresponding to the protonated molecule ($[M + H]^+$; *m/z* 282/284; Table 18.2). In APCI very little fragmentation is observed and the analysis is dependent on the SIM of *m/z* 282 (+284) and retention time. The LC-MS analysis of DCOIT is sensitive with a reported LOD of 1 ng l^{-1}.

DCOIT rapidly degrades through cleavage of the isothiazolone ring and subsequent oxidation to produce N-octyl oxamic acid, N-octyl carbamic acid and 4,5 dichlorothiazole (Thomas 2001). As far as we are aware no published methods are available for the analysis of these compounds in water samples.

18.5 Chlorothalonil

GC is the preferred analysis technique for chlorothalonil. Few studies have employed LC-MS for its analysis since GC- techniques offer greater sensitivity. The few LC-MS studies that have been performed have shown that APCI is the interface of choice, with the NI mode offering better sensitivity than the PI mode. Negative ion APCI of chlorothalonil typically produces an intense ion at *m/z* 264 $[M + OH\text{-}HCl]^-$ which is used for SIM analysis. It has been proposed that LC-APCI MS analysis of SPE extracts can detect chlorothalonil at nanogram per liter concentrations. However, GC- coupled to either ECD or MS has been more frequently used for chlorothalonil analysis (Lambropoulou et al. 2000; Albanis et al. 2002; Carbery et al. 2006). The use of GC-MS employing EI has frequently been reported with *m/z* 266 $[M + 2]^+$, *m/z* 229 $[M\text{-}Cl]^+$ and *m/z* 264 $[M]^+$ being used in SIM. Reported LODs for GC-EIMS operated in SIM are in the nanogram per liter range. Improved sensitivity can be achieved through the use of NCI particularly if used in conjunction with an ion trap MS instrument in the MS-MS mode where LODs in the sub-low nanogram per liter have been obtained (Caux et al. 1996). When using NCI for chlorothalonil analysis, methane can be used as the reagent gas which gives a simple spectrum with a base peak at *m/z* 266 $[M + 2]^+$ surrounded by a cluster generated by chlorine isotopes. NCI used in combination with MS-MS results in a product ion being formed with *m/z* 229 which corresponds to $[M\text{-}Cl]^-$.

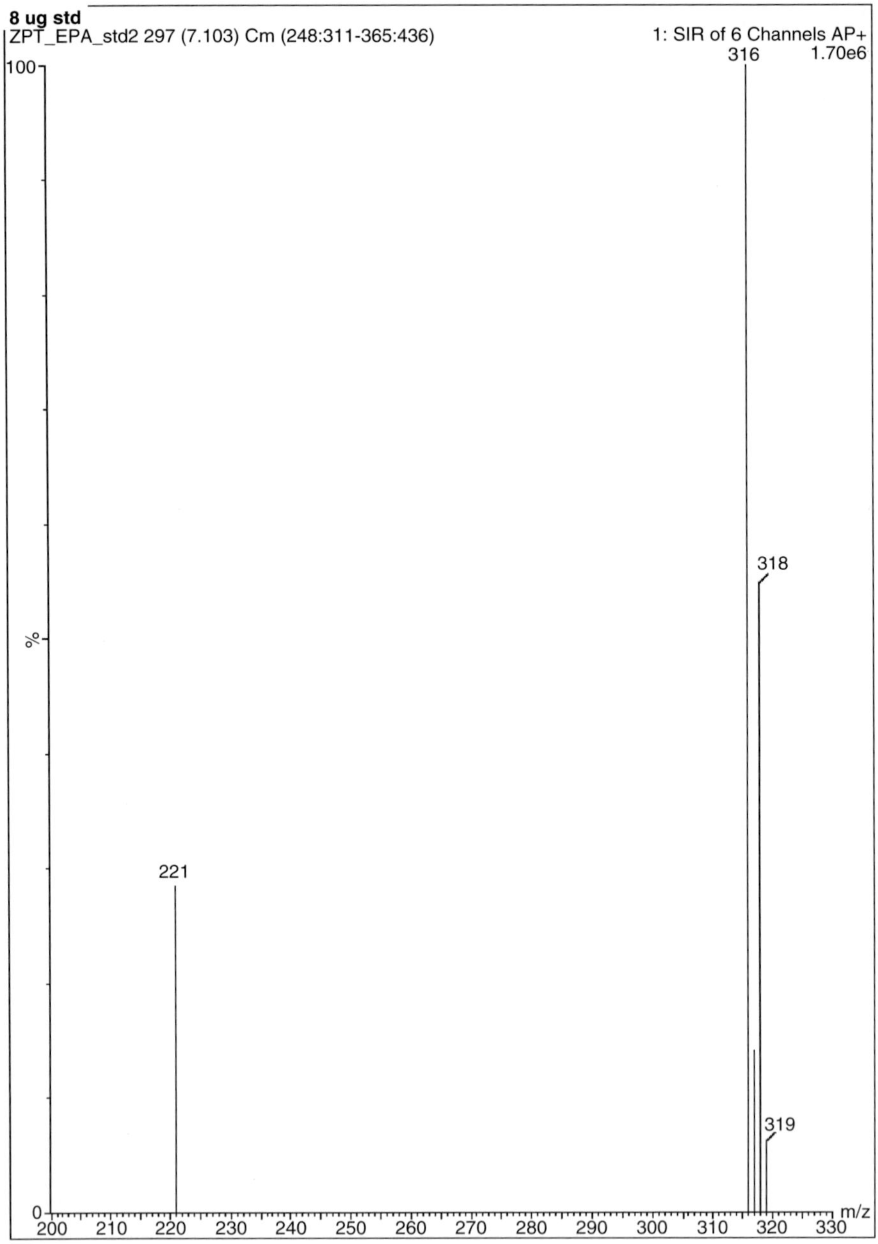

Fig. 18.4 Copper pyrithione LC/MS spectrum (APCI⁺)

18.6 Dichlofluanid

Dichlofluanid (N-dichlorofluoromethylthio-N',N'-dimethyl-N-phenylsulphamide) can be analysed by both GC- and LC-MS techniques, although GC- is favoured. GC- coupled to ECD or MS can be used to detect dichlofluanid to low nanogram per liter concentrations. The EI mass spectrum of dichlofluanid has a base peak of at *m/z* 123 $[PhNS]^+$ with fragment ions at *m/z* 224 and 167, with little signal from the molecular ion at *m/z* 232. DMSA (N'-dimethyl-N-phenyl-sulphamide), the main metabolite for dichlofluanid, has an EI mass spectrum with a base peak at 92 and two smaller peaks at 200 and 108 which can be used for quantification (Sakkas et al. 2001). It is recommended that the ions *m/z* 224 and *m/z* 200 are used for quantification of dichlofluanid and DMSA, respectively. For confirmation, the ions *m/z* 123 and 167 should be used for dichlofluanid and ions *m/z* 92 and 108 for DMSA. Care should be taken when analysing dichlofluanid by GC, since there have been conflicting reports on its occurrence based on the detection of DMSA instead of dichlofluanid itself. Dichlofluanid may also be analysed by LC-APCIMS operated in negative mode, with an LOD of around 5 ng l^{-1} (Ferrer et al. 1999; Piedra et al. 2000). Dichlofluanid fragments easily when analysed by APCI MS, with a main ion at *m/z* 199 $[M\text{-}SCCl2F]^-$.

18.7 TCMTB

The analysis of TCMTB (2-thiocyanomethylthiobenzothiazole) is possible by both GC-MS and LC-MS. TCMTB is efficiently extracted from water by SPE (Table 18.1) with recoveries of between 80% and 98% (Ferrer et al. 1999; Aguera et al. 2000; Piedra et al. 2000). GC-MS using either NCI or EI (SIM) show comparable sensitivity (1.5–3 ng l^{-1}). SIM in EI mode monitors the *m/z* 180 ion corresponding to $[M\text{-}SCN]^-$, whilst NCI-MS using methane as a reagent gas uses the *m/z* 166 and 58.

18.8 TCMS-Pyridine

TCMS-pyridine (2,3,5,6-tetrachloro-4-(methylsulphonyl)pyridine) is a little used antifouling paint biocide. The only published method for its analysis is by off-line SPE HPLC-APCI MS (Thomas 1998). TCMS-pyridine does not protonate well under PI leading to poor sensitivity, whilst it is difficult to form a de-protonated molecule in NI. What is formed in NI APCI is a characteristic cluster of ions at *m/z* 230, 232, 234, 236 corresponding to $[M\text{-}CH_3SO]^-$. TCMS-pyridine is efficiently extracted from seawater by C18 SPE (ca. 100%) with an LOD of 5 ng l^{-1}.

18.9 Zinc and Copper Pyrithione

One of the least studied antifouling paint biocides is zinc pyrithione (ZnPT). Its analysis is difficult, primarily due to its short photolytic half-life but also due to its ability to transchelate with other metals (Thomas 1999; Doose et al. 2004). The fist reported method for ZnPT in seawater used transchelation to the more stable copper pyrithione (CuPT) followed by high performance liquid chromatography-atmospheric pressure chemical ionisation mass spectrometry (HPLC-APCIMS) operated in SIM mode (*m/z* 221 and 316) (Thomas 1999). ZnPT was extracted as CuPT using DCM with adequate recovery (77% ± 17%, n = 6) from 21 water samples. The limit of detection using SIM was calculated to be 20 ng l^{-1}. The analysis of samples collected from various UK marinas showed no detectable concentrations to be present, whilst a laboratory based study confirmed that this is probably due to the rapid photodegradation of ZnPT in seawater. Further development of this method utilised an online system to minimise losses due to photolysis and sample workup (Bones et al. 2006). This method uses monolithic reversed-phase silica columns for rapid, on-line, large-volume solid phase extraction, in tandem with on-line matrix removal using sacrificial strong anion exchange (SAX) columns. This is then coupled with HPLC-APCIMS analysis, operated in full scan, using *m/z* 316 for the quantification of CuPT and *m/z* 317 for ZnPT. The LOD in spiked river water samples, using a 200 ml preconcentration volume, was determined as 18 ng l^{-1}, with a limit of quantitation of 62 ng l^{-1}. The percentage recovery from spiked river water was found to be 72 ± 9 ($n = 3$ extractions), whilst overall method precision, following ten repeat complete analyses was found to be 27% RSD at 1 µg l^{-1}. Linearity was determined over the concentration range of 0.25–10 µg l–1 and the calculated regression coefficient was $R^2 = 0.9802$. One of the advantages of this method is that it allows the simultaneous analysis of both ZnPT and CuPT.

Other similar methods have also been reported for the analysis of ZnPT or CuPT in seawater (Yamaguchi et al. 2006). Yamaguchi et al. (2006) propose a direct LC-MS method of analysis without transchelation, through the addition of ammonium acetate in mobile water phase which is reported to stabilise the ZnPT, although the method does not appear to be very sensitive (LOD = 1 mg l^{-1}). Grunnet et al. (2005) propose the use of SPE using the polymeric StrataX (Phenomenex, USA) to extract ZnPT prior to HPLC-DAD analysis. The reported LODs for this method are around 300 ng l^{-1}, with recoveries of 250 nM in seawater is 85% for ZnPT and 90% for CuPT. The method is useful for laboratory studies, however, the lack of specificity provided by the DAD suggests that MS detection should be favoured for environmental analysis. Interestingly, the chromatographic behaviour of pyrithiones has been well studied (Doose et al. 2004). The chromatographic behavior of 1-hydroxy-2-pyridinethione (pyrithione, PT), bis(2-pyridinyl)disulfide 1,1-dioxide (PT2), and the metal complexes zinc [Zn(PT)2], iron [Fe(PT)3] and copper [Cu(PT)2] pyrithione, were investigated by means of UV–vis spectroscopy, ESI-MSn, HPLC-DAD and HPLC-ESI-MSnD. This study confirmed that transformation of the analytes can occur depending on the type of stationary phase. This is particularly important when considering the direct analysis of ZnPT.

18.10 TPBP

TPBP is mainly used in Japan, with few published methods available for its analysis. No MS based methods have been reported; however, an off-line SPE HPLC-fluorescence method has been reported (Zhou et al. 2007). Extraction is performed with good recovery (96%), following the addition of pyridine, using C18 SPE and elution with acetonitrile/pyridine. Reverse phase HPLC is performed with UV detection at 220 nm to give an LOD of 500 ng l^{-1}. The method is suitable for leach-rate and laboratory measurements however a more sensitive MS-based method is required for environmental monitoring.

18.11 Multi-residue Methods

As briefly discussed above, antifouling paint biocide surveys in the environment often require the simultaneous measurement of many different biocides with different physico-chemical properties. The first multi-residue biocide method was suitable for the analysis of Irgarol 1051, diuron, DCOIT, TCMTB and TCMS-pyridine (Thomas 1998). The method was soon improved to include the metabolites of Irgarol 1051 and diuron (Martinez et al. 2001; Thomas et al. 2002). These multi-residue methods typically use off-line SPE extraction using C18, polymeric (e.g. ENV +, Isloute, UK) stationary phases or graphitized carbon black columns (ENVI-Carb). Reported recoveries for these biocides are shown in Table 18.1. Detection is carried out by reversed-phase HPLC-APCIMS operated in both negative and positive ion modes. Mass spectral data for the biocides are shown in Table 18.2 where pseudo-molecular and characteristic fragment ion(s) are presented. The limits of detection for the different biocides; vary however, LODs of between 1 and 5 ng l^{-1} can be expected.

18.12 Sediments and Biota

Most analytical methods have been developed for water samples, however, methods are available for the analysis of antifouling biocides in sediments and biota. The most commonly reported method for the analysis of biocides in sediments is solvent extraction with shaking or sonication (Table 18.3) (Voulvoulis et al. 1999; Biselli et al. 2000; Thomas et al. 2000; Albanis et al. 2002; Lambropoulou et al. 2003; Gatidou et al. 2004a,b; Hamwijk et al. 2005; Harino et al. 2005, 2006; Schouten et al. 2005). DCM, ethyl acetate or acetone, or a combination of these, is often used as the extraction solvent. The solvent extract can then be directly analysed or analysed following a clean-up step, typically using deactivated alumina or florisil. Recoveries using this procedure are generally excellent, with good limits

Table 18.3 Extraction methods for the determination of selected biocides in sediment

Compound	Extraction method	Cleanup method	Recovery (%)	LOD (ng/g)	Reference
Irgarol	Soxhlet with acetone	C18 SPE and alumina and silica GPC	61	0.05	Biselli et al. (2000)
	ASE DCM/acetone				Boxall et al. (2000)
	MAE with water	C18 SPE	94–114		Gatidou et al. (2004a)
	SFE CO2 modified with methanol	GPC XS3 biobeeds		2	Haglund et al. (2001)
	SFE CO2 modified with methanol and TFA		87	3	Carrasco et al. (2003)
	Shaker with methanol/ ethyl acetate		99	10	Thomas et al. (2000)
	Shaker with methanol/ ethyl acetate		92		Carrasco et al. (2003)
	Shaker with TFA in CO2/methanol		95		Carrasco et al. (2003)
	Shaker/sonication with methanol	C18 SPE	95	2.80	Gatidou et al. (2004a)
	Sonication with acetone	SPME extraction	90	0.50	Lambropoulou et al. (2003)
	Sonication with water	SPME extraction	67	8.00	Lambropoulou et al. (2003)
	Sonication with acetone	SPE cleanup	93		Lambropoulou et al. (2003)
M1	MAE with water	C18 SPE	85–103		Gatidou et al. (2004a)
	Shaker with methanol/ ethyl acetate		123	10	Thomas et al. (2000)
	Shaker/sonication with methanol	C18 SPE	102	3.4	Gatidou et al. (2004a)
Dichlofluanid	Mechanical shaker with acetone		101–109		Hamwijk et al. (2005)
	Sonication with acetone	SPME extraction	82	1.00	Lambropoulou et al. (2003)

(continued)

Table 18.3 (continued)

Compound	Extraction method	Cleanup method	Recovery (%)	LOD (ng/g)	Reference
	Sonication with water	SPME extraction	56	1,100	Lambropoulou et al. (2003)
	Sonication with acetone	SPE cleanup	80		Lambropoulou et al. (2003)
DMSA	Mechanical shaker with acetone		80–101	3	Hamwijk et al. (2005)
Diruon	Shaker with methanol/ ethyl acetate		102	100	Thomas et al. (2000)
	Shaker/sonication with methanol	C18 SPE	99	3.1	Gatidou et al. (2004b)
DCOIT	Sonication with acetone	SPME extraction	90	1.50	Lambropoulou et al. (2003)
	Sonication with water	SPME extraction	58	1,300	Lambropoulou et al. (2003)
	Sonication with acetone	SPE cleanup	85		Lambropoulou et al. (2003)
Chlorothalonil	Sonication with acetone	SPME extraction	75	6.00	Lambropoulou et al. (2003)
	Sonication with water	SPME extraction	36	2,500	Lambropoulou et al. (2003)
	Sonication with acetone	SPE cleanup	76		Lambropoulou et al. (2003)

Table 18.4 Methods for the extraction of Irgarol 1051 from biota samples

Material	Extraction method	Cleanup method	Recovery (%)	LOD (ng/g)	Reference
Zebra mussels	Triple shaker extraction with Hexane/ acetone	Florisil/sodium sulphate column cleanup	103	1.7 (dw)	Tóth et al. (1996)
Macrophytes	Triple shaker extraction with Hexane/ acetone	Florisil/sodium sulphate column cleanup	104	3.3 (dw)	
Algae	Triple shaker extraction with Hexane/ acetone	Florisil/sodium sulphate column cleanup	104	3.3 (dw)	
Macrophytes	Triple shaker extraction with Hexane/ acetone	Alumina/sodium sulphate column cleanup		10 (dw)	Nyström et al. (2002)
Sea grass	Soxhlet extraction with acetone/ DCM for 20h	Alumina/sodium sulphate column cleanup	97	0.3 (ww)	Scarlett et al. (1999)
Green algae	Soxhlet extraction with acetone/ DCM for 20h	Alumina/sodium sulphate column cleanup	97	0.3 (ww)	

of detection when used in conjunction with MS analysis (Table 18.3). Other extraction techniques that are suitable for the extraction of antifouling biocides from sediments include soxhlet solvent extraction, accelerated solvent extraction (ASE), microwave assisted extraction (MAE), supercritical fluid extraction (SFE) and semi-permeable micro extraction (SPME). These alternatives provide additional extraction options that provide good recovery of biocides from sediments and can be used for monitoring and fate studies.

Few methods have been reported for the analysis of antifouling biocides in biota samples. Consequently, there have been very few reports of studies relating to the biological uptake and bioaccumulation of booster biocides. Analytical methods have been reported for the analysis of Irgarol 1051 in algae, macrophytes and bivalves (Tóth et al. 1996; Scarlett et al. 1999; Nyström et al. 2002; Lambert et al. 2006) (Table 18.4). As with sediments, solvent extraction with sonication or shaking is a commonly applied technique, followed by a clean-up step. For example, Tóth et al. (1996) and later Nystrom et al. (2002), extracted Irgarol 1051 from freshwater phytoplankton and macrophytes using hexane:acetone, followed by clean-up on an alumina-Na2SO4 or florisil column. Scarlett et al. (1999) used soxhlet extraction using acetone-DCM to extract Irgarol 1051 from sea-grasses, followed by alumina clean-up, with a LOD of 0.3 ng l^{-1}. Both soxhlet and LLE appear to be suitable methods for the extraction of Irgarol 1051 from algae and macrophytes, with the methods being reproducible and sensitive when coupled to MS analysis. Methods are also available for the analysis of Irgarol 1051 in bivalves (Tóth et al. 1996).

References

Aguera A, Piedra L, Hernando MD et al. (2000) Multiresidue method for the analysis of five antifouling agents in marine and coastal waters by gas chromatography-mass spectrometry with large volume injection. J Chromatogr A 889:261–269.

Albanis TA, Lambropoulou DA, Sakkas VA et al. (2002) Antifouling paint booster biocide contamination in Greek marine sediments. Chemosphere 48:475.

Biselli S, Bester K, Huhnerfuss H et al. (2000) Concentrations of the antifouling compound Irgarol 1051 and of organotins in water and sediments of German North and Baltic Sea marinas. Mar Poll Bull 40:233.

Bones J, Thomas KV, Paull B (2006) Improved method for the determination of zinc pyrithione in environmental water samples incorporating on-line extraction and preconcentration coupled with liquid chromatography atmospheric pressure chemical ionisation mass spectrometry. J Chromatogr A 1132:157–164.

Bowman JC, Readman JW, Zhou JL (2003) Seasonal variability in the concentrations of Irgarol 1051 in Brighton Marina, UK; including the impact of dredging. Mar Poll Bull 46:444.

Boxall ABA, Comber SD, Conrad AU et al. (2000) Inputs, monitoring and fate modelling of antifouling biocides in UK estuaries. Mar Poll Bull 40:898.

Cai Z, Fun Y, Ma W-T et al. (2006) LC/MS analysis of antifouling agent Irgarol 1051 and its decyclopropylated degradation product in seawater from marinas in Hong Kong. Talanta 70:91–96.

Carbery K, Owen R, Frickers T et al. (2006) Contamination of Caribbean coastal waters by the antifouling herbicide Irgarol 1051. Mar Poll Bull 52:635–644.

Carrasco PB, Diez S, Jimenez J et al. (2003) Determination of Irgarol 1051 in Western Mediterranean sediments. Development and application of supercritical fluid extraction-immunoaffinity chromatography procedure. Water Res 37:3658.

Caux PY, Kent RA, Fan GT et al. (1996) Environmental fate and effects of chlorothalonil – a Canadian perspective. Crit Rev Environ Sci Technol 26:45–93.

Connelly DP, Readman JW, Knap AH et al. (2001) Contamination of the coastal waters of Bermuda by organotins and the triazine herbicide Irgarol 1051. Mar Poll Bull 42:409.

Doose CA, Szaleniec M, Behrend P et al. (2004) Chromatographic behavior of pyrithiones. J Chromatogr A 1052:103.

Ferrer I, Barcelo D (1999) Simultaneous determination of antifouling herbicides in marina water samples by on-line solid-phase extraction followed by liquid chromatography/mass spectrometry. J Chromatogr A 854:197–206.

Ferrer I, Barcelo D (2001) Identification of a new degradation product of the antifouling agent Irgarol 1051 in natural samples. J Chromatogr A 926:221–228.

Gatidou G, Kotrikla A, Thomaidis NS et al. (2004a) Determination of two antifouling booster biocides and their degradation products in marine sediments by high performance liquid chromatography-diode array detection. Anal Chim Acta 505:153.

Gatidou G, Zhou JL, Thomaidis NS (2004b) Microwave-assisted extraction of Irgarol 1051 and its main degradation product from marine sediments using water as the extractant followed by gas chromatography/mass spectrometry determination. J Chromatogr A 1046:41–48.

Gatidou G, Kotrikla A, Thomaidis NS et al. (2005) Determination of the antifouling booster biocides Irgarol 1051 and diuron and their metabolites in seawater by high performance liquid chromatography-diode array detector. Anal Chim Acta 528:89.

Gimeno RA, Aguilar C, Marcé RM et al. (2001) Monitoring of antifouling agents in water samples by on-line solid-phase extraction/liquid chromatography–atmospheric pressure chemical ionization mass spectrometry. J Chromatogr A 915:139–147.

González-Martínez MA, Penalva J, Puchades R et al. (1998) An immunosensor for the automatic determination of the antifouling agent Irgarol 1051 in natural waters. Environ Sci Technol 32:3442–3447.

Gough MA, Fothergill J, Hendrie JD (1994) A survey of Southern England coastal waters for the s-triazine antifouling compound Irgarol 1051. Mar Poll Bull 28:613–620.

Grunnet KS, Dahllof I (2005) Environmental fate of the antifouling compound zinc pyrithione in seawater. Environ Toxicol Chem 24:3001.

Haglund K, Pettersson A, Peterson M et al. (2001) Seasonal Distribution of the Antifouling Compound Irgarol® 1051 Outside a Marina in the Stockholm Archipelago. Bull Environ Contam Toxicol 66.

Hamwijk C, Schouten A, Foekema EM et al. (2005) Monitoring of the booster biocide dichlofluanid in water and marine sediment of Greek marinas. Chemosphere 60:1316.

Harino H, Mori Y, Yamaguchi Y et al. (2005) Monitoring of antifouling booster biocides in water and sediment from the port of Osaka, Japan. Arch Environ Contam Toxicol 48:303.

Harino H, Ohji M, Wattayakorn G et al. (2006) Occurrence of antifouling biocides in sediment and green mussels from Thailand. Arch Environ Contam Toxicol 51:400.

Konstantinou IK, Hela DG, Lambropoulou DA et al. (2002) Comparison of the performance of analytical methods based on solid-phase extraction and on solid-phase microextraction for the determination of antifouling booster biocides in natural waters. Chromatographia 56:745.

Lam KH, Cai Z, Wai HY et al. (2005) Identification of a new Irgarol-1051 related s-triazine species in coastal waters. Environ Poll 136:221.

Lambert SJ, Thomas KV, Davy AJ (2006) Assessment of the risk posed by the antifouling booster biocides Irgarol 1051 and diuron to freshwater macrophytes. Chemosphere 63:734.

Lambropoulou DA, Konstantinou IK, Albanis TA (2000) Determination of fungicides in natural waters using solid-phase microextraction and gas chromatography coupled with electron-capture and mass spectrometric detection. J Chromatogr A 893:143–156.

Lambropoulou DA, Sakkas VA, Albanis TA (2003) Determination of antifouling compounds in marine sediments by solid-phase microextraction coupled to gas chromatography-mass spectrometry. J Chromatogr A 1010:1.

Lamoree MH, Swart CP, Van Der Horst A et al. (2002) Determination of diuron and the antifouling paint biocide Irgarol 1051 in Dutch marinas and coastal waters. J Chromatogr A 970:183.

Liu DG, Pacepavicius J, Maguire RJ et al. (1999) Survey for the occurrence of the new antifouling compound Irgarol 1051 in the aquatic environment. Water Res 33:2833–2843.

Martinez K, Ferrer I, Hernando MD et al. (2001) Occurrence of antifouling biocides in the Spanish Mediterranean marine environment. Environ Technol 22:543.

Nyström B, Becker-Van Slooten K, Bérard A et al. (2002) Toxic effects of Irgarol 1051 on phytoplankton and macrophytes in Lake Geneva. Water Res 66:2020–2028.

Penalva J, González-Martínez MA, Puchades R et al. (1999) Immunosensor for trace determination of Irgarol 1051 in seawater using organic media. Anal Chim Acta 387:227–233.

Piedra L, Tejedor A, Hernando MD et al. (2000) Screening of antifouling pesticides in sea water samples at low ppt levels by GC-MS and LC-MS. Chromatographia 52:631.

Pocurull E, Brossa L, Borrull F et al. (2000) Trace determination of antifouling comopunds by on-line solid-phase extraction-gas chromatography-mass spectrometry. J Chromatogr A 885:361–368.

Readman JW, Liong Wee Kwong L, Grondlin D et al. (1993) Coastal water contamination from a triazine herbicide used in antifouling paints. Environ Sci Technol 27:1940–1942.

Sakkas VA, Konstantinou IK, Albanis TA (2001) Photodegradation study of the antifouling booster biocide dichlofluanid in aqueous media by gas chromatographic techniques. J Chromatogr A 930:135.

Sakkas VA, Konstantinou IK, Lambropoulou DA et al. (2002) Survey for the occurrence of antifouling paint booster biocides in the aquatic environment of Greece. Environ Sci Poll Res 9:327.

Scarlett A, Donkin P, Fileman TW et al. (1999) Occurrence of the antifouling herbicide, Irgarol 1051, within coastal-water seagrasses from Queensland, Australia. Mar Poll Bull 38:687.

Schouten A, Mol H, Hamwijk C et al. (2005) Critical aspects in the determination of the antifouling compound dichlofluanid and its metabolite DMSA (N,N-dimethyl-N?-phenylsulfamide) in seawater and marine sediments. Chromatographia 62:511.

Steen RJCA, Leonards PEG, Brinkman UAT et al. (1997) Ultra-trace-level determination of the antifouling agent Irgarol 1051 by gas chromatography with tandem mass spectrometric detection. J Chromatogr A 766:153–158.

Steen RJCA, Ariese F, Hattum BV et al. (2004) Monitoring and evaluation of the environmental dissipation of the marine antifoulant 4,5-dichloro-2-n-octyl-4-isothiazolin-3-one (DCOIT) in a Danish Harbor. Chemosphere 57:513.
Thomas KV (1998) Determination of selected antifouling booster biocides by high-performance liquid chromatography-atmospheric pressure chemical ionisation mass spectrometry. J Chromatogr A 825:29.
Thomas KV (1999) Determination of the antifouling agent zinc pyrithione in water samples by copper chelate formation and high-performance liquid chromatography-atmospheric pressure chemical ionisation mass spectrometry. J Chromatogr A 833:105.
Thomas KV (2001) The environmental fate and behaviour of antifouling paint booster biocides: a review. Biofouling 17:73.
Thomas KV, Blake SJ, Waldock MJ (2000) Antifouling paint booster biocide contamination in UK marine sediments. Mar Poll Bull 40:739.
Thomas KV, Fileman TW, Readman JW et al. (2001) Antifouling paint booster biocides in the UK coastal environment and potential risks of biological effects. Mar Poll Bull 42:677.
Thomas KV, McHugh M, Waldock M (2002) Antifouling paint booster biocides in UK coastal waters: Inputs, occurrence and environmental fate. Sci Total Environ 293:117.
Tolosa I, Readman JW, Blaevoet A et al. (1996) Contamination of Mediterranean (Cote d'Azur) coastal waters by organotins and Irgarol 1051 used in antifouling paints. Mar Poll Bull 32:335.
Tóth S, Becker-van Slooten K, Spack L et al. (1996) Irgarol 1051, an antifouling compound in freshwater, sediment, and biota of Lake Geneva. Bull Environ Contam Toxicol 57:426–433.
Voulvoulis N, Scrimshaw MD, Lester JN (1999a) Analytical method development for the determination of four biocides used in antifouling paints in estuarine waters and sediments by gas chromatography-mass spectrometry. Chromatographia 50:353.
Voulvoulis N, Scrimshaw MD, Lester JN (1999b) Analytical methods for the determination of 9 antifouling paint booster biocides in estuarine water samples. Chemosphere 38:3503.
Yamaguchi Y, Kumakura A, Sugasawa S et al. (2006) Direct analysis of zinc pyrithione using LC-MS. Int J Environ Anal Chem 86:83.
Zhou X, Okamura H, Nagata S (2007) Abiotic degradation of triphenylborane pyridine (TPBP) antifouling agent in water. Chemosphere 67:1904.

Section 7
Monitoring of Alternative Biocides

Chapter 19
Europe and USA

Kevin V. Thomas and Katherine H. Langford

19.1 Introduction

Antifouling paints containing active biocides are typically used on the hulls of ships and boats to prevent the growth of fouling organisms. Antifouling paint biocides are therefore released directly into surface waters following their release from painted surfaces or from the inappropriate disposal of paint related waste. The levels of biocides found in surface waters are therefore directly related to the amount released from such surfaces. Once in the water column, antifouling biocides, as with all other contaminants, are subjected to a number of environmental processes that control their environmental fate. Depending on their physico-chemical properties, biocides can partition onto sediments and accumulate in biological material. In order to measure the occurrence of antifouling biocides in water, sediments and biota analytical methods have been developed and applied. This chapter will review the data available on the occurrence of the biocides listed below in surface waters, sediments and biota for Europe and the Americas, including Canada, USA and the Caribbean (Fig. 19.1).

19.2 Irgarol 1051 and Its Metabolite M1

Since 1993, when Irgarol 1051 was reported as an aquatic contaminant (Readman et al. 1993) many studies have been performed to monitor the occurrence of Irgarol 1051 from antifouling inputs (Table 19.1). In Europe many of these studies have been performed in the UK (Boxall et al. 2000; Thomas et al. 2000, 2001, 2002;

K.V. Thomas and K.H. Langford
Norwegian Institute for Water Research (NIVA),
Gaustadalléen 21, NO-0349 Oslo, Norway

T. Arai et al. (eds.), *Ecotoxicology of Antifouling Biocides*,
DOI: 10.1007/978-4-431-85709-9_19, © Springer 2009

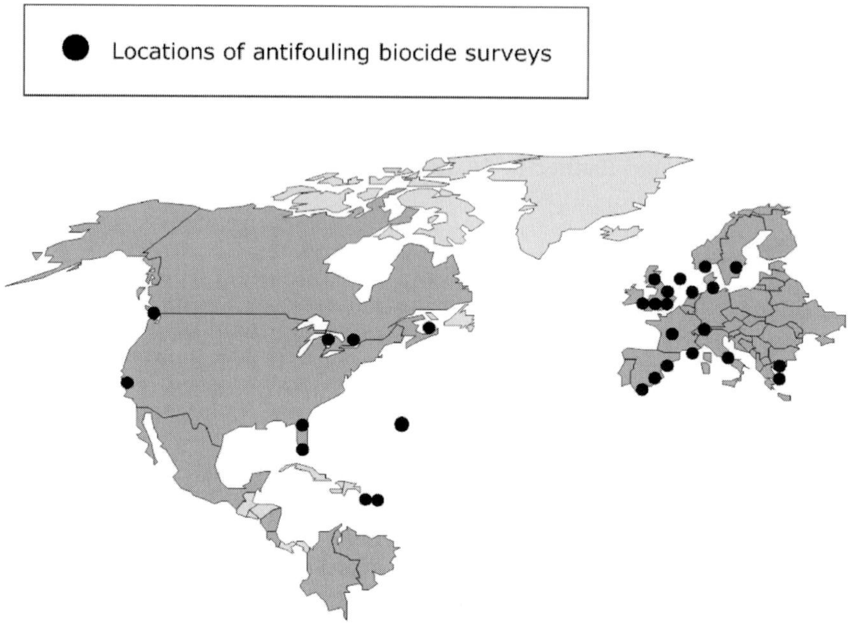

Fig. 19.1 Locations of antifouling paint biocide surveys in Europe and North America

Voulvoulis et al. 2000; Voulvoulis 2006). Early studies performed on the busy yachting areas of the Hamble estuary, UK showed Irgarol 1051 to occur in surface waters and sediments at concentrations of between 12 and 190 ng l^{-1} and 12 and 132 ng g^{-1}, respectively (Gough et al. 1994). A more widespread study in 1998, funded by the UK Department for the Environment, showed Irgarol 1051 to occur at concentrations of up to 1,421 ng l^{-1} with the higher concentrations being found early in the season in areas of high boating activity and low water exchange; such as those found in enclosed marinas (Thomas et al. 2001). Surface water concentrations in estuaries and harbours were much lower and these concentrations declined even further during the winter when most boats are removed from the water. These studies from the UK also reported that Irgarol 1051 was detected only in marina sediments where high water levels of Irgarol 1051 were measured (Thomas et al. 2000). It has been suggested that Irgarol 1051 has a low affinity for particulate matter (log K_{OC} 3.0; log K_{OW} 3.9) and in most marine waters is mainly associated with the dissolved phase (Tolosa et al. 1996; Ferrer et al. 1997). However, detectable concentrations are present in marina sediments where aqueous concentrations are at their highest and partitioning does occur (Thomas et al. 2000). Other studies on marine waters from the UK during the same period showed similar data, suggesting that Irgarol 1051 is persistent and can accumulate in areas where its release is intense. This was mirrored in data from around Europe where Irgarol 1051 painted boats were to be found (Ferrer et al. 1997; Scarlett et al. 1999; Biselli et al. 2000). Similar concentrations have been reported in other areas of Europe

Table 19.1 Irgarol concentrations in waters and sediments

Location	Description	Water concentration (ng l^{-1})	Sediment concentration (ng g^{-1})[a]	Reference
Puerto Rico, Caribbean	9 Marinas/harbours	<1–51		Carbery et al. (2006)
US Virgin Islands	5 Marinas/harbours	<1–1,300		Carbery et al. (2006)
San Diego area, California, USA	4 Marinas/harbours	1–304		Sapozhnikova et al. (2007)
Florida Keys, USA	4 Marinas	10.6–99.7		Owen et al. (2002)
Bermuda, USA	3 Harbours	37.4–294		Owen et al. (2002)
South Florida coastal waters, USA	75 Sites with varying levels of boating activities	<1–182		Gardinali et al. (2004)
Biscayne Bay, Florida, USA	2 Commercial ports	<1–2.24		Gardinali et al. (2002)
Biscayne Bay, Florida, USA	7 Sites on Miami River	<1–60.9		Gardinali et al. (2002
Biscayne Bay, Florida, USA	11 Marinas	<1–15.2		Gardinali et al. (2002)
Biscayne Bay, Florida, USA	Open bay area	<1		Gardinali et al. (2002)
Canada	Marinas/harbours	<LOD		Liu et al. (1999)
Gulf of Napoli, Italy	7 Marinas	3–22		Di Landa et al. (2006)
Gulf of Napoli, Italy	3 Harbours	3.5–8.2		Di Landa et al. (2006)
Barcelona, Spain	Marinas	<LOD–119	3–57	Ferrer and Barcelo (2001)
Almeria, Spain	Marinas	25–450		Aguera et al. (2000)
Greece	3 Ports	<LOD–14		Albanis et al. (2002)
Greece	8 Marinas	5–338		Albanis et al. (2002)
Piraeus-Elefsina, Greece	Marinas	<LOD–90	<LOD–690	Sakkas et al. (2002b)
Riviera, Monaco	Marinas/harbours	14–640		Tolosa et al. (1996)
Cote d'Azur, France	Marinas/harbours	110–1,700		Readman et al. (1993)
North Sea	6 Marinas/harbours – water samples	11–170	<LOD–14	Biselli et al. (2000)
Baltic Sea	7 Marinas/harbours – water samples	80–440	<4–220	Biselli et al. (2000)

(continued)

Table 19.1 (continued)

Location	Description	Water concentration (ng l^{-1})	Sediment concentration (ng g^{-1})[a]	Reference
The Netherlands	7 Marinas	8–90		Lamoree et al. (2002)
Western Scheldt, Netherlands	Estuary location	1.6–37		Steen et al. (1997)
Stockholm, Sweden	Marina	3–130		Haglund et al. (2001)
Stockholm, Sweden	Outside marina	<LOD–6		Haglund et al. (2001)
Fiskebackskil, Sweden	Marina	30–400		Dahl and Blanck (1996)
Southampton Water, UK	Water samples	<LOD–403	<LOD–110	Thomas et al. (2001)
Southampton Water, UK	Sediment samples		0.3–3.5	Thomas et al. (2002)
Hamble Estuary, UK	Estuary location	12–190	12–132	Gough et al. (1994)
North Sea	Sediment samples		<LOD	Thomas et al. (2000)
Conwy Marina, UK	Marina	7–543		Sargent et al. (2000)
UK	3 Marina locations	5.6–61.1		Boxall (2004)
Orwell Estuary, UK	3 Marina locations – summer water samples	24.3–201.4	<10–1,011	Boxall (2004)
Plymouth Sound, UK	Marina location	28–127		Scarlett et al. (1997)
Humber, UK	Marina location	169–682		Zhou et al. (1996)
Blackwater Estuary, UK	Estuary location	150–680	3.3–222	Voulvoulis et al. (2000)
East Anglia, UK	Rivers	<1–1,332		Lambert et al. (2006)
East Anglia, UK	Norfolk Broads	<1–1,220		Lambert et al. (2006)
UK	Coastal waters	<1–36		Cresswell et al. (2006)
UK	Coastal waters	<1–305		Thomas et al. (2002)

[a]Dry weight.

including the Mediterranean, West Coast of Sweden; Stockholm Archipelago, Sweden; Oslofjord, Norway; Western Scheldt, The Netherlands; Sas van Gent and Schaar van Ouden, The Netherlands; and the Baltic and North Sea marinas of Germany (Table 19.1).

The occurrence of Irgarol 1051 was not restricted to marine waters. Studies also showed that the use of antifouling paints formulated with Irgarol 1051 were common in freshwater lakes and waterways (Tóth et al. 1996; Lambert et al. 2006). Irgarol 1051 was detected in the waters of marinas around Lake Geneva, Switzerland and in the inland waterways of the Norfolk Broads UK. Even though the boating density in inland waterways is not as high as coastal areas concentrations as high as 1,200 ng l^{-1} were reported.

In the Americas, where Irgarol 1051 use is relatively recent, studies show that where Irgarol 1051 is in use, or where boats painted with Irgrol 1051 formulated antifouling paints are used, Irgarol 1051 occurs. An early study from Canada reported concentrations to be below the LOD when paints containing Irgarol 1051 were not in use (Liu and Pacepavicius 1999). Studies from the USA, including US territories in the Caribbean show that marina/harbour concentrations range from <1 to 1,300 ng l^{-1} (Gardinali et al. 2002, 2004; Owen et al. 2002; Sapozhnikova et al. 2007). As expected the highest concentrations are seen in the busiest boating areas such as San Diego, California (Sapozhnikova et al. 2007).

In 2001 the UK restricted the use of Irgarol 1051 in antifouling paints for small vessels (<25 m). The impact of this legislation was that these paints were removed from the marketplace for small boats. Post-restriction surveys performed in 2005 demonstrated a clear reduction in water concentrations of Irgarol 1051 (between 10% and 55% of that found during pre-restriction studies), indicating that legislation appears to have been effective (Cresswell et al. 2006).

19.3 GS26575/M1

The occurrence of the Irgarol 1051 metabolite, 2-methylthio-4-tert-butylamino-6-amino-s-triazine (GS26575 or M1) has also been reported in Irgarol 1051 contaminated surface waters and sediments (Thomas et al. 2000, 2002). Environmental degradation of Irgarol 1051 appears not to be the only source of GS26575 since it has also been reported to occur in Irgarol 1051 paint formulations (Thomas et al. 2003). The occurrence of GS26575 is therefore directly related to that of Irgarol 1051 either being released from painted surfaces with Irgarol 1051 or as a product of environmental degradation. GS26575 has been reported to occur in both surface waters and sediments (Table 19.2). Typically concentrations are lower than those detected for Irgarol 1051 in the same samples and have been reported at up to 4,000 ng l^{-1} (Ferrer et al. 1997). In the UK concentrations of between <1 and 140 ng l^{-1} have been reported for freshwater and <1 and 300 ng l^{-1} for marine samples (Thomas et al. 2002; Lambert et al. 2006). Similar levels have been recorded in both Florida and

Table 19.2 Concentrations of Irgarol 1051's main metabolite M1 in water and sediment

Location	Description	Water concentration (ng l^{-1})	Sediment concentration (ng g^{-1})	Reference
San Diego area, California, USA	4 Marinas/harbours	1–68		Sapozhnikova et al. (2007)
South Florida coastal waters, USA	75 Sites with varying levels of boating activities	<3–47		Gardinali et al. (2004)
Barcelona, Spain	Marinas	<LOD–23	0.2–3.3	Ferrer and Barcelo (2001)
Catalonia, Spain	4 Marinas	<LOD–4,000		Martínez et al. (2000)
The Solent, UK	Sediment samples		<LOD–5.7	Thomas et al. (2000)
East Anglia, UK	Rivers	<1–139		Lambert et al. (2006)
East Anglia, UK	Norfolk Broads	<1–231		Lambert et al. (2006)
Southampton Water, UK	Sediment samples		<LOD–0.3	Thomas et al. (2002)
Southampton Water, UK	Water samples	1–59		Thomas et al. (2001)
UK	Coastal waters	<1–59		Thomas et al. (2002)

California in the USA (Gardinali et al. 2002; Owen et al. 2002). GS26575 has also been found in sediments but at much lower concentrations (<1–6 ng g^{-1}) (Thomas et al. 2000)).

19.4 Diuron and Its Metabolites

Diuron, a phenylurea herbicide, has been in use since the 1950s. Predominantly this use has been associated with weed control in non-agricultural applications. Several studies have investigated the impact of non-antifouling use on diuron release into the aquatic environment (Albanis et al. 1995), however, the assessment of inputs from antifouling applications has been restricted to a few studies (Thomas et al. 2000, 2001, 2002; Lamoree et al. 2002; Gatidou et al. 2005, 2007; Lambert et al. 2006). For example Dahl and Blanck (1996) reported diuron concentrations in Swedish marinas of between 10 and 100 ng l^{-1}. A study of marinas on the Spanish Mediterranean coast reported similar concentrations (10–100 ng l^{-1}) whilst a study in the UK reported much higher concentrations in marinas (4–6,742 ng l^{-1}) probably reflecting the level of diuron use in the UK at the time. This study also reported diuron concentrations of between 5 and 226 ng l^{-1} in water samples collected from estuaries, of between 1 and 45 ng l^{-1} in coastal waters and <8 ng l^{-1} off-shore. In freshwater boating areas diuron has been shown to occur at concentrations less than 170 ng l^{-1} (Lambert et al. 2006) suggesting diuron containing paints are not so extensively used on inland waterways (Table 19.3). The principal metabolites of diuron, 1-(3-chlorophenyl)-3,1-dimethylurea (CPDU), 1-(3,4-dichlorophenyl)-3-methylurea (DCPMU) and 1-(3,4-dichlorophenyl)urea (DCPU) have also all been detected at measurable concentrations in surface water samples collected from the UK (Thomas et al. 2002), albeit at concentrations much lower than diuron itself.

19.5 DCOIT

DCOIT (4,5-dichloro-2-n-octyl-4-isothiazolin-3-one), the active ingredient in SeaNine 211, has been subject to a number of studies in Europe and the Americas, however its occurrence is rarely reported. A season long survey from the UK in the late 1990s reported concentrations < LOD in samples collected from some of the UK's busiest boating areas around Southampton Water (Thomas et al. 2001). Elsewhere in Europe, Steen et al. (2004) detected DCOIT in samples collected from a harbour which contained a ship painted with a SeaNine 211 antifouling paint. Samples collected on a transect away from the ship showed levels to be below 5 ng l^{-1} within 400 m of the ship. DCOIT has also been detected in water samples collected from marinas in Greece and Spain. DCOIT was detected at concentrations of up to 3,300 ng l^{-1} in water samples

Table 19.3 Concentrations of Diuron detected in water and sediment samples

Location	Description	Water concentration (ng l^{-1})	Sediment concentration (ng g^{-1})	Reference
San Diego area, California, USA	4 Marinas/ harbours	<2–12		Sapozhnikova et al. (2007)
Gulf of Napoli, Italy	7 Marinas	12–475		Di Landa et al. (2006)
Gulf of Napoli, Italy	3 Harbours	6.1–51		Di Landa et al. (2006)
Catalonia, Spain	4 Marinas	up to 2,000		Martínez et al. (2000)
Mediterranean Coast, Spain	Marinas/ harbours	5–200		Ferrer and Barcelo (1999)
Amvrakikos Gulf, Greece	Wetland areas	Nd–260		Albanis et al. (1995)
Lake Geneva, Switzerland		<0.3–142.5		Tóth et al. (1996)
The Netherlands	7 Marinas	90–1,130		Lamoree et al. (2002)
Sweden	Marinas	10–100		Dahl and Blanck (1996)
Southampton Water, UK	Water samples	<LOD–6,742	<LOD–1,420	Thomas et al. (2001)
Southampton Water, UK	Sediment samples		0.4–6.2	Thomas et al. (2002)
North Sea	Sediment samples		<LOD	Thomas et al. (2000)
Humber rivers, UK	11 River locations	40–8,700		House et al. (1997)
UK	3 Marina locations	10.5–786		Boxall et al. (2000)
Orwell Estuary, UK	3 Marina locations	117–768	<12–395	Boxall et al. (2000)
East Anglia, UK	Rivers	<1–169		Lambert et al. (2006)
East Anglia, UK	Norfolk Broads	<1–249		Lambert et al. (2006)
UK	Coastal waters	16–1,249		Thomas et al. (2002)

collected from marinas along Spain's Catalan coast. Lower concentrations of up to 49 ng l^{-1} were detected in Greek marinas (Sakkas et al. 2002b), whilst all other studies in Europe and the Americas have failed to detect DCOIT at concentrations > LODs. Overall, the reported occurrence of DCOIT is patchy with studies reporting reasonably high concentrations, whilst other studies report lower than LOD concentrations. DCOIT is predominantly used on large vessels (>25 m) so it is unlikely that it will occur in marinas exclusively used for pleasure craft

Table 19.4 DCOIT concentrations measured in water samples

Location	Description	Concentration range(ng l^{-1})	Reference
Puerto Rico, Caribbean	9 Marinas/harbours	<LOD	Carbery et al. (2006)
US Virgin Islands	5 Marinas/harbours	<LOD	Carbery et al. (2006)
Catalonia, Spain	4 Marinas	<LOD–3,300	Martínez et al. (2000)
Greece	Marinas/harbours	<LOD	Albanis et al. (2002)
Greece	Marinas	<LOD–49	Sakkas et al. (2002b)
UK		<LOD	Thomas (1998)
Korsør Harbour, Denmark	Harbour in immediate vicinity of ship	30–72	Steen et al. (2004)
	Harbour 2 m from ship	20–25	

(Table 19.4). In addition DCOIT is reported to rapidly degrade in water (Sakkas et al. 2002a).

19.6 Dichlofluanid and DMSA

Dichlofluanid has been reported to occur in marina surface waters and sediments from Greece and Spain, as well as UK sediments (Voulvoulis et al. 2000; Sakkas et al. 2002a,b; Readman 2006) (Table 19.5). Other extensive studies in areas of intensive boating activity in the UK have reported concentrations below detection limits for both surface waters and sediments. Interestingly the occurrence of dichlofluanid in Greek (and other) surface waters and sediments has been challenged by a study which showed that the dichlofluanid metabolite N-dimethyl-N-phenyl-sulphamide (DMSA) occurs in surface waters and sediments following rapid degradation and that previous reports of dichlofluanid may be the result of 'false positives' arising from the use of non-specific detectors or inappropriate confirmation ions when using GC-MS (Hamwijk et al. 2005; Schouten et al. 2005). The few studies performed in the Americas report dichlofluanid concentrations below LODs.

19.7 Chlorothalonil

Few studies have been performed on the occurrence of chlorothalonil in surface waters arising from antifouling inputs. Recent studies performed in the Caribbean failed to detect chorothalonil in marina or harbour waters (Table 19.6). These

Table 19.5 Concentrations of diclofluanid and DMSA detected in water and sediment samples

Location	Description	Water concentration (ng l^{-1})	Sediment concentration (ng g^{-1})	Reference
Dichlofluanid				
Puerto Rico, Caribbean	9 Marinas/harbours	<LOD		Carbery et al. (2006)
US Virgin Islands	5 Marinas/harbours	<LOD		Carbery et al. (2006)
Greece	3 Marinas	<LOD		Hamwijk et al. (2005)
Piraeus-Elefsina, Greece	Marinas	<LOD–214	12–65	Sakkas et al. (2002b)
Catalonia, Spain	Coastline	<4–600		Martínez et al. (2000)
Catalonia, Spain	Coastline		<1.6–11	Martinez et al. (2001)
Greece	3 Ports	Nd–49		Albanis et al. (2002)
Greece	8 Marinas	4–102		Albanis et al. (2002)
Greece	3 Marinas	20–205		Lambropoulou et al. (2000)
Blackwater Estuary, UK	Estuary location		7.2–688	Voulvoulis et al. (2000)
DMSA				
Greece	3 Marinas		<LOD–36	Hamwijk et al. (2005)

Table 19.6 Chlorothalonil concentrations measured in water and sediment samples

Location	Description	Water concentration (ng l^{-1})	Sediment concentration (ng g^{-1})	Reference
Puerto Rico, Caribbean	9 Marinas/harbours	<LOD		Carbery et al. (2006)
US Virgin Islands	5 Marinas/harbours	<LOD		Carbery et al. (2006)
Greece	3 Ports	Nd–49		Albanis et al. (2002)
Greece	8 Marinas	Nd–120		Albanis et al. (2002)
Greece	3 Marinas	Nd–37		Lambropoulou et al. (2000)
Greece	Marinas	<LOD–63	<LOD–126	Sakkas et al. (2002b)
Blackwater Estuary, UK	Estuary location	360–1,380	16–34.3	Voulvoulis et al. (2000)

data corroborate an extensive survey for chlorothalonil in UK marinas, estuaries and harbours where all of the samples collected contained < LOD concentrations. Chlorothalonil has been detected at one UK location in both surface waters and sediments at reasonably high concentrations (e.g. 360–1,380 ng l^{-1} in water) and at lower concentrations in Greek marinas (<LOD – 120 ng l^{-1}) (Voulvoulis et al. 2000; Sakkas et al. 2002).

19.8 Zinc Pyrithione

Zinc pyrithione (ZnPT) has been included in very few monitoring surveys in Europe and the Americas. To date we are aware of only one study whereby ZnPT concentrations were determined in surface waters collected from UK marinas (Thomas et al. 2001). None of the samples contained ZnPT at concentrations greater than the LOD of 20 ng l^{-1}. The samples were analysed by liquid chromatography coupled to APCI mass spectrometry following transchelation to copper pyrithione (CuPT) (Thomas 1999). The only reported occurrence of ZnPT in Europe is from a smaller scale study, also performed in the UK, which reported ZnPT at a concentration of 105 ± 5 nM using differential-pulse cathodic stripping voltametry (Mackie et al. 2004). These data are not surprising since ZnPT is known to rapidly photodegrade (half-life = 15 min) and also to rapidly transchelate with copper which is also commonly used in antifouling paints (Turley et al. 2000). Although ZnPT is used as an active ingredient in antifouling paints it is likely that it transchelates with copper also present in the paint to form CuPT which then acts as a biocide. As far as we are aware there are few studies on the occurrence of CuPT. Recent, unpublished work in our group has shown that ZnPT and CuPT photolysis is reduced with depth when compared to laboratory measurements. Therefore it may be that CuPT is what occurrence studies should be targeting, especially in sediments.

19.9 Summary

Extensive antifouling paint biocide occurrence data are available for Europe. Data are also available for areas of the Americas. Many of the studies performed have focused on Irgarol 1051 and reported its occurrence in marinas, harbours, estuaries and coastal areas. Diuron has also been shown to occur at elevated levels in areas where it is used as a biocide. The occurrence of elevated levels of both Irgarol 1051 and diuron is a direct effect of the large quantities of these substances released from paints and their persistence in the aquatic environment. Occurrence data are also available for other biocides such as DCOIT, dichloflaunid, chlorothalonil and ZnPT. Reported concentrations of these biocides tend to vary from study to study suggesting occurrence is localised and/or short lived. Although a number of these biocides are relatively labile (e.g. ZnPT, dichlofluanid and DCOIT) the quantities of these biocides used may be less (than Irgarol and diuron); significant changes in the volumes used should be a cause for further assessment.

References

Aguera A, Piedra L, Hernando MD et al. (2000) Multiresidue method for the analysis of five antifouling agents in marine and coastal waters by gas chromatography-mass spectrometry with large volume injection. J Chromatogr A 889:261–269.

Albanis TA, Danis TG, Hela DG (1995) Transportation of pesticides in estuaries of Louros and Arachthos rivers (Amvrakikos Gulf, N. W. Greece). Sci Total Environ 171:85–93.

Albanis TA, Lambropoulou DA, Sakkas VA et al. (2002) Antifouling paint booster biocide contamination in greek marine sediments. Chemosphere 48:475.

Biselli S, Bester K, Huhnerfuss H et al. (2000) Concentrations of the antifouling compound Irgarol 1051 and of organotins in water and sediments of German North and Baltic Sea marinas. Mar Pollut Bull 40:233.

Boxall ABA (2004) Environmental risk assessment of antifouling biocides. Chimica Oggi 22:46.

Boxall ABA, Comber SD, Conrad AU et al. (2000) Inputs, monitoring and fate modelling of antifouling biocides in UK estuaries. Mar Pollut Bull 40:898.

Carbery K, Owen R, Frickers T et al. (2006) Contamination of Caribbean coastal waters by the antifouling herbicide Irgarol 1051. Mar Pollut Bull 52:635–644.

Cresswell T, Richards JP, Glegg GA et al. (2006) The impact of legislation on the usage and environmental concentrations of Irgarol 1051 in UK coastal waters. Mar Pollut Bull 52:1169.

Dahl B, Blanck H (1996) Toxic effects of the antifouling agent Irgarol 1051 on periphyton communities in coastal water microcosms. Mar Pollut Bull 32:342–350.

Di Landa G, Ansanelli G, Ciccoli R et al. (2006) Occurrence of antifouling paint booster biocides in selected harbors and marinas inside the Gulf of Napoli: a preliminary survey. Mar Pollut Bull 52:1541.

Ferrer I, Barceló D (1999) Simultaneous determination of antifouling herbicides in marina water samples by on-line solid-phase extraction followed by liquid chromatography/mass spectrometry. J Chromatogr A 854:197–206.

Ferrer I, Barceló D (2001) Identification of a new degradation product of the antifouling agent Irgarol 1051 in natural samples. J Chromatogr A 926:221–228.

Ferrer I, Ballesteros B, Marco MP et al. (1997) Pilot survey for determination of the antifouling agent Irgarol 1051 in enclosed seawater samples by a direct enzyme-linked immunosorbent assay and solid-phase extraction followed by liquid chromatography-diode array detection. Environ Sci Technol 31:3530–3535.

Gardinali PR, Plasencia M, Mack S et al. (2002) Occurrence of Irgarol 1051 in coastal waters from Biscayne Bay, Florida, USA. Mar Pollut Bull 44:781–788.

Gardinali PR, Plasencia MD, Maxey C (2004) Occurrence and transport of Irgarol 1051 and its major metabolite in coastal waters from South Florida. Mar Pollut Bull 49:1072–1083.

Gatidou G, Kotrikla A, Thomaidis NS et al. (2005) Determination of the antifouling booster biocides irgarol 1051 and diuron and their metabolites in seawater by high performance liquid chromatography-diode array detector. Anal Chim Acta 528:89.

Gatidou G, Thomaidis NS, Zhou JL (2007) Fate of Irgarol 1051, diuron and their main metabolites in two UK marine systems after restrictions in antifouling paints. Environ Int 33:70.

Gough MA, Fothergill J, Hendrie JD (1994) A survey of Southern England coastal waters for the s-triazine antifouling compound Irgarol 1051. Mar Pollut Bull 28:613–620.

Haglund K, Pettersson A, Peterson M et al. (2001) Seasonal distribution of the antifouling compound Irgarol® 1051 outside a marina in the Stockholm Archipelago. Bull Environ Contam Toxicol 66.

Hamwijk C, Schouten A, Foekema EM et al. (2005) Monitoring of the booster biocide dichlofluanid in water and marine sediment of Greek marinas. Chemosphere 60:1316.

House WA, Leach D, Long JLA et al. (1997) Micro-organic compounds in the Humber rivers. Sci Total Environ 194–195:357–371.

Lambert SJ, Thomas KV, Davy AJ (2006) Assessment of the risk posed by the antifouling booster biocides Irgarol 1051 and diuron to freshwater macrophytes. Chemosphere 63:734.

Lambropoulou DA, Konstantinou IK, Albanis TA (2000) Determination of fungicides in natural waters using solid-phase microextraction and gas chromatography coupled with electron-capture and mass spectrometric detection. J Chromatogr A 893:143–156.

Lamoree MH, Swart CP, Van Der Horst A et al. (2002) Determination of diuron and the antifouling paint biocide Irgarol 1051 in Dutch marinas and coastal waters. J Chromatogr A 970:183.

Liu DG, Pacepavicius RJM, Lau Y L et al. (1999) Survey for the occurrence of the new antifouling compound Irgarol 1051 in the aquatic environment. Water Res 33:2833–2843.

Mackie DS, van den Berg CMG, Readman JW (2004) Determination of pyrithione in natural waters by cathodic stripping voltammetry. Anal Chim Acta 511:47–53.

Martínez K, Ferrer I, Barceló D (2000) Part-per-trillion level determination of antifouling pesticides and their byproducts in seawater samples by off-line solid-phase extraction followed by high-performance liquid chromatography–atmospheric pressure chemical ionization mass spectrometry. J Chromatogr A 879:27–37.

Martínez K, Ferrer I, Hernando MD et al. (2001) Occurrence of antifouling biocides in the Spanish Mediterranean marine environment. Environ Technol 22:543.

Owen R, Knap A, Toaspern M et al. (2002) Inhibition of coral photosynthesis by the antifouling herbicide Irgarol 1051. Mar Pollut Bull 44:623–632.

Readman JW (2006) Development, occurrence and regulation of antifouling paint biocides: Historical review and future trends. *Handbook of Environmental Chemistry, Volume 5: Water Pollution*. Springer, Berlin: 1.

Readman JW, Liong Wee Kwong L, Grondlin D et al. (1993) Coastal water contamination from a triazine herbicide used in antifouling paints. Environ Sci Technol 27:1940–1942.

Sakkas VA, Konstantinou IK, Albanis TA (2002a) Aquatic phototransformation study of the antifouling agent Sea-Nine 211: identification of byproducts and the reaction pathway by gas chromatography-mass spectroscopy. J Chromatogr A 959:215.

Sakkas VA, Konstantinou IK, Lambropoulou DA et al. (2002b) Survey for the occurrence of antifouling paint booster biocides in the aquatic environment of Greece. Environ Sci Pollut Res 9:327.

Sapozhnikova Y, Wirth E, Schiff K et al. (2007) Antifouling pesticides in the coastal waters of Southern California. Mar Pollut Bull 54:1972–1978.

Sargent CJ, Bowman JCZhou JL (2000) Levels of antifoulant Irgarol 1051 in the Conwy Marina, North Wales. Chemosphere 41:1755–1760.

Scarlett A, Donkin ME, Fileman TW et al. (1997) Occurrence of the marine antifouling agent Irgarol 1051 within the Plymouth Sound locality: implications for the Green Macroalga *Enteromorpha intestinalis*. Mar Pollut Bull 34:645–651.

Scarlett A, Donkin P, Fileman TW et al. (1999) Occurrence of the antifouling herbicide, Irgarol 1051, within coastal-water seagrasses from Queensland, Australia. Mar Pollut Bull 38:687.

Schouten A, Mol H, Hamwijk C et al. (2005) Critical aspects in the determination of the antifouling compound dichlofluanid and its metabolite DMSA (N,N-dimethyl-N?-phenylsulfamide) in seawater and marine sediments. Chromatographia 62:511.

Steen RJCA, Leonards PEG, Brinkman UAT et al. (1997) Ultra-trace-level determination of the antifouling agent Irgarol 1051 by gas chromatography with tandem mass spectrometric detection. J Chromatogr A 766:153–158.

Steen RJCA, Ariese F, Hattum BV et al. (2004) Monitoring and evaluation of the environmental dissipation of the marine antifoulant 4,5-dichloro-2-n-octyl-4-isothiazolin-3-one (DCOIT) in a Danish Harbor. Chemosphere 57:513.

Thomas KV (1998) Determination of selected antifouling booster biocides by high-performance liquid chromatography-atmospheric pressure chemical ionisation mass spectrometry. J Chromatogr A 825:29.

Thomas KV (1999) Determination of the antifouling agent zinc pyrithione in water samples by copper chelate formation and high-performance liquid chromatography-atmospheric pressure chemical ionisation mass spectrometry. J Chromatogr A 833:105.
Thomas KV, Blake SJ, Waldock MJ (2000) Antifouling paint booster biocide contamination in UK marine sediments. Mar Pollut Bull 40:739.
Thomas KV, Fileman TW, Readman JW et al. (2001) Antifouling paint booster biocides in the UK coastal environment and potential risks of biological effects. Mar Pollut Bull 42:677.
Thomas KV, McHugh M, Waldock M (2002) Antifouling paint booster biocides in UK coastal waters: inputs, occurrence and environmental fate. Sci Total Environ 293:117.
Thomas KV, McHugh M, Hilton M et al. (2003) Increased persistence of antifouling paint biocides when associated with paint particles. Environ Pollut 123:153.
Tolosa I, Readman JW, Blaevoet A et al. (1996) Contamination of Mediterranean (Cote d'Azur) coastal waters by organotins and Irgarol 1051 used in antifouling paints. Mar Pollut Bull 32:335.
Tóth S, Becker-van Slooten K, Spack L et al. (1996) Irgarol 1051, an antifouling compound in freshwater, sediment, and biota of Lake Geneva. Bull Environ Contam Toxicol 57:426–433.
Turley PA, Fenn RJ, Ritter JC (2000) Pyrithiones as antifoulants: Environmental chemistry and preliminary risk assessment. Biofouling 15:175.
Voulvoulis N (2006) Antifouling paint booster biocides: Occurrence and partitioning in water and sediments. *Handbook of Environmental Chemistry, Volume 5: Water Pollution*. Springer, Berlin: 155.
Voulvoulis N, Scrimshaw MDLester JN (2000) Occurrence of four biocides utilized in antifouling paints, as alternatives to organotin compounds, in waters and sediments of a commercial estuary in the UK. Mar Pollut Bull 40:938.
Zhou L, Fileman TW, Evans S et al. (1996) Seasonal distribution of dissolved pesticides and polynuclear aromatic hydrocarbons in the Humber Estuary and Humber coastal zone. Mar Pollut Bull 32:599–608.

Chapter 20
Asia

Hiroya Harino, Takaomi Arai, Madoka Ohji, and Nobuyuki Miyazaki

Abbreviations CuPT: Copper pyrithione (Copper, bis(1,hydroxyl-2(1H)-pyridinethionato O,S) Cas no. 14915-37-8; Dichlofluanid: N,N-Dimethyl-N'-phenyl-N'-(dichlorofluoromethylthio)sulfamide Cas no. 1085-98-9; Diuron: 3-(3,4-Dichlorophenyl)-1,1-dimetyl urea Cas no. 330-54-1; Irgarol 1051: 2-Me thylthio-4-tert-butylamino-6-cyclopropylamino-s-triazine Cas no. 28159-98-0; LOD: Limit of determination; M1: 2-Methylthio-4-tert-butylamino-6-amino-s-triazine; NOAEL: No observed adverse effect level; NOEC: No observed effect concentration; OT: Organotin; PTPB: PK (Triphenylboron-pyridine) Cas no. 971-66-4; Sea Nine 211: 4,5-Dichloro-2-n-octyl-4-isothiazolin-3-one Cas no. 64359-81-5; TBT: Tributyltin; TPT: Triphenyltin; ZnPT: Zinc pyrithione (Zinc-2-pyridinethiol-1-oxide) Cas no. 13463-41-7

20.1 Introduction

The International Maritime Organization (IMO) has adopted the International Convention on the Control of Harmful Antifouling Systems (AFS Convention), which prohibits the use of organotins (OTs) as active ingredients in antifouling systems for ships. Following the international restrictions on the use of OT-based antifoulants, paint manufacturers have developed many products as alternatives to

H. Harino
Osaka City Institute of Public Health and Environmental Sciences,
8-34 Tojo-cho, Tennoji, Osaka 543-0026, Japan

T. Arai and M. Ohji
International Coastal Research Center, Ocean Research Institute, The University of Tokyo,
2-106-1 Akahama, Otsuchi, Iwate 028-1102, Japan

N. Miyazaki
Center for International Cooperation, Ocean Research Institute, The University of Tokyo,
1-15-1 Minamidai, Nakano, Tokyo 164-8639, Japan

T. Arai et al. (eds.), *Ecotoxicology of Antifouling Biocides*,
DOI: 10.1007/978-4-431-85709-9_20, © Springer 2009

the use of OTs. More than 20 chemical substances have been used or proposed as alternative compounds. After release of these antifouling biocides from the hulls of ships, fishing nets etc. into the aquatic environment, these chemicals are distributed in water, sediment, and aquatic organisms. The coastal waters of European countries and the USA are already polluted by alternative biocide (Gough et al. 1994; Toth et al. 1996; Gardinali et al. 2002; Bowman et al. 2003).

Asia is geographically located in the eastern and northern hemispheres and is the world's largest and most populous continent. It covers 8.6% of the Earth's total surface area (or 29.4% of its land area) and, with almost four billion people, it contains more than 60% of the world's current human population. Economic growth in Asia since World War II to the present time had been concentrated in countries of the Pacific Rim such as Thailand, Malaysia, Singapore, Hong Kong, Taiwan, Japan, and Korea. Development of economies may result in marine pollution by antifouling biocides, as trade increases, as illustrated by the presence of OT pollution in Japan, Korea, Southeast Asia and India (Harino et al. 1998a,b, 1999, 2003; Kan-atireklap et al. 1997a, b, 1998; Midorikawa et al. 2004a, b; Prudente et al. 1999; Shim et al. 2002, 2005; Sudaryanto et al. 2002).

There are a few papers describing the status of contamination by alternative biocides in Japan, Malaysia, Thailand and Vietnam (Harino et al. 2005, 2006a, b, 2007). Japan is developed country and therefore has many international ports. In addition, fisheries are extensive and, consequently fishing boats are moored widely. In Japan, the sale and use of OTs has been regulated since around 1990. Malaysia is a country in which the economic condition has recovered since 1999 and because Malaysia has an abundance of mineral resource and is an area richly endowed with nature, the tourist and trade business are active. Among Southeast Asian countries, Thailand and Vietnam are the countries in which economic growth increased remarkably. Human and industrial activities in Thailand have recently increased remarkably, and trading is flourishing because of the increase of economic activity. The sailing of foreign flag vessels is therefore increasing in the Gulf of Thailand. Vietnam's ongoing economic liberalization has led to the rapid development of industries and an expansion of international trade. Thus, economic growth in these countries has increased the potential risk for antifouling biocide contaminations to occur.

In this chapter, the concentrations of alternative biocides in water, sediment and biological samples from the coastal waters of Japan, Malaysia, Thailand and Vietnam are reported and the distribution of antifouling biocides in these countries is discussed.

20.2 Concentrations in Water

Among Asian countries, contamination by alternative biocides was first described in various coastal areas around Japan (Fig. 20.1). The concentrations of alternative biocides are shown in Table 20.1.

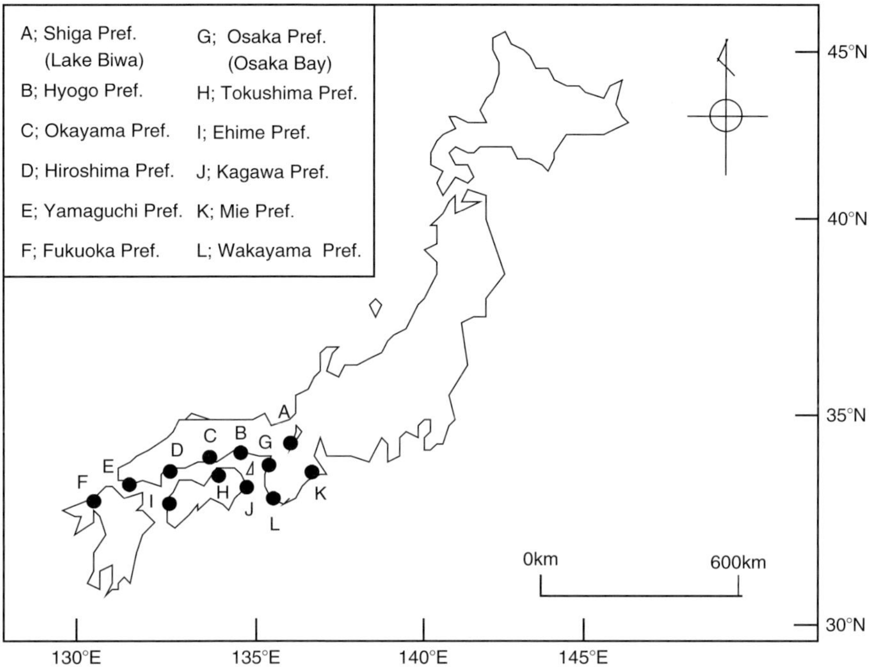

Fig. 20.1 Map of the study area in Japan. Circles indicate sampling sites

Detection frequency of Sea Nine 211 was low around coastal areas in Japan. Sea Nine 211 was detected in the Port of Osaka, Maizuru Bay and Hiroshima Bay with mean concentrations in the range of 2–28 ng l^{-1}. Thomas et al. (2002) report that Sea Nine 211 was not detected in water from Southampton, Britain; it was however detected in the Mediterranean Sea off Spain in the range of 2.6–3.7 g l^{-1} (Martinez et al. 2001). Sakkas et al. (2002) also report that concentrations of Sea Nine 211 were in the range <6.3–49 ng l^{-1} in coastal areas of Greece. The concentrations of Sea Nine 211 in water samples from Japan were similar to those in the other areas. Jacobsen et al. (1993) report degradation of Sea Nine 211 in aerobic aquatic mesocosm with a half-life in seawater of less than 1 day. Thus, despite its release from ship hulls, Sea Nine 211 may be degraded in seawater. Shade et al. (1993) reported acute toxicity to aquatic organisms. Sea Nine 211 displays relatively high toxicity to fish, with 96 h LC50 values in the range of 2.7–20.5 g l^{-1}. The concentration of Sea Nine 211 in Japan was 10^{-3} times lower than the LC50 value in fish.

Diuron was detected in various coastal waters from Japan. Mean concentrations of Diuron ranged from 26 to 372 ng l^{-1}. In Singapore, the concentration of Irgarol 1051 was in the range of 3,020–4,200 ng l^{-1}, which is higher than those in Japan. Elsewhere in the world, Diuron has been detected, for example, in the range of 0.6–305 (mean 36 ng l^{-1}) <1–334 (mean 65 ng l^{-1}), and <1–6,742 ng l^{-1} (mean 85 ng l^{-1}),

Table 20.1 The concentrations of alternative biocides in water samples from Japan and Singapore (ng l^{-1})

	Location	Sampling date	Sea Nine 211	Diuron	Irgarol 1051	M1	PTPB	Reference
Japan	Okayama Pref.	Aug 1999		<30–669 (201)	<5–143 (23)	<5–33 (6)		Okamura et al. (2003)
	Hiroshima Pref.	Aug 1999		<30–2,030 (331)	<5–148 (29)	<5–80 (14)		Okamura et al. (2003)
	Yamaguchi Pref.	Aug 1999		116–1,103 (372)	14–157 (64)	<5–49 (27)		Okamura et al. (2003)
	Fukuoka Pref.	Aug 1999		<30–471 (206)	<5–111 (33)	<5–38 (8)		Okamura et al. (2003)
	Ehime Pref.	Aug 1999		<30–373 (94)	<5–75 (12)	<5–10 (3)		Okamura et al. (2003)
	Kagawa Pref.	Aug 1999		52–424 (270)	<5–48 (18)	<5		Okamura et al. (2003)
	Tokuhshima Pref.	Aug 1999		<30–305 (191)	<5–21 (8)	<5		Okamura et al. (2003)
	Hyogo Pref.	Aug 1999		42–1,160 (547)	<5–262 (63)	<5–80 (15)		Okamura et al. (2003)
	Osaka Pref.	Aug 1999		<30–351 (158)	<5–136 (4)	<5		Okamura et al. (2003)
	Wakayama Pref.	Aug 1999		<3–2,050 (297)	<5–136 (16)	<5		Okamura et al. (2003)
	Mie Pref.	Aug 1999		<30–3,050 (56)	<5–78 (16)	<5–11 (3)		Okamura et al. (2003)
	Shiga Pref.	Aug 1999		<30–199 (39)	<5	<5		Okamura et al. (2003)
	The Port of Osaka	2002–2003	<0.3–4 (2)	<0.7–1,540 (126)	<0.8–267 (14)	<0.19–167 (17)		Harino et al. (2005)
	Maizuru Bay	2007	<1	9.6–257 (89)	<2.4–18 (6.7)	<1		Harino et al. (personal communication, 2008)
	Hiroshima Bay	May 2004	<14	<45–430 (107)	<13	<16	<25	Tsunemasa et al. (2006)
		Aug 2004	<14–92 (22)	<45–240 (108)	<13	<16–1,300 (53)	<25	Tsunemasa et al. (2006)
		Nov 2004	<14–27 (6)	<45–130 (37)	<13	<16	<25	Tsunemasa et al. (2006)
		Mar 2005	<14–100 (28)	<45–79 (26)	<13	<16	<25	Tsunemasa et al. (2006)
Singapore		Oct–Nov 2000			3,020–4,200 (2000)	<16	<25	Basheer et al. (2002)

respectively, in the River Crouch, Sutton Harbour, and Southampton, Britain (Thomas et al. 2001). In the Hamble and Orwell estuaries in Britain, the means concentrations of Diuron were 123 and 208 ng l^{-1}, respectively (Boxall et al. 2000). The levels of Diuron in Japan and Singapore were within the range of concentrations in the other coastal areas. Nebeker and Schuytema (1998) tested the chronic effects of Diuron on freshwater cladocerans (*Daphnia pulex*), amphipods (*Hyalella azteca*), midges (*Chironomus tentans*), worms (*Lumbriculus variegatus*), and snails (*Physa gyrina*). The 7- or 10-day NOAEL values in these species ranged from 1.8 to 13.4 mg l^{-1}. The concentration of Diuron in Japan was lower than the no observed adverse effect level (NOAEL) values for freshwater organisms.

Irgarol 1051 was also detected at various sites in Japan. Mean concentrations of Irgarol 1051 were in the range of 4–64 ng l^{-1}, which was lower than those of Diuron. Elsewhere, Irgarol 1051 has been detected in the range of 10–590 ng l^{-1} (means 112 ng l^{-1}) in Hamilton Harbour, Bermuda (Connelly et al. 2001). Thomas et al. (2001) measured Irgarol 1051 in three areas of Britain. The concentrations in the River Crouch, Sutton Harbour, and Southampton were in the range <1–49 (mean 23 ng l^{-1}), <1–84 (mean 8.5 ng l^{-1}), and <1–1,421 ng l^{-1} (mean 105 ng l^{-1}), respectively. The mean concentrations of Irgarol 1051 in the Hamble and Orwell estuaries were 25 and 48 ng l^{-1}, respectively (Boxall et al. 2000). In the USA, Irgarol concentrations ranging from <1 to 2,218 ng l^{-1} are reported from Chesapeake Bay marinas, nine Severn River sites, and one Mainstem Bay site (Hall et al. 2004). Irgarol 1051 in Japan is a little lower than in the other coastal areas described. The toxicity of Irgarol 1051 has been reported in a number of papers. Growth inhibition by Irgarol 1051 in spores of the green macroalga Enteromorpha intestinalis is observed at concentrations as low as 50 ng l^{-1} (Scarlett et al. 1997) and long-term effects on periphyton communities in coastal water are observable at ambient levels of between 63 and 250 ng l^{-1} (Dahl and Blanck 1996). Okamura et al. (2000) report that the no observed effect concentration (NOEC) of Irgarol 1051 in brown seaweed is 300 ng l^{-1}. Concentrations of Irgarol 1051 above these toxicity values were detected in some areas of Japan. M1 is produced by the degradation of Irgarol 1051 (Okamura et al. 1999). The concentrations of M1 were reported in some areas of Japan and, where detectable, were in the range of 3–53 ng l^{-1}. Generally Irgarol concentrations were higher than those of M1. Hall et al. (2004) reported that concentrations of M1 ranged from <1 to 332 ng l^{-1} in Chesapeake Bay marinas, nine Severn River sites, and one Mainstem Bay site. The concentrations in Japan are lower than those in Chesapeake Bay.

There are few available data on triphenylboron pyridine (PTPB). Although PTPB was detectable in Hiroshima Bay, the concentrations of PTPB were below the limits of determination.

Mean levels of each antifouling biocide detected in Japanese coastal waters decreased in the order Diuron > M1 = Irgarol 1051 > Sea Nine 211 > PTPB. Among Asian countries relatively high level of Diuron in water samples appeared to be a feature of Japanese samples.

20.3 Concentrations in Sediment

The concentrations of alternative antifouling compounds in sediment have been reported for Japan, Malaysia, Thailand and Vietnam (Harino et al. 2007, 2008). The concentrations of Sea Nine 211 in these countries were in the range of <0.02–145 g kg^{-1} dry weight (dry wt). Generally, differences in Sea Nine 211 concentrations among these countries were relatively small. However, they were higher than reported for British coastal waters (<0.1 g kg^{-1} dry wt by Thomas et al. (2002). The presence of Sea Nine 211 therefore appears to be characteristic of sediment from Japan, suggesting frequent usage on ships which sail and moor in Japanese waters.

Concentrations of Diuron in sediment from Asian countries were in the range <0.04–1,350 g kg^{-1} dry wt. and were highest in Japan. The latter values were also higher than those recorded in Britain, where Diuron was detected in the range of <0.1–13 and <0.1–1.42 g kg^{-1} dry wt in sediment from the Hamble Estuary and at offshore sites, respectively (Comber et al. 2002).

Dichlofluanid was detected in sediment from Japan and Vietnam, albeit at a low concentration. It was reported that Dichlofluanid was detected in the range <0.1–688.2 g kg^{-1} dry wt in sediment from England (Thomas et al. 2002; Voulvoulis et al. 2000). The concentrations of Dichlofluanid in sediment from Japan and Vietnam were lower than this.

Irgarol 1051 was detected in sediments from Asian countries in the range <0.04–425 g kg^{-1} dry wt. There are many reports concerning concentrations of Irgarol 1051 in sediment from European countries. Albanis et al. (2002) report a maximum Irgarol 1051 concentration of 233 g kg^{-1} dry wt in sediment from a Greek port. The concentration of Irgarol 1051 in sediment from the Hamble estuary was in the range of <10–10 g kg^{-1} dry wt (Boxall et al. 2000). Thomas et al. (2000) report Irgarol 1051 concentrations at offshore British sites at levels of <1–40 g kg^{-1} dry wt. Voulvoulis et al. (2000) report Irgarol 1051 levels of <3.1–222, <3.1–14 and 2–220 g kg^{-1} dry wt in the River Blackwater, the North Sea, and the Baltic Sea, respectively. Biselli et al. (2000) also reported the concentration of Irgarol 1051 in the range of <LOD–25 in North Sea and 4–220 g kg^{-1} dry wt in Baltic sea, respectively. The Irgarol 1051 levels from Asian countries were within the range of the results recorded at the other sites. M1 concentrations in Asian countries were in the range of 0.03–4.9 g kg^{-1} dry wt. Thomas et al. (2000) reported the concentration of M1 in sediment collected from Southampton, Britain. Irgarol was detected at only at a few locations and in the range of 0.4–5.7 g kg^{-1} dry wt, similar to levels in Asian countries.

Pyrithions were detected in Japan and Vietnam. Turley et al. (2000) reported rapid disappearance of pyrithiones from water. Nevertheless, pyrithiones were detected in sediment and it is possible that this reflects the entrainment of pyrithione enriched paint chips derived from ship hulls.

The levels of alternative biocidal compounds in sediments from Asian countries indicate differences between countries. Concentrations of Sea Nine 211 were highest in Vietnam, Diuron levels were highest in Thailand and Irgarol 1051 in

Malaysia. This reflects the fact that patterns of application of antifouling biocides vary between regions.

20.4 Concentrations in Biological Samples

There are a few papers which describe the concentration of alternative biocides in biological samples. The concentrations of alternative biocide in green mussels (*Perna viridis*) from Thailand and clams from Vietnam were reported. The concentrations in green mussels decreased in the order: Diuron > M1 > Irgarol 1051 > Sea Nine 211. The higher water solubility of Diuron (42 mg l^{-1}) has been suggested as an indication of lower bioaccumulation potential (United States Environmental Protection Agency 1989). Nevertheless, Diuron was detected in green mussels at levels of 1–10 g kg^{-1} wet wt (ww), implying relatively high concentrations of Diuron in water.

Alternative biocides were the values under the detection limit in clam (*Meretrix* spp.) from Vietnam. However, Irgarol 1051 in clams from Vietnam only detected at a trace level of about 0.05 g kg^{-1} wet wt.

20.5 Distribution in the Aquatic Environment

20.5.1 Spatial Distribution

The spatial distribution of alternative biocides has been described in the Port of Osaka and Otsuchi Bay (Harino et al. 2007). In the Port of Osaka, substantially higher concentrations of Diuron and Irgarol 1051 were observed near the moorings for small- and medium-sized vessels, situated within a poor flushing zone. Harino et al. (1998) also reported a high concentrations of tributyltin (TBT) in this region. The spatial distributions of Diuron and Irgarol 1051 thus showed similar trends to that of TBT, reflecting the input of these compounds from a combination of moored and moving vessels. The concentration levels and average values of alternative compounds detected in sediment from sites in the port of Osaka sites are also compared in Table 20.2. As in water, the concentrations of Diuron and Irgarol 1051 in sediment were highest near the poorly flushing mooring zone, indicating the continuous input of these compounds from ship hulls. The concentrations of Sea Nine 211 in this area were also a little higher than in other areas of the port of Osaka.

The relationship between Irgarol 1051 and its degradation compounds (M1) was investigated. No correlation between detected concentrations of Irgarol 1051 and M1 was observed in surface water. However, a correlation was observed between accumulated Irgarol 1051 and M1 concentration in sediment ($R^2 = 0.424$). Irgarol 1051 is reported to be rapidly degraded by sunlight in surface water, to yield M1

Table 20.2 The concentrations of alternative biocides in water samples from Japan (sediment μg kg^{-1} dw, biota mg kg^{-1} wet)

Location	Sample	Sampling date	Sea Nine211	Diuron	Dichlofluanid	Irgarol 1051	M1	Pyrithions	Reference
The Port of Osaka	Sediment	2002–2003	<0.04–2.4 (0.52)	0.64–1,350 (40.1)		<0.04–8.2 (0.64)	<0.09–2.9 (0.25)		Harino et al. (2005)
Otsuchi Bay	Sediment		<0.04–0.3 (0.06)	<0.08–1.2 (3.2)	<0.4–14 (1.0)	<0.08–21 (1.7)	<0.18–0.46 (<0.18)	<8–22 (<8)	Harino et al. (2007a)
Otsuchi Bay (Around shipyard)	Sediment		0.05–145 (12)	1.0–534 (45)	<0.4	0.14–103 (7.6)	<0.18–0.48 (<0.18)	<8–9 (<8)	Harino et al. (2007a)
Malaysia	Sediment	2007	<0.02–1.7 (0.09)	<0.04–4.8 (0.71)	<0.10	<0.04–14 (1.7)	<0.09	<2	Harino et al. (2007b)
Thailand	Sediment	Apr 2004	<0.04–0.09 (0.04)	0.07–25 (3.5)		0.03–3.2 (0.15)	0.03–4.9 (0.63)		Harino et al. (2006a)
Vietnam	Sediment	Mar, Aug 2002	<0.09–1.3 (0.45)	0.11–3.0 (1.2)	<0.10–13 (2.8)	0.05–4.0 (0.78)	<0.09–0.43 (0.16)	<2–420 (43)	Harino et al. (2006b)
Thailand	Green mussel	Apr 2004	<0.24–0.12 (0.09)	<0.64–9.6 (2.8)		<0.76–0.22 (0.15)	0.24–0.85 (0.38)		Harino et al. (2006a)
Vietnam	Clam	Mar, Aug 2002	<0.1	<0.2	<0.2	<0.1	<0.2		Harino et al. (2006b)

(Okamura 2002). As irradiation by sunlight differed at each site, the ratio of Irgarol 1051 to M1 also varied. Whilst, the constant ratio of Irgarol 1051 to M1 in sediment may be mainly due to water-sediment partitioning, the slope between Irgarol 1051 and M1 (0.274), indicate that Irgarol 1051 is the dominant species in sediment.

To help interpret the spatial distributions of contamination in Otsuchi Bay, cluster analysis was conducted, based on the concentrations of alternative biocides (Fig. 20.2). On this basis, Otsuchi Bay can be divided into three parts: the vicinity of a shipyard (St. A1), the small fishing port (St. A15), and remaining sites (Sts. A2–14). As shown in Fig. 20.3, concentrations of Sea Nine 211 and Diuron were highest in the vicinity of the shipyard (St. A1). In contrast, the concentration of Irgarol 1051 was highest at the small fishing port (St. A15), Dichlofluanid and Copper pyrithione

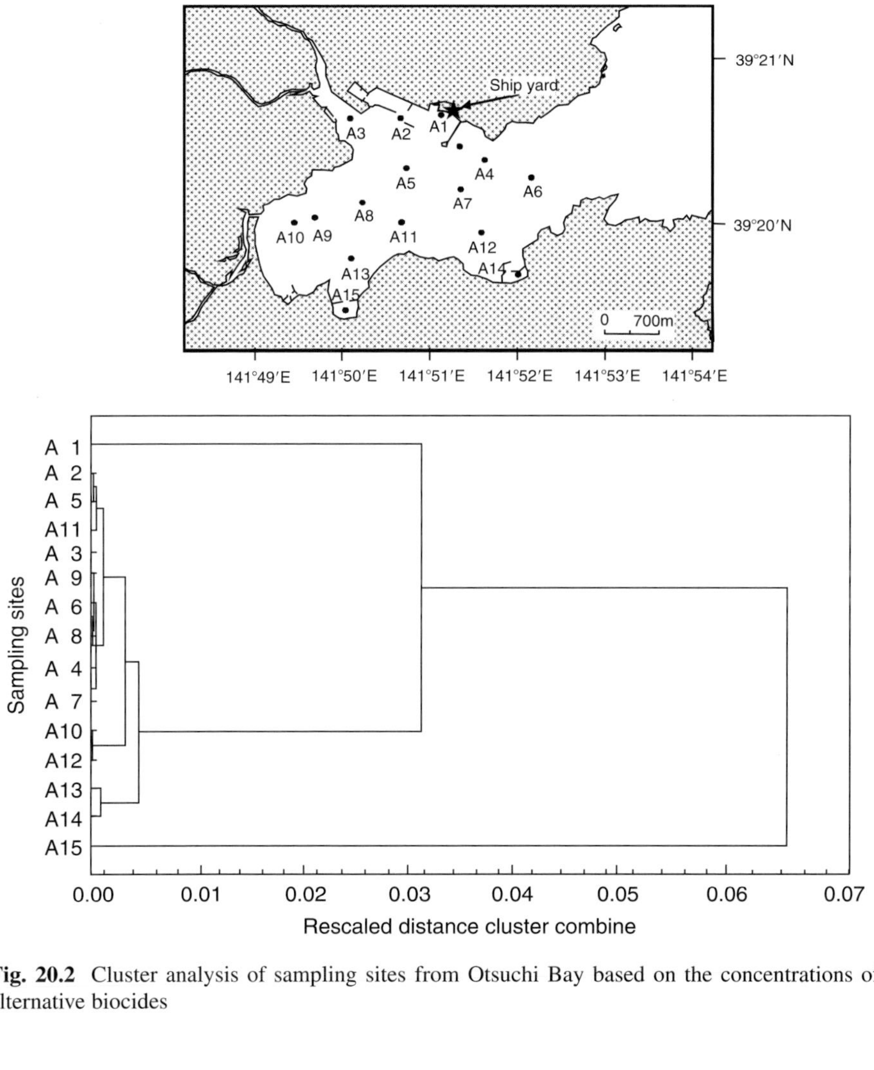

Fig. 20.2 Cluster analysis of sampling sites from Otsuchi Bay based on the concentrations of alternative biocides

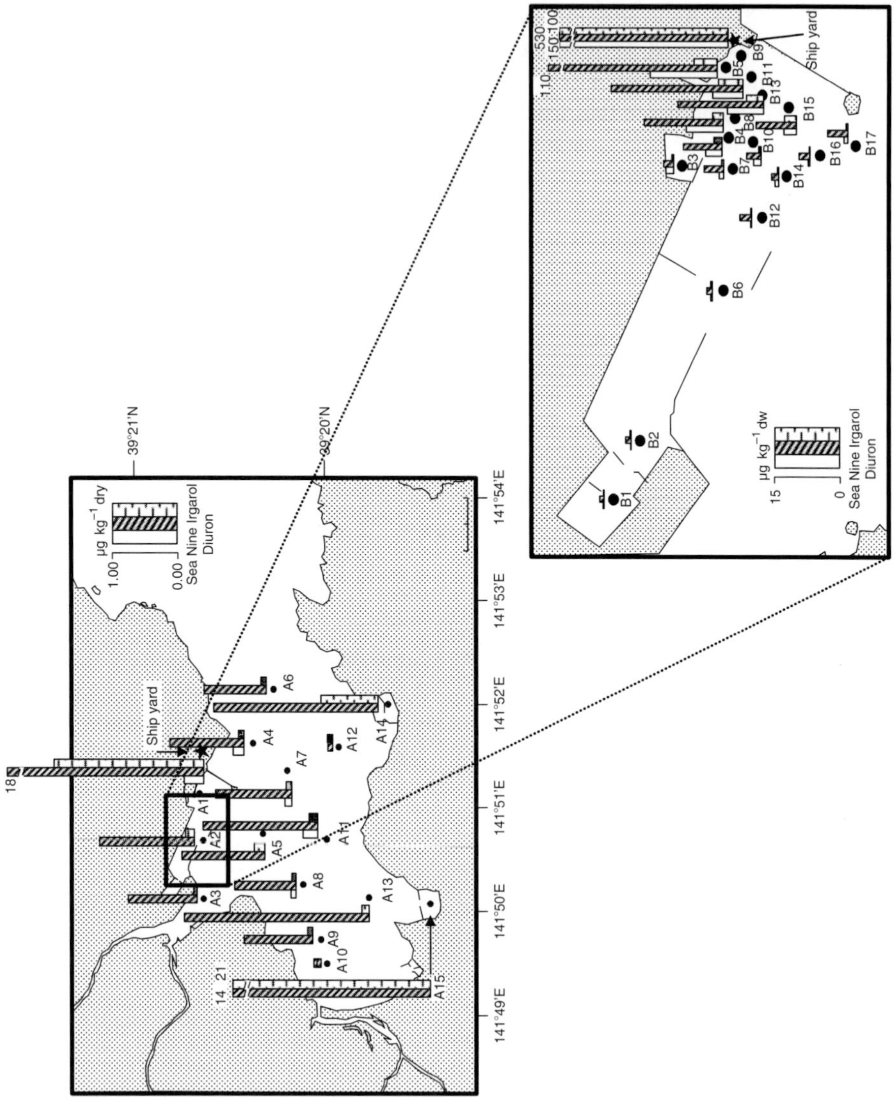

Fig. 20.3 Concentrations of alternative biocides in sediment from Otsuchi Bay

(CuPT) were detected at concentrations of 14 and 22 g kg^{-1} dry wt, respectively, at St.A15. Diuron was, numerically, the dominant compound at Sts. A2–A14.

The spatial distributions of representative alternative biocides around the shipyard in Otsuchi Bay are shown in Fig. 20.3. Concentrations of alternative biocides at St. B9 were extremely high, with the concentrations of Sea Nine 211, Diuron, and Irgarol 1051 being 150, 530, and 100 g kg^{-1} dry wt, respectively. CuPT was also detected at St. B9 at a concentration of 8.8 g kg^{-1} dry wt. The concentrations of these alternative biocides declined with distance from the shipyard. It is presumed that the alternative biocides spread widely from the pollution source in comparison with OTs, because of their lower affinity for sediments.

Liu et al. (1999) concluded from a survey of alternative biocides around coastal waters of western Japan, that Irgarol 1051 was found more frequently in commercial fishing harbours than in marinas. Thus, alternative biocides which were detected were reported to differ, according to how the port areas were used.

Alternative biocides have been used in conjunction with copper oxide in a number of antifouling preparations. Yamamoto 2007 (personal communication) measured copper in sediment from Otsuchi Bay and the correlation coefficients between various antifouling biocides were calculated (Table 20.3). Correlation coefficients between TBT, triphenyltin (TPT), Sea Nine-211, and Diuron were high. It has been reported that the kinds of alternative biocides used differ depending on application – e.g. ships or aquatic equipment such as fishing nets (Yamada and Kakuno 2002).The high correlation coefficients between TBT, TPT, Sea Nine 211, and Diuron indicate that application patterns for these compounds may be similar. Copper also significantly correlated with TBT, TPT, and Diuron, implying that the latter compounds are frequently used in combination with copper as an antifouling biocide on ships. Okamura and Mieno (2006) report that biocide mixtures of a copper (I) oxide and the other alternative biocides are used as tributyltin substitutes, generally.

Among Southeast Asian countries, the spatial distribution of alternative biocides has been described in sediment from Malaysia. The highest concentrations were observed in the southern part of Peninsular Malaysia and the Johor Strait, which have international trading ports and through which many ships sail (Harino et al. 2008).

Alternative biocide distributions in sediment from Thailand are described by Harino et al. (2006a). The kinds and levels of detected alternative biocides differed between sites, as shown in Fig. 20.4. Sea Nine 211 detected in sediment from stations C5–6, which are aquaculture areas and have offshore anchorages for large cargo vessels. The concentrations of Diuron were >5 g kg^{-1} dry wt at stations C8 and C10, which are industrial areas. Irgarol 1051 was found in the highest

Table 20.3 Correlation coefficients among various antifouling biocides

	TBT	TPT	Sea-Nine 211	Diuron	Dichlo fluanid	Irgarol 1051	CuPT	Cu	Zn
TBT	1								
TPT	0.9772*	1							
Sea-Nine 211	0.8774*	0.8405*	1						
Diuron	0.9839*	0.9861*	0.8803*	1					
Dichlofluanid	−0.0829	−0.0233	−0.0967	0.0443	1				
Irgarol 1051	0.2482	0.2629	0.2631	0.3574	0.9159*	1			
CuPT	−0.0737	−0.0303	−0.0707	0.0443	0.9323*	0.8838*	1		
Cu	0.4543*	0.4790*	0.2973	0.4193*	−0.2877	−0.1884	−0.3128	1	
Zn	0.1458	0.1077	0.1575	0.0840	−0.2690	−0.2221	−0.2562	0.0476	1

*$P < 0.05$

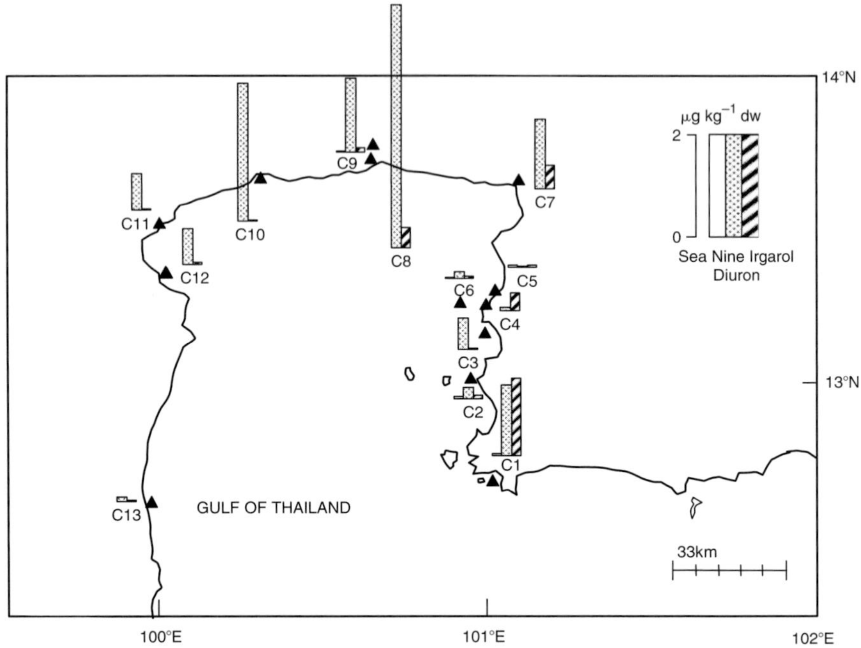

Fig. 20.4 Concentrations of alternative biocides in sediment from the coastal waters of Thailand

concentration at station C1, which is a fishing ground and the site of the Royal Thai Navy Base.

Spatial patterns of alternative biocides in northern and central Vietnam were discussed by Harino et al. (2006b). Concentrations of Sea Nine 211 (1.1–1.3 g kg^{-1} dry wt) were highest at Sts. D3 and D8 (Fig. 20.5). Diuron was also high at St. D3. A higher concentration of Dichlofluanid (12–13 g kg^{-1} dry wt) was observed at D2 and D3. Irgarol 1051 was most enhanced at St. D7 (4.0 g kg^{-1} dry wt). Pyrithiones were only detected at St.D1, at a concentration of 420 g kg^{-1} dry wt. The concentrations of Sea Nine 211, Dichlofluanid and Diuron were highest in sediment from international trading ports with poor flushing of water, such as St. D3. Highest concentrations of Irgarol 1051 were associated with fishing ports such as St. D7.

20.5.2 Temporal Trends

Figure 20.6 shows the representative seasonal variation of alternative biocides in water samples from the port of Osaka, Japan. The levels of Sea Nine 211, Diuron and Irgarol 1051 are higher in the summer season. Comber et al. (2002) also found significantly higher concentrations of Diuron and Irgarol 1051 in summer than in

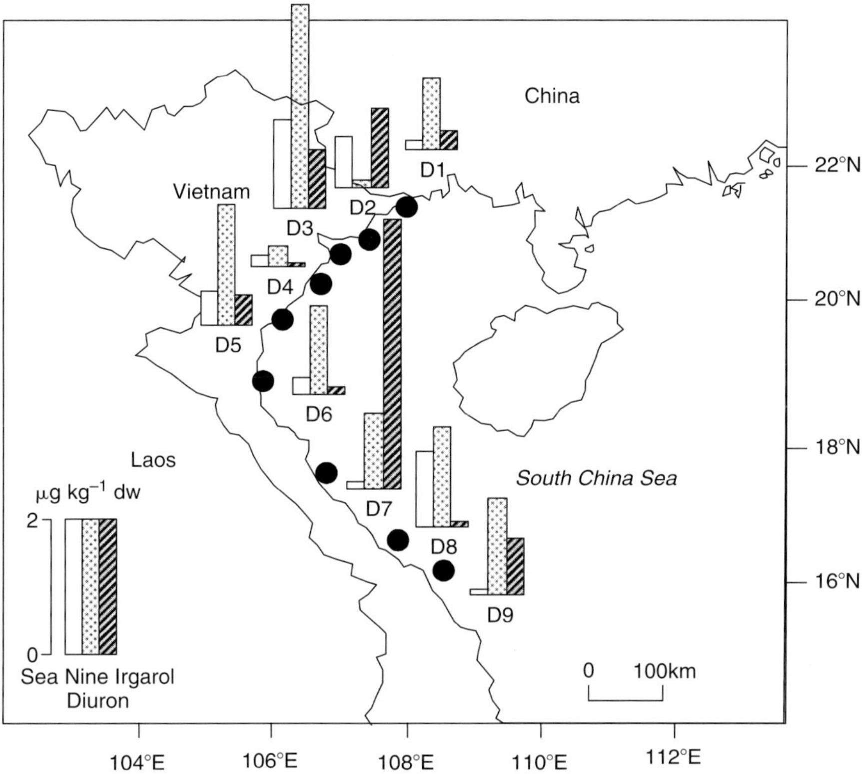

Fig. 20.5 Concentrations of alternative biocides in sediment from the coastal waters of Vietnam

winter. Thomas et al. (2001) also report that Diuron is found at much higher concentrations following the start of the yachting season than during the off-season and that peaks in Irgarol 1051 concentration are seen in mid-summer in open environments such as estuarine marinas that are well flushed by the tide. Biselli et al. (2000) reported that the concentrations of Irgarol 1051 were at a maximum during the periods March–May/July–September and at a minimum during the period December–January. The seasonal trends in Diuron and Irgarol 1051 in the port of Osaka are consistent with these patterns in European countries. The higher concentrations of Irgarol 1051 and Diuron in the summer season may be due to an increase in sailing vessels, the effect of high release rate from vessels by high water temperature, and the reapplication of antifouling paint.

The representative temporal trend for these biocides in sediment is shown in Fig. 20.7. Although the concentrations of Sea Nine 211 and Diuron were less variable than water, the concentrations of these compounds in sediment increased during the study period. These results show that the continuous input of the compounds to sediment exceeds their degradation rates.

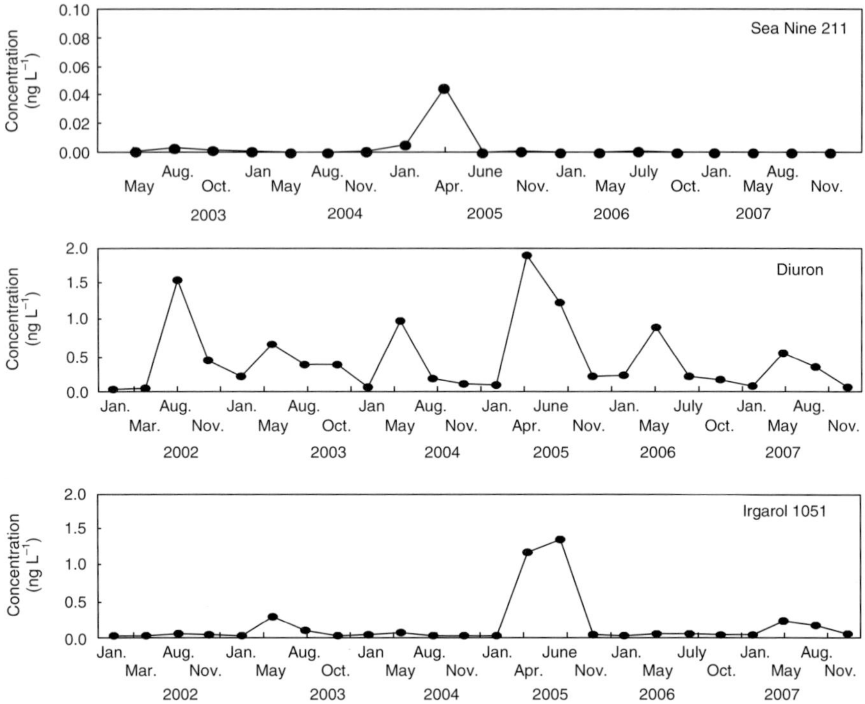

Fig. 20.6 Seasonal variation of alternative biocides in water samples

20.5.3 Partitioning in the Aquatic Environment

Based on the detected concentrations in the Port of Osaka, the sediment-water partition (Kd) values was calculated by dividing the Sea Nine 211, Diuron, Irgarol 1051, and M1 concentrations in sediment by their concentrations in water. The ratios of concentrations for Sea Nine 211, Diuron, Irgarol 1051, and M1 were 690, 2,700, 300, and 870, respectively. Bowman et al. (2003), who report Kd values calculated from Irgarol 1051 concentrations in sediment and water (0.5 m depth) sampled simultaneously at the same site, found Kd values for Irgarol 1051 ranging from 167 to 16,000 at Brighton Marina. Kd values for Irgarol 1051 may vary according to sediment type. The Kd values of alternative biocides were also compared with those of TBT. The Kd for TBT was in the range 180–2,700 according to Fent and Hunn (1991), whilst Harino et al. (1998) report a Kd of 38,000 for TBT. Generally, the Kd values of alternative compounds were lower than that of TBT, suggesting that these alternative compounds have a greater preference for the aqueous phase than TBT.

The logarithms of the octanol water partition coefficients (Pow) for Sea Nine 211, Diuron, Dichlofluanid and Irgarol 1051 were 2.8, 2.9, 3.7 and 2.8, respectively (Harino 2004). These values imply a low bioconcentration potential. In fact,

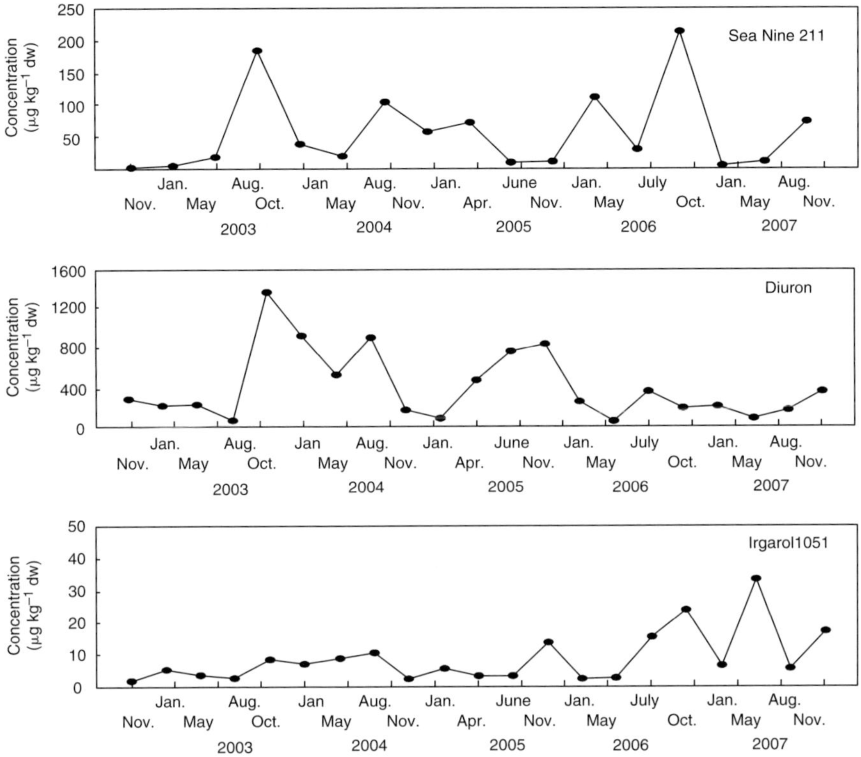

Fig. 20.7 Seasonal variation of alternative biocides in sediment samples

the concentrations and detection frequencies of these antifouling biocides in green mussel (*Perna viridis*) from Vietnam and clams (*Meretrix* spp.) from Vietnam were low (Harino et al. 2006a, b). Clam and green mussel are sold in local markets, for human consumption in Thailand and Vietnam. Although the toxicity of these antifouling biocides to humans has not been clarified, it is presumed that the effects on humans are likely to be low.

20.6 Conclusions

Asian countries are contaminated widely by alternative biocides. However, the levels and kind of alternative biocide differs between countries. For example, high concentrations of Sea Nine 211 are observed in Vietnam; Diuron in Japan and Thailand, and Irgarol 1051 in Malaysia. It is concluded that the levels of alternative biocides in Asian countries are generally lower than those which would have adverse effects on aquatic organisms. However, the elevated concentrations of alternative biocides are observed near poorly flushed moorings (for small- and medium-sized vessels). Judging from the levels of alternative biocides, Japan may already be contaminated heavily by alternative biocides. However, it is presumed that the contamination by OTs is more serious issues than that by alternative compounds in Southeast Asia.

The AFS convention which bans the use of TBT has now been ratified and a higher demand for alternative compounds seems likely in the near future. Further study is therefore needed to monitor antifouling biocides in Southeast Asian countries, to help protect against excessive contamination.

References

Albanis TA, Lambropoulou DA, Sakkas VA et al. (2002) Antifouling paint booster biocide contamination in Greek marine sediment. Chemosphere 48:475–485

Biselli S, Bester K, Huhnerfuss H et al. (2000) Concentrations of the antifouling compound Irgarol 1051 and of organotins in water and sediments of German North and Baltic Sea Marine. Mar Pollut Bull 40:233–243

Boxall A B A, Comber S D, Conrad A U et al. (2000) Inputs, monitoring and fate modelling of antifouling biocides in UK estuaries. Mar Pollut Bull 40:898–905

Bowman J C, Readman J W, Zhou J L (2003) Seasonal variability in the concentrations of Irgarol 1051 in Brighton Marina, UK; including the impact of dredging. Mar Pollut Bull 46:444–451

Comber S D W, Gardner M J, Boxall A B A (2002) Survey of four marine antifoulant constituents (copper, zinc, diuron and Irgarol 1051) in two UK estuaries. J Environ Monitor 4:417–425

Connelly D, Readman J W, Knap A H et al. (2001) Contamination of the coastal waters of Bermuda by organotins and the Triazine herbicide Irgarol 1051. Mar Pollut Bull 42:409–414

Dahl B, Blanck H (1996) Toxic effects of the antifouling agent Irgarol 1051 on periphyton communities in coastal water microcosms. Mar Pollut Bull 32:342–350

Fent K, Hunn J (1991) Phenyltins in water, sediment, and biota of freshwater marinas. Environ Sci Technol 25:956–963

Gardinali P R, Plasencia M, Mack S et al. (2002) Occurrence of Irgarol 1051 in coastal waters from Biscayne Bay, Florida, USA. Mar Pollut Bull 44:781–788

Gough M A, Fothergill G, Hendrie J D (1994) Survey of southern England coastal waters for the s-Triazine antifouling compound Irgarol 1051. Mar Pollut Bull 28:613–620

Hall Jr. L W, Killlen W D, Gardinali P R (2004) Occurrence of Irgarol 1051 and its major metabolite in Maryland waters of Chesapeake Bay. Mar Pollut Bull 48:554–562

Harino H (2004) Occurrence and degradation of representative TBT free-antifouling biocides in aquatic environment. Coast Mar Sci 29:28–39

Harino H, Fukushima M, Yamamoto Y et al. (1998a) Organotin compounds in water, sediment, and biological samples from the Port of Osaka, Japan. Arch Environ Contam Toxicol 35:558–564

Harino H, Fukushima M, Kawai S et al. (1998b) Measurement of butyltin contaminination of water and sediment in Osaka Bay, Japan. Appl Organomet Chem 12:819–825

Harino H, Fukushima M, Kawai S (1999) Temporal trends of organotin compounds in the aquatic environment of the Port of Osaka, Japan. Environ Pollut 105:1–7

Harino H, Yamamoto Y, Kawai S et al. (2003) Butyltin and phenyltin residues in water, sediment and biological samples collected from Otsuchi Bay, Japan. Otsuchi Mar Sci 28:84–91

Harino H, Mori Y, Yamaguchi Y et al. (2005) Monitoring of antifouling booster biocides in water and sediment from the Port of Osaka, Japan. Arch Environ Contam Toxicol 48:303–310

Harino H, Ohji M, Gullaya W et al. (2006a) Occurrence of antifouling biocides in sediment and green mussels from Thailand. Arch Environ Contam Toxicol 51:400–407

Harino H, Midorikawa S, Arai T et al. (2006b) Concentrations of booster biocides in sediment and clams from Vietnam. J Mar Biol Ass UK 86:1163–1170

Harino H, Yamamoto Y, Egchi S et al. (2007) Concentrations of antifouling biocides in sediment and mussel samples collected from Otsuchi Bay, Japan. Arch Environ Contam Toxicol 52:179–188

Harino H, Arai T, Ohji M et al. (2008) Occurrence of antifouling biocides in Southeast Asia-Malaysia, Thailand and Vietnam. In Miyazaki N, Wattayakorn G (ed): The ASEAN International Conference: Conservation on the Coastal Environment, Shinjusha, Tokyo, pp. 91–111

Jacobsen A, Mazza LS, Lawrence LJ et al. (1993) Fate of an antifoulant in an aquatic environment. In: Racke K. D. and Leslie A. R. (eds) ACS symposium series No.522 Pesticides in urban environments: fate and significance. American Chemical Society, San Francisco, CA, 127–138

Kan-atireklap S, Tanabe S, Sanguansin J (1997a) Contamination by butyltin compounds in sediments from Thailand. Mar Pollut Bull 34:894–899

Kan-atireklap S, Tanabe S, Sanguansin J et al. (1997b) Contamination by butyltin compounds and organochlorine residues in green mussel (Perna viridis, L.) from Thailand coastal waters. Environ Pollut 97:79–89

Kan-atireklap S, Yen N T H, Tanabe S et al. (1998)Butyltin compounds and organochlorine residues in green mussel (Perna Viridis L.) from India. Environ Toxicol Chem 67:409–424

Liu D, Pacepavicius G J, Maguire R J et al. (1999) Survey for the occurrence of the new antifouling compound Irgarol 1051 in the aquatic environment. Water Res 33:2833–2843

Martinez K, Ferrer I, Fernandez-Alba AR et al. (2001) Occurrence of antifouling biocides in the Spanish Mediterranean marine environment. Environ Toxicol 22:543–552

Midorikawa S, Arai T, Harino H et al. (2004a) Concentrations of organotin compounds in sediment and clams collected from coastal areas in Vietnam. Environ Pollut 131:401–408

Midorikawa S, Arai T, Harino H et al. (2004b) Organotin levels in bivalves in Southeast Asia. Coast Mar Sci 29:57–62

Nebeker A V, Schuytema G S (1998) Chronic effects of the herbicide diuron on freshwater cladocerans, amphipods, midges, minnows, worms and snails. Arch Environ Contam Toxicol 35:441–446

Okamura H (2002) Photodegradation of the antifouling compounds Irgarol 1051 and Diuron released from a commercial antifouling paint. Chemosphere 48:43–50

Okamura H, Mieno H (2006) Antifouling paint biocides, the handbook of environmental chemistry Vol 5 Part O. In: Konstantinou I (ed) Springer, Heidelberg, Berlin, 201–212

Okamura H, Aoyoama I, Maguire R J et al. (1999) Photodegradation of Irgarol 1051 in water. J Environ Sci Health B34:225–238

Okamura H, Aoyama I, Takami T et al. (2000) Phytotoxicity of the new antifouling compound Irgarol 1051 and a major degradation product. Mar Pollut Bull 40:754–763

Okamura H, Aoyama I, Ono Y et al. (2003) Antifouling herbicides in the coastal waters of western Japan. Mar Pollut Bull 47:59–67

Prudente M, Ichihashi H, Kan-atireklap S et al. (1999) Butyltins, organochlorines and metal levels in green mussel, Perna viridis L. from the coastal waters of the Philippines. Fish Sci 65:441–447

Sakkas V A, Konstantinou I K, Lambropoulou D A et al. (2002) Survey for the occurrence of antifouling paint booster biocides in the aquatic environment of Greece. Environ Sci Pollut Res 9:327–332

Scarlett A, Donkin M E, Fileman T W et al. (1997) Occurrence of the marine antifouling agent Irgarol 1051 within the Plymouth Sound locality: implications for the green macroalga Enteromorpha intestinalis. Mar Pollut Bull 34:645–651

Shade W D, Hurt S S, Jacobson A H et al. (1993) Ecological risk assessment of a novel marine antifoulant. Environ Toxicol Risk Assess 1216:381–408

Shim W J, Hong S H, Yim U H et al. (2002) Horizontal and vertical distribution of butyltin compounds in sediments from shipyards in Korea. Arch Environ Contam Toxicol 43:277–283

Shim W J, Hong S H, Kim N S et al. (2005) Assessment of butyl- and phenyltin pollution in the coastal environment of Korea using mussels and oysters. Arch Environ Contam Toxicol 51:922–931

Sudaryanto A, Takahashi S, Monirith I et al. (2002) Asia-pacific mussel watch: monitoring of butyltin contamination in coastal waters of Asian developing countries. Environ Toxicol Chem 21:2119–2130

Thomas K V, Blake S J, Waldock M J (2000) Antifouling paint booster biocide contamination in UK marine sediment. Mar Pollut Bull 40:739–745

Thomas K V, Fileman T W, Readman J W et al. (2001) Antifouling paint booster biocides in the UK coastal environment and potential risks of biological effects. Mar Pollut Bull 42:677–688

Thomas K V, McHugh M, Waldock M (2002) Antifouling paint booster biocides in UK coastal waters: inputs, occurrence and environmental fate. Sci Total Environ 293:117–127

Toth, S, Becker-van Slooten K, Spack L et al. (1996) Irgarol 1051, an antifouling compound in freshwater, sediment, and biota of Lake Geneva. Bull Environ Contam Toxicol 57:426–433

Tunemasa N, Hashimoto K, Yamaoka Y et al. (2006) Contamination of an alternative antifoulant in coastal waters of Hiroshima Bay. J Environ Chem 16; 201–211 (in Japanese)

Turley P A, Fenn R J Ritter J C (2000) Pyrithiones as antifoulants: environmental chemistry and preliminary risk assessment. Biofouling 15:175–182

United States Environmental Protection Agency (1989) Drinking water health advisory: pesticides. Lewis Publishers, Chelsea, MI, p364

Voulvoulis N, Scrimshaw M D, Lester J N (2000) Occurrence of four biocides utilized antifouling paints, as alternatives to organotin compounds, in waters and sediments of a commercial estuary in the UK. Mar Pollut Bull 40:938–946

Yamada H, Kakuno A (2002) Present status on the development of alternative tribytyltin-free antifouling paints and toxicity of new biocides to aquatic organisms-Review-. Bull Fish Res Agen 6:56–72 (in Japanese)

Section 8
Toxicity in Aquatic Organisms

Chapter 21
Toxicity in Plankton and Fish

Kazuhiko Mochida and Kazunori Fujii

21.1 Introduction

Organotin compounds have been strictly regulated in many countries for the last two decades because of their severe toxic effects on marine organisms (Fent 1996). Consequently, new antifouling biocides that substitute for organotin compounds have been developed. Many studies have surveyed the occurrence of these biocides in the environment. Diuron and Irgarol 1051 (herbicides used in popular slime-resistant antifouling paints) have been detected in environmental water samples from several European countries, the United States, various Caribbean countries, and Japan (Thomas et al. 2000, 2002; Konstantinou and Albanis 2004; Harino et al. 2005b; Carbery et al. 2006). The occurrence of pyrithiones (PTs), such as copper (CuPT) and zinc pyrithione (ZnPT), in the aquatic environment has only recently been reported; Harino et al. (2006) first detected CuPT in sediment from coastal northern Vietnam and, subsequently, in Otsuchi Bay, Japan (Harino et al. 2006, 2007).

Thus, it is clear that these alternative biocides are already in widespread use. The toxicity of these alternative biocides to various marine organisms has already been examined, as described below, but in view of the need to maintain wild populations of marine organisms, it is necessary to clarify the adverse effects of these alternative biocides. Indeed, many studies have been conducted to elucidate the toxic effects of alternative biocides on aquatic organisms.

Okamura and Mieno (2006) have listed alternative biocides currently registered and approved by the Japan Paint Manufacturers' Association (JPMA) (Table 21.1). In this chapter, we focus on the toxicity of several of these listed alternative biocides,

---

K. Mochida and K. Fujii
National Research Institute of Fisheries and Environment of Inland Sea,
Fisheries Research Agency, 2-17-5 Maruishi, Hatsukaichi,
Hiroshima 739-0452, Japan

T. Arai et al. (eds.), *Ecotoxicology of Antifouling Biocides*,
DOI: 10.1007/978-4-431-85709-9_21, © Springer 2009

Table 21.1 Alternative biocides approved by the Japan Paint Manufacturers' Association

Common name (abbreviation)	Chemical name	CAS no.
Dicopper oxide (Cu_2O)	Cuprous oxide	1317-39-1
Copper pyrithione (CuPT)	Copper, bis(1,hydroxy-2(1H)-pyridinethionate O,S)	14915-37-8
Zinc pyrithione (ZnPT)	Zinc-2-pyridinethiol-l-oxide	13463-41-7
PK (TPBP)	Triphenylborane-pyridine	971-66-4
Diuron (DCMU)	3-(3,4-Dichlorophenyl)-1,1-dimethyl urea	330-54-1
Sea-Nine 211 (SN211)	4,5-Dichloro-2-n-octyl-4-isothiazolin-3-one	64359-81-5
Irgarol 1051 (Irgarol)	2-Methylthio-4-tert-butylamino-6-cyclopropylamino-s-triazine	28159-98-0
Chlorothalonil (TPN)	2,4,5,6,-Tetrachloroisophthalonitrile	1897-45-6
Dichlofluanid (DCF)	N,N-Dimethyl-N´-phenyl-N´-(dichlorofluoromethylthio)ulfamide	1085-98-9
Cuprous Thiocyanate (CuSCN)	Cuprous thiocyanate	1111-67-7
PZ (Ziram)	Zinc dimethyl dithiocarbamate	137-30-4
Zineb	Zinc ethylenbis(dithiocarbamate)	12122-67-7
IT-354 (TCPM)	N-(2,4,6-Trichlorophenyl)maleimide	13167-25-4
Densil S-100 (Densil)	2,3,5,6-Tetrachloro-4-(methylsulfonyl)pyridine	13108-52-6
Naphthenic acids, copper salts (NACS)	Naphthenic acids, copper salts	1338-02-9
TT (Thiram)	Tetramethylthiuram disulfide	137-26-8

Abbreviation of each compound is described in parentheses.
Data referred as Okamura and Mieno (2006).

especially to plankton and fish, and we review progress in this research field on the basis of publicly available data.

21.2 Toxicity of Alternative Biocides to Plankton and Fish

21.2.1 Copper Pyrithione, Zinc Pyrithione, and Triphenylborane Pyridine

Both metal PTs are used in marine antifouling paints because of their broad antimicrobial activity, low water solubility, and high degradability. Many studies of the toxicological effects of PTs on aquatic organisms have been published, and the toxicity values of CuPT and ZnPT in plankton and fish species are summarized in Table 21.2. In crustaceans, the acute toxicity value of PTs is generally more than 100 μg/l, although that of CuPT in toy shrimp (*Heptacarpus futilirostris*) is exceptionally high (96-h LC50, 2.5 μg/l) (Mochida et al. 2006). The susceptibility of phytoplankton and fish species to PT toxicity, however, tends to be higher than that of crustaceans. Acute toxicity values, especially of CuPT, in some fish species (Table 21.2) are similar to those of tributyltin in fish species, such as Atlantic menhaden (*Brevoortia tyrannus,* 96-h LC50, 4.5 μg/l), sheepshead minnow (*Cyprinodon*

Table 21.2 Toxic effects of copper pyrithione (CuPT) and zinc pyrithione (ZnPT) to various aquatic organisms

Class	Test organisms	Compound	Toxicity index	Values (μg/l)	Reference
Fresh water species					
Crustacea	Waterfleas	ZnPT			Sánchez-Bayo and Goka (2006)
	Cypretta seuratti		24-h LC50	2,415 (1,298–4,495)	
	Ilyocypris dentifera			1,872 (138–25,333)	
	Chydorus sphaericus			407 (145–1,143)	
	Daphna magna			145 (2–11,151)	
	Cypretta seuratti		48-h LC50	524 (167–1,643)	
	Ilyocypris dentifera			137 (18–1,070)	
	Chydorus sphaericus			197 (73–532)	
	Daphna magna			98(1–9,602)	
Osteichthyes	Rainbow trout	CuPT	7-day LC50	7.6 (5.2–14)	Okamura et al. (2002)
	Oncorhynchus mykiss (Juvenile)		14-day LC50	3.0 (2–4.2)	
			21-day LC50	1.7 (1.0–2.4)	
			28-day LC50	1.3 (0.3–1.7)	
		ZnPT	7-day LC50	8.4 (6.6–11)	
			14-day LC50	5.6 (4.5–7)	
			21-day LC50	4.9 (4.0–6.0)	
			28-day LC50	4.6 (3.6–5.7)	
Salt water species					
Diatomeae	*Amphora coffeaeformis*	CuPT	96-h EC50 (growth)	50	Turley et al. (2005)
			96-h EC80	100	
		ZnPT	96-h EC50	30	
			96-h EC80	75	
	Chaetoceros gracilis	ZnPT	72-h IC50	3.2 (3.0–3.4)	Koutsaftis and Aoyama (2006)
	Skeletonema costatum	CuPT	72-h EC50 (growth)	1.5	Onduka et al. (2007)
		ZnPT	72-h EC50	1.6	

(continued)

Table 21.2 (continued)

Class	Test organisms	Compound	Toxicity index	Values (μg/l)	Reference
	Phytoplankton community	CuPT	EC50 (photosynthesis)	1.3–7.9	Maraldo and Dahllöf (2004)
		ZnPT		0.7–19	
	Pytoplankton community	ZnPT	LOEC(3-day, chlorophyll)	1.6[a]	Hjorth et al. (2006)
	Zooplankton community		LOEC(3-day, grazing activity)	7.9[a]	
Crustacean	*Nitocra spinipes*	ZnPT	96-h LC50 (salinity 7 ppt)	343	Karlsson and Eklund (2004)
			(20 ppt)	178	
	Toy shrimp	CuPT	96-h LC50	2.5 (1.0–6.6)[a]	Mochida et al. (2006)
	(*Heptacarpus futilirostris*)	ZnPT	96-h LC50	120 (92.3–157)[a]	
	Tigriopus japonicus	CuPT	24-h EC50 (immobilization)	23	Onduka et al. (2007)
		ZnPT		280	
Osteichthyes	Red sea bream	CuPT	96-h LC50	9.3 (8.1–10.7)[a]	Mochida et al. (2006)
	(*Pagrus major*)	ZnPT	96-h LC50	98.2 (60.5–159)[a]	
	Mummichog	CuPT			Mochida et al. (2008)
	(*Fundulus heteroclitus*)				
	2–3 days post-hatch		96-h L50	2.9–8.4[a]	
	4–5 weeks post-hatch		96-h L50	5.0–17.8[a]	
			LOEC(50-day, growth, survival)	0.37[a]	

The 95% confidence intervals are given in parentheses.
[a]Values based on acutual average toxicant concentration.

variegatus, 96-h LC50, 25.9 μg/l) (Bushong et al. 1988), and girella (*Girella punctata*, 96-h LC50, 3.2 μg/l) (Kakuno and Kimura 1987). A major difference between PTs and tributyltin (TBT) is in the length of their half-lives in seawater. CuPT and ZnPT are very light sensitive (Turley et al. 2000; Harino et al. 2005a). Under natural light conditions, the half-life of PTs is approximately 8 min (Maraldo and Dahllöf 2004) whereas that of tributyltin oxide exceeds 89 days (Maguire et al. 1983). Thus, when considering the ecotoxicological impact of PTs, their degradation characteristics must be taken into account, even although the acute toxicity values of PTs in some marine organisms are similar to those of TBT.

A histopathological study has clarified how CuPT causes lethality in a teleost, red sea bream (*Pagrus major*) (Mochida et al. 2006). The secondary lamellae of gills of fish exposed to CuPT were heavily damaged, even in the surviving fish (Fig. 21.1), and this histologically observed damage likely caused fatal hypoxemia. Ermolayeva and Sanders (1995) revealed that PT inhibits membrane transport in a fungus by affecting the primary proton pump. Thus, it is possible that PTs disrupt the membrane transport system of the fish gill causing the histologically observed alterations.

Mochida et al. (2008) also conducted an early-life stage toxicity test (ELS test) in mummichog (*Fundulus heteroclitus*) and revealed that although exposure of these fish to 2 or 4 μg/l CuPT for 50 days at embryonic to larval stages had no effect on hatchability, the survival and growth of larvae were significantly reduced. The lowest observed effect concentration (LOEC) was 0.37 μg/l, based on the actual average toxicant concentration (Table 21.2). During the ELS test, the fish showed vertebral deformity and formation of inflammatory masses in lateral muscles (Mochida et al. 2008). A possible cause of this morphological abnormality is described in Chapter 28.

Although many studies have conducted PT toxicity testing in plankton species, only a few have carried out PT toxicity testing in fish, possibly because of the difficulty of maintaining the nominal concentration in the exposure tank, even in continuous flow-through system (e.g., in ELS testing) (Mochida et al. 2006, 2008). During fish toxicity testing, PTs may be degraded by microorganisms, even

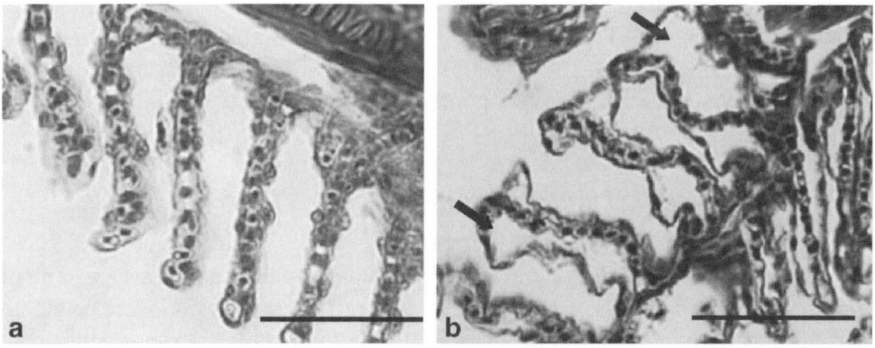

Fig. 21.1 Histological alterations in red sea bream (*Pagrus major*) gills after exposure to copper pyrithione. (**a**) Control fish, (**b**) fish exposed to the pyrithione. Arrows in (**b**) show vacuoles. Bars, 50 μm

when the tests are conducted in the dark. Metabolites of degraded PTs may thus contribute to the toxic effects revealed in these tests. To our knowledge, few studies have examined the toxicity of PT metabolites to marine fish species because quantification is difficult. Turley et al. (2000) have reported that one PT metabolite, pyridine 2-sulfonate, which is a main terminal photolysis product, is at least four orders of magnitude less toxic than the parent compound. In addition, Onduka et al. (2007) reported toxicity values of several PT metabolites to a marine phytoplankton (*Skeletonema costatum*) and a marine zooplankton (*Tigriopus japonicus*). The 72-h EC50 values of 2-mercaptopyridine-*N*-oxide, 2,2´-dithio-bis-pyridine-*N*-oxide, 2-mercpatopyridine, and 2,2´-dithio-bis-pyridine for growth of *S. costatum* are 1.1, 3.4, 730, and 65 µg/l, respectively, and those of 2-mercpatopyridine and 2,2´-dithio-bis-pyridine for *T. japonicus* are 76,000 and 550 µg/l, respectively. Because there are no data on the fates of PT metabolites, such as their half-lives, in the marine environment, it may be inappropriate to discuss the ecotoxicological risk posed by PT metabolites in this chapter. However, the elucidation of chronic toxicity of PT metabolites to marine organisms, especially during the early life stages, is worthy of additional study.

In juvenile rainbow trout (*Oncorhynchus mykiss*), the 7-day LC50, 14-day LC50, 21-day LC50, and 28-day LC50 values of triphenylborane pyridine, a pyridine-based antifouling biocide, were found to be 140, 84, 61, and 42 µg/l, respectively (Okamura et al. 2002).

21.2.2 Sea-Nine 211

Sea-Nine 211 is a compound with demonstrated high antimicrobial activity against bacteria, fungi, and algae (Shade et al. 1993). Toxicity values of Sea-Nine 211 to plankton and fish species are summarized in Table 21.3. Sea-Nine 211 is almost equally toxic to phyto- and zooplankton species and to some teleost fish. Its highest reported toxicity value is 32 µg/l (120-h EC50) for the growth of freshwater phytoplankton species, and its lowest value is 1.2 µg/l (21-day LC50) for freshwater crustaceans (Shade et al. 1993). For fish, its 96-h LC50 values range from 2.7 to 20.5 µg/l (9.6–72.7 nmol/l) (Shade et al. 1993; Okamura et al. 2002). Since the 96-h LC50 of CuPT to red sea bream is 29.4 nM, the toxicity of Sea-Nine 211 is similar to that of CuPT.

21.2.3 Irgarol 1051 and Diuron

Toxicity values of Irgarol 1051 and diuron in plankton and fish species are summarized in Tables 21.4 and 21.5, respectively. Irgarol 1051 and diuron were originally developed as herbicides, and their toxicity to phytoplankton species is drastically different from that to zooplankton or other aquatic animals. This toxicity difference can be explained by their mechanism of action. Irgarol 1051 acts by blocking electron transport in the photosynthesis system, which animal species do not have. Specifically, triazine, a component of Irgarol 1051, inhibits photosystem II

Table 21.3 Toxic effects of Sea-Nine 211 to various aquatic organisms

Class	Test organisms	Toxicity index	Values (μg/l)	Reference
Fresh water species				
Chlorophyceae	*Selenastrum capricornutum*	120-h EC50 (growth)	32.0 (28.0–36.0)[a]	Shade et al. (1993)
Crustacean	*Daphnia magna*	48-h EC50 (immobilization)	5.2 (3.9–7.0)	Shade et al. (1993)
	Daphnia magna	21-day EC50 (survival)	1.2 (1.0–1.4)	
Osteichthyes	Bluegill sunfish (*Lepomis macrochirus*)	96-h LC50	14.0 (6.5–26.0)	Shade et al. (1993)
	Rainbow trout (*Oncorhynchus mykiss*)	96-h LC50	2.7 (1.8–3.3)	Shade et al. (1993)
	Juvenile rainbow trout	7-day LC50	14 (11–17)	Okamura et al. (2002)
		14-day LC50	14 (11–17)	
		21-day LC50	14 (11–17)	
		28-day LC50	14 (11–17)	
Salt water species				
Diatomeae	*Skeletonema costatum*	96-h EC50 (chlorophyll)	20.1 (10.8–34.7)	Shade et al. (1993)
		96-h EC50 (cell growth)	13.9 (6.9–29.3)	
Osteichthyes	Sheepshead minnow (*Cyprinodon variegatus*)	96-h LC50	20.5 (17.7–23.5)[a]	Shade et al. (1993)
		NOEC (35-day, survival and growth)	6	
		LOEC	14	

The 95% confidence intervals are given in parentheses.
[a]Flow through with analytical confirmation.

Table 21.4 Toxic effects of Irgarol 1051 and its degradation products to various aquatic organisms

Class	Test organisms	Compound	Toxicity index	Values (µg/l)	Reference
Fresh water species					
Diatomae	*Navicula pelliculosa*		5-day EC50 (growth)	0.136	Hall et al. (1999)
Cyanophyceae	*Anabaena flos-aquae*			2.1	Hall et al. (1999)
Chlorophyceae	*Selenastrum capricornutum*			1.5	Hall et al. (1999)
	Selenastrum capricornutum		72-h EC50 (growth)	2.3 (2.1–2.4)	Okamura et al. (2000)
		M1	72-h EC50 (growth)	46 (42–50)	
Crustacean	*Daphnia magna*		48-h IC50	5,300	Hall et al. (1999)
	Daphnia magna		24-h LC50	16,000 (14,000–18,000)	Okamura et al. (2000)
			48-h LC50	8,300 (6,700–10,000)	
	Daphnia magna	M1	24-h LC50	17,000 (15,000–21,000)	Okamura et al. (2000)
		M1	48-h LC50	11,000 (9,000–12,000)	
	Daphnia pulex		24-h LC50	5,700 (5,100–6,300)	Okamura et al. (2000)
		M1	24-h LC50	27,000 (21,000–38,000)	
	Thamnocepharus platyurus		24-h LC50	12,000 (11,000–13,000)	Okamura et al. (2000)
		M1	24-h LC50	19,000 (17,000–21,000)	
Osteichthyes	Rainbow trout (*Oncorhynchus mykiss*)		96-hLC50	790	Hall et al. (1999)
			NOEC (hatchability)	184	
			NOEC (survival, 60-day post-hatch)	184	
			NOEC (growth, 60-day post-hatch)	4	
	Rainbow trout (juvenile)		7-day LC50	25,000 (14,000–80,000)	Okamura et al. (2002)
			14-day LC50	7,400 (6,000–10,000)	
			21-day LC50	2,500 (2,000–3,100)	
			28-day LC50	880 (460–1,300)	
	Bluegill sunfish (*Lepomis macrochirus*)	96-h LC50	2,600	Hall et al. (1999)	

Salt water species					
	Peryphiton community		LOEC (21 day, growth)	1	Dahl and Blanck (1996)
Chlorophyceae	*Dunaliella tertiolecta*		5-day IC50 (growth)	0.56	Hall et al. (1999)
Diatomeae	*Skeletonema costatum*		5-day IC50 (growth)	0.43	Hall et al. (1999)
	Skeletonema costatum	M1	5-day IC50 (growth)	16.2	
	Chaetoceros gracilis		72-h IC50 (growth rate)	1.1 (0.95–1.2)	Koutsaftis and Aoyama (2006)
Chlorophyceae	*Dunaliella tertiolecta*		96-h EC50 (growth rate)	0.7	DeLorenze and Serrano (2006)
	Dunaliella tertiolecta		96-h EC50 (growth rate)	1.1	Gatidou and Thomaidis (2007)
		M1		83	
Diatomeae	Navicula *forcipata*		96-h EC50 (growth rate)	0.6	Gatidou and Thomaidis (2007)
		M1		73	
Osteichthyes	Inland silverside (*Menidia beryllina*)		96-h LC50	1,580	Hall et al. (1999)
	Sheepshead minnow (*Cyprinodon variegatus*)		96-h LC50	3,500	

M1, 2-methylthio-4-tert-butylamino-s-triazine. The 95% confidence intervals are given in parentheses.

Table 21.5 Toxic effects of Diuron to various aquatic organisms

Class	Test organisms	Compound	Toxicity index	Values (μg/l)	Reference
Freshwater species					
Osteichthyes	Rainbow trout (juvenile)		7-day LC50	74,000 (29,000–3,681,000)	Okamura et al. (2002)
	(*Oncorhynchus mykiss*)		14-day LC50	15,000 (11,000–29,000)	
			21-day LC50	5,900 (4,700–7,700)	
			28-day LC50	230 (8.9–590)	
Salt water species					
Diatomeae	*Chaetoceros gracilis*		72-h EC50 (growth)	36.0(31.0–41.0)	Koutsaftis and Aoyama (2006)
Chlorophyceae	*Dunaliella tertiolecta*		96-h EC50 (growth)	0.7	DeLorenzo and Serrano (2006)
Chlorophyceae	*Dunaliella tertiolecta*		96-h EC50 (growth)	5.9	Gatidou and Thomaidis (2007)
		DCPMU		345	
		DCPU		<15% inhibition (at 10 mg/l)	
		DCA		6,381	
Diatomeae	*Navicula forcipata*		96-h EC50 (growth)	27	Gatidou and Thomaidis (2007)
		DCA		6,269	

DCPMU, 1-(3,4-dichlorophenyl)-3 methyl urea; DCPU, 1-(3,4 dichlorophenyl urea); DCA, 3,4-dichloroaniline. The 95% confidence intervals are given in parentheses.

and this inhibition leads to reduced CO_2 uptake by the organisms, thus reducing carbohydrate production (Hall et al. 1999). Diuron is also known to act as an inhibitor of photosystem II (Fai et al. 2007).

The toxicity of a metabolite of Irgarol 1051, such as 2-methyl-4-*tert*-butylamino-*s*-triazine (M1), has also been studied. The toxicities of Irgarol 1051 and M1 have been compared in several phytoplankton species (Hall et al. 1999; Okamura et al. 2000; Gatidou and Thomaidis 2007). In general, M1 is less toxic than Irgarol 1051. In the marine phytoplankton species *Dunaliella tertiolecta*, the following order of increasing toxicity of Irgarol, M1 and metabolites of diuron has been reported: 3,4-dichloroaniline (DCA) < 1-(3,4 dichloroophenyl urea) < 1-(3,4-dichlorophenyl)-3 methyl urea (DCPMU) < M1 < Irgarol 1051 (Gatidou and Thomaidis 2007). The toxicity of diuron to marine phytoplankton species is intermediate between that of Irgarol 1051 and M1 (Gatidou and Thomaidis 2007). As previously mentioned, Irgarol 1051 contains triazine, which has been used in combination with copper in antifouling paints on many large ships (Granmo et al. 2002). The toxic effects of triazine on cod (*Gadus morhua*), a commercially important marine fish, in its early developmental stages has been examined. Although fertilized eggs that had been exposed to triazine for 5 days exhibited higher buoyancy, the highest concentration tested (40 μg/l) did not cause significant mortality in cod larvae (Granmo et al. 2002).

21.2.4 Chlorothalonil, Dithiocarbamates, and Related Compounds

Toxicity values of chlorothalonil (TPN) and dithiocarbamates (DTCs) in plankton and fish species are summarized in Tables 21.6 and 21.7. TPN and DTCs are potent fungicides developed for agricultural uses, such as the treatment of fruit and vegetable crops (Davies and White 1985; Van Leeuwen et al. 1985a), and they are now also used as antifouling biocides. Many toxicity studies of these compounds have been conducted, especially on their effects on freshwater species. Among DTCs, zinc dimethyl dithiocarbamate (ziram) and tetramethylthiuram disulfide (thiram) exhibit strong toxic effects, particularly on rainbow trout *Oncorhynchus mykiss* in early life stages (Van Leeuwen et al. 1985a, 1986a). In addition, DTCs have been reported to be teratogenic in some teleost species (Van Leeuwen et al. 1986b; Teraoka et al. 2006; Tilton et al. 2006), and reproductive toxicity in mammals has also been reported (Mishra et al. 1993; Stoker et al. 2003). These topics are further explored in the next chapter.

21.3 Joint Toxic Effect of Antifouling Biocides

Antifouling biocides that substitute for TBT are in many cases used in combination in antifouling paints (Okamura and Mieno 2006). Among 380 paint products included in the JPMA list, 16% contain only one biocide, 52% contain two biocides, 21% contain three, and 5% contain four biocides. Only 6% of these paint products are biocide-free. Sixty products contain only one biocide. Among these, Cu_2O

Table 21.6 Toxic effects of chlorothalonil to various aquatic organisms

Class	Test organisms	Toxicity index	Values (μg/l)	Reference
Fresh water species				
Crustacean	*Ceriodaphnia dubia*	96-h LC50	156	Sherrard et al. (2002)
		LOEC (4-day, mortality)	125	
Osteoichthyes	Top minnow (*Gambusia affinis*)	48-h LC50	90	Nishiuchi (1977)
	Japanese medaka (*Oryzias latipes*)		90	Nishiuchi (1977)
	Guppy (*Poecilia reticulata*)		200	Nishiuchi (1977)
	Common carp (*Cyprinus carpio*)	48-h LC50	67	Nishiuchi (1979)
	Rainbow trout (*Oncorhynchus mykiss*)	96-h LC50 (flow through;O2, 8–9 mg/l)	14.3	Davies and White (1985)
	Common jollytail (*Galaxias maculatus*)		16.3	Davies and White (1985)
	Spotted galaxias (*Galaxias truttaceus*)		18.9	Davies and White (1985)
	Golden galaxias (*Galaxias auratus*)		29.2	Davies and White (1985)
	Rainbow trout (*Oncorhynchus mykiss*)	96-h LC50	76	Ernst et al. (1991)
	Channel catfish (Juvenile) (*Ictalurus punctatus*)	96-h LC50	52	Gallagher et al. (1992)
Salt water species				
Chlorophyceae	*Dunaliella tertiolecta*	96-h EC50 (growth)	64 (62–65)	DeLorenzo and Serrano (2006)
Osteoichthyes	Fathead minnow (*Pimephales promelas*)	96-h LC50	22.6	Sherrard et al. (2002)
		LOEC (4-day, mortality)	15	

The 95% confidence intervals are given in parentheses.

Table 21.7 Toxic effects of dithiocarbamates and related comounds to various aquatic organisms

Class	Test organisms	Compound	Toxicity index	Values (μg/l)	Reference
Fresh water species					
Chlorophyceae	Green algae (*Chlorella pyrenoidosa*)	Zineb	96-h EC50 (growth)	1,800	Van Leeuwen et al. (1985a)
		Ziram		1,200	
		Thiram		1,000	
Crustacean	*Daphnia magna*	Zineb	48-h LC50	970 (560–1,800)	Van Leeuwen et al. (1985a)
		Ziram		140 (100–180)	
		Thiram		210 (170–270)	
	Daphnia magna	Zineb	21-day LC50	89 (78–102)	Van Leeuwen et al. (1985b)
		Ziram		11 (10–12)	
		Thiram		8 (7–9)	
		Zineb	LOEC(21-day growth)	18	
		Ziram		10	
		Thiram		10	
Osteichthyes	Guppies (*Poecilia reticulata*)	Zineb	96-h LC50	7,200 (5,000–10,300)	Van Leeuwen et al. (1985a)
		Ziram		750 (560–1,000)	
		Thiram		270 (220–330)	
	Rainbow trout (*Oncorhynchus mykiss*)	Zineb	60-day LC50	211 (200–222)	Van Leeuwen et al. (1985a)
			60-day EC50 (mortality and teratogenicity)	188 (179–199)	
			LOEC	100	
		Ziram	60-day LC50	2.0 (1.8–2.1)	
			60-day EC50 (mortality and teratogenicity)	1.5 (1.4–1.5)	
			LOEC	1.8	
		Thiram	60-day LC50	1.1 (1.1–1.2)	
			60-day EC50 (mortality and teratogenicity)	0.64 (0.57–0.73)	
			LOEC	<0.32	
	Zebrafish (*Danio rerio*)	Thiram	20-h LC50 (4-hpf embryo)	40–400	Tilton et al. (2006)

Zineb, zinc ethylenbis(dithiocarbamate); Ziram, zinc dimethyl dithiocarmabate; Thiram, tetramethylthiuram disulfide. The 95% confidence intervals are given in parentheses.

is used in 52 products, followed by TPBP in five, and Sea-Nine 211, TPN, and CuSCN in one product each (Okamura and Mieno 2006). Thus, the joint toxicity of these antifouling biocides to aquatic organisms has also been studied (Hall et al. 1999; DeLorenzo and Serrano 2006; Koutsaftis and Aoyama 2006, 2007; Mochida et al. 2006; Gatidou and Thomaidis 2007).

Hall et al. (1999) investigated the joint toxicity of Irgarol 1051 and copper to three freshwater species, the green alga *Selenastrum capricornutum*, the water flea *Daphnia magna,* and rainbow trout, by using toxic units. The total number of toxic units (TU) in a mixture of two biocides is calculated by the formula $TU = C_1/R_1 + C_2/R_2$, where C_1 and C_2 represent the concentrations of components 1 and 2, respectively, and R_1 and R_2 represent the concentration of component that causes the specified response when tested alone, such as the LC50 or EC50 for growth. For example, if R1 and R2 represent LC50 values, then a mixture containing 1 TU will cause 50% mortality if the toxicity of the two components of the mixture is simply additive. If less than 1 TU or more than 1 TU causes the specified response, then the toxicities of the components are synergistic or antagonistic, respectively. Although the toxicity of the mixture of Irgarol 1051 and copper to *D. magna* and *S. capricornutum* was simply additive, the toxicity to rainbow trout was synergistic (Hall et al. 1999). Irgarol in a mixture with TPN exhibits synergistic toxicity to the marine phytoplankton, *Dunaliella tertiolecta* (DeLorenzo and Serrano 2006), but in a mixture with M1 shows additive effects on the marine phytoplankton species, *D. tertiolecta* and *Navicula forcipata*. Moreover, additive toxicity effects on both these species were also observed for mixtures of copper with either Irgarol 1051 or M1 (Gatidou and Thomaidis 2007). Although diuron exhibits synergistic toxicity in combination with its metabolites DCPMU and DCA, copper in combination with either diuron or one of its metabolites showed antagonistic toxicity to these same two phytoplankton species (Gatidou and Thomaidis 2007). Koutsaftis and Aoyama (2006) demonstrated that Irgarol 1051 in combination with Cd exhibits synergistic toxic effects on the marine phytoplankton species *Chaetoceros gracilis*, whereas ZnPT with copper showed a strictly antagonistic effect. ZnPT is readily transformed into CuPT (Nakajima and Yasuda 1990; Maraldo and Dahllöf 2004). Thus, the authors hypothesized that transchelation of ZnPT to CuPT, which has a significantly lower effect on the growth of *C. gracilis* in the presence of Cu, may explain the antagonistic effect of the mixture of ZnPT and Cu (Koutsaftis and Aoyama 2006). In contrast, Mochida et al. (2006) observed that a mixture of ZnPT and Cu exhibited more than additive toxicity to toy shrimp and red sea bream. This finding might be explained by the greater susceptibility of these organisms to CuPT toxicity because CuPT is more toxic than ZnPT to toy shrimp and red sea bream (Mochida et al. 2006).

21.4 Conclusions

Monitoring studies have revealed that the concentrations of alternative biocides, such as Irgarol 1051 and diuron, in coastal areas of some countries, including Japan, are in the order of nanograms per liter or higher (Okamura et al. 2000; Konstantinou and Albanis 2004; Harino et al. 2005b; Carbery et al. 2006). In light of the results reported to date for toxicity tests of alternative biocides, such as Irgarol 1051 and diuron, the growth of phytoplankton species, such as *S. costatum*, which has high susceptibility to these compounds (Tables 21.4 and 21.5), might be affected by these biocide concentrations. Although in the UK, diuron is no longer approved for use as an active ingredient in antifouling paints on any size of vessel (Konstantinou and Albanis 2004), it may be necessary for other countries to take measures to reduce the risk involved with the use of Irgarol 1051 as well. The importation and use of antifouling paint containing Irgarol 1051 has already been banned in Bermuda because of its acute toxicity to corals (Carbery et al. 2006). To date, however, no research had elucidated the toxicity of diuron to marine fish species, and it is essential that the toxic effects of diuron on marine fish species be examined.

Because CuPT and ZnPT are easily photolyzed (Turley et al. 2000; Harino et al. 2005a), the risk presented by PTs to marine organisms may not be very high. However, given that PTs may persist in the environments where the influence of sunlight is limited, as pointed out by Maraldo and Dahllöf (2004), and that the fate and toxicity of their metabolites remain unknown, the long-term effects of PTs and their metabolites to marine organisms deserve more study. Furthermore, Koutsaftis and Aoyama (2007) demonstrated that the combination of ZnPT and CuPT has a clear synergistic effect on the brine shrimp (*Artemia salina*). Because ZnPT is readily transformed into CuPT (Nakajima and Yasuda 1990; Maraldo and Dahllöf 2004), CuPT and ZnPT may coexist in the marine environment. Thus, the toxicity of mixtures of CuPT and ZnPT should be studied in other marine organisms.

Although the occurrence of DTCs in the marine environment has not yet been reported, Sea-Nine 211 and TPN have already been found in European waters, in a marina and a commercial estuary (Voulvoulis et al. 2000; Konstantinou and Albanis 2004), at concentrations ranging from 0.28 to 3.3 and from 0.36 to 1.38 μg/l, respectively. Although the concentration of TPN that can cause acute toxicity to some crustacean and fish species are 10–100 times those detected in the environment, the maximum detected level of Sea-Nine 211 might have toxicological effects on some plankton and fish species. In addition, as mentioned above, ziram and thiram exhibit strong toxic effects especially on teleost species in early life stages (Van Leeuwen et al. 1986a, b). However, no data on the toxicity of these compounds to marine organisms are yet available. Thus, it is necessary to gather and analyze additional data on the toxic effects of these alternative antifouling biocides on marine organisms, in addition to monitoring their occurrence in the marine environment.

Acknowledgements We are grateful to the members in the Ecotoxicology section of the National Research Institute of Fisheries and Environment of Inland Sea, Fisheries Research Agency for their kind assistance. This work was supported in part by a grant-in-aid from the Japanese Ministry of the Environment (Project: Environmental Risk Assessment of Antifouling Agents).

References

Bushong SJ, Hall Jr. LW, Hall WS et al. (1988) Acute toxicity of tributyltin to selected Chesapeake Bay fish and invertebrates. Water Res 22:1027–1032

Carbery K, Owen R, Frickers T et al. (2006) Contamination of Caribbean coastal waters by the antifouling herebicide Irgarol 1051. Mar Poll Bull 52:635–644

Dahl B and Blanck H (1996) Toxic effects of the antifouling agent Irgarol 1051 on periphyton communities in coastal water microcosms. Mar Poll Bull 32:342–350

Davies PE and White RWG (1985) The toxicology and metabolism of chlorothalonil in fish. I. lethal levels for *Salmo gairdneri*, *Galaxias maculatus*, *G. truttaceus* and *G. auratus* and the fate of ^{14}C-TCIN in *S. gairdneri*. Aquat Toxicol 7:93–105

DeLorenzo ME and Serrano L (2006) Mixture toxicity of the antifouling compound Irgarol to the marine phytoplankton species *Dunaliella tertiolecta*. J Environ Sci Health B 41: 1349–1360

Ermolayeva E and Sanders D (1995) Mechanism of pyrithione-induced membrane depolarization in *Neurospora crassa*. Appl Environ Microbiol 61:3385–3390

Ernst W, Doe K, Jonah P et al. (1991) The toxicity of chlorothalonil to aquatic fauna and the impact of its operational use on a pond ecosystem. Arch Environ Contam Toxicol 21:1–9

Fai PB, Grant A and Reid B (2007) Chlorophyll a fluorescence as a biomarker for rapid toxicity assessment. Environ Toxicol Chem 26:1520–1531

Fent K (1996) Ecotoxicology of organotin compounds. Crit Rev Toxicol 26:1–117

Gallagher EP, Cattley RC and Di Giulio RT (1992) The acute toxicity and sublethal effects of chlorothalonil in channel catfish (*Ictalurus punctatus*). Aquat Toxicol 23:155–168

Gatidou G and Thomaidis NS (2007) Evaluation of single and joint toxic effects of two antifouling biocides, their main metabolites and copper using phytoplankton bioassays. Aquat Toxicol 85:184–191

Granmo Å, Ekelund R, Sneli J-A et al. (2002) Effects of antifouling paint components (TBTO, copper and triazine) on the early development of embryos in cod (*Gadus morhua* L.). Mar Poll Bull 44:1142–1148

Hall Jr. LW, Giddings JM, Solomon KR et al. (1999) An ecological risk assessment for the use of Irgarol 1051 as an algaecide for antifoulant paints. Crit Rev Toxicol 29: 367–437

Harino H, Kitano M, Mori Y et al. (2005a) Degradation of antifouling booster biocides in water. J Mar Biol Assoc UK 85:33–38

Harino H, Mori Y, Yamaguchi Y et al. (2005b) Monitoring of antifouling booster biocides in water and sediment from the port of Osaka, Japan. Arch Environ Contam Toxicol 48:303–310

Harino H, Midorikawa S, Arai T et al. (2006) Concentrations of booster biocides in sediment and clams from Vietnam. J Mar Biol Assoc UK 86:1163–1170

Harino H, Yamamoto Y, Eguchi S et al. (2007) Concentrations of antifouling biocides in sediment and mussel samples collected from Otsuchi bay, Japan. Arch Environ Contam Toxicol 52:179–188

Hjorth M, Dahllöf I and Forbes VE (2006) Effects on the function of three trophic levels in marine plankton communities under stress from the antifouling compound zinc pyrithione. Aquat Toxicol 77:105–115

Kakuno A and Kimura S (1987) Acute toxicity of bis(tributyltin) oxide to girella (*Girella punctata*) (in Japanese with English abstract). Bull Tokai Reg Fish Res Lab 123:41–44

Karlsson J and Eklund B (2004) New biocide-free anti-fouling paints are toxic. Mar Poll Bull 49:456–464

Konstantinou IK and Albanis TA (2004) Worldwide occurrence and effects of antifouling paint booster biocides in the aquatic environment: a review. Environ Int 30:235–248

Koutsaftis A and Aoyama I (2006) The interactive effects of binary mixtures of three antifouling biocides and three heavy metals against the marine algae *Chaetoceros gracilis*. Environ Toxicol 21:432–439

Koutsaftis A and Aoyama I (2007) Toxicity of four antifouling biocides and their mixtures on the brine shrimp *Artemia salina*. Sci Total Environ 387:166–174

Maguire RJ, Carey JH and Hale EJ (1983) Degradation of the Tri-*n*-butyltin species in water. J Agric Food Chem 31:1060–1065

Maraldo K and Dahllöf I (2004) Indirect estimation of degradation time for zinc pyrithione and copper pyrithione in seawater. Mar Poll Bull 48:894–901

Mishra VK, Srivastava MK and Raizada RB (1993) Testicular toxicity of thiram in rat: morphological and biochemical evaluations. Ind Health 31:59–67

Mochida K, Ito K, Harino H et al. (2006) Acute toxicity of pyrithione antifouling biocides and joint toxicity with copper to red sea bream (*Pagrus major*) and toy shrimp (*Heptacarpus futilirostris*). Environ Toxicol Chem 25:3058–3064

Mochida K, Ito K, Harino H et al. (2008) Early life-stage toxicity test for copper pyrithione and induction of skeletal anomaly in a teleost, the mummichog (*Fundulus heteroclitus*). Environ Toxicol Chem 27:367–374

Nakajima K and Yasuda T (1990) High-performance liquid chromatographic determination of zinc pyrithione in antidandruff preparations based on copper chelate formation. J Chromatogr 502:379–384

Nishiuchi Y (1977) Toxicity of pesticides to some freshwater fish (in Japanese). Seikatsu to Kankyo 22:28–31

Nishiuchi Y (1979) Toxicity of pesticides to some fresh water organisms (in Japanese). SUISANZOSHOKU 27:36–41

Okamura H and Mieno H (2006) Present status of antifouling systems in Japan: tributyltin substitutes in Japan Antifouling Paint Biocides. In: Konstantinou I (ed) The Handbook of Environmental Chemistry Vol 5 Springer, Berlin, Heidelberg 5:201–212

Okamura H, Aoyama I, Liu D et al. (2000) Fate and ecotoxicity of the new antifouling compound Irgarol 1051 in the aquatic environment. Water Res 34:3523–3530

Okamura H, Watanabe T, Aoyama I et al. (2002) Toxicity evaluation of new antifouling compounds using suspension-cultured fish cells. Chemosphere 46:945–951

Onduka T, Mochida K, Ito K et al. (2007) Acute toxicity of pyrithione photodegradation products to some marine organisms. International Symposium on Shipbuilding Technology Proceeding 99–106

Sánchez-Bayo F and Goka K (2006) Influence of light in acute toxicity bioassay of imidacloprid and zinc pyrithione to zooplankton crustaceans. Aquat Toxicol 78:262–271

Shade WD, Hurt SS, Jacobson AH et al. (1993) Ecological risk assessment of a novel marine antifoulant. In: Gorsuch, JW, Dwyer, FJ, Ingersoll, CG, La Point, TW (eds.) Environmental Toxicology and Risk Assessment 2:381–408. American Society for Testing and Materials, Philadelphia

Sherrard RM, Murray-Gulde CL, Rodgers JHA (2002) Comparative toxicity of chlorothalonil and chlorpyrifos: *Ceriodaphnia dubia* and *Pimephales promelas*. Environ Toxicol 17:503–512

Stoker TE, Jeffay SC, Zucker RM et al. (2003) Abnormal fertilization is responsible for reduced fecundity following thiram-induced ovulatory delay in the rat. Biol Reprod 68:2142–2149

Teraoka H, Urakawa S, Nanba S et al. (2006) Muscular contractions in the zebrafish embryo are necessary to reveal thiuram-induced notochord distortions. Toxicol Appl Pharmacol 212:24–34

Thomas KV, Blake SJ and Waldock MJ (2000) Antifouling paint booster biocide contamination in UK marine sediments. Mar poll Bull 40:739–745

Thomas KV, McHugh M and Waldock M (2002) Antifouling paint booster biocides in UK coastal waters: inputs, occurrence and environmental fate. Sci Total Environ 293:117–127

Tilton F, La Du JK, Vue M et al. (2006) Dithiocarbamates have a common toxic effect on zebrafish body axis formation. Toxicol Appl Pharmacol 216:55–68

Turley PA, Fenn RJ and Ritter JC (2000) Pyrithiones as antifoulants: environmental chemistry and preliminary risk assessment. Biofouling 15:175–182

Turley PA, Fenn RJ, Ritter JC et al. (2005) Pyrithiones as antifoulants: environmental fate and loss of toxicity. Biofouling 21:31–40

Van Leeuwen CJ, Maas-Diepeveen JL, Niebeek G et al. (1985a) Aquatic toxicological aspects of dithiocarbamates and related compounds. I. Short-term toxicity tests. Aquat Toxicol 7:145–164

Van Leeuwen CJ, Moberts F and Niebeek G (1985b) Aquatic toxicological aspects of dithiocarbamates and related compounds. II. Effects on survival, reproduction and growth of *Daphnia magna*. Aquat Toxicol 7:165–175

Van Leeuwen CJ, Espeldoorn A and Mol F (1986a) Aquatic toxicological aspects of dithiocarbamates and related compounds. III. Embryolarval studies with rainbow trout (*Salmo gairdneri*). Aquat Toxicol 9:129–145

Van Leeuwen CJ, Helder T and Seinen W (1986b) Aquatic toxicological aspects of dithiocarbamates and related compounds. IV. Teratogenicity and histopathology in rainbow trout (*Salmo gairdneri*). Aquat Toxicol 9:147–159

Voulvoulis N, Scrimshaw MD and Lester JN (2000) Occurrence of four biocides utilized in antifouling paints, as alternatives to organotin compounds, in waters and sediments of a commercial estuary in the UK. Mar Poll Bull 40: 938–946

Chapter 22
Further Effects of Alternative Biocides on Aquatic Organisms

Kazuhiko Mochida and Kazunori Fujii

22.1 Introduction

In Chapter 21, we used publicly available data to review progress in research on the toxicity of several of the alternative biocides on the list (ref. Table 21.1) of those registered and approved by the Japan Paint Manufacturers' Association, especially in plankton species and teleost fish. In the first section of this chapter, we review the toxicity of several of these listed alternative biocides, particularly to marine organisms other than plankton and teleost fish. In the next section, the discussion of sublethal effects of the alternative biocides is divided into two subsections. In the first subsection, teratogenicity to fish, which is exhibited by dithiocarbamates (DTCs) as well as by zinc pyrithione (ZnPT) and copper pyrithione (CuPT) and related products is the main feature reviewed. In the second subsection, the toxic effects of herbicides such as Irgarol 1051 and diuron on corals are reviewed. Additionally, in this subsection, the physiological changes reported in some fish species exposed to these biocides are discussed.

22.2 Toxicity to Aquatic Organisms Other than Plankton and Fish

Toxicity values of the alternative biocides in several marine organisms are summarized in Table 22.1. Many toxicity data indicate that herbicide compounds such as Irgarol 1051 and diuron to exhibit markedly stronger toxicity to plant species than to animal species such as mussels, some crustacean species, or sea urchins. Whereas

K. Mochida and K. Fujii
National Research Institute of Fisheries and Environment of Inland Sea, Fisheries Research Agency, 2-17-5 Maruishi, Hatsukaichi, Hiroshima 739-0452, Japan

T. Arai et al. (eds.), *Ecotoxicology of Antifouling Biocides*,
DOI: 10.1007/978-4-431-85709-9_22, © Springer 2009

Table 22.1 Toxic effects of the alternative biocides to various marine organisms other than plankton species and teleost fish

Class	Test organisms	Compound	Toxicity index	Values (μg/l)	Reference
Rhodophyta (Florideophyseae)	*Ceramium tenuicorne* *Ceramium strictum*	ZnPT	7-day EC50 (growth)	3–6	Karlsson and Eklund (2004)
Chromophyta (Phaeophyceae)	*Eisenia bicyclis*	Irgarol	4-day EC50 (gametophyte growth)	5.9 (2.6–7.4)	Okamura et al. (2000b)
			7-day EC50 (non-dividing gametophyte)	2.2 (1.9–2.4)	
			7-day EC50 (female growth)	2.0 (1.5–2.2)	
			7-day EC50 (male growth)	2.1 (1.4–2.7)	
		M1	4-day EC50 (gametophyte growth)	>32	
			7-day EC50 (non-dividing gametophyte)	>32	
			7-day EC50 (female growth)	>32	
			7-day EC50 (male growth)	>32	
	Hormosira banksii	ZnPT	48-h EC50 (germination)	0.21 (0.16–0.28)	Myers et al. (2006)
			48-h EC50 (rhizoid growth)	0.31 (0.21–0.38)	
			72-h EC50 (germination)	0.19 (0.12–0.28)	
			72-h EC50 (rhizoid growth)	0.24 (0.15–0.36)	
	Hormosira banksii	SN	48-h EC50 (germination)	340 (280–440)	
			48-h EC50 (rhizoid growth)	430 (290–650)	
			72-h EC50 (germination)	420 (290–550)	
			72-h EC50 (rhizoid growth)	460 (340–540)	
	Hormosira banksii	Diuron	48-h EC50 (germination)	6,290 (5,930–6,830)	
			72-h EC50	6,820 (5,980–7,760)	
			48-h EC50 (rhizoid growth)	6,750 (5,930–7,590)	
			72-h EC50	7,330 (6,350–7,520)	
	Hormosira banksii	Zineb	48-h EC50 (germination)	870 (660–1,890)	
			72-h EC50	490 (300–870)	
			48-h EC50 (rhizoid growth)	2,040 (790–3,580)	
			72-h EC50	1,510 (470–3,870)	
Chlorophyta	*Enteromorpha intestinalis*	Irgarol	NOEC (144 days, growth)	0.022	Scarlett et al. (1997)
(Chlorophyceae)	*Enteromorpha intestinalis*		72-h EC50 (photosynthesis)	2.5	

(Anthopyta)	*Zostera marina*		LOEC (10 days, growth)	10	
	Porphyra yezoensis	Irgarol	4-day LC50 (conchospore survival)	5,000	Okamura et al. (2000b)
			4-day EC50 (conchospore germination)	4.1	
			4-day EC50 (conchospore growth)	0.6	
		M1	4-day LC50 (conchospore survival)	6,500	
			4-day EC50 (conchospore germination)	130	
			4-day EC50 (conchospore growth)	17	
Bivalvia	American oyster embryo (*Crassostrea virginica*)	SN	48-h EC50	6.9 (3.9–11.6)	Shade et al. (1993)
	Bay mussel embryo (*Mytilus edulis*)	SN	48-h EC50 (growth and development)	1.9 (3.9–11.6)	Shade et al. (1993)
			48-h EC10 (embryonic development)	7.1	
			48-h EC50	11	
	Mytilus edulis	ZnPT	48-h EC50 (embryonic development)	2.6	Bellas et al. (2005b)
		Irgarol	48-h EC50 (embryonic development)	1,540	Bellas (2006)
		TPN	48-h EC50 (embryonic development)	8.8	Bellas (2006)
Crustacea	*Artemia salina*	Iragarol	24-h LC50	>40,000	Okamura et al. (2000a)
		M1	24-h LC50	>40,000	
	Common prawn (*Palaemon serratus*)	Diuron	24-h LC50	3,011 (2,805–3,231)	Bellas et al. (2005a)
			48-h LC50	3,044 (2,837–3,265)	
	Amphipod (*Monoporeia affinis*)	ZnPT	EC50 (Avoidance response to sediment)	9.65	Eriksson et al. (2006)
Echinoidea	*Anthocidaris crassipina*	ZnPT	LOEC(32-h, embryonic development)	10^{-9}(1 fg/l)	Kobayashi and Okamura (2002)
		SN		10^{-7}(0.1 pg/l)	
		Irgarol		1,000	
	Paracentrotus lividus	Diuron	48-h EC50 (development)	5,600 (5,400–5,700)	Bellas et al. (2005a)
		ZnPT	48-h EC50 (embryonic development)	2.5	Bellas et al. (2005b)
		SN	48-h EC50 (embryonic development)	12.1	Bellas (2006)
			48-h EC50 (larval growth)	25	

(continued)

Table 22.1 (continued)

Class	Test organisms	Compound	Toxicity index	Values (µg/l)	Reference
		Irgarol	48-h EC50 (embryonic development)	4,021	Bellas (2006)
			48-h EC50 (larval growth)	6,032	
		TPN	48-h EC50 (embryonic development)	6.6	Bellas (2006)
			48-h EC50 (larval growth)	Nd	
		Irgarol	48–50 h EC50 (embryotoxicity)	990	Manzo et al. (2006)
			48–50 h EC50 (spermiotoxicity)	9,040	
		Diuron	48–50 h EC50 (embryotoxicity)	2,390	
			48–50 h EC50 (spermiotoxicity)	5,090	
Ascidiacea	*Ciona intestinalis*	ZnPT	24-h EC50 (larval morphological abnormalities)	72–187	Bellas (2005)
			48-h EC50 (larval settlement)	34	
		Diuron	20-h EC50 (hatchability)	24,397 (21,959–27,539)	Bellas et al. (2005a)
		SN	24-h EC50 (embryonic development)	105	Bellas (2006)
			48-h EC50 (larval settlement)	43	
		Irgarol	24-h EC50 (embryonic development)	2,115	
			48-h EC50 (larval settlement)	>6,486	
		TPN	24-h EC50 (embryonic development)	33	
			48-h EC50 (larval settlement)	42	

The 95% confidence intervals are given in parentheses.
ZnPT, zinc pyrithione; M1, 2-methylthio-4-tert-butylamino-6-amino-s-triazine; SN, Sea-Nine 211; Zineb, zinc ethylenbis (dithiocarbamate); TPN, chlorothalonil; nd, not determined.

the concentrations toxic to the growth of plant species range from approximately 1–10 μg/l, the concentrations lethal to animal species are, in most cases, more than 1 mg/l. Along with these herbicides, ZnPT also shows strong toxicity to the germination and rhizoid growth of a macroalgal species, *Hormosira banksii* (Myers et al. 2006). In addition, ZnPT and Sea-Nine 211 are extremely toxic to the embryonic development of the sea urchin *Anthocidaris crassispina* (Kobayashi and Okamura 2002). The lowest observed effect concentrations ranged from 1 to 100 fg/l. It may be necessary to elucidate whether the exceptionally high susceptibility of *A. crassispina* to these biocides is a species-specific phenomenon.

22.3 Sublethal Effects of Alternative Biocides

22.3.1 Teratogenicity

Van Leeuwen et al. (1986a) demonstrated that DTCs such as zinc dimethyldithiocarbamate (ziram), zinc ethylene-bis-dithiocarbamate (zineb), and tetramethylthiuram disulfide (thiram) have teratogenic properties. They exposed rainbow trout (*Oncorhynchus mykiss*) in embryolarval stages continuously for 60 days to these compounds and confirmed the induction of spinal and vertebral abnormalities characterized by scoliosis, lordosis, kyphosis, and irregular shrinkage of the trunk. Histological observation revealed that the vertebral deformity was due to disturbance of the notochord, and that the irregular arrangement of myomeres and myosepta was closely related to the notochordal anomalies (Van Leeuwen et al. 1986b). Extensive brain hemorrhages were also observed in fish exposed to ziram or thiram (Van Leeuwen et al. 1986b). The 60-day EC50 of ziram, zineb, and thiram for teratogenicity were 1.5, 188, and 0.64 μg/l, respectively. Thiram was the most toxic of the three: its lowest observed effect concentration (LOEC) for teratogenicity was <0.32 μg/l (Van Leeuwen et al. 1986a). Thiram is known to cause wavy distortions of the notochord in zebrafish (*Danio rerio*): exposure to thiram at 0.48–240 μg/l from 3 h post fertilization (hpf) until 24 hpf induced malformations such as wavy notochords and disorganized somites in all embryos, and the EC50 value was 1.7 μg/l (Teraoka et al. 2006). Teraoka et al. (2006) also investigated the mechanisms of the toxic effect of thiram in zebrafish embryos. Thiram-induced distortion of the notochord started at 18 hpf and was accompanied by spontaneous trunk contractions. Inhibition of spontaneous muscle contractions by using tricaine (a common fish anesthetic) and α-bungarotoxin (an inhibitor of the acetylcholine receptor) in the presence of thiram failed to induce notochord abnormalities (Teraoka et al. 2006). These results indicated that muscle activity was necessary to cause the malformations.

As an example of the teratogenic effects of the alternative biocides, ZnPT has been reported to induce skeletal deformity in zebrafish (Goka 1999) and Japanese medaka (*Oryzias latipes*) (Sánchez-Bayo and Goka 2005). In Japanese medaka, 35% of larvae exposed for 20 days to ZnPT at 3 μg/l had "wavy" vertebral columns

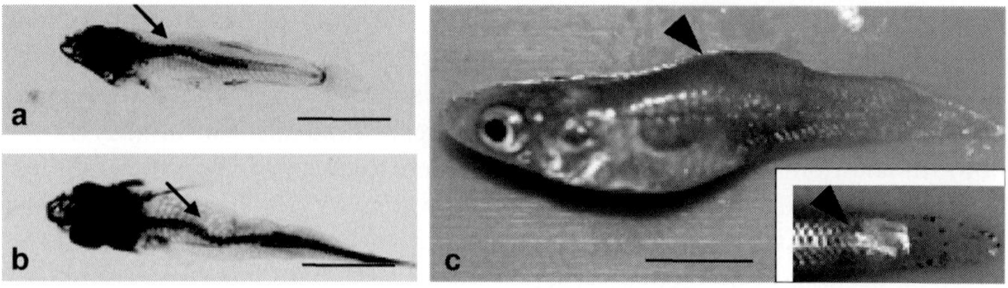

Fig. 22.1 Skeletal deformity (**a** and **b**) and inflammatory mass formation (*arrowheads* in **c**) in mummichog (*Fundulus heteroclitus*) larvae exposed to copper pyrithione. (**a**) Lateral and (**b**) dorsal views of fish with vertebral deformity. Inflammatory mass around tail region (*inset* in **c**). Arrows in (**a**) and (**b**) indicate regions of scoliosis and lordosis, respectively. Bars, 5 mm

(Goka 1999). A recent study also revealed that long-term exposure to CuPT for 50 days from embryo to larva can induce skeletal deformities and inflammatory mass formation in mummichog larvae (*Fundulus heteroclitus*) (Mochida et al. 2008) (Fig. 22.1). The frequency of occurrence of these abnormalities was higher in the groups exposed to 2 or 4 μg/l CuPT (nominal concentration) than in other groups. In the larvae of these groups, acetylcholinesterase (AChE) activity was also significantly reduced. These results suggest that the mechanism of induction of skeletal anomalies is similar to that proposed for the induction of skeletal abnormalities by organophosphorus insecticides (Meiniel 1981). In this mechanism, CuPT is thought to have neuromuscular blocking properties through its ability to inhibit AChE activity, thus causing muscular contraction that can lead to abnormal axial contortion. In addition, fish spine is particularly sensitive to mechanical forces of muscular origin (Couch et al. 1977). It is possible that these factors combine to cause the observed skeletal deformity. Electron microscopic studies suggest that skeletal muscle dysfunction is involved in the skeletal deformity induced in the mummichog after exposure to CuPT, because indicators of dysfunction, such as fibrosis and necrosis of muscle fibers, as well as swelling of the sarcoplasmic reticulum in a specific region of the lateral muscles of the abdomen, were observed (Mochida et al. 2008). These histologic observations support the notion that these anomalies form through mechanisms similar to those proposed for organophosphorus pesticide exposure (Meiniel 1981).

The inflammatory masses induced in mummichog larvae by long-term exposure to CuPT are composed mainly of infiltrating cells, such as macrophages (Fig. 22.2) and necrotic myocytes. CuPT likely first induces dysfunction of the lateral muscles through the abovementioned mechanisms, after which the myocytes undergo fibrosis and necrosis. Macrophages may infiltrate these regions to phagocytose the necrotic cells, thus forming the inflammatory masses (Mochida et al. 2008). Incidentally, exposure of rainbow trout to the DTCs such as ziram, zineb, and thiram induces formation of large nodules up to 2 mm in diameter through the muscular and integument layers (Van Leeuwen et al. 1986b). Although results of histological analyses of these nodules have not been reported, it is possible that they, too, are inflammatory masses.

Anderson et al. (2007) demonstrated that treatment with one of the metabolites of pyrithiones, 2-mercaptopyridine-*N*-oxide, at a concentration of 25 μg/l

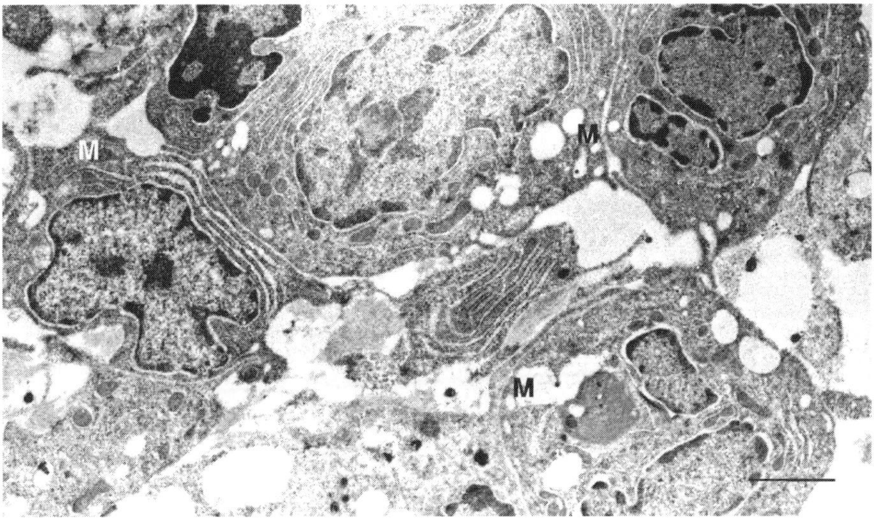

Fig. 22.2 Macrophages gathering in the region of the inflammatory mass in mummichog (*Fundulus heteroclitus*) larvae exposed to copper pyrithione. M = macrophage. Bar, 2 μm

could induce notochord malformation in zebrafish. There is a stage sensitive to the induction of notochord defects: treatment with the compound during the early (0.75–6 hpf) or late (>24 hpf) stages did not affect notochord formation, but treatment from the early sensitive phase at 12 hpf could induce defects in the posterior trunk and throughout the tail. In addition, it has been suggested that 2-mercaptopyridine-*N*-oxide inhibits lysyl oxidase, an enzyme that plays a role in the organization of collagen fibers, and thus causes malformation of the notochord (Anderson et al. 2007).

22.3.2 Other Toxic Effects

The toxic effects of Irgarol 1051 and diuron in coral are worth mentioning. The toxic mechanisms of these herbicides in coral are well documented in a review by Jones (2005). Scleractinian corals establish mutualistic symbiosis with the dinoflagellate algae *Symbiodinium*. Corals utilize the photosynthetic products of dinoflagellates, such as sugars and amino acids, as energy sources for respiration and growth (Jones 2005). The most susceptible parameter of toxicity in corals is a reduction in the photosynthetic efficiency of the dinoflagellate alga. Photosystem II (PS II), which is a membrane-protein complex, plays an important role in getting energy sources for photosynthesis through photochemical reactions. More than half of the commercially available herbicides, such as Irgarol 1051 and diuron, can reversibly bind to PS II and reduce the efficiency of photosynthesis (Jones 2005). Toxicity testing under symbiotic conditions (with the dinoflagellates still within the tissues of the coral) revealed that the 10-h EC50 values for a reduction of photosynthetic efficiency in the staghorn coral *Acropora formosa* were 1.3 μg/l for Irgarol 1051 and 2.8 μg/l for diuron. Similar results were obtained with bird's nest coral *Seriatopora hystrix* (Jones and Kerswell 2003; Jones et al. 2003). The LOEC for a reduction in efficiency of the algal symbionts were 50 ng/l for Irgarol 1051 and 200 ng/l for diuron (Jones and Kerswell 2003). Another characteristic of PS II herbicides such as Irgarol 1051 and diuron is that they have a very rapid effect on the algal symbionts. Exposure to diuron at 12 μg/l can reduce the photochemical efficiency of symbiotic dinoflagellates isolated from *Stylophora pistillata* more than 50% within 30 s (Jones et al. 2003). The same reduction was also observed within 15 min after exposure of *S. hystrix* to 3 μg/l diuron or Irgarol 1051 (Jones and Kerswell 2003). Irgarol 1051 and diuron have generally been reported to exhibit low toxicity to the coral itself. Irgarol 1051 showed no effects on fertilization in the corals *Acropora millepora* and *Montipora aequituberculata* at 1,000 μg/l (Negri et al. 2005). Metamorphosis of symbiont-free *A. millepora* larvae was significantly affected only at 300 μg/l diuron, and that of *Pocillopora damicornis* in the presence of symbiotic dinoflagellates was virtually not affected at a diuron concentration of 1,000 μg/l (Negri et al. 2005). In light of the known presence of Irgarol 1051 in coastal waters and the concentrations at which this chemical can inhibit the photosynthesis of corals symbionts (Owen et al. 2002; Carbery et al. 2006), the importation and use of antifouling paint containing Irgarol 1051 have already been banned in Bermuda because of its acute toxicity to corals; this ban also extends to diuron (Carbery et al. 2006).

A couple of studies have investigated the physiological responses of some fish species after exposure to these alternative biocides. In the liver of the freshwater teleost *Sarotherodon mossambicus* exposed to zinc dimethyldithiocarbamate at 8 μg/l for 72 h, a significant decrease in glycogen levels and an increase in lactic acid level were observed (Thangavel et al. 2004). These physiological changes are thought to be caused by the anaerobic breakdown of stored glycogen to obtain essential energy under pollutant stress. Exposure of rainbow trout to thiram at 180 μg/l for 24 h induced an increase in the lipid content of the liver and a decrease in blood hemoglobin level (Van Leeuwen et al. 1986c). These events may have been due to fatty degeneration of hepatocytes and heme degradation. In addition, osmoregulatory function was also disturbed in thiram-treated rainbow trout (Van Leeuwen et al. 1986c).

22.4 Conclusions and Outlook

As mentioned above, the alternative biocides have various toxic effects on marine organisms other than plankton and fish. Monitoring studies have detected many of these alternative biocides in both sediments and coastal waters, as described in a review article by Konstantinou and Albanis (2004), but CuPT has so far been detected only in sediments (Harino et al. 2006, 2007). The accumulated data indicate that Sea-Nine 211 and ZnPT, especially, show relatively strong toxicity to benthic organisms such as bay mussel embryos *M. edulis* and embryos of the sea urchin *A. crassispina* (Table 22.1). Additionally, some species of sea urchin are commercially important fisheries resources in Japan. Surveillance of alternative biocides in coastal environments and ecotoxicological studies of benthic organisms have to be carried out continuously.

The occurrence of the dithiocarbamates in coastal environments has not been reported until now, although the compounds exhibit teratogenicity to fish embryos at relatively low concentrations: the 60-day EC50 of thiram in rainbow trout embryos was 0.64 μg/l (Van Leeuwen et al. 1986a). Zineb and thiram have reproductive toxicity in mammals, as shown by the degeneration of seminiferous tubules (Mishra et al. 1993), inhibition of fertilization (Stoker et al. 2003; Rossi et al. 2006), and also inhibition of the action of steroid metabolic enzymes (Atanasov et al. 2003). As mentioned in the previous chapter, so far there are few available data on the toxicity of dithiocarbamates to marine organisms. We need to accumulate more data on the toxic effects of these compounds – especially on reproduction in marine organisms – and to monitor their occurrence.

Acknowledgement We are grateful to the members of the Ecotoxicology section of the National Research Institute of Fisheries and Environment of Inland Sea, Fisheries Research Agency, for their kind assistance. This work was supported in part by a grant-in-aid from the Japanese Ministry of the Environment (Project: Environmental Risk Assessment of Antifouling Agents).

References

Anderson C, Bartlett SJ, Gansner JM et al. (2007) Chemical genetics suggests a critical role for lysyl oxidase in zebrafish notochord morphogenesis. Mol Biosyst 3:51–59

Atanasov AG, Tam S, Röcken JM et al. (2003) Inhibition of 11β-hydroxysteroid dehydrogenase type 2 by dithiocarbamates. Biochem Biophys Res Commun 308:257–262

Bellas J (2005) Toxicity assessment of the antifouling compound zinc pyrithione using early developmental stages of the ascidian *Ciona intestinalis*. Biofouling 21:289–296

Bellas J (2006) Comparative toxicity of alternative antifouling biocides on embryos and larvae of marine invertebrates. Sci Total Environ 367:573585

Bellas J, Beiras R, Marinõ-Balsa JC et al. (2005a) Toxicity of organic compounds to marine invertebrate embryos and larvae: a comparison between the sea urchin embryogenesis bioassay and alternative test species. Ecotoxicology 14:337353

Bellas J, Granmo Å, Beiras R (2005b) Embryotoxicity of the antifouling biocide zinc pyrithione to sea urchin (*Paracentrotus lividus*) and mussel (*Mytilus edulis*). Mar Poll Bull 50:1382–1385

Carbery K, Owen R, Frickers T et al. (2006) Contamination of Caribbean coastal waters by the antifouling herebicide. Irgarol 1051 Mar Poll Bull 52:635–644

Couch JA, Winstead JT and Goodman LR (1977) Kepone-induced scoliosis and its histological consequences in fish. Science 197:585–587

Eriksson Wiklund AK, Börjesson T, Wiklund SJ (2006) Avoidance response of sediment living amphipods to zinc pyrithione as a measure of sediment toxicity. Mar Poll Bull 52:9699

Goka K (1999) Embryotoxicity of zinc pyrithione, an antidandruff chemical, in fish. Environ Res A 81:81–83

Harino H, Midorikawa S, Arai T et al. (2006) Concentrations of booster biocides in sediment and clams from Vietnam. J Mar Biol Assoc UK 86:1163–1170

Harino H, Yamamoto Y, Eguchi S et al. (2007) Concentrations of antifouling biocides in sediment and mussel samples collected from Otsuchi Bay, Japan. Arch Environ Contam Toxicol 52:179–188

Jones R (2005) The ecotoxicological effects of photosystem II herbicides on corals. Mar Poll Bull 51:495–506

Jones RJ, Kerswell AP (2003) Phytotoxicity of photosystem II (PS II) herbicides to coral. Mar Ecol-Prog Ser 261:149–159

Jones RJ, Muller J, Haynes D et al. (2003) Effects of herbicides diuron and atrazine on corals of the Great Barrier Reef, Australia. Mar Ecol-Prog Ser 251:153–167

Karlsson J, Eklund B (2004) New biocide-free anti-fouling paints are toxic. Mar Pollut Bull 49:456–464

Kobayashi N, Okamura H (2002) Effects of new antifouling compounds on the development of sea urchin. Mar Pollut Bull 44:748–751

Konstantinou IK and Albanis TA (2004) Worldwide occurrence and effects of antifouling paint booster biocides in the aquatic environment: a review. Environ Int 30:235–248

Manzo S, Buono S, Cremisini C (2006) Toxic effects of Irgarol 1051 and diuron on sea urchin *Paracentrotus lividus* early development, fertilization, and offspring quality. Arch Environ Contam Toxicol 51:61–68

Meiniel R (1981) Neuromuscular blocking agents and axial teratogenesis in the avian embryo. Can axial morphogenetic disorders be explained by pharmacological action upon muscle tissue? Teratology 23:259–271

Mishra VK, Srivastava MK, Raizada RB (1993) Testicular toxicity of thiram in rat: morphological and biochemical evaluations. Ind Health 31:59–67

Myers JH, Gunthorpe L, Allinson G et al. (2006) Effects of antifouling biocides to the germination and growth of the marine macroalga, *Hormosira banksii* (Turner) Desicaine. Mar Pollut Bull 52:1048–1055

Mochida K, Ito K, Harino H et al. (2008) Early life-stage toxicity test for copper pyrithione and induction of skeletal anomaly in a teleost, the mummichog (*Fundulus heteroclitus*). Environ Toxicol Chem 27:367–374

Negri A, Vollhardt C, Humphrey C et al. (2005) Effects of the herbicide diuron on the early life history stages of coral. Mar Pollut Bull 51:370–383

Okamura H, Aoyama I, Liu D et al. (2000a) Fate and ecotoxicity of the new antifouling compound Irgarol 1051 in the aquatic environment. Water Res 34:3523–3530

Okamura H, Aoyama I, Takami T et al. (2000b) Phytotoxicity of the new antifouling compound Irgarol 1051 and a major degradation product. Mar Pollut Bull 40:754–763

Owen R, Knap A, Toaspern M et al. (2002) Inhibition of coral photoshynthesis by the antifouling herbicide Irgarol 1051. Mar Pollut Bull 44:623–632

Rossi G, Palmerini MG, Macchiarelli G et al. (2006) Mancozeb adversely affects meiotic spindle organization and fertilization in mouse oocytes. Reprod Toxicol 22:51–55

Sánchez-Bayo F, Goka K (2005) Unexpected effects of zinc pyrithione and imidacloprid on Japanese medaka fish (*Oryzias latipes*). Aquat Toxicol 74:285–293

Scarlett A, Donkin ME, Fileman TW et al. (1997) Occurrence of the marine antifouling agent Irgarol 1051 within the Plymouth sound locality: implications for the green macroalga *Enteromorpha intestinalis*. Mar Pollut Bull 34:645–651

Shade WD, Hurt SS, Jacobson AH et al. (1993) Ecological risk assessment of a novel marine antifoulant. Gorsuch, J. W., Dwyer, F. J., Ingersoll, C. G., La Point, T. W. (eds.) Environmental toxicology and risk assessment 2:381–408. American Society for Testing and Materials, Philadelphia

Stoker TE, Jeffay SC, Zucker RM et al. (2003) Abnormal fertilization is responsible for reduced fecundity following thiram-induced ovulatory delay in the rat. Biol Reprod 68:2142–2149

Teraoka H, Urakawa S, Nanba S et al. (2006) Muscular contractions in the zebrafish embryo are necessary to reveal thiuram-induced notochord distortions. Toxicol Appl Pharmacol 212:24–34

Thangavel P, Nivedhitha H, Ramaswamy M (2004) Comparative study on individual and combined effects of dimecron and ziram on carbohydrate metabolites in liver, muscle, heart, and blood of a freshwater teleost, *Sarotherodon mossambicus* (Peters). Bull Environ Contam Toxicol 72:365–372

Van Leeuwen CJ, Espeldoorn A, Mol F (1986a) Aquatic toxicological aspects of dithiocarbamates and related compounds. III. Embryolarval studies with rainbow trout (*Salmo gairdneri*). Aquat Toxicol 9:129–145

Van Leeuwen CJ, Helder T, Seinen W (1986b) Aquatic toxicological aspects of dithiocarbamates and related compounds. IV. Teratogenicity and histophathology in rainbow trout (*Salmo gairdneri*). Aquat Toxicol 9:147–159

Van Leeuwen CJ, Maas-Diepeveen JL, Overbeek HCM (1986c) Sublethal effects of tetramethylthiuram disulfide (thiram) in rainbow trout (*Salmo gairdneri*). Aquat Toxicol 9:13–19

Section 9
Characteristics of Alternative Biocides in Aquatic Environments

Chapter 23
Degradation of Alternative Biocides in the Aquatic Environment

Hiroya Harino and William John Langston

Abbreviations Chlorothalonil: 2,4,5,6-Tetrachloroisophthalonitrile Cas no. 1897-45-6; $CuPT_2$: Copper pyrithione, Copper, bis(1,hydroxyl-2(1H)-pyridinethionato O,S) Cas no. 14915-37-8; Dichlofluanid: *N*-dichlorofluoromethylthio-N',N'-dimethyl-N-phenylsulfamide Cas no. 1058-98-9; Diuron: 3-(3,4-Dichlorophenyl)-1,1-dimetyl urea Cas no. 330-54-1; Irgarol 1051: 2-Methylthio-4-tert-butylamino-6-cyclopropylamino-s-triazine Cas no. 28159-98-0; Maneb: Manganese ethylene bisdithiocarbamate Cas no.12427-38-2; M1: 2-methylthio-4-tert-butylamino-6-amino-s-triazine; Sea Nine 211: 4,5-Dichloro-2-n-octyl-4-isothiazolin-3-one Cas no. 64359-81-5; Thiram: bis(dimethylthiocarbamoyl) disulphide Cas no. 137-26-8; TPBP: Triphenylborane-pyridine, Cas no.971-66-4; Zineb: Zinc ethylene bisdithiocarbamate Cas no. 12122-67-7; Ziram: Zinc bis(dimethyl thiocarbamate) Cas no. 137-30-4; $ZnPT_2$: Zinc pyrithione Zinc-2-pyridinethiol-1-oxide, Cas no. 13463-41-7

23.1 Introduction

The ideal antifouling biocide, from a marine conservation perspective, should be degraded to compounds of lower toxicity in the environment to avoid impact on non-targed organisms. On the other hand, an effective biocide needs to have high toxicity to prevent fouling. These two suppositions are, to an extent, mutually exclusive. Nevertheless, some antifouling biocides released to the marine environment undergo hydrolysis, whilst others are degraded by sunlight in the photic zone.

H. Harino
Osaka City Institute of Public Health and Environmental Sciences, 8-34 Tojo-cho, Tennoji, Osaka 543-0026, Japan

W.J. Langston
Marine Biological Association, Citadel Hill, Plymouth PL1 2PB, UK

T. Arai et al. (eds.), *Ecotoxicology of Antifouling Biocides*,
DOI: 10.1007/978-4-431-85709-9_23, © Springer 2009

Furthermore, a number of antifouling biocides are degraded by the many species of bacteria which inhabit water and sediment. Stable antifouling biocides are transported widely and can accumulate in sediment or are concentrated in aquatic organisms. Thus, in order to reduce the threat of magnification of residues, an ideal biocide should be degraded easily and rapidly to substances of lower toxicity, following their released into the aquatic environment. Information about their possible degradation mechanisms in the environment, whether by hydrolysis, sunlight or bacteria, is an important requirement in order to estimate the persistence of these compounds and to identify the factors that influence their behaviour. In this chapter, the degradation pathways and rated of representative alternative biocides in aquatic environment are reviewed.

23.2 Sea-Nine 211

There are limited data on hydrolysis of Sea-Nine 211. A review by Thomas (2001) concluded that, based on experiments with ^{14}C Sea-Nine 211 was relatively stable (half-life >720 h) in sterile water at pH 7. However, degradation occurred in alkaline and conditions: half-lives of Sea Nine 211 at pH 5 and 9 were 216 and 288 h, respectively.

Harino et al. (2005) tested the degradation of Sea-Nine 211 following irradiation with UV and sunlight. Sea-Nine 211 was degraded slowly by of sunlight and the concentrations of Sea-Nine 211 decreased to one-tenth of the initial concentration (0.1 mg l^{-1}) after 17 days of irradiation (Fig. 23.1). The half-life of Sea-Nine 211 subjected to UV treatment was <1 day. Shade et al. (1993) evaluated the rates of photo degradation in an aqueous solution at pH7. The half-lives of Sea Nine 211 were 322 h in the presence of sunlight and 1,913 h in the dark, respectively. Sakkas et al. (2002a) investigated the photochemical behaviour of the Sea-Nine 211 in a range of natural waters and reported half-lives form 131–433 h, with phototransfortilled water. The half-lives of Sea-Nine 211 exposed to simulated solar irradiation (>290 nm) were shorter, ranging from 6 to 14 h. Furthermore, the presence of humic acid and fulvic acids enhanced the rate of the photolysis. Sakkas et al. (2002a)

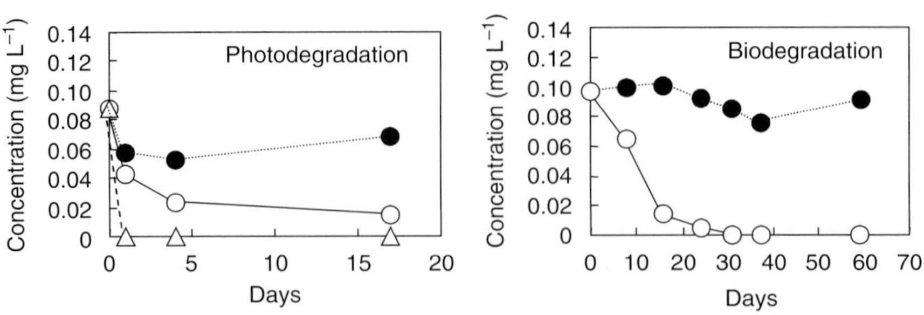

Fig. 23.1 Time course of Sea-Nine 211 concentrations in sea water, following photodegradation (*left*) and biodegradation (*right*). Open circles, sunlight; triangles, UV; black circles, dark (control)

subsequently proposed the dual degradation pathway shown in Fig. 23.2. The first pathway involves cleavage of the isothiazolone ring and subsequent oxidation of the alkyl metabolites, with the final product being n-octyl amine. The second pathway consists of the photo-transposition of Sea-Nine 211. 4,5-Dichloro-3-(n-octyl) thiazolin-2-one is produced by transposition and then N-octanal results from cleavage of the ring. Thomas et al. (2004) also support this degradation pathway.

Harino et al. (2005) have compared the degradation and disappearance of Sea-Nine 211 in sterile and non sterile seawater. Sea-Nine 211 was shown to be degraded easily by bacteria. The time-course for disappearance of the biocide, at an initial concentration of 0.1 mg l^{-1}, is shown in Fig. 23.1. Sea Nine 211 concentrations decreased rapidly during the initial 20 days and the compounds was not detected after 30 days of incubation. Callow and Finlay (1995) reported the efficient removal of Sea-Nine 211 from water by the ship fouling diatom *Amphora coffeaeformis*. Although degradation did not occur in sterile sea water, the half-life of Sea-Nine 211 was estimated to be 8.5 days in natural coastal seawater using the *A. coffeaeformis* bioassay. Jacobson et al. (1993, 2000) investigated the degradation of Sea Nine 211 in both aerobic and anaerobic microcosms consisting of marine sediment and sea water. The half-life of Sea-Nine 211 was <1 h in both aerobic and anaerobic systems. It was concluded from these findings that the influence of photolysis was stronger than that of hydrolysis and biodegradation, as a mechanism for the degradation of Sea-Nine 211. On this evidence, it seems that Sea Nine 211 is likely to be degraded quickly when released from ship hulls to the aquatic environment.

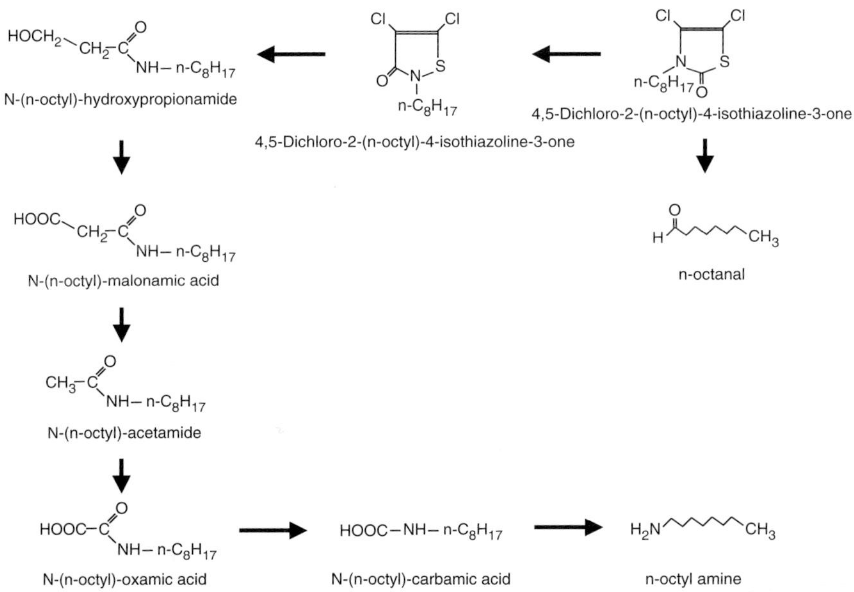

Fig. 23.2 Degradation pathway of Sea Nine 211 (Sakkas et al. 2002a)

23.3 Diuron

Harino et al. (2005) reported that changes in diuron concentrations were relatively small during 17 days of sunlight irradiation, however, degradation occurred within a day. Following irradiation with UV (Fig. 23.3). Tanaka et al. (1986) also reported that diuron was photolyzed by both UV irradiation and by natural sunlight: the six photoproducts identified were essentially the same for the two light sources tested and were 3-(3,4-dichlorophenyl)-1-methylurea, 3-(3,4-dichlorophenyl)-1-formyl-1-methylurea, 3-(chlorophenyl)-1,1-dimethylurea, 3-(3-chloro-4-hydroxyphenyl)-1,1-dimethylurea, 3-(4-chloro-3-hydroxyphenyl)-1,1-dimethylurea, trichloro-bis(N', N'-dimethylureido) biphenyl). Bonnemoy et al. (2004) researched the toxicity of degradation products of diuron arising from UV irradiation, concluding that, the toxicity increased with irradiation time.

Harino et al. (2005) also investigated the degradation of diuron by bacteria, using a 'river-die-away' method. The concentration of diuron (initial concentration of 0.1 mg l^{-1} scarcely changed after 60 days, suggesting that bacteria capable of degrading diuron are scarce in estuarine water. The time-course for 'disappearance' of diuron is shown in Fig. 23.3. Callow and Willingham (1996) monitored the breakdown of diuron by *Amphora coffeaeformis* over 8 weeks, concluding that the compound was not easily biodegraded by these fouling diatoms. Interestingly, Ellis and Camper (1982) reported that photochemical degradation of diuron was of minor importance and that most of the degradation was due to biological activity: over 80% of diuron was degraded by bacteria in seawater within 27 days in their experiments. In aerobic conditions, diuron produced 3-(3,4-dichlorophenyl)-1-methyl urea, 3,4-dichlorophenyl urea and 3,4-dichloroaniline. In anaerobic conditions, diuron was transformed to 3-(4-chlorophenyl)-1,1'-dimethylurea (Fig. 23.4).

Nevertheless, these results indicate that generally, diuron is a fairly stable compound consistent with the relatively high detection frequency seen in the aquatic environment.

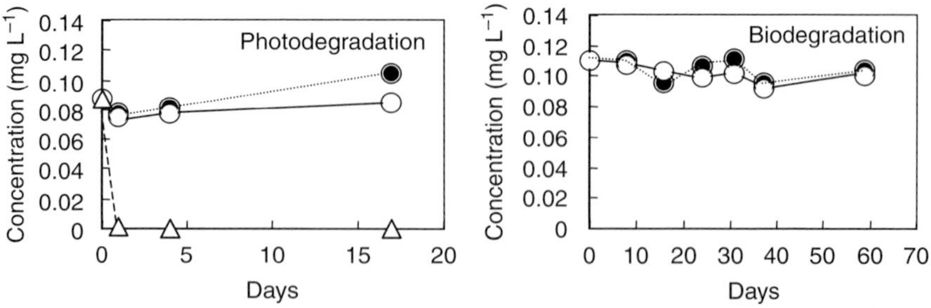

Fig. 23.3 Time course of diuron concentration in sea water following photodegradation (*left*) and biodegradation (*right*). Open circles, sunlight; triangles, UV; black circles, dark (control)

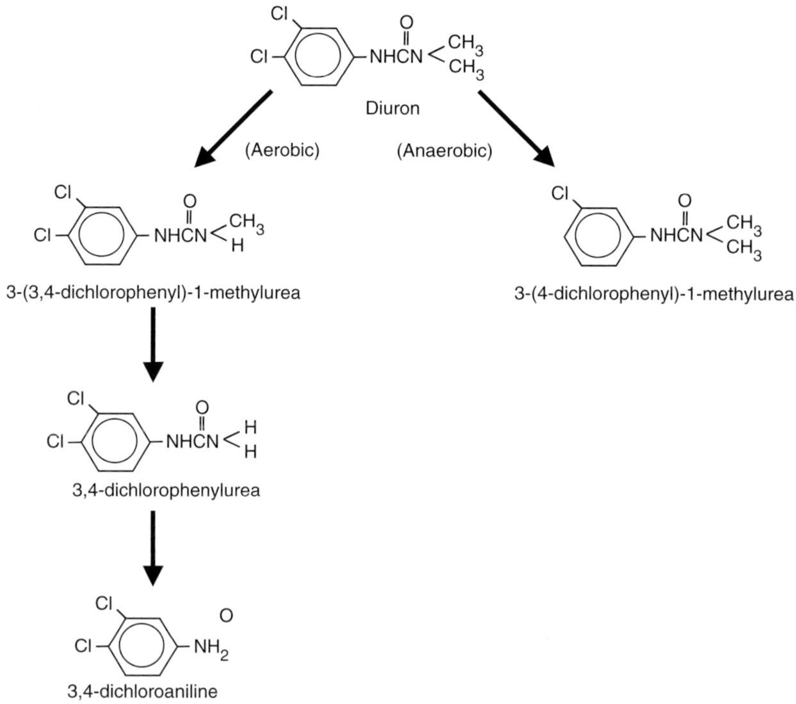

Fig. 23.4 Degradation pathway of diuron (Ellis and Camber 1982)

23.4 Irgarol 1051

Okamura et al. (1999) concluded that Irgarol 1051 is resistant to hydrolysis because no degradation of was observed in pure water, river water or sea water, or solutions buffered at pH 5, 7 and 9 following 1 week experiments at 50°C, in the dark. However, Liu et al. (1999) found that Irgarol 1051 could be hydrolysed to M1 by utilizing Hg as the catalyst (Fig. 23.5). Recently, it was also reported that M2 and M3 were produced under similar conditions (Lam et al. 2004). Furthermore, it has now been shown that these compounds exist in coastal waters (Lam et al. 2005).

Experiments conducted by Harino et al. (2005) have indicated that Irgarol 1051 was stable during 17 days of sunlight irradiation (Fig. 23.6), whilst Hall Jr. et al. (1999) reported that the half-life of Irgarol 1051 in fresh water was 36 days. However, longer half-lives were evident in fresh water with co-solvent (85 days, 0.5% methanol; 148 days, 1% acetonitrile) and in sea water with 1% acetonitrile (273 days). Sakkas et al. (2002b) have confirmed that Irgarol 1051 in seawater may be degraded by solar irradiation (half-life of 59 days) and that the photo-degradation rate increases with the addition of dissolved organic matter such

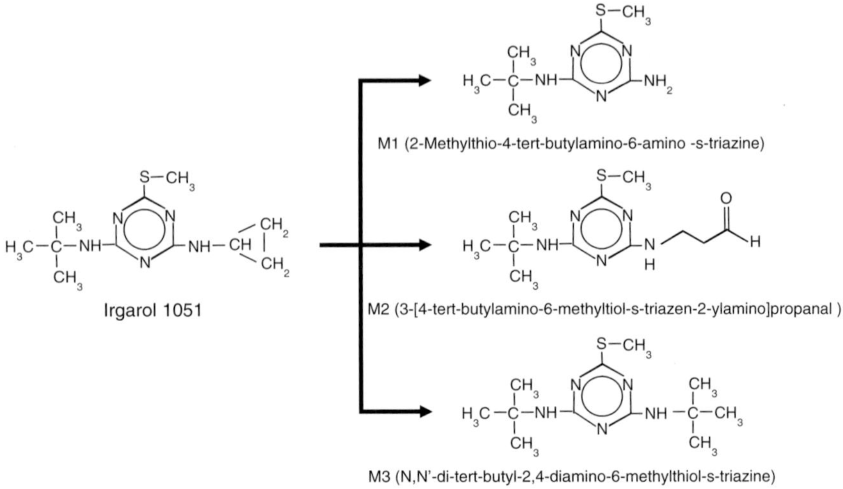

Fig. 23.5 Degradation pathway of Irgarol 1051

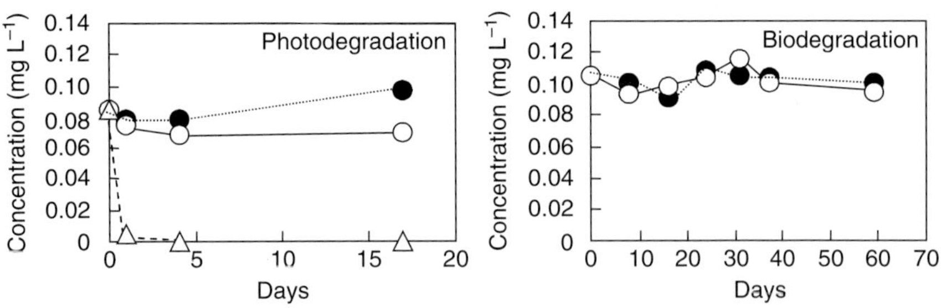

Fig. 23.6 Time course of Irgarol 1051 concentration in photodegradation and biodegradation. Open circle, sunlight; triangle, UV; black circle, dark (control). Sample sea water (Lam et al. 2004)

as humic and fulvic substances. The degradation products of Irgarol 1051 were 2-metylsulfonyl-4-terbutylamino-6-cyclopropylamino-*s*-triazine, 2-hydroxy-4-terbutylamono-6-cyclopropylamino-*s*-striazine, 2-methylthio-4-terbutylamino-6-ethylamino-*s*-triazine, 2-methylsulfonyl-4-terbutylamino-6-amino-*s*-triazine and diaminohydroxy-s-triazine.

According to Liu et al. (1997) the major route of Irgarol 1051 metabolism in the white rot fungus *Phanerochaete chrysosporium* includes N-dealkylation of a cyclopropyl group from the cyclopropylamino side-chain at the six-position of the s-triazine ring, to yield the stable metabolite M1 (2-methylthio-4-tert-butylamino-6-amino-s-triazine). M1 could not be metabolized further by the *P. Chrysosporium* culture and appeared to accumulate as a terminal metabolite. However in biodegradation tests with Irgarol 1051 in Lake Ontario water, no bacterial biodegradation or biotransformation was observed, even after incubation for >5 months (Liu-et al. 1997). The degradation of Irgarol 1051 was investigated by 'river-die-away' method (Harino et al. 2005) and the resulting kinetic of Irgarol 1051 'removal' are shown in Fig. 23.6. Irgarol 1051 concentrations (initial concentration of 0.1 mg l^{-1}) scarcely changed after 60 days. Callow and Willingham (1996) also found little evidence of significant breakdown of Irgarol 1051 over 8 weeks by the diatom (*Amphora coffeaeformis*). Thus, Irgarol 1051 does not appear to be rapidly biodegraded, or readily hydrolysed in water following leaching. Though Irgarol 1051 may be degraded slightly by bacteria and sunlight, it seems likely that the biocide will persist in the aquatic environment.

23.5 $ZnPT_2$ and $CuPT_2$

Pyrithione-containing $ZnPT_2$ and $CuPT_2$ represent a flexible group of biocides. Figure 23.7 shows some of the transformation chemistry of pyrithione (Seymour and Bailey 1981). H-pyrithione, the free form of pyrithione, is an acid with pKa of 4.4. Its salts, Na pyrithione and NH_4 pyrithione, are soluble in water, and methanol. Pyrithione combines with metal ions to form complexes, many of which have no

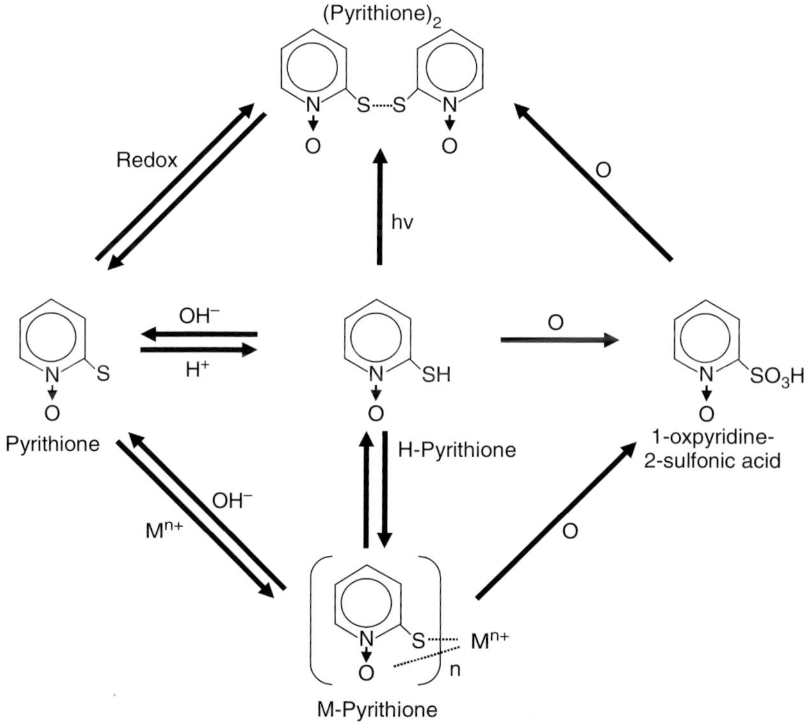

Fig. 23.7 Transformations of pyrithiones in aquatic solution (Seymour and Bailey 1981)

net charge and are insoluble in water. Some of these complexes can be extracted into chloroform and are brightly coloured such as $CuPT_2$. The stability constants of pyrithione complexes are pH-dependent and vary among different metal ions. Pyrithione may release one type of metal ion and combine with another, depending on the type and amount of ions present. Thus, following dissolution of pyrithione, various transformations are possible, depending on the characteristics of the water body. Using radiolabeled compounds, Turley et al. (2000) studied the degradation of $CuPT_2$ and $ZnPT_2$ by hydrolysis in sterilized artificial seawater (pH8) and estimated the half-lives to be 12.9 and >90 days (initial concentration 50 g l^{-1}). Grunnet and Dahllöf (2005) have studied the transchelation of $ZnPT_2$ and $CuPT_2$ and demonstrated complete transchelation of $ZnPT_2$ into $CuPT_2$ when Cu^{2+} is present at an equimolar concentration, in the absence of interfering ligands. Thus, leachates from antifouling paints containing both $ZnPT_2$ and Cu_2O, include $CuPT_2$ in addition to $ZnPT_2$.

Pyrithione, including its salts and its metal complexes are light sensitive (Seymour and Bailey 1981). When H-pyrithione is irradiated with UV light in any one of several organic solvents, it dimerizes. This compound is oxidized to 1-oxpyridine-2-sulfonic acid by reaction with molecular oxygen after further irradiation. Yamaguchi et al. (2004) identified the photodegradation products of $ZnPT_2$ as 2-mercaptopyridine-N-Oxide, Pyridine-N-oxide, 2,2-dithio-bis-pyridine, 2-mercaptopyridine, pyridine-2-sulfonic-acid, 2,2-dithio-bis-pyridine-N-oxide following irradiation with a xenon arc lamp. Using similar conditions Sakkas et al. (2007) estimated the half-life of $ZnPT_2$ by irradiation of xenon arc lamp light to range between 9.2 and 15.1 min and showed that the increasing the concentration of dissolved organic matter accelerates the photolysis reaction. The photodegradation products described were similar to those found by Yamaguchi et al. (2004). Turkey et al. (2000) have also indicated that $CuPT_2$ (50 g l^{-1}) is sensitive to light, with a half-life of 29.1 min under a filtered xenon arc lamp (154 W m^{-2}). The half-life of $ZnPT_2$ in artificial seawater (borosilicate glass vials under a filtered xenon arclamp) was 17.5 min, whilst photolytic degradation was even more effective in natural sunlight (half-life <2 min). Maraldo and Dahllöf (2004) estimated that half-lives by photodegradation were 8.3 ± 0.9 min for $ZnPT_2$ and 7.1 ± 0.2 min for $CuPT_2$, respectively, whilst Thomas (1999) reported that the half-life of $ZnPT_2$ (5 g l^{-1}) in filtered sea water by sunlight was 4 h. Grunnet and Dahllöf (2005) performed experiments with $CuPT_2$ in the field and observed photodegradation 0.5 m below the surface, however recoveries obtained at depths of 1, 2, and 3 m in the water column were not significantly different from those of nonexposed samples ($p < 0.005$). Examples of photolysis kinetics $CuPT_2$ irradiation by sunlight or artificial UV (Harino et al. 2005) are shown Fig. 23.8.

Turley et al. (2000) estimated the bacterial degradation rate of radio-labelled $CuPT_2$, using the die-away method. The observed half-life of $CuPT_2$ in sea water was approximately 4 days at an initial concentration of 52 g l^{-1}. In freshwater from the Connecticut River, the half-life of $CuPT_2$ was 7 h at ambient temperature. Examples of the kinetics of biodegradation of $CuPT_2$ investigated using the 'river-die-away' method (Harino et al. 2005), are shown in Fig. 23.8. Changes in $CuPT_2$ concentration were not observed during the initial 15 days of culture; subsequently,

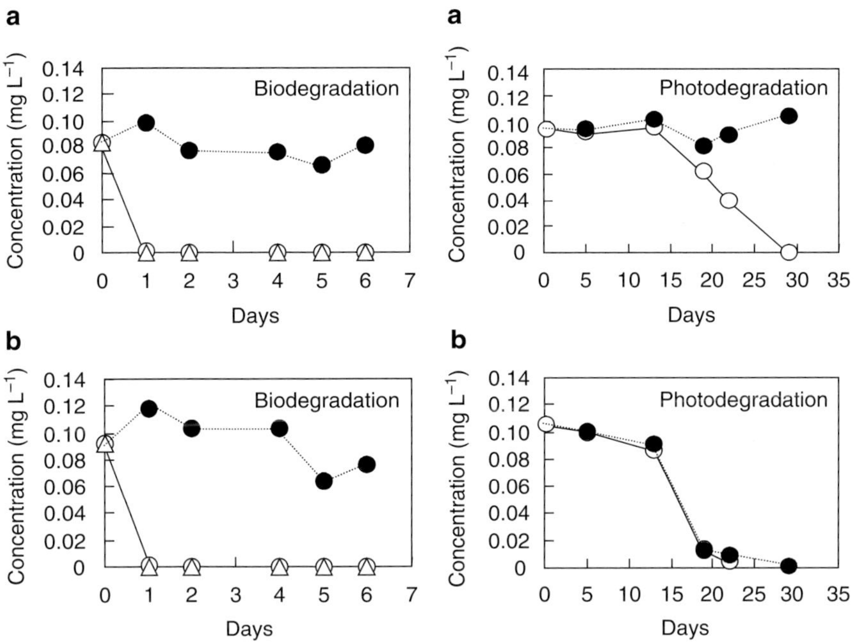

Fig. 23.8 Time course of pyrithione concentrations during photodegradation and biodegradation. Open circle, sunlight; triangle, UV; black circle, dark (control). Sample sea water. (**a**) Cupt, (**b**) znpt

degradation was rapid and $CuPT_2$ was not detected after 30 days of incubation. It is presumed from these results that degradation of $CuPT_2$ by bacteria is extensive in estuarine waters.

Turley et al. (2005) studied degradation of $CuPT_2$ in using two sediment and water systems, one of which was dosed during the day and the other at night. The pyrithione degraded in both systems – rapidly in the water phase, and with little accumulation in the sediment. 2-Pyridine sulfonic acid and carbon dioxide were the only detectable degradation products 30 days after dosing.

In summary, following the release of copper and $ZnPT_2$ into water, these compounds yield transformation products which may degrade rapidly. In order to grasp the fate of these compounds, better understanding such of transformations required and this is seen as important priority for action in the near future.

23.6 PTPB

There are relatively few available data on the hydrolysis of PTPB. Amey and Waldron (2004) reported that TPBP was hydrolyzed within 4 h in aerated artificial seawater, after which time hydrolysis approaches steady-state. The light source was

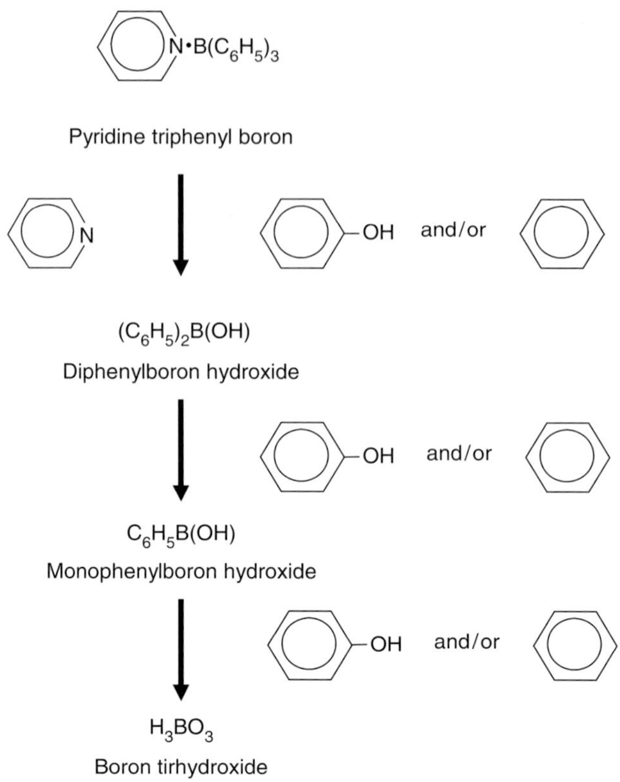

Fig. 23.9 Degradation pathway of PTPB (Amey and Waldron 2004)

a low-pressure mercury lamp providing 5.5 W of UV radiation output at 254 nm. Under these conditions, H_3BO_3 was produced in less than 1 h from PTPB by elimination of pyridine and phenol (Fig. 23.9). Samples containing PTPB, exposed to natural sunlight in mid-February in Birmingham, UK. at an average temperature of 4°C, were mostly degraded after 3 h exposure.

23.7 Dichlofluanid

The photodegradation of dichlofluanid has been studied in both natural sunlight and under conditions of artificial solar irradiation, in different types of natural water (Sakkas et al. 2001). The photodegradation rate was shown to decrease in the sequence lake water > river water > sea water > distilled water. Furthermore, the presence of humic and fulvic acids inhibited the photolysis reaction. Sakkas et al. (2001) identified dichlorofluoromethane, aniline and N,N-dimethylamino-sulfanilide as photodegradation products. The half life of dichlofluanid was

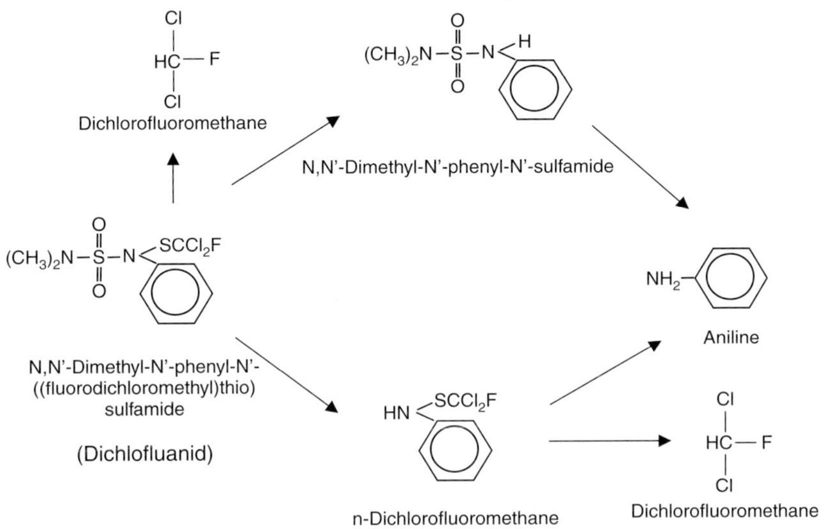

Fig. 23.10 Degradation pathway of dichlofluanid (Thomas et al. 2004)

estimated byCallow and Finlay (1995) to be 18 h in natural seawater at 25°C. The anaerobic degradation pathway is displayed in Fig. 23.10: dichlofluanid yielded N,N'-dimethyl-N'-phenylsulfamide, n-dichlorofluoromethylthion-aniline, aniline and dichlorofluoromethane (Thomas et al. 2002, 2003).

23.8 Chlorothalonil

Hydrolysis of chlorothalonil was shown to be insignificant in fresh water at $pH < 8.0$, whilst at 8.0 hydrolysis proceeded to 4-hydroxy-2,5,6-trichloroisophtalonitrile and 3-cyano-2,4,5,6-tetrachlorobenzamide (Caux et al. 1996).

Sakkas et al. (2002c) reported that chlorothalonil in natural water did not degrade in the dark, though in sunlight degradation was rapid (Table 23.1).The half life of chlorothalonil ranged from 1 to 48 h depending on the nature of the water (sea, river and lake water). Furthermore, dissolved organic matter such as humic and fulvic substances enhanced the photodegradation of chlorothalonil. Sakkas et al. (2002c) also carried experiments with UV-irradiation (>290 nm) which showed <25% degradation after 180 min: photolysis products were chloro-1,3-dicyanobenzen, dichloro-1,3-dicyanobenzene, trichloro-1,3-dicyanobenzene and benzamide.

Both biotic and abiotic degradation has been demonstrated *in vitro* using estuarine water (Walker et al. 1988). The fate in simulated marine environments appears to be similar to that in freshwater systems. However the half-life of chlorothalonil was 10 and 8–9 days in sterile water and nonsterile water, respectively, slower in comparison with photodegradation and hydrolysis studies, described above.

Table 23.1 Summary of common alternative biocides half-lives in various environmental matrices

	Study type	Condition	Half-lives (day)	Reference
Sea-Nine 211	Hydrolysis	Sterile water (pH 5)	9	Shade et al. (1993)
		Sterile water (pH 7)	>30	Shade et al. (1993)
		Sterile water (pH 9)	12	Shade et al. (1993)
	Photodegradation	Sunlight	13.4	Shade et al. (1993)
		Outdoor (distilled water)	18	Sakkas et al. (2002)
		Outdoor (sea water)	13	Sakkas et al. (2002)
		Outdoor (river water)	6.4	Sakkas et al. (2002)
		Outdoor (lake water)	5.5	Sakkas et al. (2002)
	Photodegradation	Xenon arc lamp $\lambda > 290\,nm$ (distilled water)	0.6	Sakkas et al. (2001)
		Xenon arc lamp $\lambda > 290\,nm$ (sea water)	0.4	Sakkas et al. (2001)
		Xenon arc lamp $\lambda > 290\,nm$ (river water)	0.3	Sakkas et al. (2001)
		Xenon arc lamp $\lambda > 290\,nm$ (lake water)	0.3	Sakkas et al. (2001)
	Aquatic metabolism	Aerobic seawater (25°C)	<1	Shade et al. (1993)
		Aerobic seawater (15°C)	1.9	Thomas et al. (2004)
		Anaerobic seawater (25°C)	<1	Shade et al. (1993)
		Anaerobic	<0.5	Thomas et al. (2003)
Diuron	Biodegradation	Aerobic (Over 80% of Diuron was degraded)	<27	Ellis and Camper (1982)
	Aquatic metabolism	Anaerobic (sediment)	14	Thomas et al. (2004)
Irgarol 1051	Hydolysis		>200	Hall et al. (1999)
	Photodegradation	Fresh water (no solvent)	36	Hall et al. (1999)
		Fresh water (with solvent)	85 (0.5% methanol)	Hall et al. (1999)
			148 (1% acetonitrile)	Hall et al. (1999)
		Seawater (with solvent)	273 (1% acetonitrile)	Hall et al. (1999)

Dichlofluanid	Photodegradation	Outdoor (distilled water)	1.8	Sakkas et al. (2001)
		Outdoor (sea water)	2.2	Sakkas et al. (2001)
		Outdoor (river water)	2.5	Sakkas et al. (2001)
		Outdoor (lake water)	3.5	Sakkas et al. (2001)
	Photodegradation	Xenon arc lamp λ > 290 nm (distilled water)	0.3	Sakkas et al. (2001)
		Xenon arc lamp λ > 290 nm (sea water)	0.5	Sakkas et al. (200)1
		Xenon arc lamp λ > 290 nm (river water)	0.8	Sakkas et al. (2001)
		Xenon arc lamp λ > 290 nm (lake water)	0.9	Sakkas et al. (2001)
	Aquatic metabolism	Aerobic seawater (15°C)	0.8	Thomas et al. (2002)
	Aquatic metabolism	Anaerobic (sediment)	<0.5	Thomas et al. (2003)
Chlorothalonil	Hydrolysis	Soft water (pH 6.5–7.4)	1.25	Ernst et al. (1991)
	Photodegradation	Outdoor (distilled water)	2	Sakkas et al. (2002c)
		Outdoor (lake water)	0.3	Sakkas et al. (2002c)
	Photodegradation	Outdoor (river water)	0.4	Sakkas et al. (2002c)
		Outdoor (sea water)	1.9	Sakkas et al. (2002c)
	Aquatic metabolism	Aerobic seawater (15°C)	2.8	Walker et al. (1988)
		Marine and fresh water	10	Thomas et al. (2004)
		Anaerobic (sediment)	<0.5	Thomas et al. (2003)
Dithiocarbamete	Hydrolysis	Distilled water (pH 6, 25°C)	0.08	Wesmahr and Sedlak (2000)
		Distilled water (pH 8, 25°C)	10	Wesmahr and Sedlak (2000)
Thiram	Hydrolysis	Water (pH 6, 25°C)	0.08	Wesmahr and Sedlak (2000)

23.9 Dithiocarbamates

Dithiocarbamate biocides including Maneb, Zineb, Thiram and Ziram have been used in a number antifouling paint formulation. Dithiocarbamates may degrade fairly rapidly by acid-catalyzed hydrolysis. For example, the half-life of dimethyldithiocarbamate at 25°C is estimated to range from 2 h at pH6 to 10 days at pH8 (Wessmahr an Sedlak 2000). Manab and Zineb are organic Mn and Zn complexes in which the metal ions are chelated to organic dithiocarbamate ligands. When these compounds dissolve in seawater, Cu^{2+} and Pb^{2+} ions undergo exchange reactions with Mn and Zn to form lipophilic organic Cu and Pb complexes (Phinney and Bruland 1997). These compounds also undergo hydrolysis. Maneb mainly produces ethyleneurea, ethylenethiourea, ethylenbis (isothiocyanate) sulfide, and carbimide following hydrosysis (Downing 2000).

Maneb can be degraded by sunlight and the main degradation products appear to be similar to those produced by hydrolysis. In contrast, microbial degradation represents a minor route in the breakdown of Maneb (Downing 2000).

There a few available data on the degradation of Thiram and Ziram. In water, Thiram is rapidly broken down by hydrolysis. Furthermore Thiram is degraded by photo irradiation, especially acidic conditions (Sharma et al. 2003). Thiram is degraded within 24 h by UV (>290 nm) and within 7 days by sunlight, in natural water (Samanidou et al. 1988). Furthermore, the degradation rate of Thiram is known to be affected by the characteristics of the water-body and increases in the order river water > lake water > sea water.

23.10 Conclusions

The ideal antifouling paint should contain biocides which degrade rapidly in the environment, display low partitioning affinity towards sediment and biota and represent minimal toxicity for aquatic organisms at the concentrations likely to be present in the environment. Rapid degradation was characteristic of most of the alternative biocides reviewed here. However, information on the distribution, bioconcentration and toxicity of alternative biocides, and also their many degradation products representatives a significant knowledge gap at present. Further studies on the fate and effects of these compounds should be carried out in order to enhance our understanding of risks to aquatic habitats.

Acknowledgement We would like to express thanks to Dr. Yoshitaka Yamaguchi of the National Maritime Research Institute, Japan, for his assistance in preparing this manuscript.

References

Amey RL, Waldron C (2004) Efficacy and chemistry of BOROCIDE™P triphenylboron-pyridine, a non-metal anti-fouling biocide. Pro. InSAfE 234–243

Bonnemoy F, Lavedrine B, Boulkamh A (2004) Influence of UV irradiation on the toxicity of phenylurea herbicides using Microtox® test. Chemosphere 54:1183–1187

Callow M E, Finlay J A (1995) A simple method to evaluated the potential for degradation of antifouling biocides. Biofouling 9:153–165

Callow M E, Willingham G L (1996) Degradation of antifouling biocides. Biofouling 10:239–249

Caux P-Y, Kent R A, Fan G T (1996) Environmental fate and effects of chlorothalonil: a Canadian perspective. Crit Rev Environ Sci Technol 26:45–93

Downing E (2000) http://www/odpr.ca.gov/docs/emon/pubs/fatememo/mabeb.pdf

Ellis P, Camper N (1982) Aerobic degradation of diuron by aquatic micro-organisms. J Environ Sci Health B 17:277–288

Grunnet K S, Dahllof I (2005) Environmental fate of the antifouling compound zinc pyrithione in seawater. Environ Toxicol Chem 24:3001–3006.

Hall L W Jr., Giddings J M, Solomon K R et al. (1999) An ecological risk assessment to the use of Irgarol 1051 as an algaecide for antifoulant paints. Crit Rev Toxicol 29:367–437

Harino H, Kitano M, Mori et al. (2005) Degradation of antifouling booster biocides in water. J Mar Biol Assoc UK 85:33–38

Jacobsen A H, Willigham G L (2000) Sea-nine antifoulant: an environmentally acceptable alternative to organotin antifoulants. Sci Total Environ 258:103–110

Jacobsen A H, Mazza L S, Lawrence L J et al. (1993) Fate and an antifoulant in an aquatic environment, ACS symposium series No.522 *Pesticides in Urban Environments: Fate and Significance*, Racke K D and Leslie A R (eds). 127–138

Lam K H, Lam M H W, Lam P K S et al. (2004) Identification and characterization of a new degradation product of Irgarol 1051 in mercuric chloride-catalyzed hydrolysis reaction and in coastal waters. Mar Pollut Bull 49:356–357

Lam K H, Cai Z, Lam M H W (2005) Identification of a new s-triazene species in the coastal waters originated from the use of Irgarol 1051 a booster biocide for marine antifouling coatings. Environ Pollut 136:221–230.

Liu D, Maguire R J, Lau Y L et al. (1997) Transformation of the new antifouling compound Irgarol 1051 by *Phanerochaete chrysosporium*. Water Res 31:2363–2369

Liu D, Pacepavicius G J, Maguire R J et al. (1999) Mercuric chloride-catalyzed hydrolysis of the new antifouling compound Irgarol 1051. Water Res 33:155–163

Maraldo K, Dahllöf I (2004) Indirect estimation of degradation time for zinc pyrithione and CuPT2 in seawater. Mar Pollut Bull 48:894–901

Okamura H, Aoyama I, Liu D et al. (1999) Photodegradation of Irgarol 1051 in water. J Environ Sci Health B34:225–238

Phinney J T, Bruland K W (1997) Trace metal exchange in solution by the fungicides ziram and maneb (dithiocarbametes) and subsequent uptake of lipophilic organic zinc, copper and lead complexes into phytoplankton cells. Environ Toxicol Chem 16:2046–2053

Sakka V A, Konstantinou I K, Albanis T A (2001) Photodegradation study of the antifouling booster biocide dichlofluanid in aqueous media by gas chromatographic techniques. J Chromatogr A 930: 135–144

Sakkas V A, Konstantinou I K, Albanis T A (2002a) Aquatic phototransformation stydy of the antifouling agent Sea-Nine 211: identification of byproducts and the reaction pathway by gas chromatography-mass spectroscopy. J Chromatographr A 959:215–227

Sakkas V A, Lambropoulou D A, Albanis T (2002b) Photochemical degradation study of irgarol 1051 in natural waters: influence of humic and fulvic substances on the reaction. J Photochem Photobiol A: Chem 147:135–141

Sakkas V A, Lambropoulou D A, Albanis T A (2002c) Study of chlorothalonil photodegradation in natural waters and in the presence of humic substances. Chemosphere 48:939–945

Sakkas V A, Shibata K, Yamaguchi Y et al. (2007) Aqueous phototransformation of zinc pyrithione degradation kinetics and byproduct identification by liquid chromatography-atmospheric pressure chemical ionization mass spectrometry. J Chromatogr 1144:175–182

Seymour M S, Bailey D L (1981) Thin-layer chromatography of pyrithione. J Chromatogr 206: 301–310

Samanidou V, Fytianos K, Pfister G et al. (1988) Photochemical decomposition of carbamete pesticides in natural waters of Northern Greece. Sci Total Environ 76:85–92

Shade W D, Hurt S S, Jacobson A H et al. (1993) Ecological risk assessment of a novel marine antifoulant. EnvironToxicol Risk Assess 2:381–408

Sharma V K, Aulakh J S, Malik A K et al. (2003) Thiram: degradation, applications and analytical methods. J Environ Monitor 5:717–723

Tanaka F S, Hoffer B L, Wien R G (1986) Photolysis of 3-(3,4-dichlorophenyl)-1, 1-dimethylurea(Diuron) in dilute aqueous solution. Toxicol Environ Chem 11:261–269

Thomas K (1999) Determination of the antifouling agent zinc pyrithione in water samples by copper chelate formation and high-performance liquid chromatography-atomospheric pressure chemical ionization mass spectrometry. J Chromatogr A 833:105–109

Thomas K V (2001) The environmental fate and behaviour of antifouling paint booster biocides: a review. Biofouling 17:73–86.

Thomas K, McHugh M, Waldock M (2002) Antifouling paint booster biocides in UK coastal waters: inputs, occurrence and environmental fate. Sci Total Environ 293:117–127

Thomas K V, McHugh M, Hilton M et al. (2003) Increased persistence of antifouling paint biocides when associated with paint particles. Environ Pollut 123:153–161

Thomas K V, Aldridge J, Dyer R et al. (2004) The occurrence, fate and effects of selected antifouling paint booster biocides in UK docks, harbours and marinas. Pro. InSAfE 177–195

Turley P A, Fenn R J, Ritter J C (2000) Pyrithiones as angifoulants:Environmental chemistry and preliminary risk assessment. Biofouling 15:175–182

Turley P A, Fenn R J, Ritter J C et al. (2005) Pyrithiones as antifoulant: Environmental fate and loss of toxicity. Biofouling 21:31–40

Walker W W, Cripe C R, Pritchard P H et al. (1988) Biological and biotic degradation of xenobiotic compounds in in vitro estuarine water and sediment/water systems. Chemosphere 17: 2255–2271

Wessmahr K W, Sedlak D L (2000) Effect of metal complexation on the degradation of dithiocarbamete fugicides. Environ Toxicol Chem 19:820–826

Yamaguchi Y, Kumakura A, Ishigami M et al. (2004) Spectrometric study on photodegradation of zinc and CuPT2s. Pro. InSAfE 228–233

Chapter 24
Copper Biocides in the Marine Environment

Steven J. Brooks and Mike Waldock

24.1 Introduction

Due to the restrictions on TBT usage in antifouling paints since 2003 and its complete ban on all vessels in 2008 (IMO 2001), copper has been increasingly used as the main biocide ingredient in antifouling paint coatings. Copper is toxic to a wide range of aquatic organisms, which makes it an ideal biocide, preventing the colonisation of biofouling organisms on the vessel surface. There has been much concern from regulators and scientists that copper concentrations may become elevated in areas of high boating density such as marinas and estuaries with potential damaging effects on the animal and plant communities. In certain European countries, copper has been banned from use on recreational vessels, although so far this is restricted to inland freshwaters, many countries are beginning to re-evaluate current copper risk assessments in marine coastal waters.

This chapter provides an outline of the concentrations of copper in the marine coastal environment as a result of its use as an antifouling biocide. The potential risk of copper to marine life has been evaluated with respect to copper bioavailability, speciation and toxicity. The chapter outlines some of the shortfalls of current copper risk assessment and provides some suggestions for improvement.

S.J. Brooks
Norwegian Institute of Water Research (NIVA), Gaustadalléen 21, NO-0349 OSLO, Norway

M. Waldock
Cefas Weymouth Laboratory, Fish Diseases Laboratory, The Nothe,
Barrack Road, Weymouth, Dorset, DT4 8UB, UK

T. Arai et al. (eds.), *Ecotoxicology of Antifouling Biocides*,
DOI: 10.1007/978-4-431-85709-9_24, © Springer 2009

24.2 Copper As an Antifouling Biocide

Marine antifouling paints are specialised coatings that are used to protect submerged structures such as a boat hulls and/or pier pylons from biofouling organisms. Biofouling can be described as the undesirable accumulation of biological material on the surfaces of submerged structures. The occurrence of biofouling organisms on the hulls of boats can have major environmental and economical implications. One of the main problems with biofouling is related to increased friction due to the physical presence of the organisms on the vessel surface. This can cause a reduction in vessel speed and manoeuvrability, and increased weight, resulting in an overall increase in fuel consumption and cost. The increased fuel consumption costs as a result of biofouling have been estimated to be as much as 40% (Yebra et al. 2004). Other effects of biofouling include increased frequency of dry-docking operations for cleaning and the overall deterioration of ship hulls. Consequently, there are significant financial drivers to encourage vessel companies and antifouling paint manufacturers to develop effective antifouling paint products in order to eliminate the cost of biofouling.

Environmental impacts of biofouling include increased CO_2 emissions from increased fuel consumption, contributing to global warming effects, as well as the transport of non-native species across geographical niches resulting in devastating effects on native populations and habitats (Conlan 1994). These factors have a much wider implication and support the need for reducing biofouling activities.

Conversely, the environmental implications of antifouling biocides have been well documented. For example, tri butyl tin (TBT) was once the most widely used active ingredient in antifouling paints and at peak use covered approximately 70% of the worlds shipping fleet. However, low environmental concentrations of TBT were found to cause endocrine disrupting effects in several gastropod species. For example, a condition known as 'imposex' was found in dog whelk populations, where females were found to produce male genitalia, a phenomena that has been well documented in the scientific literature (Thain 1986; Thain and Waldock 1986; Bryan et al. 1989). This eventually led to the proposed global restrictions on TBT usage as an antifouling biocide on vessels below 25 m in 2003 and a complete ban on all vessels by 2008 (IMO 2001). Despite these restrictions, TBT is still detected and the presence of imposex is still commonplace (Smith et al. 2006).

Partly due to the restrictions on TBT, copper has become one of the most widely used biocides for the protection of marine vessels. Copper is an effective biocide due to its broad spectrum toxicity to aquatic organisms, although copper based antifouling paints are often used in combination with other biocides or 'booster biocides' that can help to improve the biocidal properties of the paint to control formation of slime layers.

The amount of copper present in an antifouling paint varies typically between 4 and 30 wt %, with various attempts to reduce the amount of copper without reducing the overall effectiveness of the paint (Pérez et al. 2006). Overall, there are two main tin-free paint systems: (1) controlled depletion system (CDPs), which are an upgrade on the traditional soluble matrix paints using modern resins, exposing fresh biocidal surface at a more controlled rate; and (2) tin free self polishing copolymers (SPCs). However, paint classifications are often only based on the manufacturer's claims, mainly due to the lack of scientifically supported knowledge on the fundamental processes influencing antifouling paint behaviour.

The release rate of copper from an antifouling coating into the water column is dependent on a number of physical and biological factors such as hydrodynamics, temperature, pH and salinity, as well as the presence of biofilms on the coating surface (Thomas et al. 1999). For example, increased turbidity and shear on the vessel surface is likely to increase release rate, whilst reduced temperatures can lead to a reduction in the release rate. Valkirs et al. (2003) measured the release rate of copper from a self polishing tin free coating within the laboratory under static conditions and compared this to in situ measurements taken directly on the antifouling surface of the vessel. Higher copper leaching rates were found in laboratory studies (25–65 μg Cu/cm^2/day) compared to in situ measurements (average 8.2 μg Cu/cm^2/day). The presence of biofilms on the surface of the vessel was thought to explain the differences in leaching rate. The wide range of copper release rates within the laboratory from the Valkirs study was attributed to the different paint formulations tested.

Despite the fact that biocide leaching rates are significantly reduced by the build-up of microbial biofilms on the paint surface, current regulations for the measurement of biocide leaching rates are still based on standard laboratory tests (ASTM 2000). Consequently, current regulations provide a worst case measurement, which frequently overestimates the biocide release rate. A more realistic measurement would be achieved with better understanding of the build-up of microbial biofilms on the paint surface and the effects of boundary layers on the leaching rates in natural seawater.

The release of copper from antifouling coatings provides a significant proportion of the overall copper loading into localised areas such as harbours and marinas where high densities of vessels are moored. This can potentially lead to elevated copper concentrations in the sediment and water column.

24.3 Inputs and Concentrations of Copper in Marine Coastal Waters

Copper has a tendency to adsorb to suspended particulates within the water, which are subsequently deposited in the sediment. As a result, copper concentrations are often two to three orders of magnitude higher in the sediment than in the overlying water column. Sediment copper concentrations range between 1–10 mg/kg in relatively clean areas, reaching 3,000 mg/kg in areas receiving run-off from historical mining activity (e.g. the Fal estuary, UK, Bryan and Gibbs 1983; HSE 1999). Sediment copper concentrations can also become elevated in enclosed marinas, dockyards and harbours. For example, an extensive survey measuring sediment copper concentrations in commercial harbours, estuarine moorings and marinas of the UK found copper concentration ranges of 10–161, 4.8–30 and 9–57 mg/kg copper respectively (Jones et al. 2005).

Copper in aerobic sediments is mainly bound to metal oxides and high molecular weight organic matter, which can be released into the sediment pore waters during oxidation reactions (Skrabal et al. 1997). The movement of copper from the sediment pore waters into the water column and its contribution to the toxicity is dependent on its chemical speciation, with the free ion and inorganic species more mobile than organically bound copper. The pore waters of coastal and estuarine sediments are typically loaded with organic matter resulting in a reduction in exchangeable copper. A recent study found over 15% of sediment copper existed as an exchangeable form, ready to contribute to the copper concentration of the water column (Choi et al. 2006), although a previous study claimed that only 5% of the total sediment copper concentration was easily exchangeable (Roper 1990).

In anaerobic sediments, the formation of copper sulphides reduce copper bioavailability. In a recent study, only 0.04% of sediments sampled in areas of heavy boating activity showed copper to be potentially bioavailable (Jones et al. 2005). The presence of sulphides and the high DOC content of the pore water was thought to have resulted in the small contribution of sediment copper to the water column. However, increased copper inputs from the sediment to the water column can result from sediment disturbance events, such as dredging and storms. Typical copper concentrations in marine coastal waters are found in the range 0.5–3 μg/l (Jones and Bolam 2007).

Due to the increased usage of copper as an antifouling paint biocide, waterborne copper concentrations can become elevated in areas of high boating activity with little to no water exchange such as harbours and marinas. For example, elevated waterborne copper concentrations of up to 21 μg/l have been recorded in San Diego Bay, USA (Schiff et al. 2007). Within this study, 86% of the marina surface water area failed the environmental quality standard (EQS) of 5 μg/l. The very high density berthing, with over 17,000 recreational vessels within the bay area, was believed to be responsible for the elevated copper concentrations. Other studies have also found elevated copper concentrations above the EQS in the San Diego Bay area (Zirino et al. 1998; Noblet et al. 2002).

Despite the concentration of dissolved copper exceeding the EQS in 86% of the San Diego Bay surface waters, only 21% of these same waters were deemed toxic to the sensitive early life stages of the mussel, *Mytilus galloprovincialis* (Schiff et al. 2007). The presence of organic ligands reducing the bioavailable fraction of copper was thought to be the reason for the difference in copper toxicity from that suggested from chemical measurement. This also highlights the importance in measuring copper speciation in both field monitoring and laboratory toxicity testing.

Currently, most monitoring studies measure total dissolved copper concentration and not the different copper species. As will be discussed in detail in the next section, the speciation of copper is critical to understanding copper bioavailability and toxicity. Therefore, monitoring studies that can provide a measure of the copper species within the water column would be of great benefit to understanding copper risk within a particular water body.

Jones and Bolam (2007) measured copper species within seawater samples in an extensive survey of marinas, harbours, and ports around the UK. Copper was determined as total dissolved and labile (free ion and inorganically bound) copper, with the difference between the two reported as the organically bound copper. Total dissolved copper concentrations ranged from 0.30 to 6.68 μg/l, although only one value out of 306 measured was above the EQS of 5 μg/l. This higher value was found in an enclosed marina with little to no water exchange with the adjoining estuary. Overall, total dissolved copper tended to be higher in enclosed marinas compared to the estuarine ports where sufficient water flux was likely to have contributed to increased dilution. Labile copper concentrations ranged from 0.02 to 2.69 μg/l, with labile copper making up 10–30% of the total dissolved copper concentration, or inversely, the proportion of copper bound to organic ligands was between 70% and 90% (Jones and Bolam 2007). Even at elevated concentrations of total dissolved copper the proportion of labile copper was relatively consistent. It was suggested that this was because of the buffering capacity of the natural environment, mainly due to the presence of dissolved organic carbon within the water column.

Copper speciation measurements have also been reported within a Finnish marina at seasonal intervals (Brooks 2006). Total dissolved copper concentrations ranged from 0.62 to 3.89 μg/l, with the highest concentrations found in the summer. This was attributed to the increased boating traffic during this sampling season. By contrast, the labile copper concentration remained stable below 1 μg/l irrespective of sampling season or locality to the marina. The high concentration of dissolved organic carbon (DOC, 4.9–10.6 mg/l) in the water column was suggested as being responsible for buffering the labile copper concentration. The labile copper concentration within the Finnish marina was similar to that found in the UK monitoring study, and was overall, approximately 10–30% of the total dissolved copper concentration.

24.4 Copper Speciation, Bioavailability and Toxicity

Fundamentally, copper is an essential ion that is required by all organisms for cellular processes. For example, it is involved in the functioning of a variety of proteins including the enzymes carbonic anhydrase (Henry 1996), cytochrome oxidase (Hassall and Dangerfield 1990) and the respiratory protein haemocyanin (Dallinger 1977). As a result, many organisms have developed methods for the uptake and excretion of copper (e.g. the metal binding protein, metallothionein), enabling copper to be maintained within cell tolerance limits. It is only when these tolerance limits are exceeded that copper toxicity is reported (US EPA 1985).

It is widely acknowledged that the bioavailability of copper is highly dependent on its chemical form, which is influenced by a wide range of physicochemical factors, such DOC, pH and ion concentration of the media (Erickson et al. 1996). The free ion is the most bioavailable form able to pass through biological membranes where it can then elicit a toxic effect. However, the free copper ion has a strong tendency to form complexes with both inorganic and organic ligands, which results in a reduction in copper bioavailability and toxicity. Copper inorganic ligand complexes are considered to be bioavailable to a certain extent, although to what extent is not fully understood (MacRae et al. 1998). Copper bound to organic ligands is believed to be almost entirely non-toxic, while there is some evidence to suggest that the strength of the copper-ligand complex is important when assessing organic copper bioavailability (MacRae et al. 1999). In the coastal marine environment, a high concentration of both inorganic and organic ligands ensures that the free copper ion only makes up a small percentage of the total dissolved copper concentration in seawater.

Certain studies have reported weakly bound organic copper complexes to be bioavailable and toxic (Florence et al. 1992; Erickson et al. 1996). Copper binding studies with fish gills found organic copper complexes to be bioavailable and to bind to the gill ligand when the copper binding affinity to the gills was greater than that of certain organic ligands (Playle et al. 1993a, b; MacRae et al. 1999). This suggests that copper speciation is not the only consideration when predicting copper toxicity, but the binding characteristics of the individual ligands may also need to be considered.

A list of copper binding characteristics is shown in Table 24.1. These are expressed as stability constants (Log K), with the value of K determined by Equation (24.1), where M is the metal and L the ligand.

$$K = [ML]/[M][L] \tag{24.1}$$

Table 24.1 A selection of stability constants for copper-ligand interactions. A Log K value below that of the biological ligand would suggest bioavailability

Ligand type	Specific ligand	Stability constant (Log K)	Source
Biological	Rainbow trout gill	6.4–7.5	MacRae et al. (1999)
	Brown trout gill	7.25	MacRae et al. (1999)
	Daphnia	8.02	Karel et al. (2001)
	Fathead minnow gill	7.4	Playle et al. (1993b)
Inorganic	Cu(OH)	7.66	van den Berg (1984)
	$Cu(CO_3)_2^{2-}$	8.92	Smith and Martell (1976)
	$CuCO_3$	5.75	Smith and Martell (1976)
	$Cu(OH)_2$	12.66	van den Berg (1982)
Organic	Nitrilotriacetic (NTA)	10.3	MacRae et al. (1999)
	2,6-pyridine-dicarboxcylic	8.6	MacRae et al. (1999)
	Ethylenediamine (EDA)	6.9	MacRae et al. (1999)
	Citric acid	6.4	MacRae et al. (1999)
	Malonic	5.6	MacRae et al. (1999)
	Tartaric	4.2	MacRae et al. (1999)
	Fulvic acid	10	MacRae et al. (1999)
	EDTA	18	MacRae et al. (1999)
	Natural Irish Sea organics	10–10.4	Van den Berg (1984)
Exudates of	Cyanobacteria	9.2–9.5	Gouvêa et al. (2005)
	Synurophycea	8.42	Lombardi and Vieira (1998)
	Synechococcus	12.3–13.8	Moffett and Brand (1996) Croot et al. (2000)
	Diatoms	8.6–8.8	Gouvêa et al. (2005)
	Thalassiosira weissflogii	10.6	Croot et al. (2000)
	Skeletonemacostatum	11.6–12.3	Croot et al. (2000)
	Coccolithophore		
	Hymenomonas carterae	10.8	Croot et al. (2000)
	Dinoflagellates		
	Amphidinium carterae	11.8–12.2	Croot et al. (2000)

A higher K value suggests stronger binding (Stumm and Morgan 1996). Copper-ligand stability constants (Log K) less than those calculated for the biological ligand (gill) (ranging from 7.25 to 8.02) are potentially bioavailable. The majority of the organic ligands listed in the table, particularly those derived as macroalgae exudates, have a stability constant greater than that of the biological ligand. Only in a few instances is this not the case. In theory, if the composition of DOC in the natural environment were composed of only weakly bound copper complexes then potentially all organically bound copper present would be bioavailable. However, the organic content (i.e. DOC) of marine coastal waters is made up of a multitude of different organic ligands with up to 60% composed of humic and fulvic substances with high copper binding affinities (Ma et al. 1999). Therefore, copper present in these waters is likely to become bound to these organic ligands. It is only when the dissolved copper concentration exceeds the concentration of these strong organic ligands (in relation to binding characteristics of the biological ligand) that organically bound copper will potentially contribute to the overall toxicity. Since the DOC concentration in marine coastal waters is relatively high, typically ranging between 2 and 4 mg/l (Jones and Bolam 2007) (up to 10 mg/l in the Baltic), the contribution of organically bound copper to organism toxicity in the natural environment is likely to be negligible.

The presence of biofilms on the vessel surface can act as a source of DOC and may result in the rapid complexation of copper as it leaves the paint surface. Although the presence of the biofilm is likely to have some effect, the extent of this interaction is uncertain and is still poorly researched. Copper leaching from a vessel surface and the likely metal seawater interactions with respect to speciation and bioavailability is shown diagrammatically in Fig. 24.1.

In controlled laboratory studies, using measured copper concentrations, increasing the DOC concentration of the test media was found to increase the copper EC/LC50 concentration in fish (Playle et al. 1993a, b), echinoderms (Lorenzo et al. 2006), bivalves (Brooks et al. 2007), macroalgae (Brooks et al. 2008), and unicellular organisms (Florence and Stauber 1986). An increase in the proportion of organically bound copper with a concomitant decrease in the free ion or labile copper concentration was suggested by all authors as the reason for the reduction in toxicity. In the mussel (*Mytilus galloprovincialis*) a good correlation was found between DOC concentration and copper toxicity that could be explained by Equation (24.2; Arnold 2004).

$$\mathrm{EC50} = 11.53\mathrm{DOC}^{0.54} \tag{24.2}$$

This correlation was found to exist over a wide range of DOC concentrations from 0.3 to 10 mg/l. In addition, correlations between the concentration of the free copper ion and toxic effects in marine species have also been documented (Zamunda and Sunda 1982; Sanders et al. 1983; Lorenzo et al. 2006). Equally good correlations have been found between the labile copper concentration (Cu^{2+} and inorganic Cu) and the toxic effects to oyster embryo larvae (Brooks et al. 2007) and *Fucus* germlings (Brooks et al. 2008).

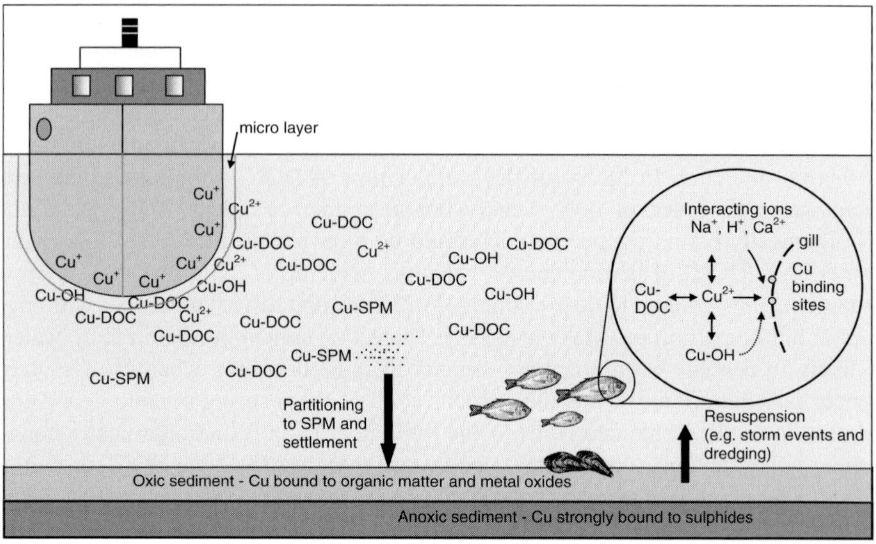

Fig. 24.1 Copper leaching from a vessel surface painted with copper antifouling paint. As copper enters the water in its free ion state (Cu^{+}), it is immediately oxidised to Cu^{2+} and forms complexes with inorganic and organic ligands. This is thought to occur within the first few micrometers of the painted surface. The presence of biofilms on the vessel surface can act as a source of dissolved organic carbon (DOC). Copper can also adsorb to suspended particulate matter (SPM), which will lead to copper deposition to the sediment. Copper can return to the water column from the sediment through resuspension events, such as storms and dredging activities. Copper bioavailability and toxicity is dependent on chemical speciation with the Cu^{2+} the most bioavailable, inorganic copper is only partly bioavailable, whilst organically bound copper is not. The presence of seawater ions at the gill prevents copper binding through competition and can reduce copper uptake. Note: Cu-OH refers to inorganically bound copper (e.g. CuOH, $CuCO_3$, CuCl)

Due to the importance of chemical speciation in influencing the bioavailability and toxicity of copper to aquatic organisms, toxicity studies that report nominal copper concentrations have lost much of their scientific credibility. This has been acknowledged by many laboratories who now report measured copper concentrations to support toxicity measurements. Toxicity data expressed as no observable effect concentration (NOEC) are summarised in Fig. 24.2. These were taken from 23 separate studies on 21 marine species in which measured copper concentrations were reported. A large range of chronic NOEC values can be seen, from 4.4 to 223 μg/l total dissolved copper. Although algal species were found to be the most sensitive group, two algae species were also found to be the least sensitive. Most mollusc species were shown to be sensitive, whilst fish were particularly tolerant to copper exposure.

Marine toxicity data for copper taken from the US EPA database have been plotted in Fig. 24.3. Only studies that reported measured copper concentrations have been included. According to this data the embryo larvae of all marine species were the most sensitive compared to juvenile and adult life stages. In some cases

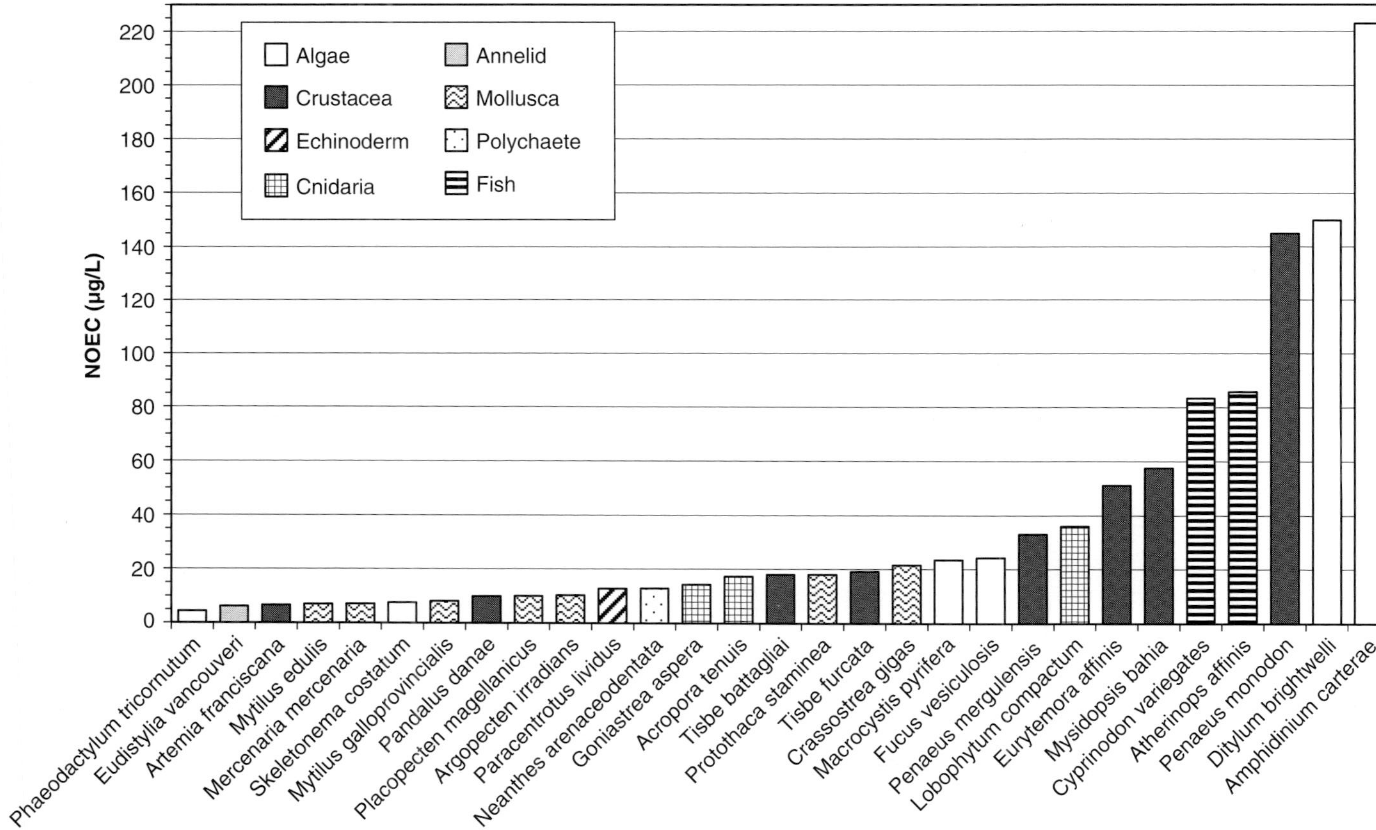

Fig. 24.2 Ranked copper No Observable Effect Concentrations (NOECS) for the main groups of marine organisms. Data taken from chronic toxicity studies in which measured copper concentrations were reported. (66 copper NOEC values of 21 marine species from 23 studies, data obtained from EU risk assessment report, in press)

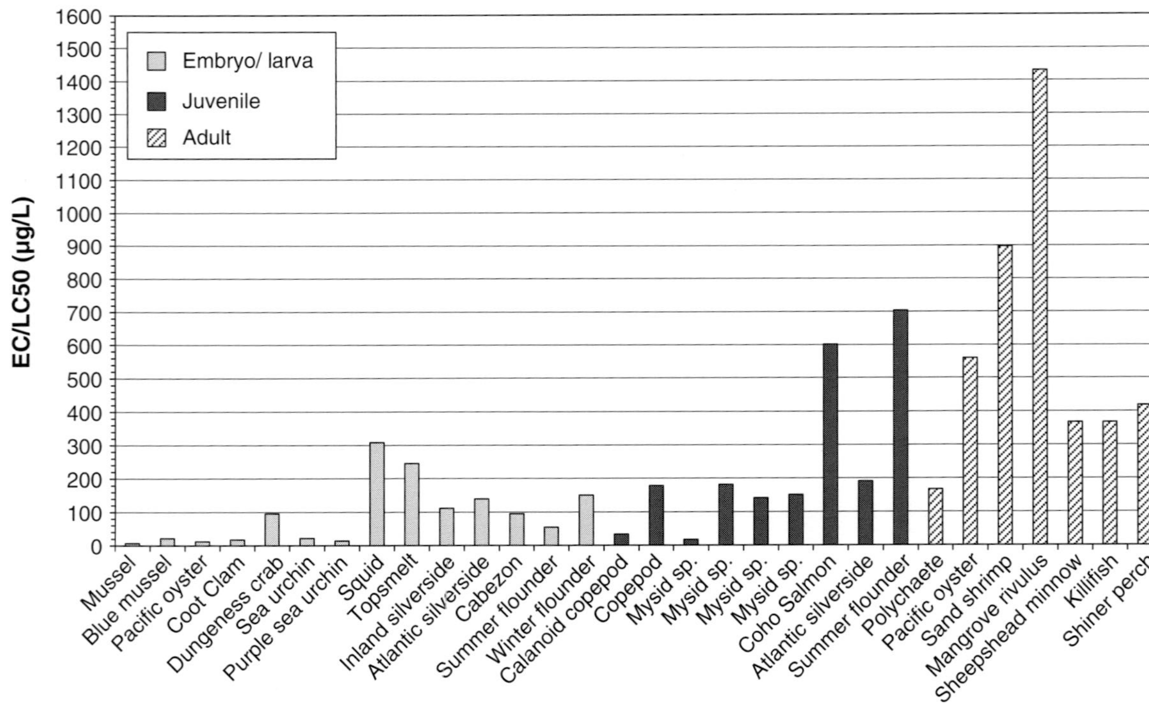

Fig. 24.3 Copper toxicity data expressed as EC or LC50 for a range of marine organisms. Data obtained from the United States Environmental Protection Agencies (US EPA) database on acute toxicity of copper to salt water animals. Toxicity values were only used in tests that reported measured copper concentrations in the exposure media. Http://www.epa.gov/waterscience/criteria/copper/2003/index.htm

there was an approximate 10–50-fold difference in toxicity to the early life stage compared to the adult (e.g. Pacific oyster 12.1 and 560 μg/l respectively; Summer Flounder 54.4 and 703.7 μg/l respectively). The embryo/larval stages of the mussels, oysters and sea urchins were found to be the most sensitive with copper EC50 concentrations of 6.8, 12.1 and 14.3 μg/l respectively. These animals are regularly used in water quality assessment due to their high level of sensitivity to environmental contaminants such as copper. Overall, with the exception of the highly copper tolerant squid embryo, the fish were the most tolerant group to copper exposure with higher EC/LC50 concentrations in all life stages.

Of all the marine species tested cyanobacteria have been found to be the most sensitive group to copper toxicity. Chronic effects on growth in the genus *Synechoccus* have been detected at a free copper concentration of approximately 0.63 ng/l (Moffett and Brand 1996), although adverse effects to copper were found at 0.2 μg/l. These values are much less than the background concentrations measured within marine coastal waters (typically 1–2 μg/l). For this reason there are conflicting views as to whether cyanobacteria should be included in water quality risk assessments in coastal marine waters.

24.5 Copper Risk and Regulation

The assessment of contaminant risk in the environment is often carried out using the simple Equation (24.3):

$$\frac{\text{Predicted Environmental concentration (PEC)}}{\text{Predicted No Effect concentration (PNEC)}} = \text{Risk Quotient (RQ)} \qquad (24.3)$$

If the PEC is greater than the PNEC then the RQ will be greater than one. In this case there would be potential for environmental harm and the need for further investigations to take place. Conversely, an RQ <1 would suggest no environmental harm. However, the accuracy of the assessment is entirely dependent on the PEC and PNEC values and therefore it is important that the calculations are based on reliable data and reflect what is happening in the real environment. In many instances, PECs have been inferred from either measured total copper concentrations or from predictive models, while the PNECs have been taken from toxicity tests in clean filtered seawater with very different copper binding characteristics than those in the natural environment. In this case the copper PNEC value would be significantly lower than that expected in the natural environment. Since it is fairly well established that copper toxicity is a measure of the bioavailable form, it is important that effect levels based on measured concentrations are used in risk assessment.

Partly due to the difficulties in measuring copper speciation in seawater, the majority of copper risk assessments have been based on the total or dissolved copper concentration rather than the free copper ion or labile copper concentration (Allen and Hansen 1996). Less than 30% (typically <10%) of dissolved copper is considered bioavailable (i.e. labile copper – free copper ion and inorganic copper, Jones and Bolam 2007). This proportion is reduced to as little as 0.1% when only the free copper ion is considered as the bioavailable form (HSE 1999; Elzvik and Hanze 1992). Consequently, current risk assessments based on total dissolved copper concentrations provide a very conservative estimate of copper risk in the marine coastal environment.

An assessment of copper risk in the marine coastal environment has been carried out using the risk quotient approach (Hall and Andersen 1999). The PEC data were taken from measured copper concentrations from eight separate studies of harbours and marinas of the Mediterranean Sea, Baltic Sea and North Sea between 1986 and 1997; copper PNEC values were derived from acute toxicity tests on 65 marine species from algae to fish. Overall the calculated PNEC that provided 95% protection was 5.6 µg/l total dissolved copper. Risk quotients greater than one were found in three out of 101 marine coastal water sites, and led the authors to conclude that the ecological risk of copper in European marine environments was generally low (Hall and Andersen 1999).

Copper models for the prediction of copper toxicity (e.g. Biotic Ligand Model) and the prediction of environmental fate, transport and concentration (e.g. MAMPEC) for marine coastal waters have been developed with the ultimate aim to improve assessments of environmental risk. The Biotic Ligand Model (BLM) which itself incorporates three models ((1) – gill surface interaction model (GSIM, Pagenkopf 1983); (2) – chemical equilibrium model (CHESS, Santore and Driscoll 1995); and (3) – metal dissolved organic matter interaction model (WHAM v.5, Tipping 1994)) was first developed to predict copper toxicity to freshwater organisms. The model takes into account the water parameters that can influence copper bioavailability, including the concentration of DOC, suspended particulate matter (SPM) and the main ions (e.g. Na^+, Ca^{2+}, H^+) as well as water chemistry (pH, alkalinity, hardness). The freshwater BLM was found to predict known copper toxicity within a factor of ±2 (Di Toro et al. 2001; Santore et al. 2001; US EPA 2003) and is currently being considered by the US EPA for the direct calculation of copper criteria in freshwaters.

The success of the freshwater BLM has led to the development of a marine BLM for the prediction of copper toxicity. The marine mussel (*Mytilus edulis*) was used to test the suitability of the marine BLM for predicting copper toxicity (Arnold et al. 2005). A strong correlation was found between the measured and the BLM predicted copper EC50s, although limitations in the model were found when predicting EC50s below 10 µg/l copper. Despite the limitations of the marine BLM, it has shown potential for the prediction of copper toxicity to marine organisms. However, further validation of the model is required before it can be incorporated into future environmental risk assessments for the prediction of PNECs.

Models for predicting the fate, transport and concentration of copper in the coastal marine environment have been developed and used in risk assessments to determine the PECs of antifouling products, such as copper. An example of such a model is the Marine Antifoulant Model to Predict Environmental Concentrations (MAMPEC). The MAMPEC model can be applied to a range of marine scenarios (e.g. open sea, estuary and marina) and takes into account many physicochemical factors and hydrodynamic processes of these environments, as well as parameters specific to antifouling (e.g. leaching rate). Validation studies of the MAMPEC model have shown good agreement between predicted and measured concentrations of copper in the marine environment (van Hattum et al. 2002). However, further validation is required to increase confidence in the predictions. Once validated such models could potentially provide regulators with a fast and cost effective method for predicting copper risk and improving protection.

24.6 Conclusions

This review has outlined the environmental concentrations and biological effects of copper in the coastal marine environment with respect to inputs from its use in antifouling paints. The speciation of copper is fundamental for predicting its biological impact in the coastal marine environment. Particularly as these coastal waters are typically composed of high DOC, SPM and ion concentrations that reduce the proportion of bioavailable copper to between 10% and 20% of the total dissolved copper concentration. This has been confirmed in the limited number of monitoring studies that have reported copper speciation concentrations and seawater parameters within harbours and marinas (Jones and Bolam 2007; Brooks 2006; Hall and Andersen 1999). These studies suggest that the overall impact of copper in the marine environment is low. However, there have been other cases, particularly in the US, that have reported copper concentrations to be continually higher than the copper EQS with potential for environmental harm. High berthing densities with little water mixing have been suggested as reasons for these elevated copper concentrations. In such cases restrictions on copper usage could be recommended. Overall, copper toxicity problems tend to occur in areas of high boating density with little or no water mixing, in these cases copper can build up in the water column and underlying sediment, and can reach toxic levels. However, in most cases (as suggested from monitoring studies) copper effects are limited and the buffering capacity of the natural environment appears to be dealing with the increased inputs from antifouling usage. Alternative environmentally friendly antifouling paints should be encouraged in order to reduce copper inputs in isolated water bodies, however according to current knowledge the benefits of copper as an antifouling biocide generally outweigh its environmental risk.

References

Allen HE, Hansen DJ (1996) The importance of trace metal speciation to water quality criteria. Water Environ Res 68:42–54.

Arnold WR (2004) Effects of dissolved organic carbon on copper toxicity: implications for saltwater copper criteria. Integr Environ Assess Manag 1:34–39.

Arnold WR, Santore RC, Cotsifas JS (2005) Predicting copper toxicity in estuarine and marine waters using the biotic ligand model. Mar Pollut Bull 50:1634–1640.

ASTM (2000) Standard test method for copper release rates of antifouling coating systems in seawater. ASTM Method D6442–6499.

Brooks SJ (2006) Copper speciation in samples collected from a Finnish marina. Cefas contract report CEFAS/PRO/C2415.

Brooks SJ, Bolam T, Tolhurst L et al. (2007) The effects of dissolved organic carbon on the toxicity of copper to the developing embryos of the pacific oyster, *Crassostrea gigas*. Environ Toxicol Chem 26:1756–1763.

Brooks SJ, Bolam T, Tolhurst L et al. (2008) Dissolved organic carbon reduces the toxicity of copper to germlings of the macroalgae, *Fucus vesiculosus*. Ecotoxicol Environ Safety 70: 88–98.

Bryan GW, Gibbs PE (1983) Heavy metals in the Fal Estuary, Cornwall: a study of long-term contamination by mining waste and its effects on estuarine organisms. Mar Biol Assoc UK (occ publ 2):1–112.

Bryan G. W, Gibbs PE, Huggett RJ et al. (1989) Effects of tributyltin pollution on the mud snail, *Ilyanassa obsoleta*, from the York river and Sarah creek, Chesapeake bay. Mar Pollut Bull 20:458–462.

Choi SC, Wai OWH, Choi TWH et al. (2006) Distribution of cadmium, chromium, copper, lead and zinc in marine sediments in Hong Kong waters. Environ Geol 51:455–461.

Conlan KE (1994) Amphipod crustaceans and environmental disturbance: a review. J Nat Hist 28:519–554.

Croot PL, Moffett JW, Brand LE (2000) Production of extracellular Cu complexing ligands by eucaryotic phytoplankton in response to Cu stress. Limnol Oceanogr 45:619–627.

Dallinger R (1977) The flow of copper through a terrestrial food chain III. Selection of an optimum copper diet by isopods. Oceologia 30:273–277.

Di Toro DM, Allen HE, Bergmann HL et al. (2001) Biotic Ligand Model of the acute toxicity of metals. 1. Technical basis. Environ Toxicol Chem 20:2383–2396.

Elzvik A, Hanze K (1992) Ecotoxicological evaluation of copper in antifouling paints. Copper cuprous oxide, cuprous thiocyanate. National Chemicals Inspectorate.

Erickson RJ, Benoit DA, Mattson VR et al. (1996) The effects of water chemistry on the toxicity of copper to fathead minnows. Environ Toxicol Chem 15:181–193.

EU Risk Assessment – [Copper, Copper II Sulphate pentahydrate, Copper(I)oxide, Copper(II) oxide, Dicopper chloride trihydroxide] CAS [7440–50–8, 7758–98–7, 1317–3–1, 1317–38–0, 1332–65–6]. Voluntary risk assessment. European copper Institute, in press.

Florence TM, Stauber JL (1986) Toxicity of copper complexes to the marine diatom *Nitzschia closterium*. Aquat Toxicol 8:11–26.

Florence T, Powell H, Stauber J et al. (1992) Toxicity of lipid-soluble copper(II) complexes to the marine diatom *Nitzschia closterium*: amelioration by humic substances. Water Res 26: 1187–1193.

Gouvêa SP, Vieira AAH, Lombardi AT (2005) Copper and cadmium complexation by high molecular weight materials of dominant microalgae and of water from a eutrophic reservoir. Chemosphere 60:1332–1339.

Hall LW, Anderson RD (1999) A deterministic ecological risk assessment for copper in European saltwater environments. Mar Pollut Bull 38:207–218.

Hassall M, Dangerfield JM (1990) Density dependant processes in the population dynamics of *Aramadillium vulgare* (Isopoda: Oniscidae). J Anim Ecol 59:941–958.

Henry RP (1996) Multiple roles of carbonic anhydrase in cellular transport and metabolism. Annu Rev physiol 58:523–528.

Health and Safety Executive (HSE) (1999) Evaluation on: Copper compounds. 1st review of their use in antifouling products. Prepared by: The Health and Safety Executive. Biocides and pesticides assessment unit. Magdalen House, Stanley Precinct, Bootle, Merseyside, L20 3QZ. Advisory committee on pesticides. Issue No 183.

International Maritime Organisation (IMO) (2001) International conference on the control of harmful anti-fouling systems on ships, adoption of the final act of the conference and any instruments, recommendations and resolutions resulting from the work of the conference 18 October 2001. IMO Headquarters, London, UK.

Jones B, Bolam T (2007) Copper speciation survey from UK marinas, harbours and estuaries. Mar Pollut Bull 54:1127–1138.

Jones B, Bolam T, Waldock M (2005) The speciation of copper in samples collected from the marine environment. Cefas Report study number: CEFAS/PRO/C1385, 127 pp.

Karel AC, de Schamphelaere Janssen CR (2001) A biotic ligand model predicting acute copper toxicity for *Daphnia magna*: the effects of calcium, magnesium, sodium, potassium, and pH. Environ Sci Technol 36:48–54.

Lombardi AT, Vieira AAH (1998) Copper and lead complexation by high molecular weight compounds produced by *Synura* sp. (Chrysophyceae). Phycologia 37:34–39.

Lorenzo JI, Nieto O, Beiras R (2006) Anodic stripping voltammetry measures copper bioavailability for sea urchin larvae in the presence of fulvic acids. Environ Toxicol Chem 25:73–81.

Ma H, Kim SD, Cha DK et al. (1999) Effect of kinetics of complexation by fulvic acid on toxicity of copper to *Ceriodaphnia dubia*. Environ Toxicol Chem 18:828–837.

MacRae et al. (1998) In: Syke, environmental hazard assessment of copper in antifouling paints. Finnish Environmental Institute, Sanna Koivisto. pp 66.

MacRae RK, Smith DE, Swoboda-Colberg N et al. (1999) Copper binding affinity of rainbow trout (*Oncorhynchus mykiss*) and brook trout (*Salvelinus fontinalis*) gills: Implications for assessing the bioavailable metal. Environ Toxicol Chem 18:1180–1189.

Moffett JW, Brand LE (1996) Production of strong, extracellular Cu chelators by marine cyanobacteria in response to Cu stress. Limnol Oceanogr 41:388–395.

Noblet JA, Zeng EY, Baird R et al. (2002) Regional monitoring program: VI Sediment Chemistry, Southern California Coastal Water Research Project, Westminster, CA.

Pagenkopf GK (1983) Gill surface interaction model for trace-metal toxicity to fishes: role of complexation, pH, and water hardness. Environ Sci Technol 17:342–347.

Pérez M, Blustein G, García M et al. (2006) Cupric tannate: A low copper content antifouling pigment. Prog Org Coat 55:311–315.

Playle RC, Dixon DG, Burnison K (1993a) Copper and cadmium binding to fish gills: Modification by dissolved organic carbon and synthetic ligands. Can J Fish Aquat Sci 50:2667–2677.

Playle RC, Dixon DG, Burnison K (1993b) Copper and cadmium binding to fish gills: Estimates of metal-gill stability constants and modelling of metal accumulation. Can J Fish Aquat Sci 50:2678–2687.

Roper WL (1990) Toxicological profile for copper. Sponsor – US agency for toxic substances and disease registry. Contract House – Syracuse Research Corporation, published by National Technical Information Service, Springfield, VA, USA.

Sanders BM, Jenkins KD, Sunda WG et al. (1983) Free cupric ion activity in seawater: effects on metallothionein and growth in crab larvae. Science 222:53.

Santore RC, Driscoll CT (1995) The CHESS model for calculating equilibria in soils and solutions, chemical equilibrium and reaction models. In: Loeppert R, Schwab AP, Goldberg S. (Eds), Chemical Equilibrium and Reaction Models. American Society of Agronomy, Madison, pp. 357–375.

Santore RC, Di Toro DM, Paquin PR et al. (2001) Biotic ligand model on the acute toxicity of metals. 2. Application to the acute copper toxicity to freshwater fish and *Daphnia*. Environ Toxicol Chem 20:2397–2402.

Schiff K, Brown JB, Diehl D et al. (2007) Extent and magnitude of copper contamination in marinas of the San Diego region, California, USA. Mar Pollut Bull 54:322–328.

Skrabal SA, Donat JR, Burdige DJ (1997) Fluxes of copper complexing ligands from estuarine sediments. Limnol Ocenaogr 42:992–996.

Smith RM, Martell AE (1976) Critical Stability Constants. IV. Inorganic Complexes. Plenum, New York.

Smith AJ, Thain JE, Barry J (2006) Exploring the use of caged *Nucella lapillus* to monitor changes to TBT hotspot areas: a trail in the River Tyne estuary (UK). Mar Environ Res 62:149–163.

Stumm W, Morgan JJ (1996) Aquatic Chemistry: Chemical Equilibria and Rates in Natural Waters. Wiley, New York.

Thain JE (1986) Toxicity of TBT to bivalves: effects on reproduction, growth and survival. Proceedings of the organotin symposium of the oceans'86 conference. Washington, DC, September 23–25, 1986, vol. 4. IEEE, Piscataway, NJ, 1306–1313.

Thain JE, Waldock MJ (1986) The impact of tributyltin (TBT) antifouling paints on molluscan fisheries. Water Sci Technol 18:193–202.

Thomas KV, Raymond K, Chadwick J et al. (1999) The effects of short-term changes in environmental parameters on the release of biocides from antifouling coatings: cuprous oxide and tributyltin. Appl Organomet Chem 13:453–460.

Tipping E (1994) WHAM-a chemical equilibrium model and computer code for waters, sediments, and soils incorporating a discrete site/electrostatic model of ion-binding by humic substances. Comput Geosci 20:973–1023.

United States Environmental Protection Agency (US EPA) (1985) Ambient water quality criteria for copper. Office of water regulations and standards, criteria and standards division. Washington, DC, EPA 440/5-84-031.

United States Environmental Protection Agency (US EPA) (2003) Draft update of ambient water quality criteria for copper. US Environmental Protection Agency, Office of Water, Washington, DC, EPA-822-R-03-026.

Valkirs AO, Seligman PF, Haslbeck E et al. (2003) Measurement of copper release rates from antifouling paint under laboratory and in situ conditions: implications for loading estimates to marine water bodies. Mar Pollut Bull 46:763–779.

van den Berg CMG (1982) Determination of copper complexation with natural organic ligands in seawater by equilibration with MnO2 II. Experimental procedures and application to surface seawater. Mar Chem 11:323–342.

van den Berg CMG (1984) Organic and inorganic speciation of copper in the Irish Sea. Mar Chem 14:201–212.

van Hattum B, Baart AC, Boon JG (2002) Computer model to generate predicted environmental concentrations (PECs) for antifouling products in the marine environment. 2nd edition accompanying the release of Mam-Pec version 1.4, Report number E-02-04/Z 3117.

Yebra DM, Kiil S, Dam-Johansen K (2004) Antifouling technology – past, present and future steps towards efficient and environmentally friendly antifouling coatings. Prog Org Coat 50:75–104.

Zamunda CD, Sunda WG (1982) Bioavailability of dissolved copper to the American oyster *Crassostrea virginica*. I. Importance of chemical speciation. Mar Biol 66:77–82.

Zirino A, Belli SL, Van der Weele DA (1998) Copper concentration and CuII activity in San Diego Bay. Electroanalysis 10:423–427.

Part IV
General Summary

Chapter 25
Overview

Hiroya Harino

Organotin compounds are amongst the worst environmental pollutants in history because of their unforeseen stability in the environment and adverse effects on marine organisms, including both direct toxicity and indirect consequences such as endocrine disruption. Many studies are addressed the spread of organotin contamination and the mechanisms for toxicity in aquatic organisms; these are summarized in dedicated sections on organotin compounds in this volume.

As a consequence of such research, the application of organotin compounds has now been banned by IMO and inputs to the marine environment will be reduced in the near future. However, to what extent an how quickly will organotin pollution be improved by this prohibition? The answer is uncertain. Many issues remain concerning the timescales of recovery of the marine environment from organotin contamination. Probably the most serious of these is the legacy organotin loadings in sediment. Organotin compounds may be reversibly adsorbed on to sediment, or entrained as paint particles, and are likely to remain stable for long time, with a strong possibility of gradual re-elution from sediment to the water column. Close to shipping channels, harbours and ports, dredging of heavily contamination sediment may be needed to maintain access, promoting the likelihood of enhanced remobilization. The spread of organotin contamination to the deep sea environment confirms concerns over long-distance transport and even indicates potential for recirculation back to coastal waters by currents. Furthermore, some fish back forth between deep sea and shallow coastal waters, crossing and extending contamination boundaries. These phenomenon present additional risks for marine organisms in shallow water, and, potentially, human consumers. The elucidation of contaminant transport processed to deep sea environments in an important prerequisite if steps are to be taken to reduce such pathways. Our closing concern is over organotins is the presence of relatively high concentrations in marine mammals compared other marine organisms. This is an emerging, and potentially serious, characteristic of organotin pollution and one which requires better understanding of the mechanisms of organotin metabolism in different taxonomic groups.

H. Harino
Osaka City Institute of Public Health and Environmental Sciences,
8-34 Tojo-cho, Tennoji, Osaka 543-0026, Japan

T. Arai et al. (eds.), *Ecotoxicology of Antifouling Biocides*,
DOI: 10.1007/978-4-431-85709-9_25, © Springer 2009

Increasingly, restrictions on the use of organotin compounds will lead to a rise in the incorporation of alternative biocides in antifouling preparations. The environmental effects of these alternative biocides are evaluated in various countries. Unfortunately, the evaluations of environmental effects can differ significantly. One of the reasons for this difference in opinion is that the PEC (predicted environmental concentration) which is used in the risk assessment process, is estimated using modelling. These predicted values by modelling are difficult to validate for a number of alternative biocides, because the detection limits of current analytical methods are often higher than the predicted values. Therefore the development of highly sensitive analytical methods for these compounds is an important requirement. The unstable nature of several alternative biocide presents additional challenges since it is difficult to assess the threat of toxicity for marine organisms given the likelihood of transitions of many of these compounds in the environment. Established risk assessment protocols, as applied, for example to Persistent Organic Pollutants (POPs) may not be appropriate for some alternative biocides. The development of new risk assessment methods which consider the instability of the parent compounds is desirable.

Biocide-free antifoulants have been developed recently but there is no available information concerning the risks for marine life. Whilst extensive ecotoxicological impact would seem unlikely, the possibility of some environmental effects can not be ignored given that such coatings are designed to prevent the settlement of biota. We hope that these antifoulants undergo rigorous assessment of risks and benefits and that the threat is shown to be acceptably low, before widespread application: False assumptions made about the harmless nature of organotin antifouling have taught us expensive lessons about the folly of such actions without appropriate and environmentally relevant testing.

Without wishing to sound emotive, but on a more personal note, Dr. Ohji, one of the editors of this volume, became a parent in March, 2008. The eyes of the baby are clear and innocent. We guess from baby's eyes that there is anticipation to experience new world and hopefully a safe and comfortable world. This expectation applies to all animals on the earth. Our job as parents and guardians is to help fulfil these expectations. We hope that this book will make a useful contribution towards understanding and managing the specific issue of antifouling would like to appeal to all for "continuous efforts help restore and sustain a clean marine environment" for the benefit of future generations.

Finally, we would like to thank Springer for their encouragement and help in publishing this book.

Index